Rachid Alami, Raja Chatila, Hajime Asama (Eds.)

Distributed Autonomous Robotic Systems 6

Rachid Alami, Raja Chatila,
Hajime Asama (Eds.)

Distributed Autonomous Robotic Systems 6

With 220 Figures

Springer

Rachid Alami, Ph.D.
Senior Scientist,
LAAS CNRS,
7 Avenue du Colonel Roche, 31077 Toulouse Cedex 4, France

Raja Chatila, Ph.D.
Senior Scientist,
LAAS CNRS,
7 Avenue du Colonel Roche, 31077 Toulouse Cedex 4, France

Hajime Asama, Ph.D.
Professor, Research into Artifacts, Center for Engineering,
The University of Tokyo,
5-1-5 Kashiwanoha, Kashiwa, Chiba 277-8568, Japan

Library of Congress Control Number: 2006928856

ISBN 978-4-431-35869-5 Springer Tokyo Berlin Heidelberg New York

Springer is a part of Springer Science+Business Media
springer.com

Typesetting: Camera-ready by the editors and authors

Printed on acid-free paper

Preface

The DARS (Distributed Autonomous Robotic Systems) symposia series have spun around the globe every two years since 1992.

The 2004 edition was held on June 23-25 at LAAS-CNRS in Toulouse, France, and attended by over 75 international participants from 16 countries in Asia, the Americas and Europe.

The 46 papers selected by a peer review process provide an excellent coverage of the latest progress in the state of the art in multi-robot systems. The main topics at DARS 2004 addressed the challenges facing the distribution of embodied intelligence, the interaction of intelligent machines and the complex dynamics emerging from interacting agents.

Papers reported on latest research results on several frontier problems. One of the exciting issues is reconfigurability, be it software or hardware, and its relation to growth. Another flourishing research topic is the design of architectures providing for distributed control among (sometimes heterogeneous) robots, while preserving coherence of their behavior. Mobility and motion coordination among multiple robots is another central issue. Other reported work focuses on the relationship between mobility and intelligence, where cooperative behaviors emerge from interaction. Cooperation and coordination are however much wider than mobility and several papers address these issues for the accomplishment of tasks by multiple robots. One of those tasks is perception and mapping, in which cooperation poses difficult problems of information sharing. Several papers addressed the interaction of large numbers of entities in swarms or in groups, a very rich interdisciplinary question on which robotics and ethology share common research agendas.

Transversal to all the issues is the communication problem: how much is necessary? what to exchange, how and when? How to preserve the communication link? This was the topic of several papers as well as the keynote on "Communication-sensitive Planning and Behavior for Multi-Robot Teams" by Ronald C. Arkin.

The conference would not have taken place without the contribution of sponsors, whom we would like to thank: the French National Research Center

(CNRS), the Robotics Society of Japan (RSJ), The French Ministry of Research, the Midi-Pyrénées Region, the European Robotics Network EURON, The Society of Instrument and Control Engineers of Japan (SICE), the Federation of Research in Computer and Control Sciences in Toulouse (FERIA) and LAAS.

The success of the conference is based on the active contribution of the members of the organizing committee, and on the quality and thoroughness of the reviews provided by the Program Committee members whom we would like to thank. We would like also to thank Ms. Jackie Som, Ms. Dominique Daurat, Ms. Marie-José Fontagne, Ms. Marie Dervillers and Jean-Michel Pons for their invaluable help with local arrangements, document edition and the conference website.

Raja Chatila
Rachid Alami
Hajime Asama

Contents

Part I

Reconfigurable Robots I

Self-Reconfiguration Using Directed Growth

K. Stoy[1] and R. Nagpal[2]

[1] The Maersk Mc-Kinney Moller Institute for Production Technology, University of Southern Denmark, Denmark, kaspers@mip.sdu.dk
[2] Division of Engineering and Applied Sciences, Harvard University, USA rad@eecs.harvard.edu

Abstract

Self-reconfigurable robots are built from modules which are autonomously able to change the way they are connected, thus changing the overall shape of the robot. This process is difficult to control because it involves the distributed coordination of large numbers of identical modules connected in time-varying ways.

We present an approach to self-reconfiguration where the desired configuration is grown from an initial seed module. Seeds produce growth by creating a recruitment gradient, using local communication, which spare modules climb to locate the seed. The growth is guided by a novel representation of the desired configuration, which is automatically generated from a 3D CAD model. This approach has two salient features: (1) the representation is concise, with a size proportional to the global shape rather than the number of modules and (2) there is a clean separation between the goal and the local, goal independent rules used by the modules. We demonstrate three implementations of the local rules for recruitment, and show how one can trade-off the number of moves and messages, against time taken to reconfigure.

1 Introduction

Reconfigurable robots are built from modules and can be reconfigured by changing the way the modules are connected. If a robot is able autonomously to change the way the modules are connected, the robot is a self-reconfigurable robot. Self-reconfigurable robots are versatile because they can adapt their shape to fit the task. They are also robust because if a module fails it can be ejected and replaced by a spare module. Potential applications for such robots include search and rescue missions, planetary exploration, building and maintaining structures in space, and entertainment. Challenges exist both in the development of the hardware for the modules, as well as their control. This paper focuses on the challenge of controlling reconfiguration in a robot with many identical modules.

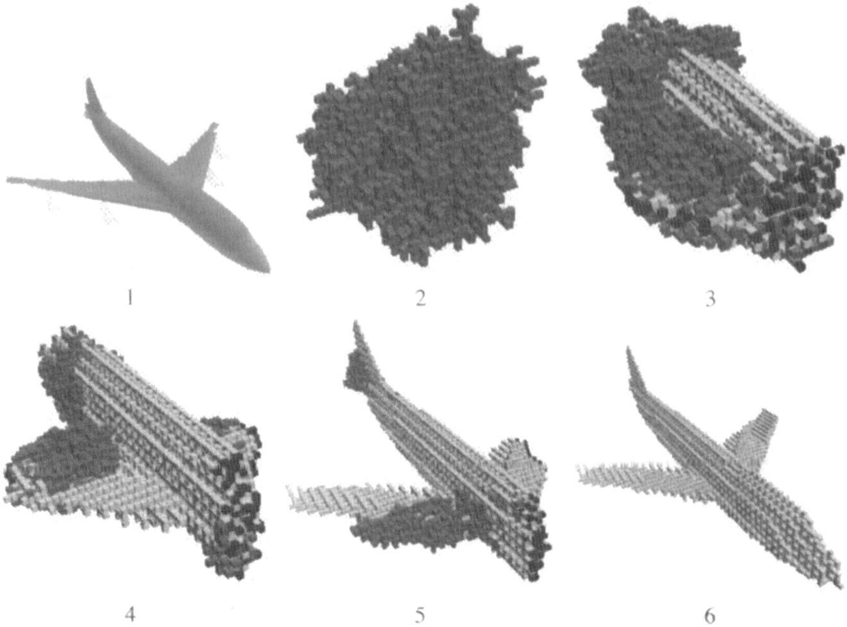

Fig. 1. 1: A brick based representation (grey) is generated based on a CAD model (light grey). **2-6**: This representation is used to control a reconfiguration process.

In this paper we present an approach for reconfiguration that consists of two steps. First, a 3D CAD model representing the desired configuration is transformed into a geometric representation based on overlapping bricks of different sizes. The representation is supplemented with a scaffold structure which removes local minima, hollow, or solid sub-configurations from the configuration. The second step is the actual reconfiguration process. The desired configuration is grown by choosing an arbitrary initial seed module. The seed module uses the brick representation to determine if a neighbour module is needed at an unfilled neighbour position, and if so creates a recruitment gradient in the system. Spare modules climb this gradient to reach the unfilled position and may become seeds themselves if further construction is needed. Figure 1 shows an example of this self-configuration approach.

This approach has several salient features. The representation is automatically generated, is independent of initial configuration, and is concise with a size proportional to the global shape rather than the number of modules. The local rules for the module remain the same, irrespective of the goal shape. This separation, between goal and local rules, allows one to easily optimise or retarget the representation and explore alternate local rules. We demonstrate and compare three different implementations of recruitment that trade-off the extent of the gradient. The general method for recruitment using gradients was first introduced in [12], which used global gradients that cover the entire robot. Two alternate implementations introduced here are:

1) the range of the gradient is increased linearly until the unfilled neighbour position is filled and 2) the range is increased exponentially. We compare the number of moves, messages and time steps taken to complete reconfiguration. The simple global recruitment gradient is more time-efficient than the two new approaches, but the linear strategy uses fewer moves.

2 Related Work

The self-reconfiguration problem is: given a start configuration, possibly a random one, how to move the modules in order to arrive at the desired final configuration. It is computational intractable to find the optimal solution (see [4] for a discussion). Therefore, self-reconfiguration planning and control are open to heuristic-based methods.

One type of approach is planning based, where a path is determined for each module in the original configuration. Chirikjian and others have proposed heuristic methods based on finding a suboptimal sequence of moves from initial to final configuration, which is then optimised by using local searches [4, 10]. Rus et al. simplify the planning problem by using an intermediate chain configuration, which is easy to configure into and out of [11]. Several papers have proposed hierarchical planners, where at the high level some of the hardware motion constraints are abstracted away to facilitate efficient planning. Based on the high-level plans, the lower level then produces the detailed sequence of actions [6, 15]. Another approach is to use meta-modules consisting of a small number of modules [6]. By planning at the meta-module level there are no or few motion constraints; on the other hand, meta-modules make the granularity of the robot larger. A related approach is to maintain a uniform scaffolding structure, facilitating planning [14]. Butler implemented the distributed Pacman algorithm on the Crystalline robot, which has very few motion constraints making the planning problem easier [3, 16]. The advantage of the planning approach is that it can accommodate motion constraints and minimise unnecessary moves; the disadvantage is that plans are often comparable in size to the number of modules and depend on knowing the initial configuration.

A different approach is to rely on common local rules as far as possible and then add randomness to deal with the problems that could not be solved using local rules. This was true in early work such as the work on Fracta [7] and also later work [18, 13]. The problem tended to be that even though the robot often ended up in the desired configuration, it did not always converge. This problem was also present in the work of Yim et al [17], however local communication was used to increase the probability of converging to the final shape. One solution to convergence, proposed by Bojinov et al. [2], is not to focus on a specific configuration. Instead, the idea is to build something with the right functionality. Using this approach it is acceptable if a few modules are stuck as long as the structure maintains its functionality. Alternatively, Jones et al. insist on a specific configuration, but achieve convergence by enforcing a specific sequence of construction [5]. In the work presented here, scaffolding is used to guarantee convergence by making sure that the configurations do not contain local minima, hollow, or solid sub-configurations.

Our system can be thought of as combining the two approaches: the global representation is a plan for constructing a shape from simpler shapes (bricks), while the local rules allow modules to recruit nearby modules to form bricks. This approach is similar to approaches for self-assembly used in Amorphous Computing, such as [8, 9, 1]. There a global goal is specified as a construction which is then compiled into biologically-inspired local rules for agents, resulting in self-assembly that is scale-independent, robust and space efficient. The representation we use is inspired by the circle-network proposed by Kondacs for 2D self-assembly, however the agent model and local rules are completely different [9]. Instead we use local rules proposed by Støy [12] to control module movement.

3 Simulated Robot Model

In our simulation, we use modules which are more powerful than any existing hardware platforms but do fall within the definition of a Proteo module put forward by Yim *et al.* [17]. The modules are cubical and when connected they form a lattice structure. They have six hermaphrodite connectors and can connect to six other modules in the directions: east, west, north, south, up, and down. Modules directly connected to a module are referred to as neighbours. A module can sense whether there are modules in neighbouring lattice cells. In this implementation we do not control the actuator of the connection mechanism, but assume that neighbour modules are connected and disconnected appropriately. A module can only communicate with its neighbours. It is able to rotate around neighbours and to slide along the surface of a layer of modules. Finally, we assume that coordinate systems can be transformed uniquely from one module to another. This is necessary to propagate the gradients and the coordinates used to guide the growth process.

The simulator is programmed in Java3D. The simulation uses discrete time steps. In each time step all the modules are picked in a random sequence and are allowed: 1) to process the messages they have received since last time step, 2) to send messages to neighbours (but not wait for reply), and 3) to move if possible.

In the simulation one module moves at a time so the problem of two modules moving into the same position at the same time is not addressed. However, a solution may be to have modules move back to their original position if a collision occurs and retry a random period of time later.

4 From CAD Model to Representation

It is difficult and time-consuming to hand-code local rules which result in a desired configuration being assembled. Therefore, we need an automatic way of transforming a human-understandable description of a desired configuration into a representation we can use for control.

In our system, the desired configuration is specified using a connected three-dimensional volume in the VRML 1997 or Wavefront .obj file format, which are industry standards produced by most CAD programs. In earlier work we transformed the model into a cellular automaton, which represents relationships between neighbour modules in the desired configuration [12]. This representation has the disadvantage that it scales linearly in the number modules and has to be completely recompiled if the number of modules is changed.

Here we introduce a representation whose size instead scales with the complexity of the three-dimensional model and does not require recompilation if the number of modules changes. We approximate the input shape using a set of overlapping bricks of different sizes. This choice is fairly arbitrary and other basic geometric shapes, such as spheres or cones, could be used as well. The set of bricks is generated by starting at a user specified point inside the CAD model. The algorithm then fits as large a brick as possible which contains this point and does not intersect the CAD model. This is done recursively for all points just outside this brick, but inside the CAD model. This process continues until the volume has been filled with overlapping bricks. Figure 2 shows a simple example of a shape and its brick representation. The fewer bricks needed, the more concise the representation.

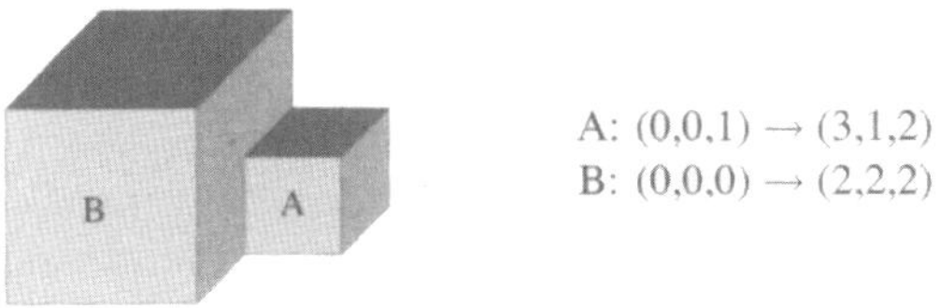

Fig. 2. This figure shows how a volume can be approximated with two overlapping bricks and how we represent this.

In order to control the resolution of the approximation a parameter r is supplied. The points and the corners of the bricks are then constrained to be positioned at coordinates equaling an integer times r. Table 1 shows the number of bricks needed to approximate a model of a Boing 747 at different resolutions. The size of representation scales with the complexity of the input shape and the resolution of the approximation. Furthermore, the brick based representation can at run-time be scaled to match a specific number of modules.

5 From Representation to Self-Reconfiguration Algorithm

Starting from a random configuration the robot needs to reconfigure into the desired configuration as described by the representation. The self-reconfiguration algorithm consists of three components: a coordinate propagation mechanism, a mechanism to create gradients in the system, and a mechanism the modules use to move without disconnecting from the structure. We will look at these in turn.

Resolution	low	medium	high
Modules	32	4512	34493
Bricks	3	168	1152

Table 1. This table shows the number of modules and bricks needed to approximate a CAD model of a Boing 747 at three different resolutions.

5.1 Coordinate Propagation

All the modules are initially connected in a random configuration, have a copy of the representation of the desired configuration, a scale parameter, and start in the wandering state. An arbitrary module is given a random coordinate contained in the representation. The idea is to grow the configuration from this seed module. The seed can detect whether a module is needed in a neighbour position based on its coordinate and the representation. If this is the case, the seed attracts a wandering module to the unfilled position. When a module has reached an unfilled position and is given its coordinate it also may act as a seed if further construction is needed at this position. A module stops attracting modules and stops acting as a seed when all neighbour modules, specified by the representation and the seed's coordinate, are in place.

In order to simplify the reconfiguration problem a scaffold structure is enforced on the desired configuration; neighbour modules are only needed at positions which are contained in the brick representation and belong to the scaffold. The introduction of the scaffold sub-structure into the desired configuration simplifies the reconfiguration problem because during reconfiguration it can be assumed that the configuration does not contain local minima, hollow, or solid sub-configurations. This simplification means that the system is convergent by design as described in [12].

5.2 Creating a Recruitment Gradient Using Local Communication

In this section, we will describe how seed modules attract wandering modules by creating a gradient in the system. A seed module acts as a source and sends out an integer, representing the concentration of an artificial chemical and the strength of the source, to all its neighbours. A non-source module calculates the concentration of the artificial chemical at its position by taking the maximum received value and subtracting one. This value is then propagated to all neighbours and so on. When the concentration reaches zero, the gradient is not propagated further. Note, that the source can control the range of the gradient by changing the source strength. We will explore different strategies for deciding this range in the experimentation section.

Also, since messages take one time step to travel between neighbours, it can take many time steps for gradients to be propagated in the system.

If wandering modules have to rely on the basic integer based gradient to locate the source, they would have to move around randomly for a while to detect the direction of the gradient. Instead we introduce a vector gradient which makes direction information available locally, thereby eliminating unnecessary moves. The basic gradient implementation is extended with a vector indicating the local direction of the gradient. This vector is updated by taking the vector from the neighbour with the highest concentration, adding a unit vector in the direction of this neighbour and renormalising the result. The paths to the unfilled positions always go through or on the surface of the structure. The structure does not contain local minima, because of the scaffold structure. Therefore, the paths to unfilled positions never contain local minima.

5.3 Staying Connected

Wandering modules climb the vector gradient to reach unfilled positions. Unfortunately, the wandering modules cannot move independently of each other, because they depend on each other for connection to the robot. The problem is then to keep the system connected while allowing wandering modules to move. In our solution finalised modules in the configuration emit a *connection gradient* and wandering modules only move if they do not perturb the gradient. Detailed rules for movement and proofs were presented in [12].

6 Experiments

In this section we investigate and compare three different strategies for implementing the recruitment gradient. In global recruitment, a module needing a neighbour creates a gradient throughout the entire configuration using a high source concentration. This, in effect, means that wandering modules always go toward the closest unfilled position even though other wandering modules may already be on their way there. This may result in three problems: 1) poor performance, because many modules are attracted to the same unfilled position and therefore many move in vain; 2) interference between modules because of overcrowding - a well-known problem in literature, see for instance [17]; 3) the amount of communication needed to maintain global gradients increases with the size of the configuration and limits the system's scalability.

We address these problems by investigating two alternative recruitment strategies. In the first strategy, a source linearly increases its strength and therefore the range of the recruitment gradient. In the second strategy, the source increase its strength exponentially. The motivation behind these two recruitment strategies is that they recruit as locally as possible and therefore address the three problems mentioned above. We compare the strategies based on three criteria: time taken to reconfigure, number of module moves, and number of messages.

	time steps	moves	messages
global	535±39.3	17800±1690	572000±41700
linear	21300±7070	13500±993	1250000±225000
exponential	1610±466	30000±13200	1280000±536000

Table 2. This table shows how the total number of time steps, moves, and messages depend on the recruitment strategy. Mean and standard deviation of 20 trials are shown.

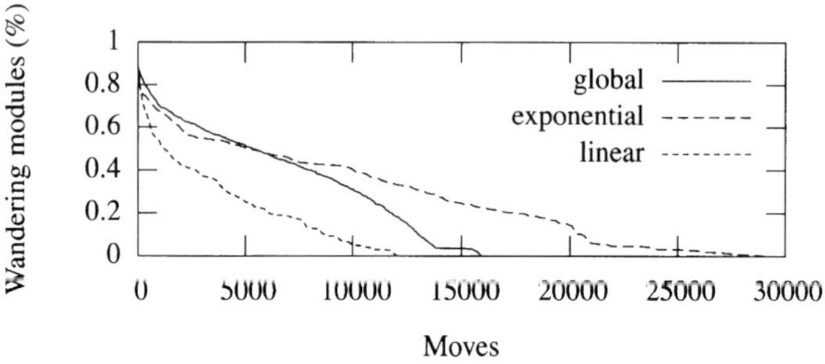

Fig. 3. This figure shows how moves for each strategy are distributed over a reconfiguration process.

The task is to self-reconfigure from a randomly connected configuration of 618 modules to one, which resembles a Boing 747 aeroplane. The representation of the configuration is built by the generator based on a CAD model. The representation is then downloaded into the modules of the simulation, and the self-reconfiguration process is started.

In Table 2, we can see that the global recruitment strategy outperforms the linear and exponential strategies in terms of the number of time steps and messages needed to reach the desired configuration. It can also be seen that the linear strategy outperforms the other two in terms of moves needed to complete a configuration. Therefore, the choice of strategy depends on which constraint is the most important.

Communication, in our system, is time-consuming and as can also be seen in Table 2 the global strategy uses significantly fewer messages than the other two strategies. This could lead one to conclude that the poor time efficiency of the linear strategy relies on the fact that it uses more messages. However, this is not the entire explanation. Figure 3 shows in more detail how the three strategies use their moves during construction. The global strategy recruits aggressively initially, making many modules move in parallel. This is indicated by the fact that initially the global strategy uses more moves per percent completed compared to the other strategies. However, this pays off later when fewer moves are needed per percent completed compared to the other strategies. Aggressive recruitment seems to improve time efficiency without increasing the number of moves significantly, because it takes advantage of a *heuristic*: if one module is needed more will be needed later. Where the global strategy is a parallel process, the linear essentially is a sequential process: instead of recruiting n modules in parallel, n modules are recruited one at a time. This makes the linear approach inefficient in terms of time even without factoring in the cost of communication.

The exponential strategy uses many more moves compared to the other two strategies. This is the case because sources tend to compete for the same modules. This causes the wandering modules to be trapped in the middle causing many unnecessary moves. Therefore, the global strategy is always preferable compared to the exponential.

For the Boing shape the strategies perform differently, however, we do not yet know whether some shapes have significantly different behaviour. A hypothesis is that the performance of the global strategy depends on the degree to which the heuristic holds and what the performance penalty is if it fails. The performance of the other two strategies may be less dependent on the shape. We plan to evaluate this in future work.

7 Conclusion

We have explored an approach to the control of self-reconfiguration which consists of two steps. In the first step a generator takes as input a 3D CAD model of a desired configuration and outputs a set of overlapping bricks which represent this configuration. In the second step this representation is combined with a control algorithm to produce the final self-reconfiguration algorithm. This algorithm controls the self-reconfiguration process through a growth process: seed modules create recruitment gradients in the configuration that wandering modules climb to locate the seed.

In this paper we demonstrate that a representation based on geometric shapes is efficient in terms of space and is independent of the number of modules. We also show that a global recruitment strategy is more efficient in terms of time and messages, while a linear strategy is more efficient in the number of moves. This highlights a key feature of our approach, which is that one can separately optimise (or even change) the global representation and the local rules for module movement. Overall, the proposed system represents a step toward systematic and efficient control of self-reconfigurable robots.

8 Acknowledgements

This research is in part funded by the EU contract IST-2001-33060, the Danish Technical Research Council contract 26-01-0088, and NSF grant EIA-0130391.

References

1. H. Abelson, D. Allen, D. Coore, C. Hanson, G. Homsy, T. Knight, R. Nagpal, E. Rauch, G. Sussman, and R. Weiss. Amorphous computing. *Communications of the ACM*, 43(5):74–82, 2000.

2. H. Bojinov, A. Casal, and T. Hogg. Emergent structures in modular self-reconfigurable robots. In *Proc., IEEE Int. Conf. on Robotics & Automation (ICRA'00)*, volume 2, pages 1734 -1741, San Francisco, California, USA, 2000.
3. Z. Butler, S. Byrnes, and D. Rus. Distributed motion planning for modular robots with unit-compressible modules. In *Proc., IEEE/RSJ Int. Conf. on Intelligent Robots and Systems (IROS'01)*, volume 2, pages 790–796, Maui, Hawaii, USA, 2001.
4. G. Chirikjian, A. Pamecha, and I. Ebert-Uphoff. Evaluating efficiency of self-reconfiguration in a class of modular robots. *Robotics Systems*, 13:317–338, 1996.
5. C. Jones and Maja J. Matarić. From local to global behavior in intelligent self-assembly. In *Proc. of the IEEE Int. Conf. on Robotics and Automation (ICRA'03)*, pages 721–726, Taipei, Taiwan, 2003.
6. K. Kotay and D. Rus. Algorithms for self-reconfiguring molecule motion planning. In *Proc., IEEE/RSJ Int. Conf. on Intelligent Robots and Systems (IROS'00*, volume 3, pages 2184–2193, Maui, Hawaii, USA, 2000.
7. S. Murata, H. Kurokawa, and S. Kokaji. Self-assembling machine. In *Proc., IEEE Int. Conf. on Robotics & Automation (ICRA'94)*, pages 441–448, San Diego, California, USA, 1994.
8. R. Nagpal. Programmable self-assembly using biologically-inspired multiagent control. In *Proc., 1st Int. Joint Conf. on Autonomous Agents and Multi-Agent Systems (AAMAS)*, pages 418–425, Bologna, Italy, 2002.
9. R. Nagpal, A. Kondacs, and C. Chang. Programming methodology for biologically-inspired self-assembling systems. In *Proc., AAAI Spring Symposium on Computational Synthesis: From Basic Building Blocks to High Level Functionality*, 2003.
10. A. Pamecha, I. Ebert-Uphoff, and G.S. Chirikjian. Useful metrics for modular robot motion planning. *IEEE Transactions on Robotics and Automation*, 13(4):531–545, 1997.
11. D. Rus and M. Vona. Self-reconfiguration planning with compressible unit modules. In *Proc., IEEE Int. Conf. on Robotics and Automation (ICRA'99)*, volume 4, pages 2513–2530, Detroit, Michigan, USA, 1999.
12. K. Støy. Controlling self-reconfiguration using cellular automata and gradients. In *Proc., 8th int. conf. on intelligent autonomous systems (IAS-8) (to appear)*, Amsterdam, The Netherlands, 2004.
13. K. Tomita, S. Murata, H. Kurokawa, E. Yoshida, and S. Kokaji. A self-assembly and self-repair method for a distributed mechanical system. *IEEE Transactions on Robotics and Automation*, 15(6):1035–1045, Dec 1999.
14. C. Ünsal and P.K. Khosla. A multi-layered planner for self-reconfiguration of a uniform group of i-cube modules. In *Proc., IEEE/RSJ Int. Conf. on Intelligent Robots and Systems (IROS'01)*, volume 1, pages 598–605, Maui, Hawaii, USA, 2001.
15. C. Ünsal, H. Kiliccote, and P.K. Khosla. A modular self-reconfigurable bipartite robotic system: Implementation and motion planning. *Autonomous Robots*, 10(1):23–40, 2001.
16. S. Vassilvitskii, M. Yim, and J. Suh. A complete, local and parallel reconfiguration algorithm for cube style modular robots. In *Proc., IEEE Int. Conf. on Robotics and Automation (ICRA'02)*, volume 1, pages 117–122, Washington, DC, USA, 2002.
17. M. Yim, Y. Zhang, J. Lamping, and E. Mao. Distributed control for 3d metamorphosis. *Autonomous Robots*, 10(1):41–56, 2001.
18. E. Yoshida, S. Murata, H. Kurokawa, K. Tomita, and S. Kokaji. A distributed reconfiguration method for 3-d homogeneous structure. In *Proc., IEEE/RSJ Int. Conf. on Intelligent Robots and Systems (IROS'98)*, volume 2, pages 852–859, Victoria, B.C., Canada, 1998.

A hardware/software architecture for the control of self reconfigurable robots

Claude Guéganno[1] and Dominique Duhaut[2]

[1] Université de Bretagne Sud Lorient – France claude.gueganno@univ-ubs.fr
[2] Université de Bretagne Sud Lorient – France Dominique.Duhaut@univ-ubs.fr

Summary. Self-reconfigurable robots are built on a set of elementary modules. We focus here on a part of the realization of such system: the control system. This control is based on a processing unit developed around a CPU + FPGA computing system communicating through *bluetooth*. On this hardware architecture we build a software architecture that we describe and discuss the advantages and limitations.

1 Introduction

In the general field of modular robotics research, we can notice that on-board control systems are very different.

We can mention [1]

- Basic Stamp Parallax for MEL Fructum, USC Conro, AIST Mtran
- 68HC11 Motorola for Stanford Polypo
- PowerPC 555 for Xerox Park Polybot [6].

In our case we propose a new approach to control architecture based on a CPU coupled to a FPGA. This coupling allows to simplify the software architecture because the FPGA integrates the low level control loops (for actuators and sensors) and the CPU is just used for the task level control and communication of the system.

The second feature of the proposed architecture is to allow local or global communication. As shown in this paper the low level sensors loops can be used for a one/one communication between elementary modules, while a *bluetooth* connection on the CPU module is used for global communication.

In this paper we'll first describe the requirement for a control system for modular self reconfigurable robot. In a second part we'll present the `maam` project which is our contribution. In a third part we present our hardware/software architecture designed for the `maam` project.

2 Requirements for a control system

Controlling modular robots needs at least three basic functionalities in each module, classical control of motors (at least one up to twelve in our case), input of low level sensors (again the number of sensors can be very different), a communication between modules (with two classical levels: local communication between connected modules or a broadcast system).

The modular robot `maam` is composed with a set of atoms and a host–system whose purpose is to operate the atoms by remote control and interface the whole with the world (figure 1).

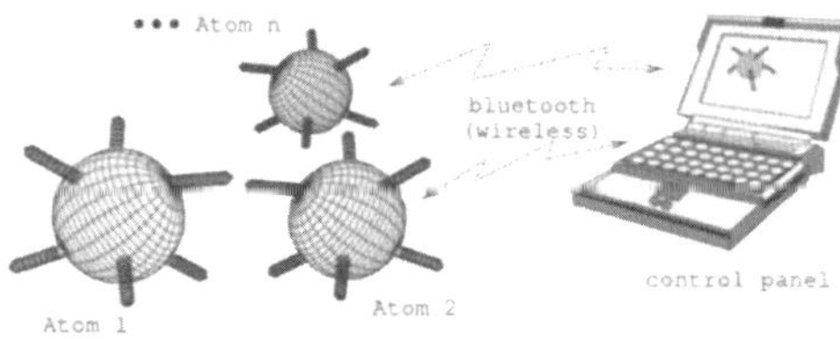

Fig. 1. Overview : some atoms an the host.

An atom is composed of six legs which are directed towards the six orthogonal directions of space. They allow the atom move itself and/or couple to another one.

This leads to the following functional analysis :

1. **control 12 axis** (2 for one leg) : each leg is driven by two servo–motors and a servo–motor is controlled by a PWM (Pulse Width Modulation) signal.
2. **control the coupling of two legs** : the mechanic system under consideration provides a *flip-flop* control. The same control must alternatively couple then uncouple the two atoms.
3. **identify the legs at the touch of the ground** : an atom may have 3 or 4 legs touching the ground at the same time. The presence of pincers at the tip of the leg make the installation of a sensor hard. But, in our case, it is possible to extract this information from the inside of the servo by reading control-signals.
4. **line up 2 legs** : the mechanical connection between two atoms require the lining up of two legs. We propose an infrared transmitter/receiver system. The search for an optimal position with gradient calculus needs the use of 6 analog–to–digital converters for each atom. It may be useful to activate or desactivate the transmitter if necessary: that leads to add 6 digital outputs in our system.
5. **communicate** with another atom or with a host computer: this aspect is discussed in subsection 3.3.

We also have the following general constraints for robotic and embedded systems:

- **mechanical:** the electronic is embedded in a robotic atom; it must fit in a cube which edges $\leq 50mm$.
- **adaptation:** emergence of new requirements due to unforeseen problems during the development of robotic atom must not question the general architecture.

The architecture represented by the diagram in figure 2 takes the previous enumeration of functions and constraints into account. We will now present and discuss this architecture.

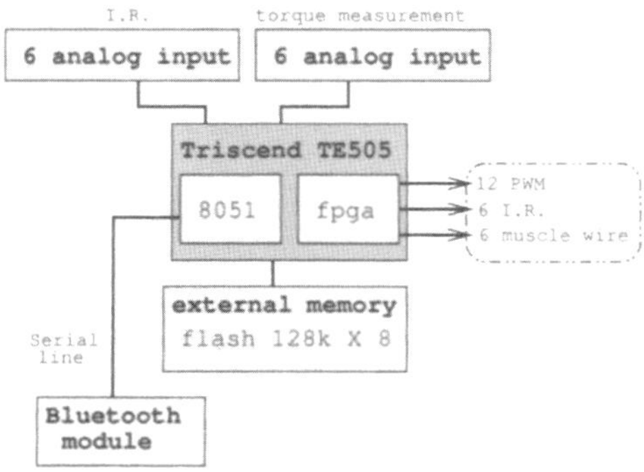

Fig. 2. Embedded electronics. We can see the TE505 CSoC with external memory, AD convertor card and external *bluetooth* module for radio–communication

3 Hardware and software

3.1 CPU of atom

Our proposal is a configurable system on chip (CSoC), which integrates a micro–controller and a FPGA[1] in a single component. This solution gives a suitable answer for previous constraints. The micro-controller provides usual functions of a computing architecture: central unit, serial line, timers, internal memory...

With the FPGA we can realize the equivalent of an input/output card with low level functionalities. It provide most of classical combinatory and sequential circuits (latches, counters, look–up–tables, comparators...).

We've opted for the Triscend TE505 CSoC. This component integrates a CPU 8051, a FPGA with 512 cells and an internal 16KB RAM [7]. Figure 4

[1] Field Programmable Gate Arrays

shows the synoptic diagram of TE505 and figure 3 shows the footprint of the prototype CPU card build around it. The card includes the CSoC, 128 KB of external memory, IO ports, JTAG interface and serial line.

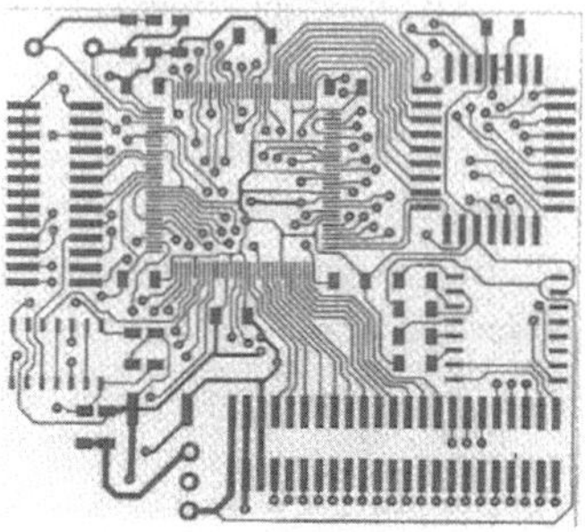

Fig. 3. Footprint of CPU card (true size).

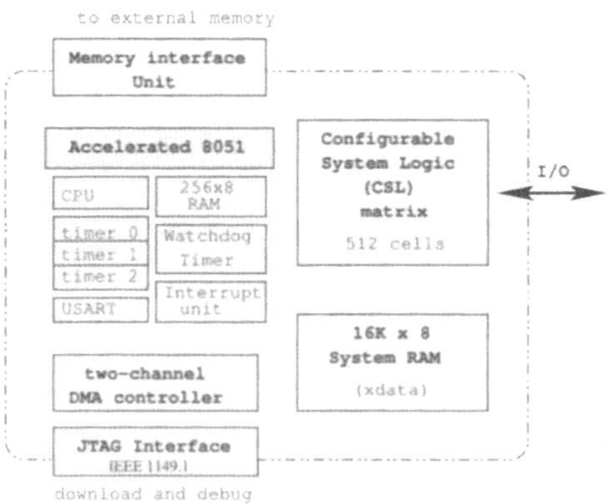

Fig. 4. TE505 central processing unit. We get a micro–controller and a FPGA in one component

3.2 Inputs an outputs

All IO functions are distributed among FPGA and external cards. As many as possible functions are embedded in the FPGA.

PWM control

Position control of servo–motors is obtained by *pulse width modulation* (PWM). The position is proportionate to the width of a periodic pulse. The period is about $20ms$, the range of the pulse width is from $1.9ms$ to $2.1ms$ The servo performs the position loop.

The FPGA provides the 12 PWM control signals.

From the CPU point of view, *controlling the position of one axis amount to write in one byte-register.* The principle of pulses generation is shown on the schematic diagram of figure 5. The corresponding logical circuitry is entirely build in the FPGA.

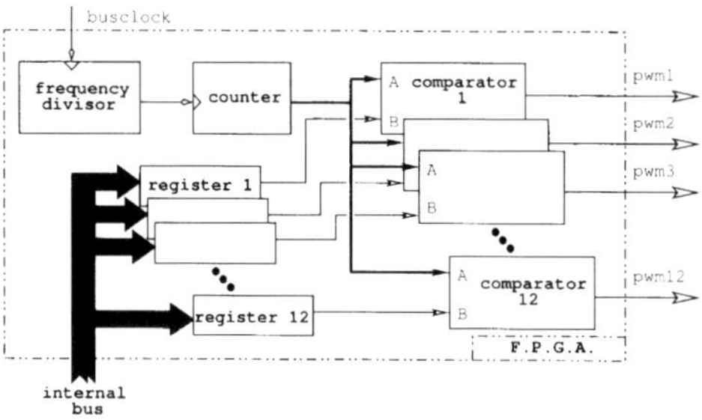

Fig. 5. PWM position control build in FPGA. The FPGA provides 12 independent pulses generations. The output signal of the 16 bits-counter has a period of about $20ms$. The comparison of this signal with the register i acts on the duty cycle of the pwm_i output.

Contact detection

Using a sensor for detecting the contact with the ground should be possible, but difficult because of the pincers. However, we can take this information from the internal circuitry of the servo–motor. Looking at the inputs of the H-bridge, we know if the motor is active or not. When $\neq 0$, they mean that the motor is active. They are held while there is a resistant torque. We get a boolean information.

Infrared positioning

coupling one atom to another involves lining up two legs. The accuracy of alignment will be fixed by mechanic constraints of the system under development. The solution which has been carried for lining up two legs implements infrared transmitter/receiver. The location of the transmitter by the receiver is got by scanning the space (quickest algorithm than a simple scanning should be studied later in the project). The lining up involves a dialog between the two atoms to be coupled, each of them becoming alternately transmitter then receiver. To do that, we need at least 6 channels digital-to-analog converters, and 6 outputs (taken on the CSoC) to control the transmitters.

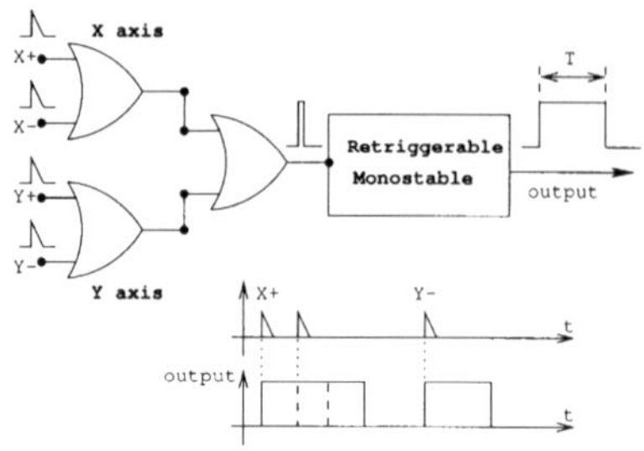

Fig. 6. Contact detection. The inputs are the control signals of inner H-bridge. This schematic diagram takes the two axes of a leg into account, each of them with 2 inputs for 2 directions. The chronogram shows the output state when motors are in motion.

Coupling control

for coupling one atom to another, we have to activate a pincer. Due to size and weight feature, we propose to drive the pincer with a *muscle wire*[4] [8]. The muscle wire is a type of shape memory alloy that contracts in length when electrically heated. Compared to motors, they have some advantages: very small size, light weight, high strength-to-weight ratio, simple activation... On the other hand, they can be stretched by only 4% for a normal use.

Controlling this motorless motion is elementary: we just have to provide a current $I_0 = 400mA$ during coupling or un-coupling two atom's leg. So, 6 ouputs of the FPGA are used for this purpose.

3.3 Communication

An atom must be able to communicate with one or more of its neighbors, and/or with a host-system. Moreover, in the final application atoms must communicate inside little independent groups (for example by pairs of atoms when coupling or un-coupling). That looks like a network including sub-networks.

Among the wireless technologies, *Bluetooth* gives us suitable responses for noise constraint, miniaturization of modules, and low cost.

All layers from radio to `HCI`[5] are implemented in an industrial module: `Bluebird` created by `Inventel` [9]. This module is connected to the CPU with a serial line. One can drive it with the `HCI` commands. The host system uses exactly the same module.

In a *Bluetooth* network, all the units are identical for the hardware and also for the software interface. The only difference is the 6-byte address of each one. When a module establishes a connection with another one, it becomes the *master* during the communication. A master can have up to seven open

[4] *Muscle Wire* is a registered trademark of `Mondo--Tronics`

[5] Host Control Interface

links at the same time The set of the slaves and the master is called a *piconet*. One slave of a *piconet* can be the master of another *piconet* (see figure 7).

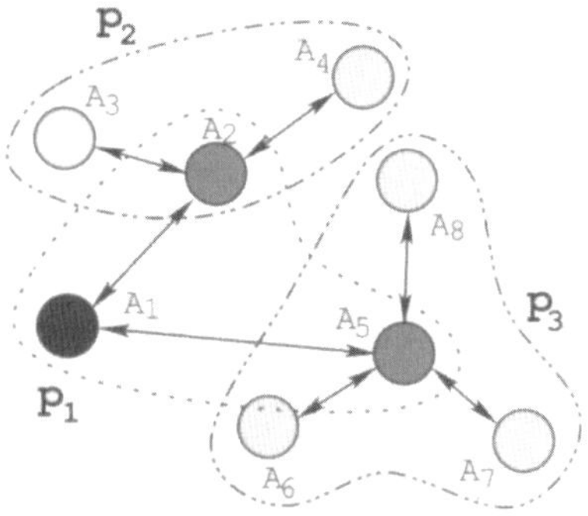

Fig. 7. A network of atoms with *Bluetooth*. The atoms A_1, A_2 and A_5 make up a *piconet* p_1 whose master is A_1. Slaves A_2 and A_5 are the respective masters of *piconets* p_2 and p_3. The 8 atoms make up a *scatternet*.

3.4 Software

At the present time, the embedded software in the atom is written above the `HCI` layer of the *Bluetooth* stack. It is a program in C language which waits for a connection request, accepts the connection and finally exchanges data through the open link. The exchanged data encapsulates low level command for the atoms. In the future, the data should encapsulate a real command language for the atoms.

In the host system, we use the `Java` language. The program first identifies dynamically all the active atoms in the local area and then requests to connect with all the identified atoms in accordance with the scenario of figure 8. This program links with the Johan Eker's *Bluetooth* stack called `"Harald"` [10]. This free stack seems to work properly after a few changes.

The goals of the host-program are, first to validate the communication between the host and several atoms, and second to help the learning of the atom's motion. Low level functions (*i. e.* separate commands of 12 axis) are embedded in atoms and can be operated by remote control from the host. This allows to finalize more easily high-level algorithms on the host-system and then embed them in the atom software. As shown in the example of figure 9, a new algorithm embedded give us a new instruction for the command-language of atoms.

A screen-shot of the present control-panel is shown in figure 10.

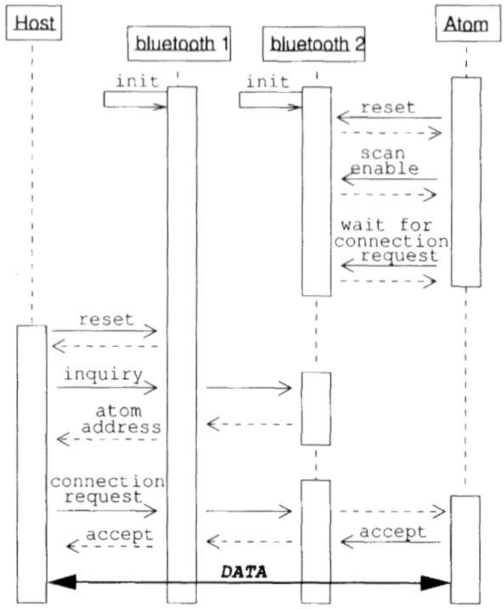

Fig. 8. Dialog between an atom and the host-system. In this diagram, the objects called *Bluetooth i* are the lower layers of the *Bluetooth* stack implemented in the industrial *Bluebird* module. The result of the inquiry-command is an array list of remote modules. Then the host tries to establish links with one or more detected modules.

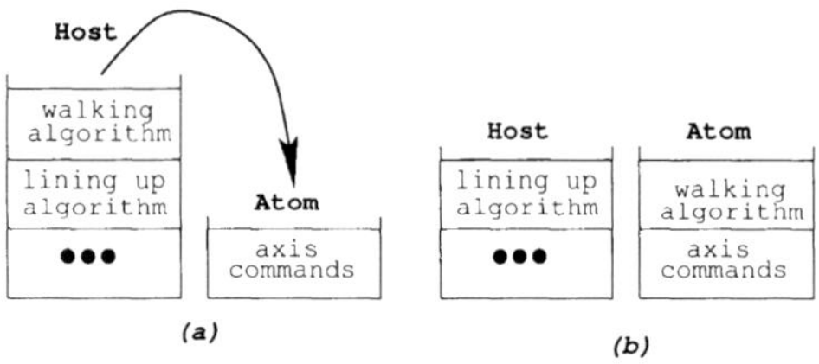

Fig. 9. Software development. When finalized, an algorithm can be embedded in the atom and becomes accessible through an instruction of the command language encapsulated in the communication-protocol

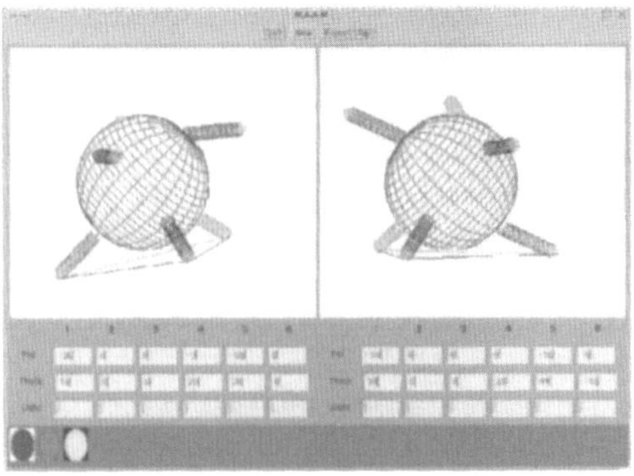

Fig. 10. Control panel In this example, 3 *Bluetooth*-modules were detected by the inquiry-instruction. After a connection request, the atom send a synchronization frame to the host. The host program give the user a 3D interface to drive the atom.

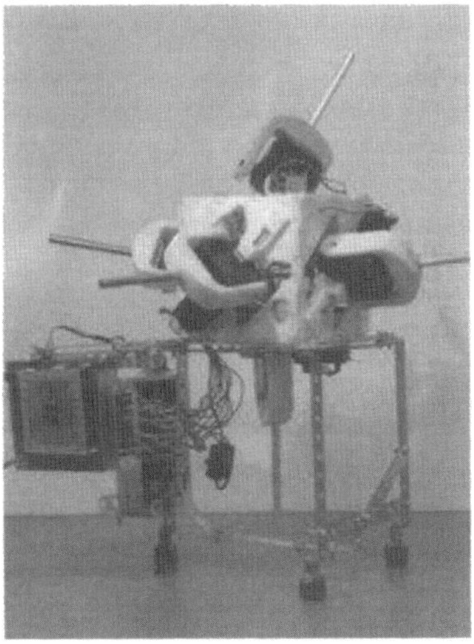

Fig. 11. First version of the atom. One can see the CPU and *Bluetooth* module near the atom.

4 A versatile architecture

Beyond our application, the proposed architecture can be brought into general uses.

Most of robotic applications need to drive motors or other ouputs, according to information from some sensors (feedback), to communicate with a remote system (indeed many systems for a modular robot). Mechanical solutions for robots may be very different from one to another. But, the software and hardware architecture presented in this paper can be applied in most cases.

Robotic systems are distinguished by their outputs, inputs and low level loops. Using a technology that integrates a FPGA in a low level layer is a suitable way to take these differences into account. The association of the FPGA with a micro-controler in the same reconfigurable component increases the versatility of the architecture, allowing flexible connexion between intelligent peripherals (timers, UART ...) and embedded internal logic of the FPGA. Developing and then using this architecture will allow us to treat our project with more efficiency and to share a technical knowledge between a set of applications.

5 Future direction

At the end of 2003, we have a wireless atom which understands simple orders from the host through *Bluetooth*. This atom is shown in figure 11.

The second version of atom is under development. It will embed all the hardware (in particular, the industrial version of the CPU card). It will allow us to study dynamic aspects (moving, walking ...). The result should be a library of moving functions which glues to the command language.

At the same time, solutions must be achieved for coupling and un-coupling atoms. That involves mechanic, infrared lining–up and communication between two atoms.

6 Conclusion

This paper has presented a hardware control architecture based on the coupling of a CPU and a FPGA on a single board. To this basic module, we add a wireless communication based on *Bluetooth*. We've presented here how we use this for the control of self-reconfigurable module.

In fact, by changing the low level loops in the FPGA, we can implement PID controler to control more classical robots. So this basic architecture can be extended to different kinds of robots.

Acknowledgment

The `maam` project is supported by the Robea project of the CNRS. All references to people participating to this work can be found in [2].

References

1. http://staff.aist.go.jp/e.yoshida/test/top-e.htm A. Kaminura & all, *"Self reconfigurable modular robot"*, IEEE/RSJ, IROS conference, Maui, Hawaii, USA, Oct 29-Nov 03, 2001 p 606-612
2. http://www.univ-ubs.fr/valoria/duhaut/maam
3. D. Duhaut *"Robotic Atom"* CLAWAR, Paris, septembre 2002
4. *The DARPA Knowlege Sharing Initiative. "Specification of the KQML agent communication language : Draft"*. June 1993
5. *IEEE/ASME Transactions on Mechatronics Volume: 7*, Issue: 4, Dec 2002
6. *Polybot* http://www2.parc.com/spl/projects/modrobots/chain/polybot/
7. *Triscend E5 Configurable System-on-Chip Platform*, Triscend Corporation, February 2001.
8. *Muscle wires* Mondo Tronics Inc. San Rafael U.S.A.
9. *Bluebird 2* Datasheet, Inventel, 2003.
10. Johan Eker, *Harald Bluetooth Stack* Department of Automatic Control, Lund Institute of Technology, Lund, Sweden, 2000.

Emergent Morphology Control of a Modular Robot by Exploiting the Interaction between Control and Mechanical Dynamics

Masahiro Shimizu[1], Masayasu Takahashi[1], Toshihiro Kawakatsu[2], and Akio Ishiguro[1]

[1] Dept. of Computational Science and Engineering, Nagoya University, Japan
{shimizu/masayasu/ishiguro}@cmplx.cse.nagoya-u.ac.jp
[2] Dept. of Physics, Tohoku University, Japan
kawakatu@cmpt.phys.tohoku.ac.jp

Abstract: This paper discusses a fully decentralized algorithm able to control the morphology of a modular robot, consisting of many identical modules, according to the environment encountered. One of the significant features of our approach is that we explicitly exploit an "emergent phenomenon" stemming from the interaction between control and mechanical dynamics in order to control the morphology in real time. To this end, we particularly focus on a "functional material" and a "mutual entrainment", the former of which is used as a connection mechanism between the modules, and the latter of which plays as the core of the control mechanism for the generation of locomotion. Simulation results indicate that the proposed algorithm can induce "protoplasmic streaming", which allows us to successfully control the morphology of modular robot in real time according to the situation without losing the coherence of the entire system.

1 Introduction

Recently, a modular robot (or called reconfigurable robot), consisting of many mechanical units (hereinafter called modules), have been attracting lots of concern. Since the relative positional relationship among the modules can be altered actively according to the situation encountered, a modular robot is expected to show significant abilities, *e.g.*, adaptability, fault tolerance, scalability, and flexibility, compared with a robot on a fixed-morphology basis [1-5]. Under these circumstances, so far various morphology control methods have been proposed for modular robots. Most of these studies, however, have the following problems:

- Morphological alteration is discussed in some studies, but is usually resolved by turning into a module rearrangement problem in a centralized-planning manner.
- Modules are normally connected mechanically and/or electromagnetically by highly rigid mechanisms.

In order to fully exploit the advantages mentioned above, (1)each module should be controlled in a fully decentralized manner, and (2)the resultant morphology of the entire system should be *emerged* through the module-to-module and module-to-environment interactions.

In light of these facts, this study is intended to deal with an emergent control method which enables a modular robot to change its morphology in real time according to the situation encountered without the use of any global information as well as without losing the coherence of the entire system. Since there still remains much to be understood about how such emergent systems can be created, in this study, we employ the following *working hypothesis*: well-balanced coupling between control and mechanical dynamics plays an essential role to elicit interesting emergent phenomena, which can be exploited to increase adaptability, scalability, and so on.

Based on this working hypothesis, here we particularly focus on the exploitation of a *functional material* and a *mutual entrainment* among nonlinear oscillators, the former of which is used as a connection mechanism between the modules, and the latter of which plays as the core of the control mechanism for the generation of locomotion. In what follows, we will explain these in more detail. As mentioned before, most modular robots developed so far have their modules connected mechanically and/or electromagnetically by highly rigid mechanisms. Under this kind of connection mechanism, however, the control algorithm required usually ends up to be extremely complicated and intractable since it has to always specify which modules should be connected physically as well as how each module should be moved. In addition, module connections done by such a highly rigid mechanism may impair some of the advantages expected, particularly the flexibility against environmental changes. In order to alleviate this problem, we focused on a functional material. More specifically, we used *Velcro strap* as a practical example, since this intrinsically have an interesting properties: when the male and female halves of Velcro contact each other, they are connected easily; and when the halves are disconnected by a force greater than the yield strength, they come apart automatically. Exploiting the property of this material itself as a part of the mechanical dynamics is expected not only to reduce the computational cost required for the connection control dramatically, but also to induce emergent properties in morphology control[1].

For efficient morphology control of a modular robot with this kind of material, the induction of *protoplasmic streaming* is considered in this study. Protoplasmic streaming is a many-body behavioral phenomenon widely observed in nature. We expect that this contributes to control the morphology of an entire module group in an emergent manner from the interactions between the modules and between the module group and its surrounding environment. Here, a mutual entrainment plays an essential role to elicit protoplasmic streaming, which will be discussed later.

[1] Due to the automatic disconnection by a force beyond the yield strength, this material is expected to absorb the conflict between a modular robot and its environment. When a highly rigid connection mechanism is employed, one has to always control precisely, which will lead to the huge computational cost. Therefore, a functional material can be viewed as a mechanism which autonomously controls the connection and disconnection among the modules by exploiting its intrinsic properties.

Since the study is still in the initial stage, this paper deals with the morphology control of a modular robot placed two dimensionally. More specifically, we attempt to construct a control method able to induce protoplasmic streaming inside the modular robot. Simulation results indicate that the proposed method is highly promising.

2 Proposed Method

2.1 The Mechanical Structure

A modular robot considered in this study consists of many identical modules, each of which has a mechanical structure like the one shown in Fig. 1. Each module is equipped with telescopic arms and a ground friction control mechanism (explained later). Each module is also equipped with two types of light-detecting sensor: one is for detecting the goal; and the other is for ambient light. Note that the module is covered with Velcro straps with different polarities, *i.e.*, male and female halves of Velcro. The dynamics of the connection mechanism is specified by the yield stress of Velcro employed: connection between the modules is established spontaneously where the arms of each module make contact; disconnection occurs if the disconnection stress exceeds the yield stress. We also assume that local communication between the connected modules is possible, which will be used to create phase gradient inside the modular robot (discussed below). In this study, each module is moved by the telescopic actions of the arms and by ground friction. Therefore, each module itself does not have any mobility but can move only by the collaboration with other modules.

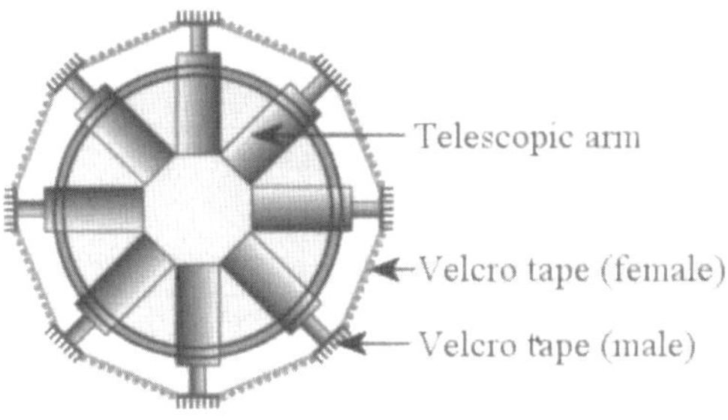

Fig. 1. Mechanical structure of the modular robot employed (top view).

2.2 The Control Algorithm

Under the above mechanical structure, we consider how we can generate stable and continuous protoplasmic streaming inside the modular robot. As observed in *slime mold* and other organisms, the generation of an appropriate phase gradient inside the modular robot is indispensable in order to induce protoplasmic streaming. To this end, a nonlinear oscillator is implemented onto each module with which we expect to create an appropriate equiphase surface suitable for generating protoplasmic streaming through the mutual entrainment among the oscillators. In what follows, we will give a detailed explanation of this algorithm.

Active mode and passive mode

Here, the basic operation of each module is defined. Each module in the modular robot can take one of two exclusive modes at any time: *active mode* and *passive mode*. As shown in Fig. 2, a module in the active mode actively contracts the connected arms, and simultaneously reduces the ground friction. In contrast, a module in the passive mode increases the ground friction[4], and return its arms to their original length. Note that a module in the passive mode does not move itself but serves as a supporting point for efficient movement of the module group in the active mode.

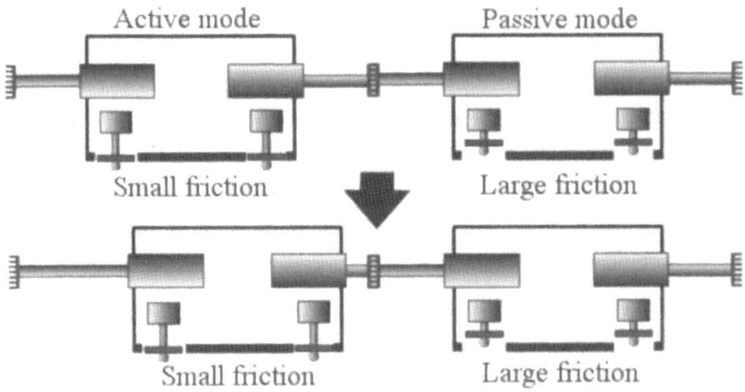

Fig. 2. A schemtic of the active mode and the passive mode (A side view of the connected modules is shown for clarity).

Configuration of the phase gradient through mutual entrainment

As mentioned before, the modes of each module should be switched appropriately in order to induce a protoplasmic streaming phenomenon. Therefore, the configuration of an equiphase surface is extremely important as a guideline for the mode switching. In this study, the creation of an equiphase surface effective for generating the protoplasmic streaming phenomenon is attempted only through the local communication. To do so, we focused on a *mutual entrainment phenomenon* created through the interaction among nonlinear oscillators. In the following, we will explain this in more detail.

As a model of a nonlinear oscillator, *van der Pole oscillator* (hereinafter VDP oscillator) was employed, since this oscillator model is widely used for its significant

[4]Slime molds do really use this mechanism. This is called a *pseudopod*.

entrainment property. The equation of VDP oscillator implemented on module i is given by

$$\alpha_i \ddot{x}_i - \beta_i(1 - x_i^2)\dot{x}_i + x_i = 0, \tag{1}$$

where the parameter α_i specifies the frequency of the oscillation. β_i corresponds to the convergence rate to the limit cycle.

The local communication among the physically connected modules is done by the local interaction among the VDP oscillators of these modules. This is conducted by referencing the study of Kakazu *et al.* [6], which is expressed as[5] :

$$x_i(t+1) = x_i(t) + \varepsilon \left(\frac{1}{N_i(t)} \sum_{j=1}^{N_i(t)} x_j(t) - x_i(t) \right), \tag{2}$$

where $N_i(t)$ represents the number of modules neighboring module i at time t. The parameter ε specifies the strength of the interaction.

When VDP oscillators interact according to Equation (2), significant phase distribution can be created effectively by varying the value of α_i in Equation (1) for some of the oscillators [6]. In order to create an equiphase surface effective for the generation of protoplasmic streaming, we set the value of α_i as:

$$\alpha_i = \begin{cases} 0.7 \text{ (if the goal light is detected)} \\ 1.3 \text{ (if the ambient light is detected)} \\ 1.0 \text{ (otherwise)} \end{cases} \tag{3}$$

Note that except the modules detecting the goal light, the modules on the boundary, *i.e.*, the outer surface, have the value of $\alpha_i = 1.3$. This allows us to introduce the effect of *surface tension*, which is indispensable to maintain the coherence of the entire system. Figure 3 shows the phase distribution when the modules are arranged circularly. The top and bottom of the figure corresponds to the front and rear of the modular robot, respectively. In the figure, arrows – each of which represents the direction of gradient vector at the corresponding point – are also depicted for clarity.

Generation of the protoplasmic streaming

Here, we consider a control algorithm able to generate protoplasmic streaming exploiting the phase distribution created from the aforementioned mutual entrainment among the VDP oscillators. To do so, the two possible modes, *i.e.*, the active and passive modes, of each module should be appropriately altered corresponding to the emerged phase distribution. In this study, therefore, we first divide one period T,

[5] In this study, the mutual entrainment among the VDP oscillators adopted by Kakazu *et al.* [6] is employed to create an appropriate phase gradient. The point of this study, however, is to induce the protoplasmic streaming inside the modular robot by exploiting the interaction between the dynamics of the functional material and the distribution of the velocity vectors created from the resultant shape of the equiphase surface. This is totally different from the study of Kakazu *et al.*, which is aimed at controlling a swarm of motile elements, *i.e.*, autonomous mobile robots.

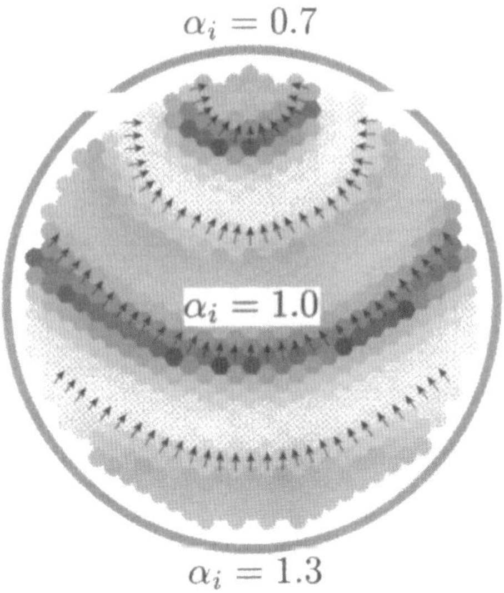

Fig. 3. Phase distribution created through the mutual entrainment among the VDP oscillators in a circular arrangement. The gray scale denotes the value of the phase at the corresponding point.

i.e., the period of $(n-1)\pi \leq \theta_i(t) < (n+1)\pi$, of the VDP oscillator equally into N_p sections. Then, in each phase section corresponding to time T/N_p the two mode is altered according to the duty ratio γ expressed as:

$$\gamma = \frac{T_a}{T_a + T_p}, \tag{4}$$

where T_a and T_p are the period of the active mode and passive mode in time T/N_p, respectively. Fig. 4 illustrates how the active and passive modes are altered according to the phase of the oscillation. The vertical and horizontal axes are the phase of the VDP oscillator (hereinafter denoted by θ_i) and the time step, respectively. For clarity, N_p is set to two in the figure.

When the duty ratio is set as above under the phase distribution shown in Fig. 3, the timings of the mode alternation are propagated from the front to the rear inside the modular robot as traveling waves. In this study, the extension/contraction of each arm of module i in the active mode is determined according to the phase difference with the neighboring module. This is given by

$$F_i^m(t) = -k\{\theta_j(t) - \theta_i(t)\}, \tag{5}$$

where, $F_i^m(t)$ is the force applied for the extension/contraction of the m-th arm of module i at time t. k is the coefficient. $\theta_j(t)$ represents the phase of the neighboring module physically connected to module i. Due to this, the degree of arm extension/contraction of each module will become most significant along the phase gradient (see Fig. 3).

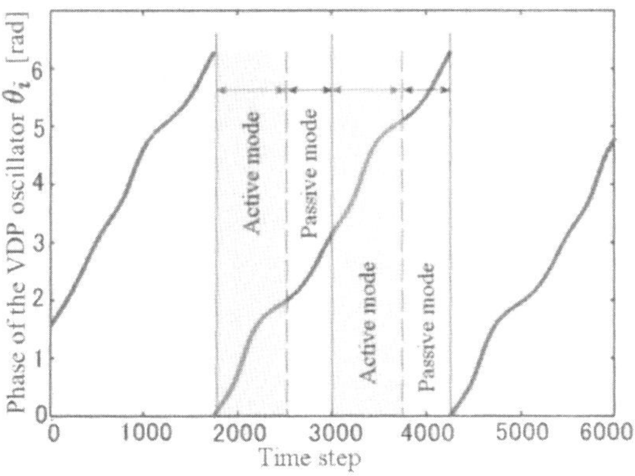

Fig. 4. Mode alternation.

What should be noticed here can be summarized as: (1) the motion-direction vectors of the modules along the midline connecting from the front to the rear are oriented almost in the same direction; (2) the others are heading inward, the latter of which induces the effect of surface tension (see the arrows in Fig. 3). This enables the entire system to advance forward while maintaining its coherency.

3 Simulation Results

3.1 Problem Setting

In this study, a *phototaxis behavior* is adopted as a practical example: the task of the modular robot is to move toward the goal without losing the coherece of the entire system. In the simulation discussed below, the light from the goal is given from the top of the figure, and thus the modular robot moves upward. The simulation conditions employed are as follows:

Initial arrangement: Circular (each module is placed so as to be the most densely filled structure, as shown in Fig. 3).
Parameters of the VDP oscillator: $\beta_i = 1.0$; $\varepsilon = 1.0$; α_i is varied according to equation (3).
Duty ratio: $\gamma = 0.6$.

3.2 Verification of the Creation of Protoplasmic Streaming

In order to confirm the validity of the proposed method, simulations were performed under the above problem settings. Figure 5 (a) and (b) show representative results obtained under the condition where the number of modules was set to 92 and 563, respectively. The thick circles in the figures denote obstacles. These snapshots are in the order of the time transition (see from the left to right in each figure). As in the figures, the modular robot can successfully negotiate the environmental changes

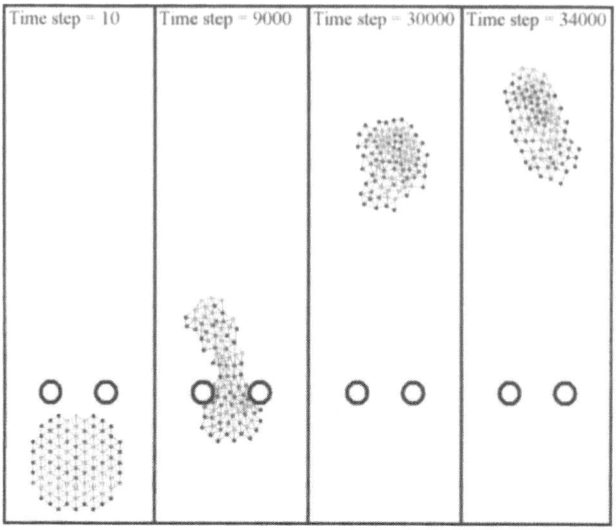

(a) The number of the modules: 92

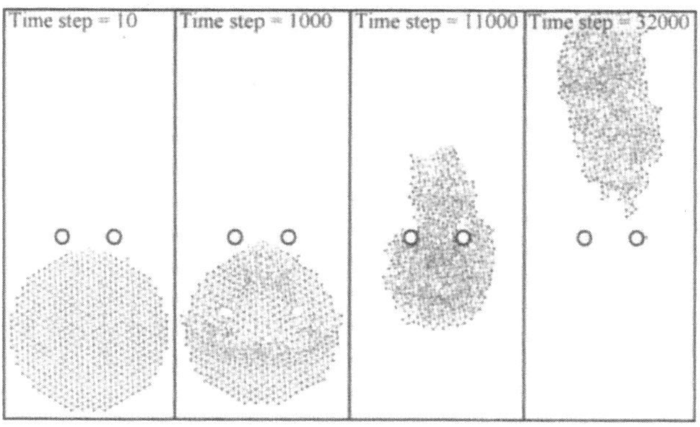

(b) The number of the modules: 563

Fig. 5. Representative data of the transition of the morphology (see from the left to right in each figure). The thick circles in the figures are the obstacles.

without losing the coherence. These results provide us the following three points that have to be noted. First, as we clearly see from the figure (b), the traveling wave stemming from the phase distribution created through the mutual entrainment gradually becomes conspicuous (see time step 1000 in the figure), and the right and left outer sections in the module group start moving toward the center. As a result, the protoplasmic streaming is emerged by causing the connection and disconnection among the modules. It should be noted that the dynamics of the connection mechanism provided by the functional materials is fully exploited in the process. Second, the way of negotiating the environment seems significantly different: the modular robot in the figure (a) passes through the obstacles by narrowing the width of the entire system, whilst the one in the figure (b) negotiates its environment by enclosing the obstacles. Note that these behavior are not pre-programmed, but are totally emergent. Third and finally, the effect of surface tension contributes to maintain the coherence of the entire system. Around the time step of 3000 in the figure (a), we temporarily turned off the goal light. As we see from the figure, the modular robot starts to form a circular shape. This is due to the effect of surface tension. We have observed that the modular robot cannot maintain the coherence without this effect.

3.3 Verification of the Spontaneous Connectivity Control

The control method discussed the above does not *explicitly* control the connectivity among the modules by fully exploiting the dynamics of the functional material. To verify the feasibility of this idea, we have measured the number of spontaneous disconnection occurred. In the following experiment, we have employed exactly the same condition as shown in Fig. 5 except that the number of modules was set to 64. The result is illustrated in Fig. 6.

As the figure explains, the number of spontaneous disconnection varies depending on the situation encountered. In other words, the number of disconnection occurred has strong correlation with the process of negotiation: we can observe the significant increase when the modular robot is passing through the obstacles by narrowing the width of its embodiment (see from 2000 to 8000 time steps); once the entire system has almost converged to a shape expanding along the moving direction (see around 8000 time steps), the number of disconnection immediately starts to decrease. This strongly supports that the proposed control method allows us to fully exploit emergent phenomena during the deformation of morphology.

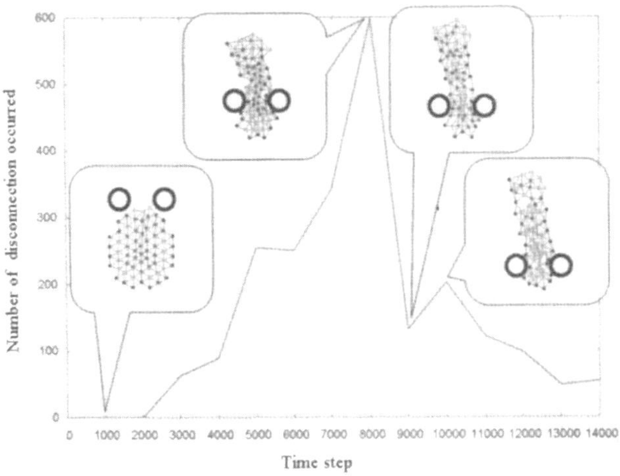

Fig. 6. Time evolution of the number of spontaneous disconnection among the modules.

4 Conclusion and Further Work

This paper discussed a decentralized control method enabling a modular robot to control its morphology in real time by explicitly exploiting an emergent phenomena stemming from the interplay between the control and mechanical dynamics. To this end, we focused on the functional material and the phase distribution created through the mutual entrainment among the VDP oscillators by utilizing the former as a mechanism for inter-module connection control and the latter as a core mechanism for locomotion pattern generation. Simulations conducted indicate that the proposed algorithm can induce a stable protoplasmic streaming inside the modular robot, which allows us to successfully control the morphology in real time according to the situation encountered without losing the coherence of the entire system. It should be noted that the control method discussed here does not control the connection/disconnection among the modules explicitly. This is totally an emergent phenomenon.

In order to control the morphology of a modular robot having a great degree of freedom in real time, the concept of exploiting the protoplasmic streaming inside the modular robot derived from the mutual entrainment among the VDP oscillators and the dynamic characteristics of the functional material introduced in this paper is considered to play an extremely important role in simplifying the necessary control algorithm. In addition, the protoplasmic streaming discussed here is emergent as the number of modules increases. This satisfies one of the important aspects of emergent phenomena: "a quantitative change leads to a qualitative change". To our knowledge, this is a first study explicitly based on this idea in the field of modular robots. To verify the feasibility of our proposed method, an experiment with a real physical modular robot is significantly important. This is currently under investigation.

References

1. T. Fukuda and Y. Kawauchi (1990), Cellular robotic system (CEBOT) as one of the realization of self-organizing intelligent universal manipulators, Proc. of IEEE ICRA, pp.662–667
2. M. Yim, C. Eldershaw, Y. Zhang, and D. Duff (2003), Self-Reconfigurable Robot Systems: PolyBot, Journal of the Robotics Society of JapanCVol.21, No.8, pp.851–854
3. Hydra Project (EU Project): http://www.hydra-robot.com
4. A. Castano, W.-M. Shen, and P. Will (2000), CONRO: Towards Miniature Self-Sufficient Metamorphic Robots, Autonomous Robots, pp.309–324
5. H. Kurokawa, E. Yoshida, A. Kamimura, K. Tomita, S. Yoshida, and H. Kokaji (2003), Autonomous Modular Robot M-TRAN for Metamorphosis and Locomotion (in Japanese), Journal of the Robotics Society of Japan, Vol.21, No.8, pp.855–859
6. Y. Kakazu and N. Takahashi (2003), Biomimetics and Modular Robotic System (in Japanese), Journal of the Robotics Society of Japan, Vol.21, No.8, pp.839–842

HydroGen: Automatically Generating Self-Assembly Code for Hydron Units

George Konidaris, Tim Taylor, and John Hallam

Institute of Perception, Action and Behaviour, University of Edinburgh.
{gkonidar, timt, jhallam}@inf.ed.ac.uk

Abstract. This paper introduces HydroGen, an object compiler system that produces self-assembly instructions for configurations of Hydron units. The Hydron is distinct from other self-reconfigurable robotic units in that it operates under water, and can thus move without being constrained by gravity of connectivity requirements. It is therefore well suited to self-assembly as opposed to self-reconfiguration, and faces similar control problems to those expected in nanotechnology applications.

We describe the first version of the Hydron Object Compiler and its supporting software. The object compiler uses a basic instruction set to produce instructions for the distributed self-assembly of any given connected configuration of Hydron units. We briefly outline the implementation of a preliminary interpreter for this instruction set for Hydron units in a reasonably realistic simulated environment, and demonstrate its operation on two example configurations.

1 Introduction

The HYDRA project aims to develop self-synthesising robot systems based on the use of simple robotic building blocks. One approach to this is the development of *object compilers*, which generate instructions for the distributed synthesis of objects from some initial configuration. Tomita et al. [9] have shown that such a process is possible in 2D using a recursive self-assembly process for some (but not all) configurations of Fractum units. More recent work by Støy [7] has shown that such a process is possible for 3D self-reconfiguration using proteo modules [10], which can be simulated using three of the HYDRA project's Atron units [5]. However, this work requires restrictions on the form of the target object in order to make local decision-making sufficient because Atron units must crawl over each other to move, and must remain connected to a configuration as it changes.

This paper therefore introduces a complementary system and accompanying object compiler for the Hydron, the HYDRA project's other primary robotic platform. Hydron units operate while suspended in water, and are thus free of the constraints of gravity. This system, called HydroGen, is therefore able to assemble objects from units dispersed in water, rather than transforming one robot configuration to another, and provides a self-assembly system complementary to Støy's self-reconfiguration system.

2 The Hydron Unit

A Hydron unit prototype is show in Figure 1. Each unit is roughly circular with a slightly narrowed equator, giving an approximate height of 12cm and width of 10cm. The Hydron is suspended in water, and actuated in the horizontal plane by four nozzles which expel water drawn through an impeller at the bottom of the unit when activated, and which are selected by a rotating collar. A syringe draws or expels water through the bottom of the unit to control unit buoyancy, and thereby actuate the unit along the vertical axis. Each unit's hull will also support a small set of switchable optical sensors and emitters capable of transmitting data over short ranges. Optical sensors and transmitters were chosen because they provide a simple and flexible underwater communication mechanism. Experiments to determine the range and effectiveness of this scheme are currently underway.

Fig. 1. A Hydron Unit Prototype

Although previous research has assumed an alternate optical sensor and transmitter placement [8], in this paper the optical sensors and transmitters are assumed to be located at each nozzle, and directly on top of and underneath the unit. Each transmitter unit is surrounded by four sensor units, forming a diamond. We assume that each sensor and emitter is active at a maximum angle of just less than $\frac{\pi}{4}$ to the normal, so that sensors at a docking site can receive only signals from emitters at a compatible docking site on another Hydron.

We term the the area around the transmitter a *docking site* or *binding site*, and it is at these sites that the Hydrons are to align themselves with each other during assembly. During the assembly process, units that wish to bind at a particular site switch on the relevant transmitter; free units move toward these lights, and two units are considered docked when their transmitter and sensor sites are sufficiently close together. The sensor and transmitter placing thus simplifies control because each quartet of sensors allow precise alignment against a particular transmitter.

At the time of writing, two prototypes with functional propulsion systems have been successfully designed and built at Edinburgh. The third prototype will include the optical communication system, and be batch produced.

Since the Hydron is still at the prototype stage, current research takes place in simulation. The simulator models the physical characteristics of the Hydron unit and its light sensors and emitters, as well as the drag properties of water, although it does not as of yet fully model its fluid dynamics. This should not present a major problem since the prototypes move fairly slowly (at an average of about 1.4cm per second), thus hopefully minimising turbulence.

One serious potential problem that will likely affect docking in the physical robots is the fact that they are unactuated about the vertical axis. They may thus obtain docks skewed by up to $\frac{\pi}{4}$ radians if they encounter turbulence that rotates them about the vertical axis. Although control code could be added to allow individual units to constantly shift their positions to compensate for the effects of turbulence and minor position fluctuation, rotation about the vertical axis may result in the deformation of the target object, and such a situation would be difficult to detect and rectify. We solve this problem in the simulator through the use of electromagnets that are switched on to affect alignment during docking, although the eventual physical solution may differ.

Because Hydron units are suspended in water and can move about freely within it, they avoid having to perform planning (e.g., [2]) or imposing structural requirements on the assembled configuration (e.g., [7]). During assembly, Hydron units do not have to move across the surface of an already constructed body – instead, they can float around until they notice an optical signal from a site searching for a binding unit and then follow this light until they reach their intended position. Although this may require sufficiently many units initially placed in useful positions to achieve rapid assembly, and thus more units than strictly necessary, it may bring a measure of redundancy to the system.

Because the units would start in a completely disconnected initial configuration, self-assembly with Hydron units would be true self-assembly, rather than self-reconfiguration. This process has more in common with morphogenesis than other models, which rely on robots which start in one (arbitrary) configuration and crawl over each other to reach a second one.

A self-assembly system using the Hydron units as a basis might then be able to draw more inspiration from and provide more insight into the developmental stages of an organism than reconfiguration approaches, thus providing a bridge between reconfigurable robotics and computational development [3], and would provide a proof-of-concept complementary to Støy's self-reconfiguration system [7] (also developed under the HYDRA project).

In addition, the knowledge gained from such a system may be useful in the future since at very small scales air is viscous, and robots employed in nanotechnology applications will likely encounter similar control problems.

3 The HydroGen Process

The HydroGen design process is depicted in Figure 2. First, either an already available CAD model of the object is converted to a "modularised" Hydron unit representation (following Støy [7]), or design takes place at the unit level.

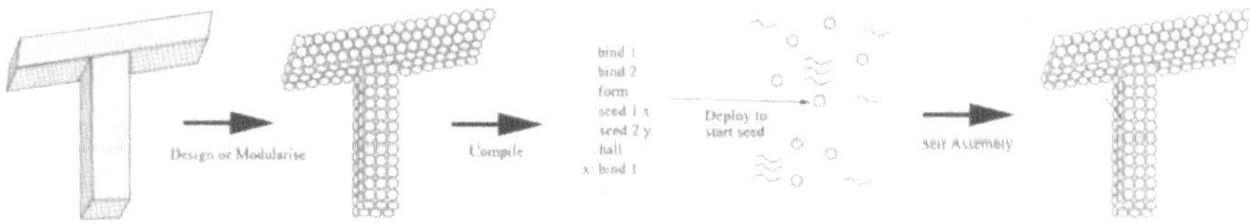

Fig. 2. The HydroGen Design Process

Given the unit representation, an object compiler then produces a list of instructions for its assembly, and these instructions are uploaded into a single initial *seed unit*, which floats in water with sufficiently many other units to complete the assembly process. This seed unit is responsible for initiating assembly and propagating the instruction list.

The code would operate under a simple seeded assembly scheme, where there are two types of Hydron units: free-floating units, and seeded units. Seeded units are already bound to the assembled structure, and can open their remaining binding sites. A free-floating unit then attaches to such a binding site, and is transformed to a seeded unit. No free-floating units execute any instructions until they dock and become seeded, whereupon the instruction list is transmitted along with an instruction label to begin execution with. Control and instruction propagation are thus both completely local and distributed, with the instructions specifying which binding sites each unit should open, and which instruction numbers the units bound to each should be seeded with. Thus, each unit carries an entire copy of the target representation, and the target is built using the progressive transfer of assembly position between units, as in Tomita et al. [9], resulting in a recursive assembly process.

A basic instruction set capable of accomplishing this is given in Table 1, along with a brief description of each instruction's function. In addition to the list given in Table 1, instructions may be given labels which are used as the second parameter to the `seed` instruction.

These instructions are clearly sufficient to reach any connected object configuration, provided that some Hydron unit can reach each binding site where necessary. In natural systems, cells and structures can be grown where they

Table 1. The Basic Hydron Construction Set

`bind` x	Bind another Hydron unit to site x.
`form`	Wait for all sites to complete binding.
`seed` x y	Transform the Hydron bound at site x to a seed, starting execution there at the instruction labelled y.
`halt`	Stop processing at this unit.

are required, but in a system consisting of seeded and free-floating Hydron units, a bind point may become unreachable because the path to it from all points outside of the seeded region is blocked.

One way to solve to this would be to enforce a breadth-first ordering on binding and seeding; however, for some combinations of structure, seed choice and controllers, blocking may still occur. The `form` instruction has been included as a point where some form of synchronisation behaviour can take place if necessary. This could take the form of signal propagation (where units that have not completed forming broadcast a growth suppression signal), or a simple timed seeding pulse.

In later work, we intend to expand the instruction set to potentially include explicit signal and gradient propagation, variable access instructions, etc., outlined in section 6.

4 Generating Self-Assembly Code

The object compiler is the heart of the HydroGen process. This section describes the first HydroGen compiler and its supporting programs.

The Hydron Object Designer provides an intuitive interface allowing users to construct objects out of Hydron molecules, by manipulating the user's viewpoint and adding and removing individual units. Although object descriptions will eventually likely be discretisations of CAD object models (as in Støy [7]), an object designer that works at unit level is initially a more useful prototyping tool. The Object Designer produces a description of the object as a simple list of Hydron coordinate triplets, suitable as input to the object compiler.

The Hydron Object Compiler then generates a set of self-assembly instructions that can be used to build the required structure in a distributed fashion. These instructions are required to be interpreted either by a control program running on an individual Hydron unit, or by the Hydron Object Interpreter, described below.

The compiler generates the instruction list as follows. First, the Hydron coordinate list is read into memory, and sorted on x coordinate, breaking ties first on y and then on z coordinate values. The first unit in the unit list is then placed in a unit queue, and marked as considered. The algorithm enters a loop that removes the first unit in the queue, and performs a binary search for each of the coordinate tuples that would be occupied by units docked at each site to determine which ones are present. Each unit present at a binding site generates a `bind` instruction; those that have not been marked as considered generate `seed` instructions and are added to the end of the unit queue. A `form` instruction is placed between the unit's list of `bind` and `seed` instructions and a `halt` instruction is placed at the end of the unit's instruction list. This list is then given the label hp and output, where p is the unit's position in the input list, and the loop repeats. The entire compilation process takes $\Theta(n \log n)$ time, where n is the number of Hydrons in the configuration, and $\Theta(n)$ space.

The use of a queue rather than simple recursive method for code generation results in an instruction list that seeds units in breadth-first order, with the aim of increasing the amount of binding that can occur in parallel without synchronisation. We consider this a good compromise between allowing for unordered growth, which maximises the amount of parallel docking possible but potentially blocks many docking sites, and a serial docking schedule, which removes the potential for parallel construction but allows for a completely predictable development process. The controller implementation could either rely on a breadth-first growth order happening naturally because of this, or employ a synchronisation method to ensure it, as in Tomita et al. [9].

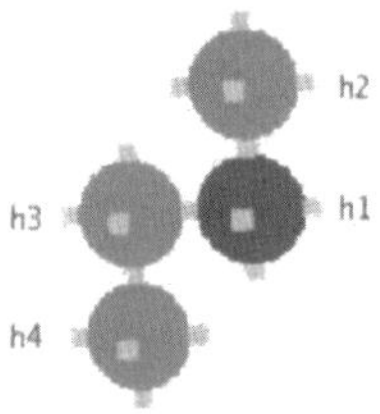


```
h1:  bind 1          h3:  bind 0
     bind 5               bind 4
     form                 form
     seed 1 h2            seed 0 h4
     seed 5 h3            halt
     halt            h4:  bind 1
h2:  bind 0               form
     form                 halt
     halt
```

Fig. 3. A Sample Configuration and its Self-Assembly Code.

A simple object configuration (with the seed unit (h1) in darker gray than the others) along with the resulting compiler output is given in Figure 3. At present the compiler generates instructions for each unit, and since no code optimisation is performed, the instruction list length is proportional to the number of units in the target configuration.

The Hydron Object Interpreter is used to verify the code produced by the the Object Compiler, or to rapidly evaluate the object configurations produced by some other process (e.g., a genetic algorithm) without requiring a physically realistic Hydron unit simulation.

The interpreter starts with the seed unit and (if necessary) creates a Hydron with the required coordinates whenever a `bind` instruction is executed. Figure 4 shows four frames of the development of a simple Hydron structure in the Object Interpreter.

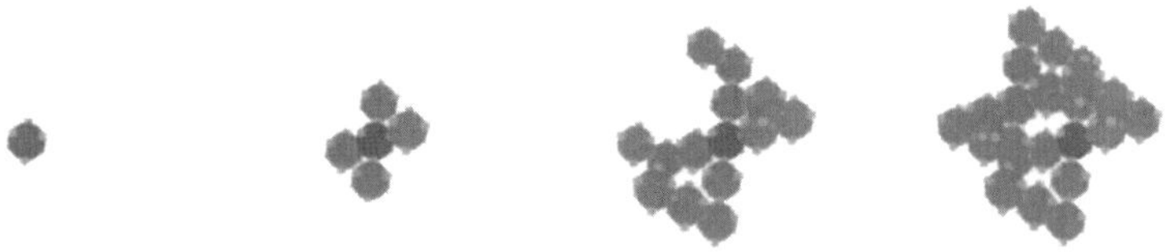

Fig. 4. The Object Interpreter Assembling a Hydron Structure

5 Self-Assembly in a Physically Realistic Simulator

This section outlines a preliminary implementation of the HydroGen instruction set in a reasonably realistic physical simulator. The implementation does not tackle the problem of breadth-first seeding co-ordination, or the problem of multiple light sources confusing a free-floating unit. In addition, it assumes that seeded modules bound to the existing structure are sufficiently rigidly attached as to not be knocked away by docking modules.

Here, free-floating modules pick the quartet of optical sensors with the highest overall reading, and use a proportional control scheme (with water resistance providing a differential damping component) to align the appropriate docking port with the light source by attempting to equalise the readings across the sensor quartet. Experience with the simulator indicates that this control method reliably brings the floating unit sufficiently close to the signalling unit for it to activate the appropriate electromagnet and complete the dock. The `form` instruction does not complete until all of the required binding sites have docked. This implementation is sufficient for constructing some structures but will fail when the structure has a hole in it because some docking sites will be blocked.

Figure 5 shows two assembly sequences. In the first, six free-floating Hydrons bind to all the sites on a seed Hydron, demonstrating docking and simple structural formation. This sequence required approximately 1030 simulated seconds.

The second sequence shows a slightly more complex configuration in development. Although in this case the Hydrons had to be placed so that they did not physically interfere with each other, the instructions are clearly being propagated, and the structure was assembled correctly in 3600 seconds.

6 Future Work

The first version of the Hydron Object Compiler system presented here is intended to form the basis for further work that will further develop the system using ideas from compiler optimisation and cell biology. This section outlines the major directions that future research is expected to take.

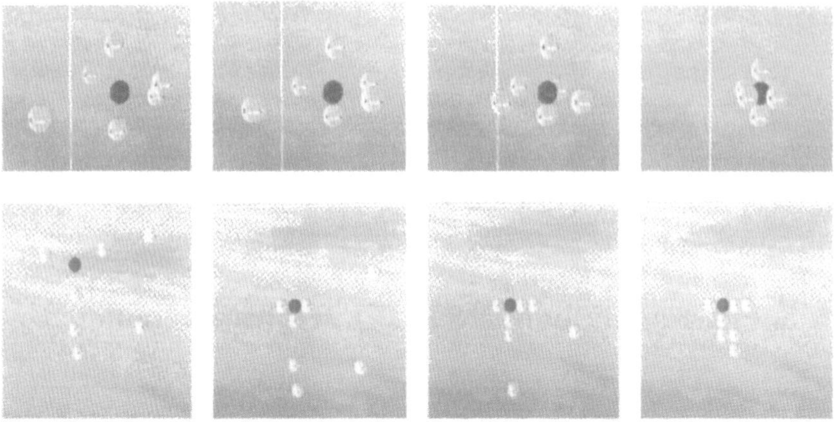

Fig. 5. Simulated Self-Assembly Sequences

6.1 Hardware and Simulation Implementation

Although the implementation given in section 5 can perform rudimentary assembly, it is not yet complete. It does not implement any form of breadth-first signal propagation, nor does it tackle the problems that could occur when two adjacent Hydrons switch on transmitters pointing in the same direction (although a form of local signal suppression would probably be sufficient here).

More importantly, it cannot reach all connected object configurations, because of the simplistic method used for Hydron docking and the narrow range of the Hydron's sensors. Other controller implementations (e.g., where a binding site notices that its way is blocked, lights up an orthogonal docking site, and instructs an arriving Hydron to move around) may solve this problem, or it may require preemptive blockage handling by the compiler; we expect it to require some combination of the two.

Further work is thus required to develop methods that can assemble all HydroGen instruction sequences in real and realistically simulated robots. The development of these methods and the further research detailed below (which is concerned exclusively with the compiler) will proceed concurrently, since the two processes are mutually informing.

6.2 Growing Disconnected Components using Cell Death

Another immediate limitation of the basic instruction set is that it cannot be used to assemble objects that consist of two or more disconnected pieces. One way to resolve this would be to construct the object with extra scaffolding units connecting the separate components of the object. A form of timed (or programmed) cell death, which occurs during development in natural systems for fine feature formation [4], could then be used to remove the scaffolding units once assembly is complete. The system could then reach any Hydron unit construction, provided the scaffolding units could escape from the configuration or perhaps find somewhere else to bind. This could also be used to solve blocking conflicts by initially construction portions of the object as solid and then removing Hydrons to obtain the intended structure.

6.3 Variables and Common Code Segments

The code generated by this version of the object compiler is linear in the size of the target configuration. One way to reduce code size and increase modularity would be to reuse of common code segments [1] and use variables to express repeated structures. This would require the addition of simple branching and arithmetic operators to the instruction set and extra functionality to the compiler, but would result in more concise code.

6.4 Obtaining Symmetry through Cellular Gradients

Another way to increase the expressiveness of the instruction set would be the use of a cellular gradient to specify binding site numbering. For example, if the start seed sent out a gradient and the other units numbered their binding sites starting at the site receiving it, then common code would produce radial symmetry. This would not require major changes to the compiler but there may be other interesting symmetry generating mechanisms (e.g., having multiple symmetric origins) that would require more instructions and further compiler functionality.

6.5 Evolving Hydron Configurations

Finally, it would be useful to attempt to bridge the gap between explicitly designed structures and those developed by evolutionary techniques, especially since the instruction set augmented with the additions described above represents an expressive genomic language. The use of an Object Interpreter may also allow for extremely fast genome evaluation, while preserving the ability of evolved solutions to be expressed in realistic environments.

Another possibility might be the translation of either the basic instruction set or a later variant of it to a Genetic Regulatory Network [6, 8]. The HydroGen system could then be used to seed such systems, rather than requiring evolution to start from scratch.

7 Summary

This paper has introduced the HydroGen object compiler, a system that produces instructions for the self-assembly of Hydron unit configurations, and briefly described the Hydron unit. It has also presented the first implementation of the Hydron Object Compiler, supporting a basic instruction set capable of expressing any connected unit configuration, and a controller that demonstrates that the system is capable of assembly in a reasonably realistic simulator. This system is intended as a platform for the further development of self-assembly methods, through the future research areas given in section 6. The aim of this research programme is to shed some light on the characteristics and functional requirements of natural and synthetic cellular self-assembly systems.

Acknowledgments

This work forms part of the HYDRA project (EU grant IST 2001 33060, `http://www.hydra-robot.com`). Facilities for this research were provided by the Institute of Perception, Action and Behaviour in the School of Informatics, University of Edinburgh. We would also like to thank the anonymous referees for their helpful comments and suggestions.

References

1. A.V. Aho, R. Sethi, and J.D. Ullman. *Compilers: Principles, techniques and tools.* Addison-Wesley, 1986.
2. K.D. Kotay and D.L. Rus. Algorithms for self-reconfiguring molecule motion planning. In *Proceedings of the International Conference on Intelligent Robots and Systems*, 2000.
3. S. Kumar and P.J. Bentley. An introduction to computational development. In S. Kumar and P.J. Bentley, editors, *On Growth, Form and Computers*, chapter 1, pages 1–44. Elsevier, 2003.
4. H. Lodish, A. Berk, S.L. Zipursky, P. Matsudaira, D. Baltimore, and J. Darnell. *Molecular Cell Biology.* W.H. Freeman & Co., New York, NY, 4th edition, 1999.
5. E. Østergaard and H. H. Lund. Evolving control for modular robotic units. In *Proceedings of the 2003 IEEE International Symposium on Computational Intelligence in Robotics and Automation*, Kobe, Japan, 2003.
6. T. Reil. Dynamics of gene expression in an artificial genome – implications for biology and artificial ontogeny. In D. Floreano, J.-D. Nicoud, and F. Mondada, editors, *Proceedings of the Fifth European Conference on Artificial Life (ECAL99)*, 1999.
7. K. Støy. Controlling self-reconfiguration using cellular automata and gradients. In *Proceedings of the 8th International Conference on Intelligent Autonomous Systems (IAS-8)*, March 2004.
8. T. Taylor. A genetic regulatory network-inspired real-time controller for a group of underwater robots. In *Proceedings of the 8th International Conference on Intelligent Autonomous Systems (IAS-8)*, March 2004.
9. K. Tomita, S. Murata, H. Kurokawa, E. Yoshida, and S. Kokaji. Self-assembly and self-repair method for a distributed mechanical system. *IEEE Transactions on Robotics and Automation*, 15(6):1035–1045, December 1999.
10. M. Yim, Y. Zhang, J. Lamping, and E. Mao. Distributed control for 3D metamorphosis. *Autonomous Robots*, 10(1):45–56, 2001.

Part II

Emergence of Intelligence Through Mobility

Local Obstacle Avoidance with Reliable Goal Acquisition for Mobile Robots

Chomchana Trevai[1], Ryota Takemoto[1], Yusuke Fukazawa[1], Jun Ota[1], and Tamio Arai[1]

The Univ. of Tokyo, School of Eng., Dept. of Precision Eng. Arai&Ota Lab.
{ctrevai,takemoto,fukazawa,ota,arai}@prince.pe.u-tokyo.ac.jp

Summary. In this paper, we propose a method of obstacle avoidance and goal acquisition for mobile robots in unknown environments. We have modified the navigation method, Vector Field Histogram(VFH) by Borenstien *et al.*. Our method, Sensor Based Vector Field Histogram(SBVFH), designs for more sensor-reactive method. Our method concerns the situation that several mobile robots travelling in the environment. The mobile robot is goal-directed while trying to avoid static and moving obstacles. The results of simulation and experiment show that the algorithm is efficient with relatively cheap sensors.

Key words:mobile robot, obstacle avoidance, navigation

1 Introduction

Obstacle avoidance is one of fundamental key to successful applications of autonomous mobile robot systems. Obstacle avoidance means detection of obstacle and stop or change direction of the mobile robot in order to avoid the collision. Also there are sophisticated algorithms , which enable the robot to avoid the obstacle and proceed toward its goal. In this paper, we focus at the problem of obstacle avoidance and goal acquisition for mobile robots operating in unknown or partially known environments. The problem has been studied in previous works[1],[2],[3], [4], [5].

Our method is the extended version of Vector Field Histogram(VFH) method[1]. Our method utilizes raw sonar sensor readings to rapidly avoid obstacles. The method is called Sensor Based Vector Field Histogram(SBVFH). Our approach concerns the situations that there are moving obstacles (e.g. other mobile robots) in the environments. The method navigates a mobile robot safely, even with a noisy sensors(such as sonar sensor). The mobile robot is goal-directed while trying to avoid obstacles. The mobile robots is able to avoid collision with each other. Our mobile robots can communicate through wireless LAN. However, the mobile robots are not use the communicated information to plan priority for each mobile robot to move. The goal acquisition is done using map, which is built as the mobile robot moving. Our method works by choosing velocity and direction that satisfies all constraints and maximizes an objective function. Constraints derive from physical limitations of the mobile robot, and from the sensor readings that indicate the presence of obstacles that also be used to build map. Our method also interprets the sensor readings to obtain acceleration and deceleration in the selected approaching direction. The deceleration of the mobile robot prevents the deadlock situation when encounter other mobile robots.

To achieve real-time performance, our method simply searchs approximated direction to goal in the map. Several simple extensions make the basic method more robust to sensor noise and reduce the possibility of the robot getting stuck. The experiments on our actual mobile robot demonstrate that our method provides safe and reliable navigation for indoor mobile robot systems.

2 Obstacle Representation

To represent the obstacles in an environment, the mobile robot could build a model of the environment from the sensor readings. In our research, an occupancy based model is used for representation of the model[8],[6],[9]. The model is especially suited to path planning, navigation and obstacle avoidance because it explicitly models free space.

In occupancy based model, the mobile robot's environment is represented by a two-dimensional array of square occupancy grid. The sensor readings are fused into the model using the Dempster-Shafer inference rule[9]. The model becomes map, which represents the environment of the mobile robot. The sensor model of the sonar sensors equipped on our mobile robot is shown in Fig. 1. The assumption made from the sensor reading is that the echo is generated somewhere on an arc at range R, within the sensor beam $\pm\beta$.

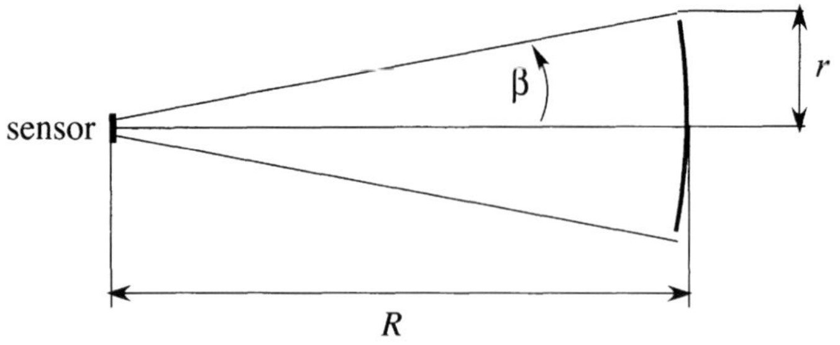

Fig. 1. Sonar sensor model

3 Sensor-Based Vector Field Histogram Algorithm

Our method is inspired by Vector Field Histogram(VFH)[1] method originally developed by Borenstien and Koren for real-time local obstacle avoidance. Our method utilizes histogram directly from raw sonar sensor readings. The traversibility to the goal is evaluated using occupancy map[6].

Our mobile robot is omnidirectional mobile robot. The mobile robot was equipped with k sonar sensors. The sensing area of sensor $1, 2, \ldots, k$ is shown in Fig. 2a. The n sector of candidate moving direction for the mobile robot is depicted in Fig. 2b. Our mobile robot exploits the raw sensor readings to avoid obstacles while determines the velocity to approach it goal.

3.1 Sensor Reading Interpretation

The sensor reading is evaluated to calculate traversibility for each sector. The value of traversibility is called Sector Value(SV). At each sampling cycle, the SV is classified into 4 cases. Each case indicates the possibility of the mobile robot to collide with obstacles as follows.

- S_1 : no collision possibility at maximum speed in the direction
- S_2 : deceleration is needed to avoid the collision if the obstacle is moving
- S_3 : deceleration can avoid only a static obstacle
- S_4 : cannot approach at any speed in the direction

For each case, following condition is obtained respectively.

- if($\tau_H \leq d$) then S_1
- if($\tau_M \leq d \leq \tau_H$) then S_2
- if($\tau_L \leq d \leq \tau_M$) then S_3
- if($d \leq \tau_L$) then S_4

Here, d is the distance from sensor reading and τ is the distance indicates the borderline in each case of collision possibility. Therefore, each value of the borderline can be defined as

$$\tau_H = 2 \cdot V_{max} \cdot \Delta t + d_{safe} \tag{1}$$

$$\tau_M = (\rho + 1) \cdot V_{max} \cdot \Delta t + d_{safe} \tag{2}$$

$$\tau_L = \rho \cdot V_{max} \cdot \Delta t + d_{safe} \tag{3}$$

Here, V_{max} is the maximum speed of the mobile robot, Δt is the sampling rate of the sensor readings, ρ is the deceleration rate of the mobile robot and d_{safe} is the allowable distance from the mobile robot to obstacles. Collision possibility of each SV is determined using configuration obstacles[7]. Fig. 3 shows the configuration obstacles have influence on the sectors.

The priority of S_1 is higher than S_2, S_2 than S_3 and S_3 than S_4 at the sectors that was effected by multiple sensor readings.

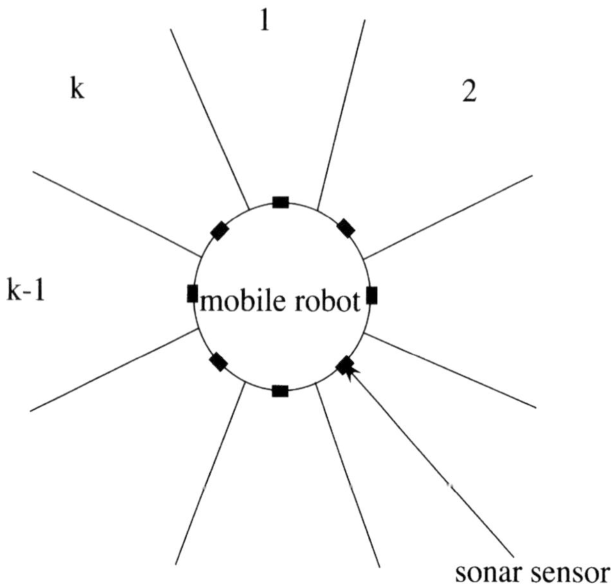

a)

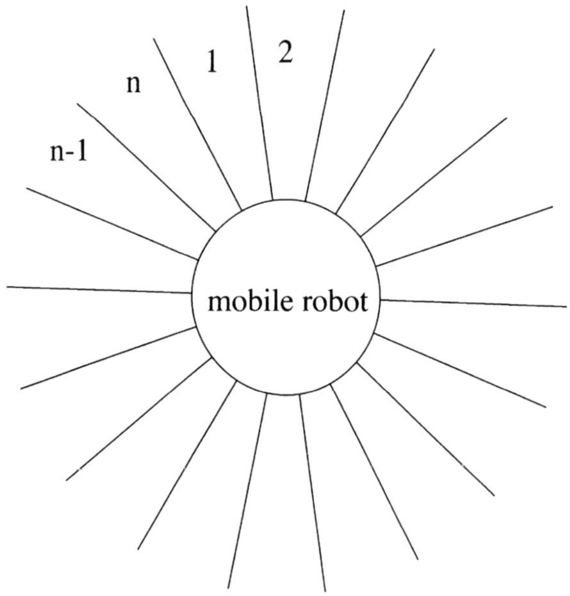

b)

Fig. 2. Sensing area (a) and Sector (b)

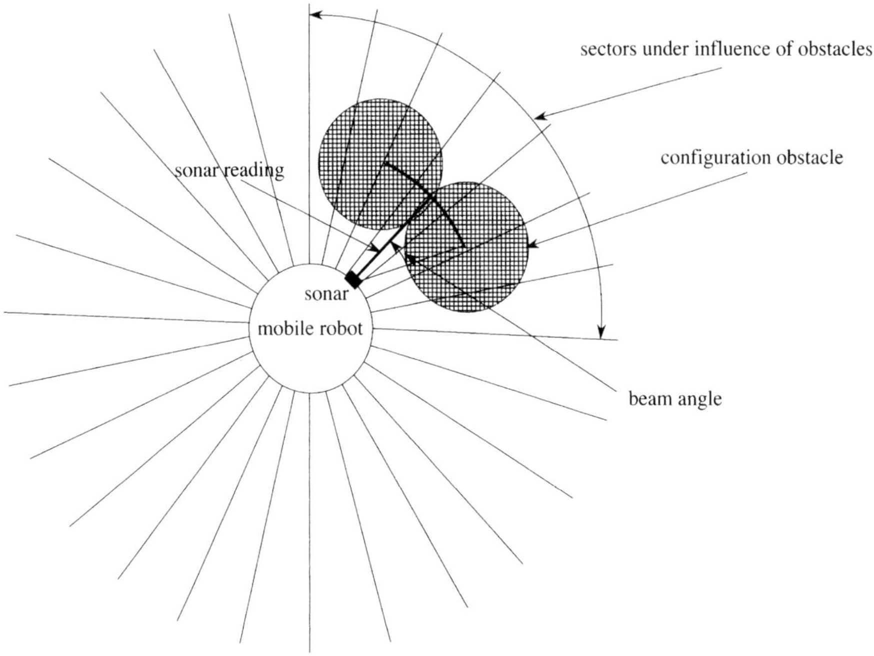

Fig. 3. Configuration obstacles over the sectors

3.2 Constructing Histogram

The polar histogram is built from SV of each sector as shown in Fig. 4. The directions$(\theta_1, \theta_2, \ldots, \theta_n)$ in the polar histogram is corresponding to the momentary position of the mobile robot. H_k shows the proper action that the mobile robot should take during traverse to the direction. H_k for each direction is derived from the condition in Section 3.1. H_k indicates the following actions.

- $H_k = 0$: acceleration
- $H_k = 1$: maintain speed
- $H_k = 2$: deceleration
- $H_k = 3$: not traversable

The mobile robot has to choose the approaching direction from the directions, which histogram are $H_k < 3$.

3.3 Goal Acquisition and Navigation Cost Factor

In the next stage, our algorithm has to compute the actual approaching direction that leads the mobile robot to the goal. The occupancy based map of the environment is used here to determine the proper approaching direction for the mobile robot. The direction to goal α is determine using map information. An instance of the navigation function family of algorithms [7]

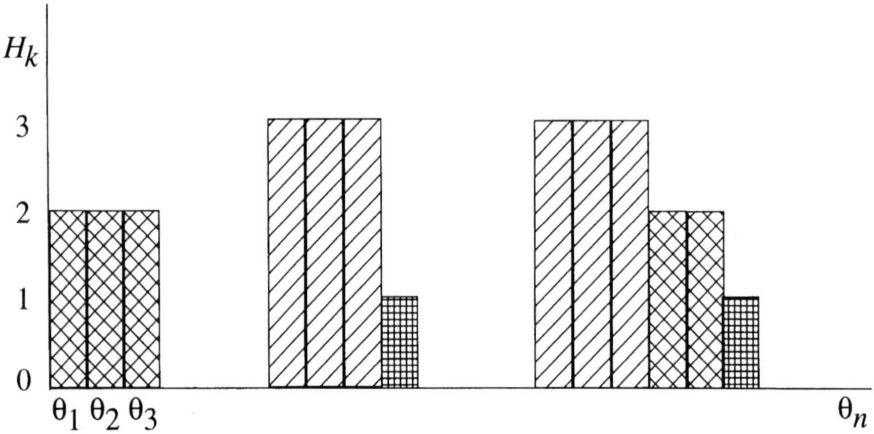

Fig. 4. Graph of histogram from the *SV* of each sector

is used to calculated the cost of driving from the current position to the proposed destinations based on the information in the occupancy grid map. α is the approximated direction, which has the lowest cost to the goal(Fig. ??). The driving cost is calculated from goal point to the current position of the mobile robot. By backtracking the search in the limited number of step the approximated direction to goal position can be obtained.

The navigation direction is evaluated from function $f(\zeta)$.

$$f(\zeta) = |\zeta - \alpha| + |\zeta - \beta| + \gamma \tag{4}$$

Here, ζ is the desired direction that should minimize $f(\zeta)$. α is the direction to goal. β is the current direction of the mobile robot and γ is the deceleration rate randomly selected from 0.3, 0.4, 0.5, 0.6, and 0.7. Randomly selecting deceleration rate also prevent the mobile robots to be trapped in the deadlock situations.

4 Simulation and Experiment

We have performed simulations to validate the SBVFH method. Our method is implemented in Player/Stage simulator[10],[11],[12]. The simulation is run on 750[MHz] PC. Fig. 5 shows the navigation and mapping simulation of a mobile robot. The mobile robot is approximated to be a circle with 230[mm] in radius. The mobile robot is equipped with 8 sonar sensors. The maximum range of the sonar sensor is 2[m]. The number of sector is 24. The parameters of SBVFH are as follows:

- $V_{max} = 0.5$[m/s]
- $\Delta t = 0.1$[s]

- $d_{safe} = 0.3$[m]
- $\rho = 0.5[\mathrm{m/s^2}]$

The sampling time Δt 0.1[s] is the value measured from the simulation. This sampling rate can be told to be fast enough for actual mobile robot to perform in actual environment.

The experiment with actual mobile robot has been done using two omnidirectional mobile robots. The drive mechanism of both mobile robot is developed by RIKEN [13]. The on-board processor of the mobile robot is the industrial 650[MHz] computer. Each mobile robot is equipped with 8 sonar sensors. The maximum range of the sonar sensor is 3[m] and the beam angle is ±45[deg]. The mobile robots maintain their odometry using encoders. The motion controller of the mobile robot receives desired velocities from the processor and sends velocity command to external PID controllers.

In the experiment, the mobile robots have a difficulty to accurately detect each other when encounter in close range. The difficulty happens due to the direct hit of another mobile robot's sonar sensor. To avoid the fault detection, in the experiment, the mobile robot broadcast their position through wireless LAN. The broadcasted position information is used only when mobile robots are too close to each other. The position information is used to recover the fault sensor readings.

The goal of the mobile robots in the experiment is to switch their position. There is a static obstacle in the middle of the environment. Fig. 6 shows the experiment of two actual mobile robots. In Fig. 6c, the mobile robot No.1 avoids the static obstacle and meet the mobile robot No.2. The mobile robot No.2 detects No.1 and changes direction to the other side of the environment. At the end, the mobile robots avoid each other and reach their goals successfully.

5 Conclusions and Future Works

We have presented the sensor based vector field histogram method for local obstacle avoidance. The method utilizes raw sensor readings and build map for safe navigation and goal acquisition in environment. SBVFH achieves real-time performance by approximating direction to achieve a goal from online built map. The method needs low communication cost and does not has to explicitly determine the priority of every mobile robots moving in the environment.

The method has been implemented on actual omnidirectional mobile robots. The method provides safe navigation in the environment with several mobile robots moving together.

The future work based on this method is to apply the method to more complex task. We would also like to investigate the method to apply to navigation and/or planning problems. Also the situations, which are people walk-

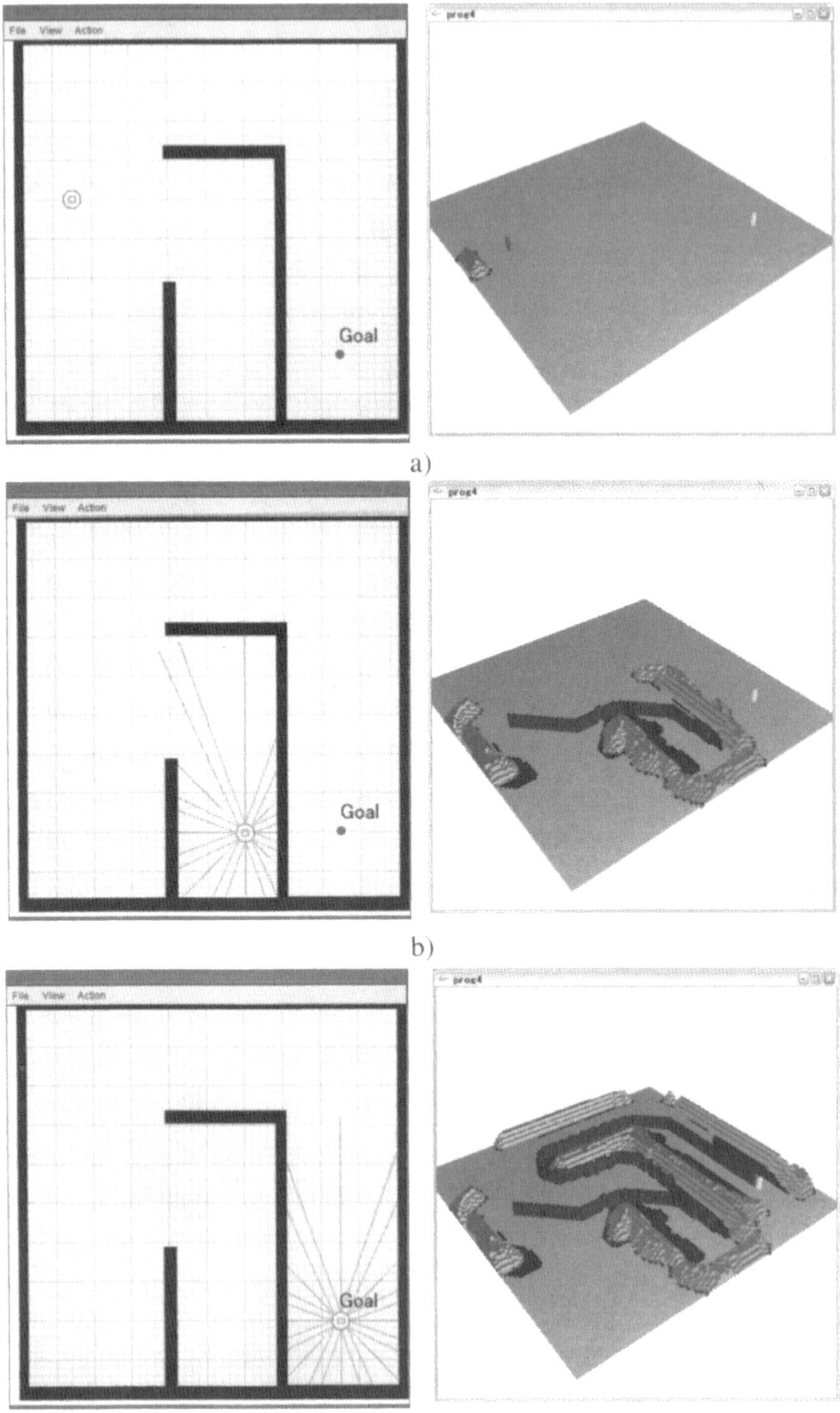

Fig. 5. Navigation and mapping simulation

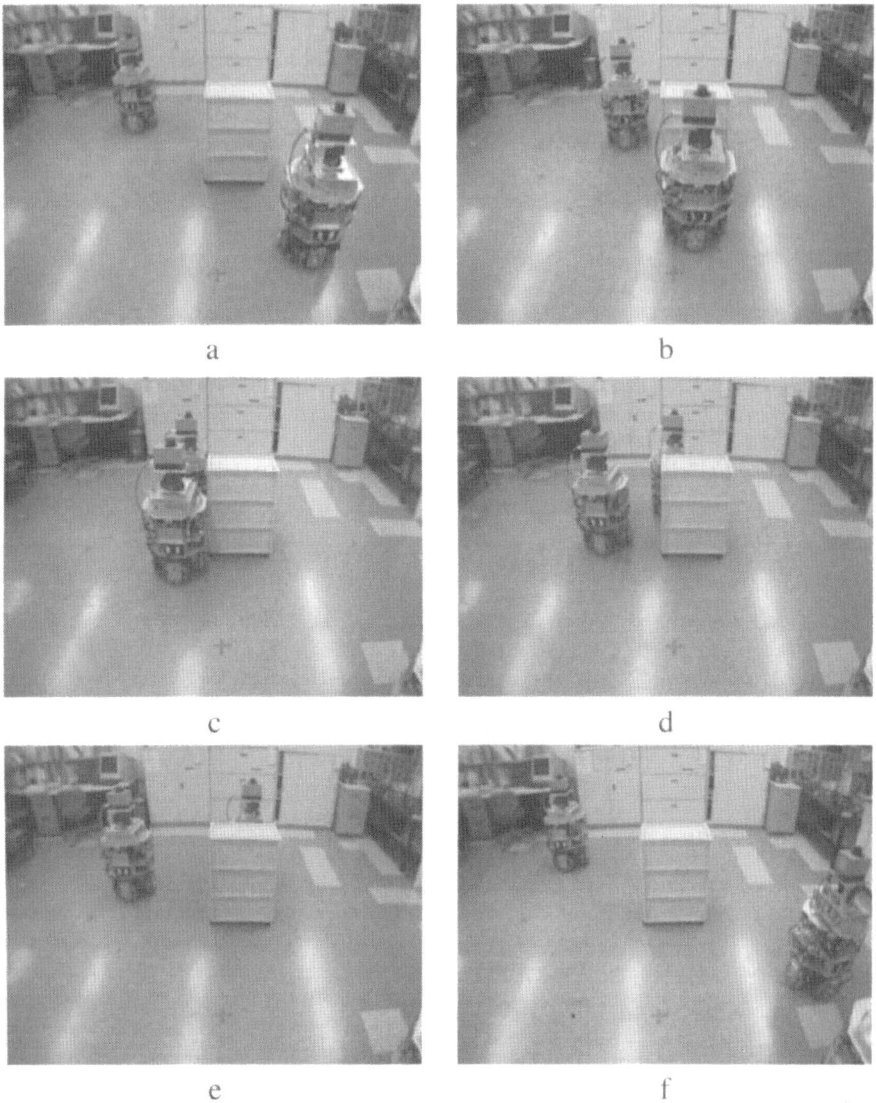

Fig. 6. Experiment with two actual mobile robot

ing in the environment are challenging. The extension and modification for non-holonomic should be made in the near future.

Acknowledgement

The experiment using omnidirectional drive machanism in this paper has been done under advisory of Prof. Hajime Asama of the Univ. of Tokyo.

References

1. J. Borenstein and Y. Koren, The vector field histogram-fast obstacle avoidance for mobile robots. IEEE Transactions on Robotics and Automation. 7(3),pp.278–288, June 1991.
2. D. Fox, W. Burgard and S. Thrun. The dynamic window approach to collision avoidance, IEEE Robotics and Automation Magazine, Vol. 4, No. 1, pp.23-33, March 1997.
3. R. Simmons, The Curvature-Velocity Method for Local Obstacle Avoidance, IEEE Int. Conf. on Robotics and Automation, pp. 3375–3382, April 1996.
4. O. Khatib, Real-time Obstacle Avoidance for Manipulators and Mobile Robots, IEEE Int. Conf. on Robotics and Automation, pp. 500-505, March 1985.
5. K. Konolige, A gradient method for realtime robot control. IEEE/RSJ Int. Conf. on Intelligent Robots and Systems, 2000.
6. H.P. Moravec, Sensor Fusion in Certainty Grids for Mobile Robots. AI Magazine, pp.61–74, 1988.
7. J.-C. Latombe, Robot Motion Planning, Kluwer Academic Publishers, Boston, 1991.
8. A. Elfes, Sonar-based real-world mapping and navigation, IEEE Journal of Robotics and Automation, Vol.RA-3, No. 3, pp.249–265, 1987
9. D. Pagac, E. M. Nebot and H. Durrant-Whyte, An Evidential Approach to Map-Building for Autonomous Vehicles, IEEE Transaction on Robotics and Automation, Vol.14, No. 4, August, 1998
10. R. T. Vaughan, B. P. Gerkey, and A. Howard, On device abstractions for portable, reusable robot code. In Proc. of the IEEE/RSJ Int. Conf. on Intelligent Robots and Systems (IROS), pp.2121–2427, Las Vegas, Nevada, October, 2003.
11. B. P. Gerkey, R. T. Vaughan, and A. Howard, The Player/Stage Project: Tools for Multi-Robot and Distributed Sensor Systems, In Proc. of the Int. Conf. on Advanced Robotics(ICAR), pp. 317–323, Coimbra, Portugal, July 2003.
12. B. P. Gerkey, R. T. Vaughan, and K. Stϕy, A. Howard, M. J. Mataric and G. S. Sukhatme, Most Valuable Player: A Robot Device Server for Distributed Control. In Proc. of the IEEE/RSJ Int. Conf. on Intelligent Robots and Systems(IROS),pp. 1226–1231, Vailea, Hawaii, October 2001.
13. H. Asama and M. Sato and L. Bogoni and H. Kaetsu and A. Matsumoto and I. Endo, Development of an omni-directional mobile robot with 3 dof decoupling drive mechanism, Proceeding of IEEE International Conference on Robotics and Automation, April, 1996

Adaptive Routing System by Intelligent Environment with Media Agents

Takenori Matsuoka, Daisuke Kurabayashi, and Katsunori Urano

Tokyo Institute of Technology, 2-12-1 Ookayama, Meguro-ku, Tokyo, 152-8550

Summary. In this paper, we consider a distributed robotic system that includes special agents that convey the information. We address the issue of selecting one course from two;a long one-way detour or a short two-way path on which traffic jams may occur. We consider a system in which the environment, instead of mobile agents, learns feasible parameters for task execution. To correct problems with this system and improve it we introduce media agents that carry data for the learning. They adjust information flow. We formulate the system and evaluate its performance.

1 Introduction

With the continuing development of robots, higher-level work by cooperative robots is becoming possible in various situations;well-defined places like plants, uncertain and dangerous environments such as disaster areas, planets and so on. When cooperative robots work in the areas described above, there are many problems. Assembling in small areas, they must avoid collisions with each other. Scattered over an area, they have to have some method of transmitting information. One of these problems is a physical routing problem. If most robots that configure as a swarm concentrate on the same route, their efficiency in moving is decreased. So at least some of them should select another route. To solve this problem various researchers have considered using learning techniques. Some researchers have viewed robots as learning actors. Ota proposed a learning method[1] to make one-way roads autonomously. Yoshimura proposed a method[2] to select a detour or a direct route depending on the crowds. The other researchers considered environment as teaching the actor a system similar to ants that exploit their pheromone to form a line of ants[3][4][5]. But they have not sufficiently taken account of mutual interference, which Beckers indicated[6], and have not evaluated the performance of those systems quantitatively. Kurabayashi compared two learning actors [7]—robots and environment— that learn the strategy about an issue of selecting paths. He quantitatively indicated the performance of the two as learning actors, and showed that environments are more suitable than robots.

In this paper we analyze a distributed robotic system that gradually optimizes a strategy to select courses for various parameters of environment. We address the issue of whether to select a detour or a straight course. The straight course is a two-way path which is the shortest distance between two points and on which traffic jams may occur, meanwhile the detour is a one-way path whose distance is longer than the straight course,but there is no problem of traffic jams. Because of the work of [7], we consider junctions in environment as learning actors, building up a system that maximizes the efficiency of movement in the environment by optimizing the choice of courses. By the way, it is said that a swarm of ants has "media ants" to convey information. So we introduce "media agents" analogous to media ants to convey data between junctions that have no tools to transmit information to each other. We formulate the system with media agents and estimate the advantage of them quantitatively.

This paper consists of 6 sections as follows. Section 2 describes the environment model. Section 3 formulates the optimal condition of the model to estimate the performance of other conditions, and shows the defects of reinforcement learning by environment. Section 4 introduces media agents to cover the defects of reinforcement learning. Section 5 gives the conclusion.

2 Environment model

We set a network-like environment that has several routes and junctions(for example Fig. 1). The black circles in Fig. 1 represent junctions and the line segments represent routes. Although routes physically link junctions, they can't transmit information to each other by themselves. Robots move on these routes. Each of the routes consists of "course A" which is a one-way detour and "course B" which is the shortest way between two junctions. But course B is two-way(Fig. 2). To move between ${}^iG_1$ and ${}^iG_2$ in Fig. 2, course A of route i costs "${}^ia$", and course B of route i costs "${}^ib_1(<^i a)$" and "$n{}^ib_2$" when there are n robots on the same course. b_2 represents the width of course B, which means b_2 also represents the degree of traffic jams. Hereinafter "$({}^ia,\ {}^ib_1,\ {}^ib_2)$" represents an i-th route cost.

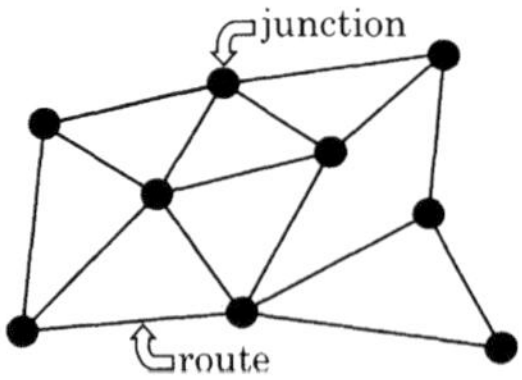

Fig. 1. Environment

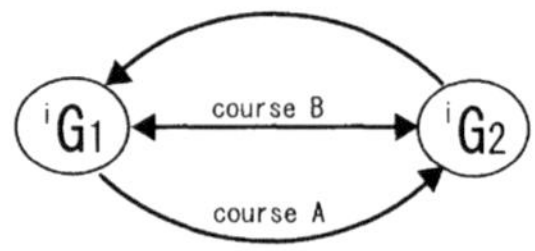

Fig. 2. Courses in route i

The efficiency of movement for robots in the environment changes with the choice of course A or B. In this paper we optimize the probability of choosing course A "p_a" as a strategy for moving on the environment.

We define the robot model as follows;

- Robots start moving from a junction with probability "p_G". The route that robots move on is random.
- The choice of course is determined according to p_a at junction ${}^iG$. (Fig. 2)
- Robots move at 1 cost per 1 step.

- Robots are distributed autonomous systems.
- Robots don't have the map of the environment.

Under this model we consider the following situations.

(i) Junctions keep and learn p_a as part of the strategy. They learn p_a by information from robots and point out to the robots which way to go according to p_a. We refer to these junctions as "Intelligent junctions".
(ii) To improve the efficiency of (i) we introduce "media agents" that carry information about routes that they took. They compensate for the inadequacy of junction's communication ability.

We ran computer simulations for a specific time and evaluated the accumulated number of movements the robots execute as "the number of achievements".

3 Optimal condition and reinforcement learning

The movement cost must be minimized to maximize the number of achievements. When p_a satisfies the condition that minimizes the movement cost, we refer to it as the optimum probability of course A "p_{opt}". In this section we formulate the system with a probabilistic model and derive p_{opt}. And we compare p_{opt} and p_a obtained by reinforcement learning to show the limit of its performance. In this section we point out some of its defects.

3.1 Formulation of optimal condition

When a junction retains p_a, then a route that has a junction at both ends has two values of p_a. So we assume that a route has the average value of the two p_a.

Consider the following condition; the environment has "m" routes and "N" robots. There are "${}^i n_B$" robots on course B of route i. Route i has cost "$({}^i a,\ {}^i b_1,\ {}^i b_2)$". And the probability of course A of route i is "${}^i p_a$".

The expected movement cost of route i "${}^i f({}^i p_a)$" is represented by

$$ {}^i f({}^i p_a) = S_G + \ {}^i p_a \ {}^i a + (1 - \ {}^i p_a)({}^i b_1 + \rho \ {}^i n_B \ {}^i b_2) \tag{1} $$

where $\rho = (N-1)/N$, and S_G is the expected waiting time until a junction orders a robot to start moving. This is derived from p_G. "$f({}^1 p_a, {}^2 p_a, \cdots, {}^m p_a)$", which is the expected movement cost considering all routes becomes equation(2).

$$ f({}^1 p_a, {}^2 p_a, \cdots, {}^m p_a) = S_G + \frac{1}{m} \left\{ \sum_{i=1}^{m} {}^i p_a^i a + \sum_{i=1}^{m} (1 - {}^i p_a)({}^i b_1 + \rho^i n_B^i b_2) \right\} \tag{2} $$

When we assume that the robots are evenly distributed on the routes, the number of robots on route i " ${}^i n$" and the number of robots on course B on the same route " ${}^i n_B$" have the relationship given bellow.

$$\frac{{}^if({}^ip_a)}{{}^in} = \frac{(1 - {}^i p_a)({}^ib_1 + \rho \, {}^in_B \, {}^ib_2)}{{}^in_B} \tag{3}$$

When we set the value of ${}^in$, we can determine a unique optimal ${}^ip_a(= {}^ip_{opt})$ to minimize ${}^if({}^ip_a)$. Then ${}^if({}^ip_{opt})$ can be expressed as ${}^if({}^in)$(the function of ${}^in$). Route i and $j(\in m)$ have the relationship of equation(4). The sum of the robots on all routes corresponds to N as equation(5).We can derive ${}^in$ and ${}^ip_{opt}$ with equations(2)-(5).

$$\frac{{}^if({}^in)}{{}^in} = \frac{{}^jf({}^jn)}{{}^jn} \quad (i, j \in m) \tag{4}$$

$$\sum_{i=1}^{m} {}^in = N \tag{5}$$

We show the comparison of the derived p_{opt} and the searched p_{opt} in Fig. 3 for the following parameters; $m = 3$, $N = 20$, $({}^1a, {}^1b_1, {}^1b_2) = (6, 1, 2)$, $({}^2a, {}^2b_1, {}^2b_2) = (6, 2, 4)$, $({}^3a, {}^3b_1, {}^3b_2) = (12, 1, 4)$. The average error is 0.047, and the variance is 2.6×10^{-3}. We have successfully formulated the system.

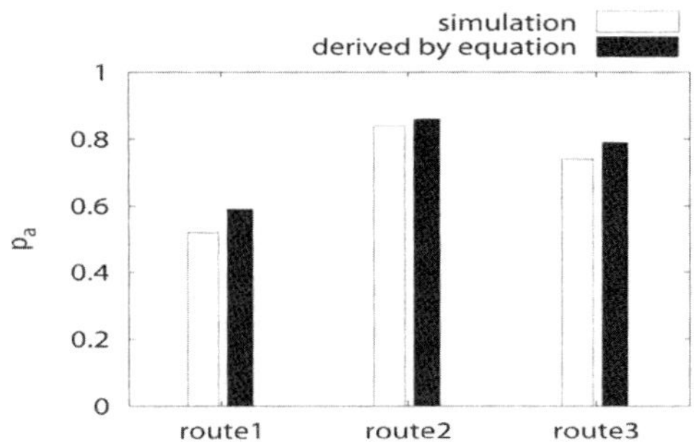

Fig. 3. The comparison of p_{opt}

3.2 Learning of p_a by Intelligent junction

We try to optimize the strategy "p_a" by reinforcement learning with intelligent junctions. We employ often-used reinforcement learning, the same as [7].Each of the junctions renews p_a to optimize it with the following algorithm.

I Junction G_1 which is at one end of route i points out course A or B according to ${}^ip_a$ to a robot which comes into route i. We call this behavior "trial".

II Junction G_2 which is at another end of route i gets ${}^is$ and the course information(A or B) from a robot which comes from G_1. ${}^is$ is the number of steps from G_1 to G_2. Junction G_2 estimates course X(X=A, B) with the function $e^{-0.1^is}$.

III Junction G_2 changes the expected gain ${}^iE_X$ with the following equation. K_f $(0 < K_f < 1)$ is the coefficient which influences the amount of change.

$${}^iE_X = K_f \, {}^iE_{X\_old} + (1 - K_f)e^{-0.1^is}$$

IV Junction G_2 renews the strategy "p_a" with the following equation. K_d $(0 < K_d)$ is the changing coefficient and $p_{min}(0 < p_{min} < \frac{1}{m})$ is the minimum probability to guarantee a course selection.

$$
{}^{i}p_{X\_tmp} = max \left\{ \begin{array}{l} {}^{i}p_{X\_old} + K_d \left({}^{i}E_X - \frac{{}^{i}E_A + {}^{i}E_B}{2} \right) \\ p_{min} \end{array} \right.
$$

$$
{}^{i}p_X = \frac{({}^{i}p_{X\_tmp} - p_{min})(1 - 2p_{min})}{{}^{i}p_{A\_tmp} + {}^{i}p_{B\_tmp} - 2p_{min}}
$$

Note that junction G_1 does not obtain the result of the trials which junction G_1 performed . Junction G_2 obtains the results of the trials which junction G_1 performed.

Next we compare p_{opt} derived in section 3.1 and p_a obtained by reinforcement learning in Fig. 4(under the following condition; $m = 2$, $({}^1a, {}^1b_1, {}^1b_2) = (6, 1, 2)$, $({}^2a, {}^2b_1, {}^2b_2) = (6, 2, 4)$). Both routes p_a cannot reach each p_{opt} because p_a converges at the point that the movement cost of course A is equal to that of course B when we use the algorithm described above.

To see how intelligent junctions adapt to changes in parameters, we show the results of simulations in Fig. 5. It shows the number of achievements per robot for a change in the number of robots. They get better strategies by reinforcement learning, keeping achievements at a higher level than for the case where junctions do not do reinforcement learning(p_a is fixed on 0.0 or 1.0). But the difference between "optimal" and "learning" is large, because junctions cannot get the results of trials that were done by them. So, if the robots convey the results of trials to the correct junctions, the junctions can learn more effectively. In the next section we introduce "media agents" to solve this problem.

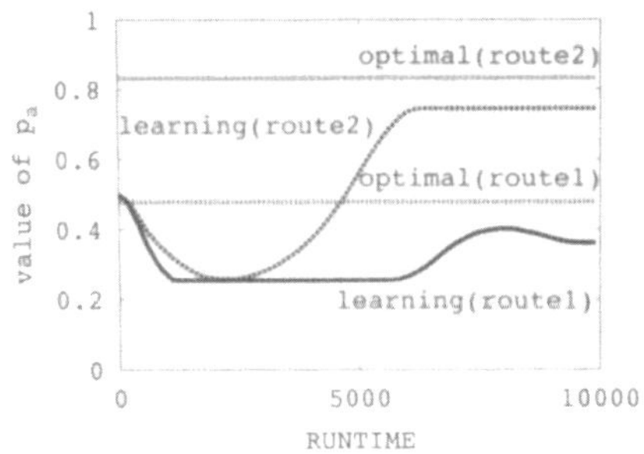

Fig. 4. Learning of p_a by Intelligent junction

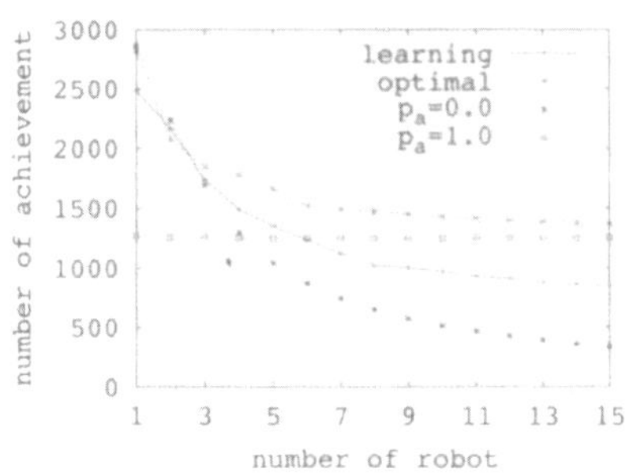

Fig. 5. Comparison of achievements

4 Learning with media agents

We discussed the difficulties of learning by intelligent junction in the previous section. To improve the learning we introduce "media agents" as carriers of information. A media agent conveys the results of trials to the correct junction.

4.1 Introduction of media agents

Junctions fixed on the environment can estimate the strategies statistically by observing robots for a certain period of time. Therefore we can introduce the following algorithm to search for the optimal probability of p_a with the media agents.

(i) A junction samples and accumulates evaluations of $e^{-0.1^{r_s}}$ for a certain number of times at the present probability ${}^b p_a$.
(ii) The junction does the same action as (i) at probability ${}^b p_a \pm \Delta p$.
(iii) The junction compares three values of evaluation(${}^b p_a, {}^b p_a \pm \Delta p$) and shifts the present probability ${}^b p_a$ to the best probability of the three.

We refer to this algorithm as "p_a *search*". Media agents follow the steps given below(with reference to Fig. 6 and 7);

1. Junction G_1 appoints some robots to be media agents according to p_m(the ratio of media agents to the number of robots which were ordered to start according to p_G).
2. A media agent moves to junction G_2 as a normal robot.
3. The media agent records the evaluated value and goes back to the start point G_1 immediately, not obeying the order of junction G_2. This movement from G_2 to G_1 is not counted as an achievement.
4. The media agent gives the information(the evaluated value) to Junction G_1. Junction G_1 optimizes p_a according to the algorithm described previously.

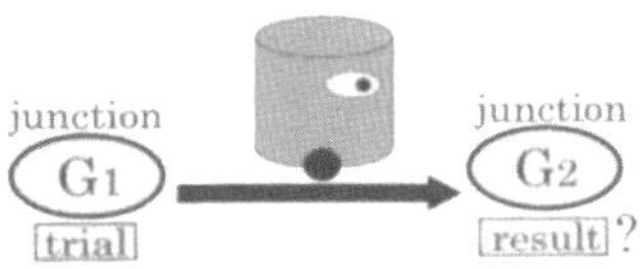

Fig. 6. Action of normal robots

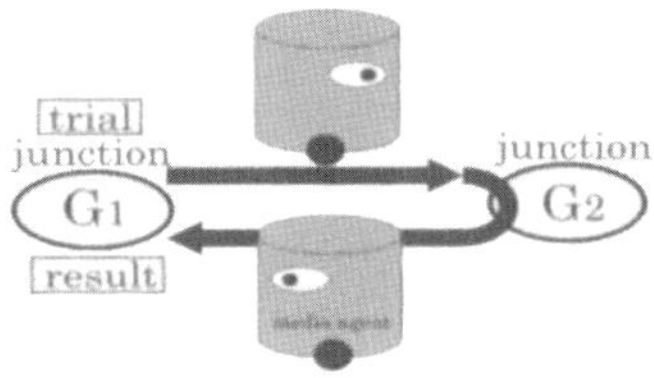

Fig. 7. Action of media agents

4.2 Task efficiency

Media agents immediately go back to the start point junction after they have reached the opposite junction of the route. So media agents leave junctions without the orders of junctions when they go back to the start point junction. Therefore p_G is changed to $\hat{p}_G$. $\hat{p}_G$ per step is as follows under the condition that all robots act synchronously.

$$\hat{p}_G(0) = p_G$$
$$\hat{p}_G(1) = p_G(1 - p_G p_m) + p_G p_m = p_G + (1 - p_G) p_G p_m$$
$$\hat{p}_G(2) = p_G + p_G p_m (1 - p_G) - p_G^2 p_m^2 (1 - p_G)$$
$$\vdots$$
$$\hat{p}_G(n) = p_G - (1 - p_G) \sum_{i=1}^{n} (-p_G p_m)^i$$

Therefore $\hat{p}_G$ is expressed as equation(6)

$$\hat{p}_G = \lim_{n\to\infty} \hat{p}_G(n) = \frac{p_G(1+p_m)}{1+p_G p_m} \tag{6}$$

In a similar way S_G also changes to $\hat{S}_G = (1+p_G p_m)/(p_G(1+p_m))$ because S_G is derived from p_G. The optimal probability of route i "${}^i p_{opt}$" also changes to "${}^i \hat{p}_{opt}$" as S_G changes to $\hat{S}_G$. But we treat it as "$p_{opt} = \hat{p}_{opt}$", because the difference between p_{opt} and $\hat{p}_{opt}$ is small.

Media agents do not work(their movements are not counted as an achievement) when they go back to the start point junction. The more media agents we use to acceralate the optimization, the lower the task efficiency becomes. The ratio "p_w"of working robots to robots which start from junctions becomes $p_G/(1+p_m p_G)$. Therefore task efficiency "q" is represented by

$$q = \frac{p_w}{\hat{p}_G} = \frac{1}{1+p_m} \tag{7}$$

4.3 Formulation of adaptation with media agents

Media agents accelerate the optimization but cause a decrease in task efficiency(equation(7)). We have to determine the optimal ratio of media agents to the number of robots which were ordered to start "p_{m_opt}" paying attention to both the speed of optimization and the task efficiency. We formulate the system with media agents and derive p_{m_opt}.

The initial condition of p_a is "p_0", the target p_a is "p_d", and "$\bar{p} = |p_0 - p_d|$". We formulate the connection between "t" and p_m. "t" is the time required for p_a to shift from p_0 to p_d by optimization.

When $p_m = 1.0$, we define "s_1" as the minimal time to shift from p_0 to p_d. A junction needs "$M = \bar{p}/(\Delta p) \cdot SMP$" data to optimize p_a from p_0 to p_d."SMP" is the number of samplings((ii) in algorithm).

A junction receives an expected number of data "E_M" from media agents per step.

$$E_M = \frac{1}{2m} \cdot \frac{N - n_G}{\frac{1}{m}\sum_{i=0}^{m}({}^i a + {}^i b_1 + \rho\, {}^i n_B\, {}^i b_2)} \tag{8}$$

where $n_G = N \cdot S_G / f({}^i p_a)$."$n_G$" is the number of robots that stay in junctions. The latter part's numerator of equation(8) represents the number of media agents on the routes because all robots are media agents($p_m = 1.0$).The latter part's denominator of equation(8) is the expected cost that a media agent requires when it goes and returns between junctions.

A junction does not receive any data from media agents over a time interval at each trial until the first media agent comes back to the start point junction. We estimate its expected time with ${}^i f((p_0+p_d)/2)$. We refer to this time as "T_{no}". T_{no} is represented by

$$T_{no} = \frac{\bar{p}}{\Delta p}\frac{1}{m}\sum_{i=0}^{m} {}^i f\left(\frac{p_0+p_d}{2}\right) \tag{9}$$

Therefore s_1 is expressed in equation(10).

$$s_1 = \frac{M}{E_M} + T_{no} \tag{10}$$

Next we define "d" as the average step time to shift from p_0 to p_d. The average step time is the amount of time for the average cost. That is to say, "d" is time when we consider the average cost as a time unit. (The relation of average step time, average cost and time integral value of movement cost corresponds to that of transit time, average velocity and moving distance.)To simplify analysis, we normalize d by $\bar{p} = 0.1$.

We define $s_2|_{\bar{p}=0.1}$ as the amount of time to change from p_0 to $p_d|_{\bar{p}=0.1}$. And the change of p_a in the same period is approximately linear(${}^i p_a = p_0 - (p_0 - p_d)t/s_2|_{\bar{p}=0.1}$).Then the time integral value of the expected movement cost "I_c" between $t = 0$ and $t = s_2|_{\bar{p}=0.1}$ is equation(11).

$$I_c = \int_0^{s_2|_{\bar{p}=0.1}} {}^i f({}^i p_a)dt \tag{11}$$

The average cost of moving between two junctions "C_a" corresponds to the movement cost at halfway p_a between p_0 and $p_d|_{\bar{p}=0.1}$ in equation(12).

$$C_a = {}^i f\left(\frac{p_0 + p_d|_{\bar{p}=0.1}}{2}\right) \tag{12}$$

Therefore d is represented by $d = I_c/C_a$.

As we formulate the system by a probabilistic model, differences from the expected values emerge. So s_1 may become bigger than the value calculated by equation(10). We introduce coefficient "K" for s_1 to represent the influence of the probabilistic model.

The evaluation is conducted based on the movement cost((i)(ii) in algorithm). a and b_1 are constant values because they represent the distance of courses. But $n_B b_2$ depends on the number of robots on course B "n_B". If the actual n_B is very different from the expected n_B, a junction may mistakenly evaluate at step(iii) in the algorithm. So we formulate the condition of variance of n_B as a Gaussian distribution.

We consider two conditions for route i; ${}^i p_a = p_1$ and ${}^i p_a = p_2 (= p_1 \pm \Delta p)$. The variances of n_B under each set of conditions are represented by Gaussian distribution $g_1(n_1)$ and $g_2(n_2)$

$$g_x(n_x) = \frac{1}{\sqrt{2\pi\sigma_x^2}} exp\left(-\frac{(n_x - m_x)^2}{2\sigma_x^2}\right) \quad (x = 1, 2) \tag{13}$$

m_x is the expected value of n_B, σ_x is the variance of n_B. We set an appropriate value for σ_x. When $n_1 = n_{tmp}$ and $m_1 > m_2$, the area that is surrounded by $g_2(n_2)$, $n_B = n_{tmp}$ and horizontal axis represents the probability of the untruthful evaluation. In this case, the probability of failure becomes $g_1(n_{tmp}) \int_{n_{tmp}}^{\infty} g_2(n_2)dn_2$. Therefore, the probability of failure to estimate condition"${}^i p_{f_{(1,2)}}$" is represented as follows.

$$
\begin{aligned}
{}^{i}p_{f(1,2)} &= \int_{-\infty}^{\infty} g_1(n_1) \int_{n_1}^{\infty} g_2(n_2) dn_2 dn_1 \\
&= \frac{1}{2\sqrt{1+\frac{\sigma_1}{\sigma_2}}} exp\left(\frac{(m_1-m_2)^2}{2(\sigma_1^2+\sigma_2^2)}\right)
\end{aligned} \tag{14}
$$

In consideration of all conditions, "p_f", which is the probability of failure for the environment, becomes as follows.("n" is the number of conditions and "m" is the number of routes.)

$$
p_f = \frac{1}{m}\sum_{i=1}^{m}\left(\frac{1}{n}\sum_{j=0}^{n-1} {}^{i}p_{f(j,j+1)}\right) \tag{15}
$$

The relation of K and p_f is represented as $K = 1/(1-p_f)$.Because s_1 increases with the increase in the probability of failure. As long as we can set the appropriate values of σ, we can derive K.

The connection between t and p_m is formulated as equation(16) with coefficients described above. Because "d" is normalized by $\bar{p} = 0.1$, we multiply "d" by "$10\bar{p}$" in the numerator of equation(16) to extrapolate the value of each case. Even if $p_m = 1.0$, the optimization of p_a needs minimum time"Ks_1".So the denominator of equation(16) is described below.

$$
p_m = \frac{10\bar{p}d}{t - Ks_1} \tag{16}
$$

Next we derive p_{m_opt} by using equation(16). We have to consider both the speed of optimization and the decrease of task efficiency due to the use of media agents.

${}^{i}f({}^{i}p_a)$ becomes ${}^{i}f(t, p_m)$ by equation(16). The time integral value of ${}^{i}f(t, p_m)$ "$F(t : 0 \rightarrow T)$" becomes a function of p_m(T:simulation time). We define "Y(equation(17),a function of p_m)" as the total cost, which includes the movement cost and the task efficiency. p_{m_opt} is p_m which minimizes Y. Finally we can derive p_{m_opt}.

$$
Y = \frac{F}{q} = \frac{\frac{1}{m}\sum_{i=0}^{m}\int_0^T {}^{i}f(p_a)dt}{\frac{1}{1+p_m}} \tag{17}
$$

4.4 Evaluation of media agents

We compare $(\alpha)p_{m\_opt}$ derived from equation(17) and $(\beta)p_{m\_opt}$ searched in the simulation under various conditions(a-i) in Tab. 1. The common condition is as follows; $m = 1, p_0 = 1.0, T = 100,000$.

The average error is 0.015, and the variance is 4.7×10^{-4}. We succeeded in formulating the system with media agents.

Next we show the result of p_a search with media agents in Fig.8. We compare the number of achievements of 2 patterns;reinforcement learning by intelligent junctions(RL), p_a search by media agents(MA)(p_m is fixed at the value derived in section 4.3). Each value is normalized by the achievement of optimal condition. Conditions are the same as a-i described above. In all cases MA does not reach optimal condition but exceeds RL. The average of RL and MA are 0.612 and 0.881. MA improves the system by 44%.

Table 1. Comparison of p_{m_opt}

	condition cost	N	(α)	(β)
a	(9,1,2)	10	0.070	0.05
b		20	0.022	0.02
c	(9,2,4)	10	0.012	0.03
d	(12,1,2)	10	0.110	0.08
e		20	0.037	0.02
f	(12,2,4)	10	0.045	0.04
g	(15,1,2)	10	0.143	0.11
h		20	0.054	0.03
i	(15,2,4)	10	0.066	0.04

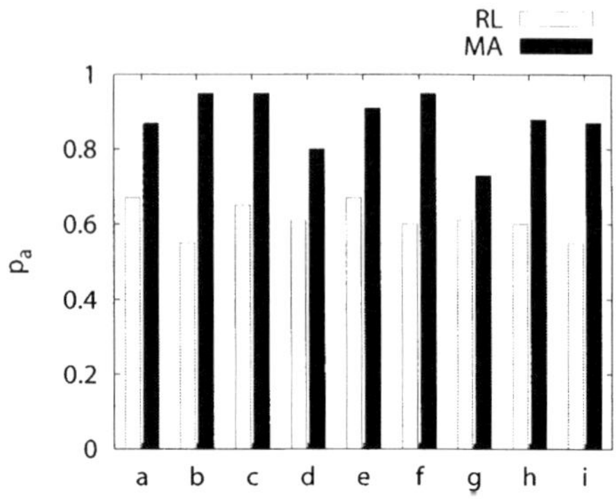

Fig. 8. Comparison of achievements

5 Conclusion

In this paper we described our work addressing the issue of the optimal routing problem. We arranged the environment so that the efficiency of movement changed depending to the course selected. We formulated the optimal condition by a probabilistic model and confirmed the consistency of formulation by means of comparing the derived p_{opt} with the searched p_{opt} in simulation. Next we estimated the performance of reinforcement learning by intelligent junctions, revealing its problem. Then we proposed media agents to solve this problem. We formulated the system with media agents and confirmed its consistency. Finally we showed the availability of media agents quantitatively. The performance of this system significantly exceeds systems using reinforcement learning.

References

1. Jun Ota et al.:Distributed Strategy-Making Method in Multiple Mobile Robot System, DARS, 123-133, 1994.
2. Yuji Yoshimura et al: Iterative Transportation Planning of Multiple Objects by Cooperative Mobile Robots, DARS2, 171-182, 1996.
3. Luc Steels: Cooperation Between Distributed Agents Through Self-Organization, Decentralized AI, Elsevier Science publishers, pp.175-196,1990.
4. Alexis Drougoul and Jacques Ferber: From Tom Tumb to the Dockers: Some Experiments with Foraging Robots, Animals to Animats 2, pp.451-459, 1992.
5. Daisuke Kurabayashi et al. :Distributed Guidance Knowledge Management by Intelligent Data Carriers, Int. robotics & automation, Vol. 16, No. 4, pp. 207-216, 2001
6. R. Beckers, et al.:From Local Actions to Global Tasks:Stigmergy and Collective Robotics, Artificial Life IV, pp.181-189, 1994.
7. Daisuke Kurabayashi, et al.: Performance of Decision Making: Individuals and an Environment, IEEE Int. Conf. on Intelligent Robots and Systems, 2831/2836, 2002.

Multi-Robot Concurrent Learning in Museum Problem

Zheng LIU[1], Marcelo H. ANG Jr.[2], and Winston Khoon Guan SEAH[3]

[1]Dept. of Electrical & Computer Engineering, National University of Singapore. g0202255@nus.edu.sg. [2]Dept. of Mechanical Engineering, National University of Singapore. mpeangh@nus.edu.sg. [3]Institute for Infocomm Research, Singapore. winston@i2r.a-star.edu.sg.

Abstract:
Multi-robot concurrent learning on how to cooperatively work through the interaction with the environment is one of the ultimate goals in robotics and artificial intelligence research. In this paper, we introduce a distributed multi-robot learning algorithm that integrates reinforcement learning and neural networks (weighting network). By retrieving continuous environment state and implicit feedback (reward), the robots can generate appropriate behaviors without deliberative hard coding. We test the learning algorithm in the "museum" problem, in which robots collaboratively track moving targets. Simulation results demonstrate the efficacy of our learning algorithms.

Key words:
Multi-robot, reinforcement learning, neural networks, concurrent learning, tracking.

1. Introduction

The multi-robot system has been one of the focuses of robotics research in the last two decades. It includes a wide range of research topics such as multi-robot cooperative transportation, exploration and mapping, distributed sensing, robot soccer, etc [1]. The multi-robot system is not simply an extension of the single-robot system by increasing the performance owing to parallel operation; it can accomplish tasks impossible to a single-robot system through "cooperation" [2].

Normally, the cooperation in multi-robot systems is concentrated on the task level [3], whereby the mission is broken down into tasks, and robots choose different tasks (roles) according to the state and behave differently. To achieve mission decomposition, task allocation, and conflict coordination, the designer needs to predict all possible scenarios and preset corresponding actions for each robot to react accordingly. Such development and coding work is undesirable and sometimes extremely difficult, especially when the mission is very complex and the robot group is heterogeneous. One possible solution is to let the robot learn how to cooperatively work through the interaction with the environment and other robots, hence generating appropriate behaviors without human design or coding.

In this paper, we introduce typical reinforcement learning and its constraints in Section 2, and present our learning algorithms that integrate reinforcement learning and neural networks in Section 3. Following which, we introduce how to implement our learning algorithms for the museum problem in Section 4 and show the simulation and results in Section 5. Finally, Section 6 concludes this paper.

2. Reinforcement Learning

Emergent generation of multi-robot cooperation is one of the ultimate goals of robotics and artificial intelligence research. While there are dozens of basic learning algorithms in machine learning research [4], only reinforcement learning (RL) is extensively studied for behavior based control (cooperation in the task level) in multi-robot systems [5]. An explanation is that compared with other learning algorithms, reinforcement learning has the following advantages [6]:

- Model free – can learn the control policy even if the model of the environment is unknown.
- Not strictly supervised – no need for explicit human training, implicit reward is sufficient.
- Optimal – subject to user defined criteria.
- Practical – simple and real-time.

The basic concept of reinforcement learning is to find the optimal control policy that chooses the appropriate action under any given state; in other words, to find the optimal link/mapping from states to actions. Usually, reinforcement learning can sufficingly solve the control problems that execute in discrete state/action space. For instance, for path planning, if the map is divided into grids, the agent/robot can find the path to approach to the destination by reinforcement learning.

Because of above advantages, reinforcement learning is predominant in both single- and multi-robot system researches. However, the limitation of discrete/finite input (state) and output (action) constrains the application of reinforcement learning. For example, to implement reinforcement learning in robot behavior based control, the designer must define discrete/finite state and action space first (as in [7]). Obviously, this is not realistic because some tasks and missions can hardly be discretized. Furthermore, even if the state and action spaces are discrete, the huge size of the space will badly affect the learning process, which requires clustering (grouping) to reduce the space. Reasonably, one important question arises: can the robot do the state/action discretization and clustering by itself without human intervention, or even more, perform reinforcement learning without discretization or clustering? In this paper, we address this problem by integrating reinforcement learning with neural networks (weighting network).

Besides the limitation of finite and discrete input (state) and output (action), other critical research issues of the reinforcement learning include the Markov Decision Process (MDP) and stationary environment assumptions, reward definition and assignment, state-action link value update, and action selection. However, they are not the main topic of this paper. Related work on these issues can be found in [8-11].

3. Our Learning Algorithms

As introduced before, the aim of our research is to address the problem of discrete and finite input/output space in the reinforcement learning. A reasonable solution is to modify the architecture of reinforcement learning or integrate it with some control algorithms that can deal with continuous and infinite input/output space.

In robotics and artificial intelligence research, the neural network (NN) is a well known control and learning algorithm that has been extensively studied for decades. Neural networks can deal with continuous and infinite input/output spaces; however, the learning in neural networks is normally supervised. As shown in Figure 1, in a typical Back Propagation (BP) neural networks, a "trainer" is needed to generate desired output according to the input, and then some algorithms are used to adjust the parameters/weights inside the neural networks by the error between the desired output and real output of the controller.

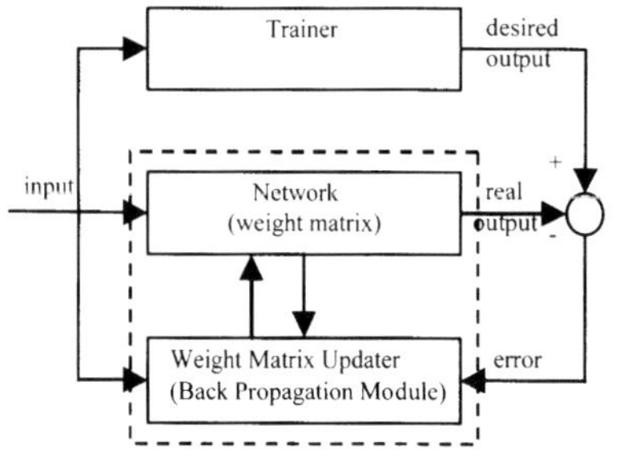

Figure 1. Back Propagation Neural Networks

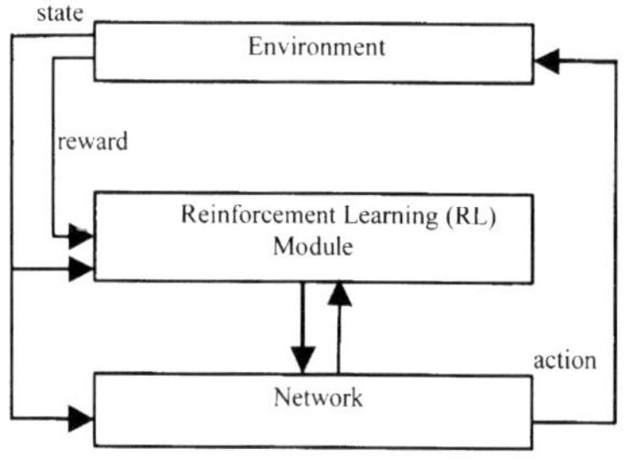

Figure 2. Our Learning Architecture

The design of our learning architecture is shown in Figure 2. In this new architecture, the back propagation module in the BP neural networks is replaced by the reinforcement module, and no trainer is needed to generate the explicit desired output. Instead, the reinforcement learning module adjusts the weight inside the network through the interaction with the environment. By integrating reinforcement learning with neural networks (weighting network), we combine the advantages of both neural networks and reinforcement learning. The neural networks (without the BP module) can retrieve continuous and infinite input, and then generates continuous and infinite output by the weight matrix (weighting network). On the other hand, the reinforcement module which replaces the BP module can find the "best" weight inside the networks through the interaction with the environment. It should be noted that the input/output spaces of the reinforcement learning module are still discrete and finite. However, since the learning module is only used to adjust the weight inside the networks, the input and the output of the overall controller/system is now continuous and infinite.

Because the learning architecture is a framework of a learning methodology, and the learning algorithms are highly coupled with the mission, we need to test our learning algorithms in real applications. For this purpose, museum problem is a good choice. We will introduce the museum problem and show how to implement our learning algorithms in the museum problem in the next section.

4. Learning in Museum Problem

4.1. Museum Problem

Museum problem is the research on multi-robot tracking of multiple moving targets. The assumptions and descriptions of the museum problem are as follows:

- The environment is a large bounded plain area.
- Several targets move in the environment.
- Several mobile robots are in the environment. Each robot has a 360 degree view within a certain range. When an object is inside this circle, the robot can differentiate it as obstacle, target, or robot. The summation of the sensible area of all robots is far less then the size of the environment.
- The targets are mobile and the sensor range of the robot is limited, hence the robot needs to track targets to maintain observation.
- For the robots, the number and motion pattern of the targets are unknown. Localization and intercommunication are unavailable.
- The objective is to maximize the number of targets being observed.

In current research for the museum problem, Artificial Potential Field (APF) based control is mostly used. The concept of APF is very simple: map the targets as attractive force sources and map the robots and obstacles as repulsive force sources; then, let the robot move under the vector sum of the attractive and repulsive forces. Artificial potential field based control can be seen as a kind of competitive neural networks (weighting network) in which the attractive forces compete with the repulsive forces. It is simple and can be used in real-time applications.

The pure potential field based control just adds the repulsive and attractive forces. However, purely summing the repulsive and attractive forces may not achieve desired cooperation in most cases. When two robots find the same target, intuitively, the best cooperation is to let one robot track the target, and the other robot leave to search for other targets, so as to maximize the use of the resource (robots force). This target selection is a kind of high level cooperation. However, pure potential field based control cannot guarantee such cooperation in that case. Neither of the two robots will leave.

A solution to achieve better target selection is to modify the pure potential field based control by adding a weight to the attractive forces. This weight of the attractive forces represents the preference that a robot tracks targets. If the weight value is high, the robot is likely to keep tracking detected targets; if the weight value is low, the robot is likely to leave detected targets if other robots are already around it. In previous research [12-14], some algorithms have been designed to adjust this weight. However, it is difficult to find the best weight value for each robot, especially when the scenario is very complex and the robot team is heterogeneous. A natural thought is to let the robot get the best weight value through learning. Hence the museum problem is well suited to the implementation and testing of our learning algorithms.

4.2. Implementation of Learning in Museum Problem

To implement reinforcement learning, we need to design the functions and set the parameters for state-action definition, reward generation and allocation, state-action link value update and action selection.

Firstly, we need to define the states and actions for the reinforcement learning module. To make the learning simple, yet not lose its generality, we define the input state of the learning as the number of targets and robots detected.

Secondly, we need to define the rewards for reinforcement learning. Since the objective of museum problem is to maximize the observation of moving targets, the reward should be given to the robot that tracks target or cooperates with other robots. For this purpose, we define three kinds of rewards:

- *Reward_TT*: track target reward (positive) – if target(s) is within the sensor range.
- *Reward_NR*: near robot reward (negative) – if other robot(s) is nearby.
- *Reward_SC*: state change reward (positive/negative) – if the new state has less neighbor or more targets, the reward is positive, else negative.

Because the learning process is distributed and there is no intercommunications among the robots. For each individual robot, these three kinds of rewards are all generated by its local sensing.

Thirdly, in reinforcement learning, the learning process needs to update the value of the state-action links based on the reward received. In our learning algorithms, the state-action link value is updated in every simulation step. The new value of the state-action link is the summation of the previous value and the rewards.

Finally, the robots need to select the action (weight) according to state and the value of the state-action links. Every time the state changes, the robot will reselect the action (weight). However, if the state is unchanged for a long period of time (N simulation step), we also let the robot reselect the action (weight) because we want to accelerate the learning speed. In reinforcement learning, for action selection, the learning process needs to both explore and exploit the action space. Therefore, when selecting action, we add an exploration factor to the real state-action link value, and then choose the action that has the highest resultant value.

5. Simulation and Results

5.1. Simulation Scenario

We set a scenario where two robots track one target. If pure potential field based control is used, the two robots will both track the target. Obviously, this is a waste of robot force and we expect that through learning, one of the two robots will learn to neglect the target tracked by other robots. To test our learning algorithms, we simulate three kinds of control mode in both homogeneous and heterogeneous robot groups:

- Pure Artificial Potential Field based controller.
- All-adjust heuristics of APF controller: if a robot detects a target and finds that some other robots are near to that target, it will decrease the weight of the attractive force to the target. [12, 13]
- Robot concurrent learning controller by our learning algorithms.

For above three kinds of controllers, we aim to find the following results:

- Waste time length – represent the cooperation level.
- Learning results – the difference in learned weight for tracking targets.

5.2. Simulation Parameters

The parameters and settings of the simulation are as follows:

- The simulations are run on Webots, a differential-wheel robot simulator.
- Museum: 4m * 4m square plain area with no obstacles inside.
- Each learning episode is 20000 simulation step long. For each scenario, 100 episodes is run to get the average of the simulation results.

- In the heterogeneous robot group, one robot is 30% faster than the other(s).
- For the all-adjust heuristics of pure potential field based control mode, the All-adjust Weight Decrease Ratio (AWDR) is 0.95.
- For the learning mode, the input state of the learning is *(robot number, target number)*, e.g., state *(1, 1)* means there are one neighbor robot and one target detected. The output action (weight) space is {0.5, 1.0, 1.5}.
- For the learning mode, the initial value of all state-action links is 10.
- For the learning mode, *Reward_TT* = 0.005, *Reward_NR* = - 0.01, *Reward_SC* = (m-a)*0.5 – (n-b)*2.0 (m/n is the current target/robot number; a/b is the previous target/robot number).
- For the learning mode, if the state changes or if the state has been unchanged for N = 100 simulation steps, the robot will reselect the action (weight). When reselecting the action, an exploration factor (uniformly distributed in [-1, 1]) is added to the real state-action link value, then the action having the highest resultant value will be chosen.

5.3. Simulation Results and Discussion

Figure 3 shows the waste time length. When two robots are simultaneously tracking a target, it is a waste of resource since one of the robots can leave and search for other targets. Obviously, short waste time length means high level cooperation. The results demonstrate the efficacy of our learning algorithms that the waste time length is greatly shortened, even better than the deliberative coded control mode (all-adjust heuristics of pure potential field based control). It should be noted that the performance of the heuristic controller is highly dependent to the value of weight decrease ratio. If an optimal value is selected, the performance may be better. However, considerable human effort is needed to find this optimal value; while the learning controller can learn the best weight value without such work.

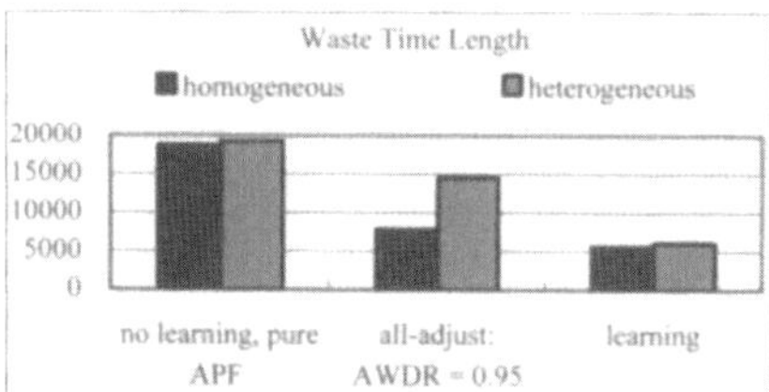

Figure 3. The Waste Time Length in Different Scenario

Since the objective of the learning is to generate cooperative behaviors between the two robots when they meet the same one target, the most meaningful learning results are for the state *(1, 1)* (one neighbor robot and one target detected). Figures 4 and 5 show the learning results (the x-axis represents the difference of the learned weights between two robots; the y-axis represents the probability of getting such result). In the end of the simulation, each robot will find a preferred weight for this state *(1, 1)* as low, mid, or high. Here we show the "difference in learned weight" instead of the "value of learned weight". This is because the difference in the weights between two robots is the key for cooperation. The robot with high weight will keep tracking and the robot with low weight will leave.

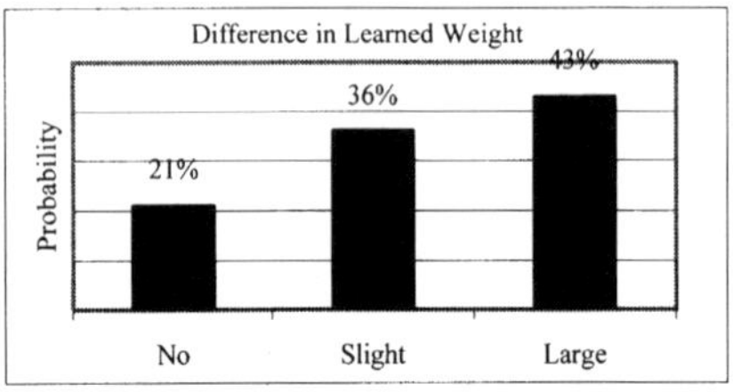

Figure 4. Learning Results of Homogeneous Robot Group

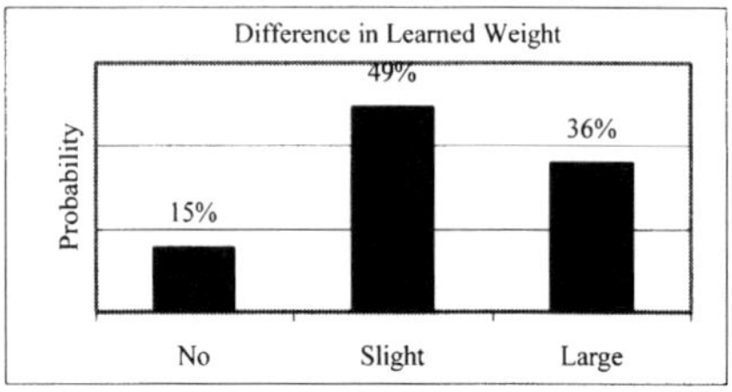

Figure 5. Learning Results of Heterogeneous Robot Group

As shown in Figures 4 and 5, for both homogeneous and heterogeneous robot groups, in most cases, the two robots can learn different weights. This is the key for the cooperation. However, in some cases (homogeneous 24%, heterogeneous 15%) the two robots may learn the same weight in the end. This draw case is undesired because we require the robots to have different weights for cooperation. This problem may be explained by the following reasons:

- In the learning environment, there is no localization and intercommunication available for the robots. The only input of the robots is the local sensed data. Since the sensor ability is quite limited, the partial observation may badly affects the Markov Decision Process (MDP) and stationary environment assumptions.
- Furthermore, since the two robots are learning concurrently, their learning process may interfere with each other, therefore fall in local minima or change control policy cyclically. This problem further affects the two assumptions.

For the homogeneous and heterogeneous robot groups, the learning results are different. The heterogeneous group has less draw cases. This may be due to the fact that the difference in functionality catalyzes the role differentiation between the robots. Besides, the homogeneous robot group prefers highly different weights; while the heterogeneous robot group prefers slightly different weights. An explanation is that in the heterogeneous robot group, the two robots are already quite different that one robot is 30% faster than the other. Therefore a slight weight difference is enough for them to generate cooperative behaviors.

To further validate the efficacy of our learning algorithms, we extend the learning to three targets and three robots scenario using the same parameters as before. The simulation results of this "three plus three" scenario is consistent to the previous "one plus two" scenario. For the pure potential field based controller, all adjust heuristic controller, and the learning controller, the average number of targets being tracked during the simulation are almost the same (2.66, 2.68 and 2.59). However, if the robots can cooperate, fewer robots are needed to track the targets because the situation that several robots track one same target is avoided. Therefore, large average free robot number means high level cooperation. In the learning mode, 0.26 robots are free (no need for tracking) and they can search targets in the environment; while in the all adjust heuristic control mode and the pure potential field based mode, only 0.15 and 0.10 robots are free. (The above numbers are the average of homogeneous and heterogeneous robot groups.) The results of the extended scenario also show that for our learning environment, the cooperation is mostly happened in the situation that two robots meet one same target. This is possibly due to the limitation of the robot sensor ability that it can only cover a small region around the robot.

Simulation videos can be found in http://guppy.mpe.nus.edu.sg/~mpeangh/kevin.

6. Conclusion and Future Work

Multi-robot concurrent learning on how to cooperatively work is one of the ultimate goals of robotics and artificial intelligence research. Reinforcement learning has achieved great success for this purpose. However, typical reinforcement learning cannot deal with continuous and infinite inputs and outputs. In this paper, we address these problems by integrating reinforcement learning with neural networks (weighting network). The efficacy of our learning algorithms is proved by simulation results.

For multi-robot concurrent learning, the Markov Decision Process and stationary environment assumptions, reward generation and allocation, action selection, and state-action link value update, are critical research issues that may greatly affect the learning process and results. By carefully designing and choosing the functions and parameters of our algorithms for the museum problem, the learning results are satisfactory.

Integrating reinforcement learning with neural networks (weighting network) is a good solution to solve the discrete and finite input/output problem. However, in our learning architecture, the reinforcement module still needs to retrieve discrete input state and perform discrete actions (weights). A more challenging work is to design a totally continuous and infinite learning algorithm, or at least, let the robot do state/action definition or discretization by itself through learning. The answer is to be found in the future research.

Reference

[1] Balch, T., and Parker, L. E. (2002), Robot Teams: From Diversity to Polymorphism, Natick, Massachusetts, A K Peters Ltd.
[2] Cao, Y. U., Fukunaga, A. S., Kahng, A. B., and Meng, F. (1995), Cooperative Mobile Robotics: Antecedents and Directions, in proceedings of IEEE/RSJ International Conference on Intelligent Robots and Systems, Vol:1, pp:226-234.
[3] Tangamchit, P., Dolan, J. M., and Khosla, P. K. (2002), The Necessity of Average Rewards in Cooperative Multirobot Learning, in proceedings of IEEE International Conference on Robotics and Automation.
[4] Mitchell, T. M. (1997), Machine Learning, McGraw Hill.
[5] Mataric, M. J. (1997), Reinforcement Learning in the Multi-Robot Domain, in Autonomous Robots, Vol:4(1), pp:73 - 83.
[6] Sutton, R. S., and Barto, A. G. (1998), Reinforcement Learning: An Introduction, MIT Press, Cambridge, MA.
[7] Chu, H. T., and Hong, B. R. (2000), Cooperative Behavior Acquisition in Multi Robots Environment by Reinforcement Learning Based on Action Selection Level, in proceedings of IEEE/RSJ International Conference on Intelligent Robots and Systems, Vol:2, pp:1397-1402.
[8] Kaelbling, L. P., Littman, M. L., and Moore, A. W. (1996), Reinforcement Learning: A Survey, in Artificial Intelligence Research, Vol: 4, pp237-285.
[9] Kawakami, K. I., Ohkura, K., and Ueda, K. (1999), Adaptive Role Development in a Homogeneous Connected Robot Group, in proceedings of IEEE International Conference on Systems, Man, and Cybernetics, Vol:3, pp:254-256.
[10] Uchibe, E., Asada, M., and Hosoda, K. (1998), Cooperative Behavior Acquisition in Multi Mobile Robots Environment by Reinforcement Learning Based on State Vector Estimation, in proceedings of IEEE International Conference on Robotics and Automation, Leuven, Belgium, Vol:1, pp:425 - 430.
[11] Michael Bowling, and Manuela Veloso (2003), Simultaneous Adversarial Multi-Robot Learning, in proceedings of the Eighteenth International Joint Conference on Artificial Intelligence, Acapulco, Mexico.
[12] Parker, L. E. (2002), Distributed Algorithm for Multi-Robot Observation of Multiple Moving Targets, Autonomous Robots, vol. 12(3), May.
[13] Parker, L. E., and Emmons, B. A. (1997), Cooperative Multi-robot Observation of Multiple Moving Targets, in proceedings of IEEE International Conference on Robotics and Automation, vol. 3, pp. 2082-2089, Albuquerque, New Mexico, USA.
[14] Liu, Z., Ang, M. H., and Seah, W. K. G. (2003), A Searching and Tracking Framework for Multi-Robot Observation of Multiple Moving Targets, in International Journal of Advanced Computational Intelligence & Intelligent Informatics (JACI3), 8-1, 2004.

How a Cooperative Behavior can emerge from a Robot Team

A. D'Angelo[1], E. Menegatti[2], and E. Pagello[2]

[1] DIMI, via delle Scienze 206, University of Udine antonio@dimi.uniud.it
[2] DEI, via Gradenigo 6, University of Padua {emg,epv}@dei.unipd.it

In this paper, we suggest an hybrid architecture where the deliberative part takes advantages from the reactive one and vice versa, to make a multi-robot system to exhibit some assigned cooperative task. We explain our architecture in terms of schemas and a set of firing conditions. To experiment our approach, we have realized an implementation that tries to exploit the resources of our robot team participating to Middle-size RoboCup tournaments. Each individual exhibits both reactive and deliberative behaviors which are needed to perform cooperative tasks. To this aim we have designed each robot to become aware of distinguishing configuration patterns in the environment by evaluating descriptive conditions as macroparameters. They are implemented at reactive level, whereas the deliberative level is responsible of a dynamic role assignment among teammates on the basis of the knowledge about the best behavior the team could perform. This approach was successfully tessted during the Middle-size Challenge Competition held in Padua on last RobCup2003.

1 Introduction

The most recent robotics applications have shown a growing interest in developing colonies of robots within industrial and civil environments, switching robot design from the goals of controlled speed, high accuracy and repeatability toward new targets of flexibility and reliability. A key issue to successfully perform such kind of advanced tasks is figuring out how to make emerging cooperative abilities within this context.

The *differentiating* robot societies[15] show a large number of homogeneous individuals with limited abilities, whereas the *integrating* societies are usually characterized by a small number of heterogeneous and specialized individuals. Both societies include individuals with well-distinguishing skills referred to the role to play inside the group or the aptitude to modify dynamically its behavior while performing an assigned task.

A group of robots is a robotics team only if it exhibits the ability to perform cooperative tasks, providing better performance for their individual components and taking advantages from distributed sensing and acting. Soccer-robots International Games, like RoboCup [8], are very useful testbeds to experiment various approaches to coordinate multiagent systems operating in real environments. The solution, implemented at the IAS Laboratory of Padua University evolves from those successfully adopted in the ART Team on RoboCup-2000 [16], and now in the new Team Artisti Veneti [14] and [13].

2 Multi-robot Systems

A multi-robot system is characterized by attributes like its size, composition, and reconfigurability as well as its communication topology, availability, and range [11]. Also its collective intelligence [9] and agent redundancy are important features. Thus, a group of mobile robots gives rise to an intelligent multi-robot system if they cooperate to solve a given complex task by communicating among individuals and allowing dynamic group reconfigurability.

Robotics team design addresses issues such as the specification if each individual robot share or not a common goal [10] or the choice between distributed and centralized control. Nevertheless, communication among individuals cannot be ignored and, depending on *explicit* or *implicit* one, or a combination of both, the group exhibits very distinguishing behaviors.

In the next sections we shall insist between *explicit communication*, where signals are intentionally shared between two or more individuals, and *implicit communication*, by observing other robot actions. Despite what appears at a first glance, intelligent cooperation doesn't necessarily require an explicit communication among robots. For example, in our preceding papers [7], [12] and [14], we have exploited the case of forcing collective behaviors through implicit communication. There, the idea of perceptual patterns, recognizable by evaluating a set of scalar quantities, termed *macroparameters*, has been introduced. Every agent was equipped with a set of basic behaviors and, moreover, each behavior was defined with its *complementary* [6].

3 Behavior-based Approach

Developing robot agents includes both the design of physical components and the implementation of new software architectures with the aim of investigating the issues that arise from the integration of different software components which support the *decide-sense-adapt behavior cycle* and which, starting from the pioneeristic work of Brooks [5], is controlled by a set of behaviors.

This architecture, known as *behavior-based approach* but also termed *reactive control*, has become very popular along the time. It refers to the direct coupling of perception to action as specific technique which provides time-bound responses to robots moving in dynamic, unstructured and partially

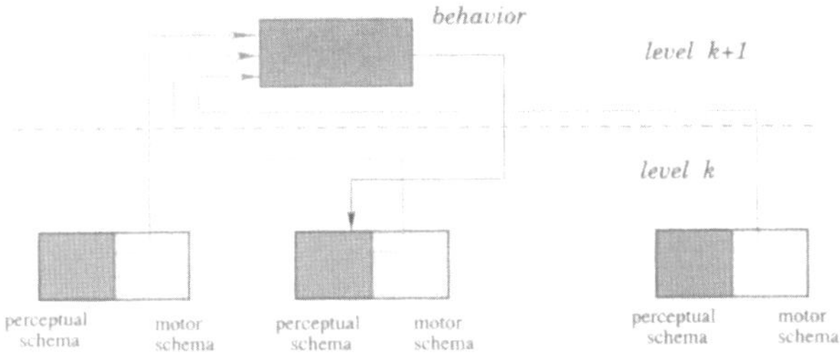

Fig. 1. Controlling the underline level

unknown environments. A behavior is defined to be a control law for *achieving* and/or *maintaining* a given goal. Usually, robot agents have multiple goals, including at least one achievement goal and one or more maintenance goals.

3.1 Schema Approach

The behavior-based approach assumes a robot to be *situated* within its environment. This means that a robot interacts with the world on its own, without any human intervention, namely, its perspective is different from observer's. Moreover, since robots are not merely information processing systems, its *embodiment* requires that both all acquired information and all delivered effectors command must be transmitted through their physical structure. Different research areas like biology, ethology and psychology, have contributed to the design of robot control. Among them, *schema*-based theories have been adapted by Arbib [2] to build the basic blocks of robot behaviors.

Originally, when they appeared during eighteenth century, they provided a mechanism of understanding sensory perception in the process of storing knowledge. Such a philosophical model to explain behaviors has also become an useful abstraction to implement behaviours taking advantage from the object-oriented programming. In this perspective, a *schema* is a generic template for doing some activity which is parameterized and created like a class (*schema instantiation*). Following Arbib [1] we implement a *behavior* with one **motor schema**, representing the physical activity, and one **perceptual schema** which includes sensing.

3.2 Implementing Schemas

Schema-based methodologies are largely used in robotics. So, *motor schemas*, as they were proposed and developed by Arkin [3], are the *basic units of behavior from which complex actions can be constructed; they consist of the knowledge of how to act or perceive as well as the computational process by which they are enacted* [4]. Each schema operates as a concurrent, asynchronous process initiating a behavioral intention by reacting to sensory information.

In Arbib and Arkin, all schemas are always active producing outputs as action vectors which are summed up, whereas our implementation assumes

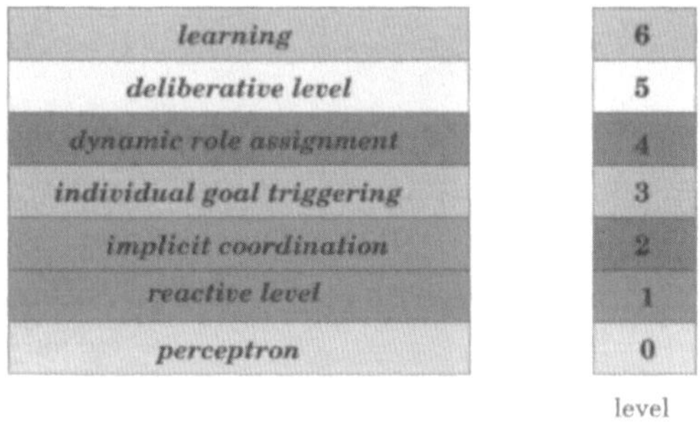

Fig. 2. Levels of Control

only one schema to be active at a time, in a winner-take-all fashion. Moreover, the output is not a continuous signal but either a motor command to feed some servo or an evaluated condition affecting the activation/inhibition mechanism for another schema.

In this perspective, the governor's unit of each robot is organized at many levels of abstraction, the lowest one being directly coupled with the environment by the robot servos. They are implemented as C routines which access sensor and effector devices of the robot. Each level is populated by a set of control units, which are schema-based *behaviors* receiving sensor information by monitoring the lower level and acting on some *releaser*. Fig. 1 makes explicit how conditional activations propagate through levels.

For example, simple behaviors like *defendArea* or *carryBall* are implemented in C as motor schemas accessing directly robot effectors. Now, we are able to build the two basic behaviors *playDefensive* and *chaseBall* by simply appending the two perceptual schemas as explained by the following behavior constructing rules:

$$\begin{array}{ll} playDefensive & : seeBall \rightarrow defendArea \\ chaseBall & : haveBall \rightarrow carryBall \end{array}$$

Also the perceptual schemas *seeBall* and *haveBall* are implemented in C by accessing virtual sensor devices like *senseBall* and *touchBall* which are fed by robot physical sensors. At any level, the primitive control component is a behavior with only a perceptual and motor schemas. By releasing a behavior we mean an *activation-inhibition* mechanism built on some given *evaluating-condition* basis. Thus, a primitive behavior results in appending just one perceptual schema to one motor schema so that, at reactive level, we have the *sensorimotor coordinations* the individual robot is equipped with.

The reactive level is the lowest control level because it uses only information coming from sensors and feed motors with the appropriate commands. Compound behaviors appear only at higher levels while they are receiving more abstract information about the environment as they are filtered by lower behavior functioning. As suggested by fig. 2, individual control has been organized into different layers, each of which represents a different level of abstraction such that an upper level risults in a more abstract handling of the environment. So, the second layer assumes that some perceptual patterns represent events generated by other individuals, either opponents or teammates.

Moreover, the corresponding schemas can control the underline reactive behaviors but, at the same time, they are also triggered by the individual goals every robot should pursue. The higher layers refer to the cooperation capabilities that any robot could exhibit as teamate while a cooperative behavior is going to emerge. We shall describe them in the next sections.

As a matter of implementation, we want to sketch some solutions we have devised to actually implement such an architecture. All schemas are executed as threads in the so called **ade** runtime environment, expecially designed for real-time systems over an Unix/Linux kernel. Also the arbitration module is executed as a thread; more exactly, three different threads have been committed to select a behavior for its execution on the winner-take-all basis.

Looking at fig. 3, it can easily understood how the governor's unit operates to control robot behaviors. First of all, sensor information, coming from different sources are piped towards the *sensor drivers* which work as input controllers. They provide all perceptual schemas with the required sensing, also feeding the C-implemented motor schemas which demand immediate sensor data for triggering. The modules labelled *brain*, *ruler* and *teamplay*, implemented as threads, are committed to select the most suitable motor schema to gain exclusive control of the robot. The thread *brain* evaluates all the possible activating conditions, implemented as **and-or** networks but organized to cover levels of abstraction. The *ruler* affects robot behaviors by adapting their execution to the constraints which stem from soccer play rules avoiding, as far as possible, violating situations.

At last, the module *teamplay* provides the necessary coordination a single teammate must exhibit to eventually make emerging a collective behavior inside the group, for example, a *triangulation* while passing the ball. We deal with this topic in the next section.

4 Building an Hybrid Architecture

As previously stated a schema is the building block of our architecture where perceptual components are organized into an hierarchy of abstraction levels. They feed motor schemas acting as either a control mechanism or a delivery device towards robot effectors, namely, the *wheel-driving motors* and the *kicker*. No effector is needed to control vision as it is implemented by a camera monitoring the environment with an omnidirectional mirror. At reactive level (cfr. fig. 2) schemas are true behaviors whereas at higher levels they work as triggering mechanism to modulate the whole behavior of any individual.

The actual implementation rearranges perceptual schemas in a network of **and-or** nodes, generated at startup by executing appropriate scripts de-

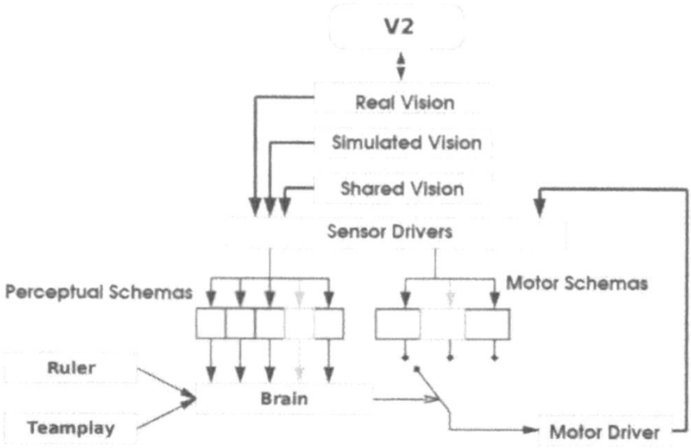

Fig. 3. General Architecture of a single Robot

scribing that hierarchy and which can be easily changed. So many different configurations have been tested during experimental trials in our Lab.

4.1 Integrating Deliberation

If we build robots only considering the reactive level depicted in fig. 2, we couldn't endow cooperation capabilities in our team because of the lack of any mechanism which allows a robot behavior to take into accounts the behavior of other robots. So, an *implicit coordination* is required to trigger individual robot behaviors in such a way some *actions that are a part of an agent's own goal-achieving behavior repertoire, but have effects in the world, help other agents to achieve their goals* [10].

However, a group of robots whose coordination is based on some *stigmergic*[1] property, may exhibit no collective behavior because stigmergy doesn't guarantee cooperation. To force an emergent collective behavior we could need to endow deliberation into the group of robots. Integrating deliberation within a behavior-based architecture is a current topic of research and, moreover, it is a matter of an active debate because the *reactive/deliberative* trade-off depends on how many representational issues are endowed and how much reasoning process is made available to the system.

Considering that any deliberative process slows down the *decide-sense-adapt behavior cycle*, a solution to this problem could suggest a different priority levels to be assigned to the different layers appearing in fig. 2 and which are mapped into different levels of abstraction. So, we have devised an hybrid architecture growing up from a level to the next level in such a way *the more is the number of levels the more are the deliberative capabilities of an individual*

[1] this term, commonly used in biological literature, refers to the animal capabilities to coordinate without explicit communication.

robot. At the top we have the learning process which, at the present, we have not yet implemented. Just a level below there is the effective *deliberative level* but this property becomes less feasible as we approach the reactive level.

There are also two intermediate levels which cope with the communication capabilities of the robot. The lower implements stigmergy whereas the higher deals with the dynamic role exchange, which is needed if we want an effective control on cooperation to be triggered by internal and external firing conditions. We could say that our robots are featured with both *reactive and deliberative communication*, the latter implying some form of negotiation.

In our preceding works we have tried to have the same result only using stigmergy, avoiding any form of negotiation. Macroparameters and quality functions have been the tools we have used to this aim. At the present we want better evaluating the two solutions.

4.2 Implementing Coordination

As previously stated, coordination has been implemented at two stages: the former, dealing with the reactive level, provides the necessary conditions to be verified to start an activation cycle of cooperations. Such conditions are evaluated acquiring information from the environment by testing for specified patterns. The latter is involved in coordination properly by examining and scheduling the behaviors which are the best candidates to cooperate with.

As an example, let us address the coordination between two robots with the task of carrying the ball towards the opponent goal, eventually passing and defending it from opponents' attacks. A number of conditions must be be continuously tested if we want such a cooperative task to become effective. First of all, at any stage of cooperation we should require the two robots to play well-specified roles. So, if we assign the *master* role to the robot chasing the ball, the latter can be considered the *supporter* of the former. Any role is played at different levels; let's call them *canbe*, *assume,acquire*, *grant*, *advocate* so that we can build the following coupled behaviors.

behavior **clampmaster**
$haveBall(\mathbf{me}) \& \neg haveBall(\mathbf{mate}) \rightarrow acquire(Master);$
$acquire(Master) \& Notify(Master) \rightarrow advocate(Master);$
$grant(Master) \rightarrow advocate(Master);$
behavior **clampsupporter**
$\neg acquire(Master) \& canBe(Supporter) \rightarrow assume(Supporter);$
$assume(Supporter) \& Notify(Supporter) \rightarrow acquire(Supporter);$

Here, because the role assignment depends on *ball possess*, we have used the condition *haveBall* to discriminate the robot which is really carrying the ball. It can be understood as a *macroparameter* [7], [6] in the style of our preceding works, that describes the characteristic of the environment and because it can be evaluated by different robots. Moreover, it resyncronizes the activation of a new cooperation pattern.

They are complementary behaviors and they must be arbitrated in such a way they must be assigned to the robots with the right referring roles. The basic rule is that a *master* role must be **advocated** whereas the *supporter* role should be **acquired**. To this aim, we require two reciprocity rules where a role is switched either from *acquire* to *advocate* or from *assume* to *acquire* provided that a **notification** is made to the referred teammate. Such rules imply a direct communication between teammates to negotiate the role on the *first notified/first advocated* basis as depicted below.

$$
\begin{aligned}
&notify(\mathbf{role})\\
&\quad Supporter(\mathbf{mate}) \rightarrow reply(\mathbf{role}, \mathbf{mate});\\
&\quad Master(\mathbf{mate}) \rightarrow request(\mathbf{role}, \mathbf{mate});
\end{aligned}
$$

In such a way, the robot carrying the ball suggests a teammate to become supporter by advocating the master role and forcing the latter to acquire the supporter role. By so doing the former issues a behavior of *chaseBall* whereas the latter exhibits a behavior of *approachBall* and they work as a reinforcement to maintain or exchange these roles.

5 Experimental Results

Implicit coordination and role exchange are necessary tools for activating and tailoring cooperative behaviors. To tell the truth only implicit coordination is strictly necessary as it has been repeatedly pointed out in literature. The problem is how many times the interaction patterns are detected by different robots to initiate a cooperation task. In the case of simulated soccer games, we have shown [6] that a continuous evaluation of environmental patterns could fire ball exchange between teammates during attack. The number of succeded cooperations was made high by increasing the circumstances of positive activations by a kind of *brownian motion* among teamates. The situation becomes more difficult in the case of Middle-size robot competitions where the evolving dynamic of teammates cannot provide such a satisfactory number of active interactions. So, role assignment becomes a very important feature to be endowed into a soccer team. In a preceding implementation [14] we have forced cooperations by evolving *macroparameters* into *quality functions*. On the contrary, the current approach would exploit a different point of view by considering an active engagement of teammates during the phase of role assignment. As shown in fig. 4[2], for the last RoboCup International Competition, held in Padua on July 2003, we have developed some testing programs, that were effectively exhibited during the Middle-size Challenge tournament.

We had to show a cooperative behavior of two companion robots. As it can be easily understood looking at the figure, two soccer robots were involved

[2] at the location http://www.dei.unipd.it/~robocup/video/tenaglia.avi it is accessible the full video

Fig. 4. Two attacking robots chasing the ball in a clamp

in a cooperative task which results in carrying the ball towards the opponent goal to score safely. The first robot is chasing the ball, whereas its companion is approaching to protect it. Its *approachingBall* behavior is a consequence of a sharp negotiation implemented by a low number of *short message exchanges.*

Thus, the two robots are moving in a strict coordinate behavior: the former carries the ball, the latter protects it. Moreover, depending on the actual environmental conditions, the roles can be swapped. Then, when the two robots are near the goal, the first robot exchanges the ball with the second robot, because the two robots have evaluated that the former can score more easily. The resulting complex emergent behavior (*exchangingBall*) seems to emphasize a deliberative aptitude of soccer robots to activate a cooperative cycle of actions. The sequence of actions reported in fig. 4, took place during the Challenge competitions, and was evaluated for the final Team score.

Conclusions

In this paper we have illustrated our current research, aimed to understand how much deliberative process should be endowed in the distributed control of a middle size team to play a soccer game cooperatively. Our current work evolves from our past experience to design behavior arbitration which triggers and it is triggered on the basis of purely *stigmergic* mechanism, namely, implicit coordination. Considering the inherent difficulty to force coordination when the dynamics of the game is not quite fast, we have tried to drive a cooperative task by a dynamical role assignment, switching from the implicit team assessment, given by the evaluation of quality functions, to the explicit negotiation on the *first notified/first advocated* basis. The actual implementation has been made possible with the heavy collaboration of the students at the Eng. School of Padua University who participate to the Research Project on RoboCup at IAS-Laboratory.

References

1. M.A. Arbib. Schema theory. In M.A. Arbib, editor, *The Handbook of Brain Theory and Neural Networks*, pages 830–834. MIT Press, Cambridge, 1995.
2. M.A. Arbib. Perceptual structures and distributed motor control. In *Handbook of Physiology-The Nervous System II: Motor Control*, pages 1449–1480. 1981.
3. R.C. Arkin. Motor schemas based navigation for a mobile robot. In *IEEE Conference on Robotics and Automation*, pages 264–271. 1987.
4. R.C. Arkin. Modeling neural function at the schema level. In *Biological Neural Networks in Inverterbrate Neuroethology and Robotics*, pages 383–410. 1993.
5. R. Brooks. A robust layered control system for a mobile robot. *IEEE Jour. of Robotics and Automation*, 2(1):14–23, 1986.
6. A. D'Angelo, E. Pagello, F. Montesello, and C. Ferrari. Implicit coordination in a multi-agent system using a behaviour-based approach. In *Distributed Autonomous Robot Systems 4 (DARS98)*, Karlsruhe (D), 1998.
7. E. Pagello A. D'Angelo F. Montesello F. Garelli C. Ferrari. Cooperative behaviors in multi-robot systems through implicit communication. *Robotics and Autonomous Systems*, 29(1):65–77, 1999.
8. M. Asada A. Birk E. Pagello M. Fujita I. Noda S. Tadokoro D. Duhaut P. Stone M. Veloso T. Balch H. Kitano and B. Thomas. Progress in robocup soccer research in 2000. In *Proceedings of the 2000 International Symposium on Experimental Robotics*. Honolulu, Dec 2000.
9. D. Kurabayashi. Toward realization of collective intelligence and emergent robotics. In *IEEE Int. Conf. on Sys., Man and Cyb. (IEEE/SMC)*, pages 748–753. Tokyo, 1999.
10. M. Mataric. Issues and approaches in the design of collective autonomous agents. *Robotics and Autonomous Systems*, 16(2-4):321–331, 1995.
11. G. Dudek M. Jenkin E. Milios and D. Wilkes. A taxonomy for swarm robotics. *Autonomous Robots*, 3(4):375–397, 1996.
12. E. Pagello C. Ferrari A. D'Angelo F. Montesello. Intelligent multirobot systems performing cooperative tasks. In *Special Session on Emergent Systems - Challenge for New System Paradigm, IEEE/SMC*, pages 754–760. Tokyo, 1999.
13. E. Pagello, A. D'Angelo, and E. Menegatti. Artisti veneti 2002 evolving an heterogeneous robot team for the middle-size league. In A. Birke and G. Kamilka, editors, *RoboCup2002*, Fukuoka (J), June 2002.
14. E. Pagello and A. D'Angelo et al. Emergent behaviors of a robot team performing cooperative tasks. *Advanced Robotics*, 17(1):3–19, 2003.
15. L. Parker. From social animals to teams of cooperating robots. In *IEEE/RSJ Int. Conf. on Intelligent Robots and Systems (IROS'97)*, Workshop on Multi-robot cooperation: Current trends and key issues. Grenoble (France), 1997.
16. G. Adorni A. Bonarini G. Clemente D. Nardi E. Pagello M. Piaggio. Art'00 - azzurra robot team for the year 2000. In P. Stone T. Balch and G.Kraetszchmar, editors, *RoboCup-2000: Robot Soccer World Cup IV*, number 2019 in Lecture Notes in Artificial Intelligence. Springer, Berlin, 2001.

Vehicle Guidance System using Local Information Assistants

Kuniaki Kawabata[1], Madoka Doi[2], Daisuke Chugo[3], Hayato Kaetsu[1] and Hajime Asama[4,1]

[1] DARS Research Unit, RIKEN, 2-1, Hirosawa, Wako, Saitama, Japan, 351-1098
kuniakik, kaetsu@riken.jp
[2] Tokyo University of Science, 2641, Yamazaki, Noda, Chiba, 278-8510
doi@dars.riken.go.jp
[3] Saitama University, 255, Shimo-okubo, Sakura, Saitama, Saitama, 338-8570
chugo@riken.jp
[4] RACE, University of Tokyo, 4-6-1, Komaba, Meguro, Tokyo, 153-8904
asama@race.u-tokyo.ac.jp

Summary. In this paper, we propose a vehicle guidance system using local information assistants. The information assistant device which is embedded into the environment, realizes to exchange and manage local information related to the environment and the situation. Therefore, proposed system can realize to provide the information without direct communications among the vehicles and global communications. We develop local information device for vehicle guidance and attempt experiment using the electrical vehicle. Also, we discuss a simple guidance method based on local information manaement such devices.

1 Introduction

In recent years, AGVs (Automated Guided Vehicles)[1] are introduced to products transportation system in the factory and so on. Such technologics are expected to apply and extend to general environment. Practically, in the golf course, the AGV technologies are utilized as human transportation system. For example, towards barrier free society, ITS(Intelligent Transportaion Systems) technologies are also applying to construct safety and comfortable transportation for the elderly and handicapped people. In most case of transportation system using AGVs, supervisory management approach is utilized based on global communication. However, such system has the problems related to high calculation load because the amount of information increases according to the number of the vehicles. Therefore, practical guidance systems of AGV utilize the guidance line (magnetic, electrical and so on) and fixed command devices (magnet and so on) that are distributed at the side of the guidance line. The devices realize to send control command to the vehicles, locally. In such systems, the fixed information is handled by such distributed signal devices and it is hard to correspond to change the course layout and the traffic situation.

On the other hand, it is developing that the research related to the technologies for autonomous AGVs using on-mount sensors without the guidance line. However, in such system, the system should equip plural sensors to realize to improve the reliability of the recognition in general environment[2]. The control system of such vehicle becomes highly complex and the cost also becomes high. Therefore, it is required to develop a new guidance system that introduces distributed information management and realizes flexible course construction. Chiba *et al* proposed a method for classifying the transportation route of AGVs[3]. Higashi *et al* proposed a self-organizing control method for AGV system[4]. Moreover, in conventional technologies of AGV, the main scope is to transport the object in the factory with small number of AGVs and it is not the scope to apply to the general traffic road enviroment.

In this research, we are developing a guidance system for applying AGV to human transportation on the general road. Our system realizes to construct un-supervised vehicle control system based on local infomrmation management. In this paper, we develop an electrical outdoor vehicle guidance system with local information devices. Also, we examine to extend the system for realizing effective guidance and propose simple guidance algorithm based on combining distributed information.

2 Conventional Vehicle Guidance System

Generally, in AGV guidance system, the tape or electric wire for the guidance is on the ground and each AGV equips the sensor to detect such guidance line. Figure 1 shows a commercial electric vehicle and we utilize this platform in this research[5].

Standard guidance system for this commercial electric vehicle utilizes the electric wire as the guidance line and the permanent magnets for the control

Fig. 1. An Electric Vehicle

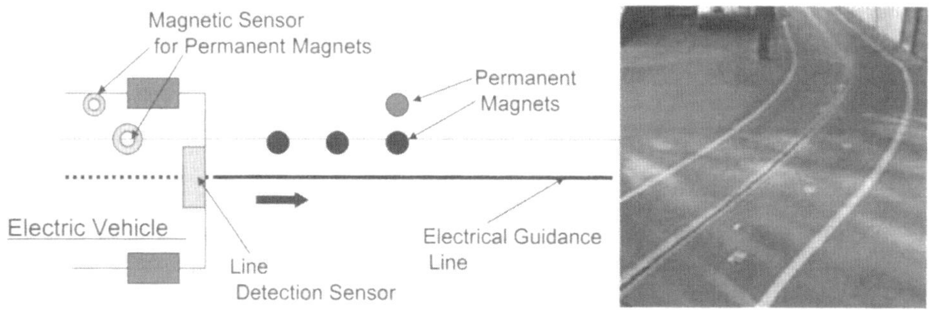

Fig. 2. Conventional Guidance System

command are placed on the side of the guidance line(Figure 2). The arrangement pattern of permanent magnets expresses the control command to the vehicle utilizing the combination of S and N pole (moving speed change, stop and so on.) However, such conventional guidance system only manages fixed information and operates small kinds of the control commands. Therefore, flexibility for system construction is acquired to extend this system to the other applications. In next section, we explain our local information management devices and to apply them to vehicle guidance system.

3 System Configuration

Here, we explain our developing vehicle guidance system based on Intelligent Data Carrier.

3.1 Intelligent Data Carrier (IDC)

In our current work, we proposed and are developing *Intelligent Data Carrier* (IDC) system. Figure 3 shows the concept of the IDC system. IDC tag is a portable electronic data carrier device with functionality of information storage (rewritable nonvolatile memory), information processing (MPU), local wireless data exchange (RF-ID weak radio communication), power supply (battery - optional), and external ports (I/O interface, optional).

As shown on Figure 3 , IDC system consists of a number of IDC tags embedded in the environment(wall, floor, obstacles, objects, etc) and reader/writer devices. Each IDC tag manages and processes local information depending on specific place or objects. The agents can communicate with the IDCs via a reader/writer device provided with each agent, and extract/add/update the local information in IDC tags through local radio communication. Or, the IDC tags inform the agents of the necessary local information for guidance or task execution, and mediate the knowledge or commands stored by an agent to the other agents. With IDC system, it becomes possible to make the environment

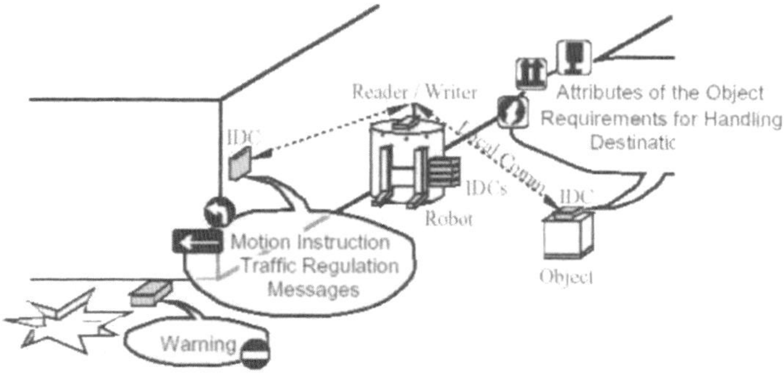

Fig. 3. Concept of the IDC system

intelligent and implement information structure in the environment, namely the infrastructure for mobile agents to acquire the necessary information. In other words, the IDC system facilitates realization of a ubiquitous computing environment [6], which makes autonomous operation of mobile robots feasible and task execution efficient and flexible. Moreover, with the interactions between robots and the environment, the agents can share knowledge via environment without mutual direct communication. Consequently, the IDC system provides not only the utility for agents as their infrastructure, but also means for emergent adaptiveness of cooperating agents constructing active and dynamic affordance[7]. The global order of the intelligent system is expected to emerge through the local interactions by using IDC system [8].

In this research, we developed an IDC system for electric vehicle control. Table 1 indicates the speccification of the developed system. Figure 4 shows IDC tags and IDC Reader/Writer, respectively.

3.2 Guidance Experiments using IDC system

Our developing system is shown on Figure 5. The vehicle equips the computer(PC) for its own controller, the antenna and IDC Reader/Writer(IDC R/W) which communicates with the control PC via RS-232C(serial communication).

Table 1. Specifications of IDC system

media	electromagnetic wave
communication	RF-ID
communication rate	read:4.8[kbps]/write:0.96[kbps]
communication range	150[mm]
memory capacity	110[byte]

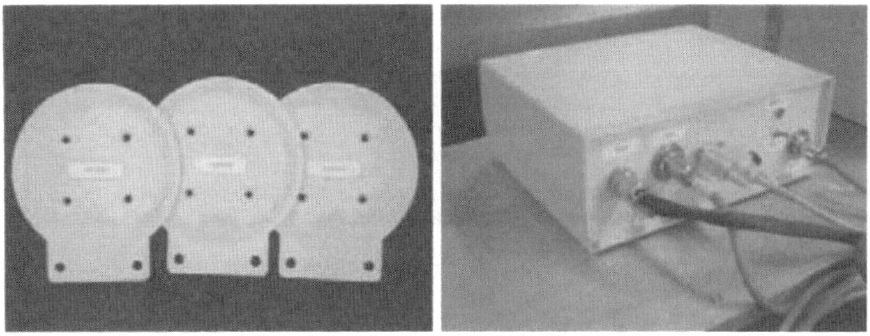

Fig. 4. IDC Tags and IDC Reader/Writer

Figure 5 also shows IDC R/W with antenna and it works to read/write the information from/to IDC tags on the road(Figure 6). Using developed system, we have some basic experiments related to running speed control, branching off control and so on. The control commands are stored in each IDC tag and the tags are placed on the side of the guidance line. We confirmed that the vehicle can adjust its own running speed from 1.0[km/h] to 8.0[km/h] at every

Fig. 5. Overview of the Electrical Vehicle and Communication Antenna

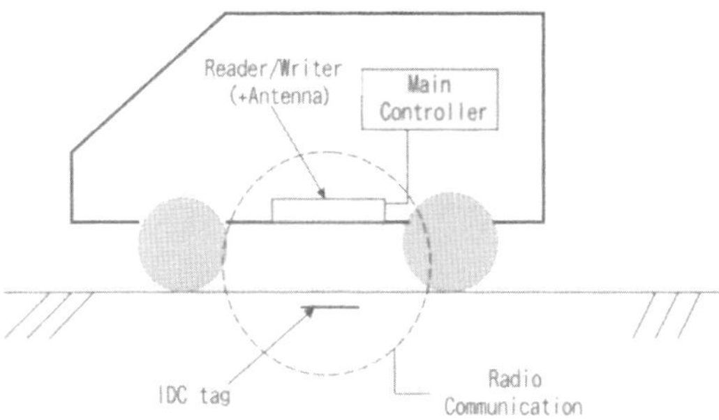

Fig. 6. Communication between IDC tag and Vehicle System

Fig. 7. Overview of Running Experiment

1.0[km/h] based on information on IDC tags. Also, the vehicle can branch off to right or left blanch line, and run through the junction of two branch lines. Figure 7 shows the overview of running control experiments using developed system. In this experiment, the vehicle starts at 3.0[km/h] and accelerates to 8.0[km/h] after passing first IDC tag. Passing the second IDC tag, the vehicle slows down to 2.0[km/h] and finally stops after passing the third tag.

As the result, we can confirm that developed system can work to guide and control the vehicle. Although the system can locally manage static or semi-static information based on RF-ID communication, it can not treat dynamic information. Here, semi-static information include the map information, vehicle control command and the data which are stored when previous vehicles passed at that point.

We must consider to improve our system for effective guidance system based on the distributed manner.

4 Guidance Method with Local Information Assistant and IDC

In this section, we examine a vehicle guidance method based on local information management using Information Assistant(IA) which is a sort of extended IDC system. IA equips simple calculation capacity and local communication (based on like a wireless LAN devices) with the other IA. At each intersection, there is one IA and plural IDC tags. IDC tags are utilized to provide semi-static infomation as velocity command, map and so on to the vehicle. IAs also support to provide dynamic situation change including traffic condition. IAs support that the vehicles decide the path based on mutual information exchange.

4.1 Environment Setup and Selection of Path Candidate

We assume a simple road layout as Figure. 8. and it consists of straight roads and the intersections. The intersections are connected to straight roads(bi-directional paths) each other and the distance between the intersections is equal to the road length.

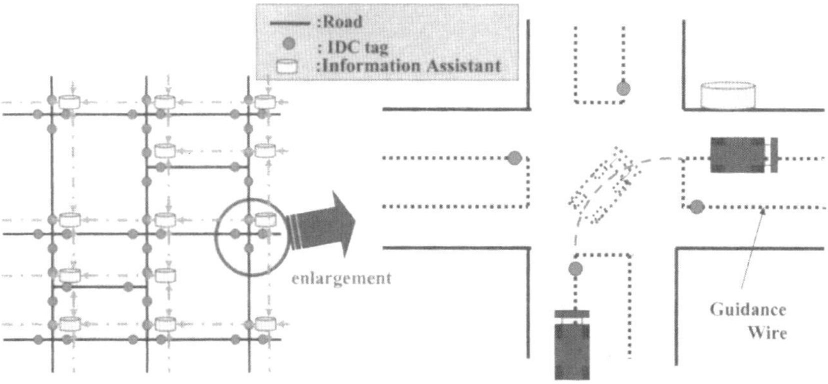

Fig. 8. Experimental Setup

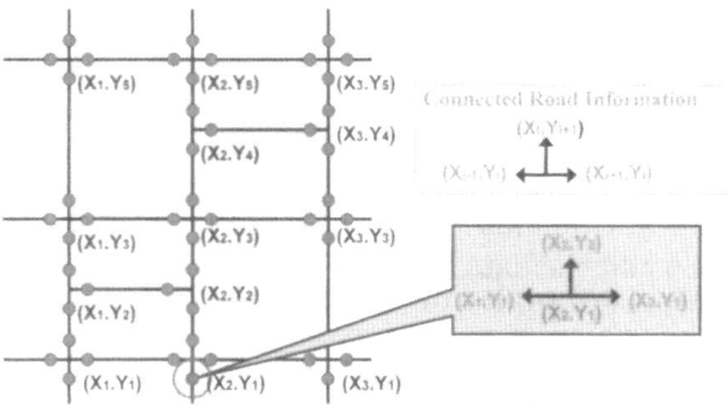

Fig. 9. Stored Information in IDC Tags

The guidance line is set up to the road part and IDC tags are placed at every branches of the intersection. As guidance at the intersection, we proposed a laser guidance system using Information Assistant (IA)[9]. IA can communicate to the IAs at the neighboring intersections, locally. IDC tags, which are placed at the intersection, supervise semi-static information including the map, command and so on. The example layout is shown on Figure 9 and there are IDC tags at each intersection in proportion to the number of the roads. Each intersection is identified using the position in the world coordination and it is stored in IDC tags.

For example, when the vehicle arrives at the intersection $P_{ij} = (X_i, Y_j)$, the vehicle can read the position of the neighboring intersections $P_{(i-1)j}$, $P_{(i+1)j}$, $P_{i(j-1)}$ and $P_{i(j+1)}$ from IDC tag at the intersection P_{ij}. Therefore, the vehicle has the destination position $P_{dst} = (X_{dst}, Y_{dst})$ and can calculate each path length to the destination based on IDC information. Each IDC tag, which is placed at the intersection, has the local map information related to the position of connected branches. Here, the distance $L_{\alpha\beta}$ between the destination P_{dst} and the next branch position $P_{\alpha\beta} = (X_\alpha Y_\beta)$ is calculated by following equation.

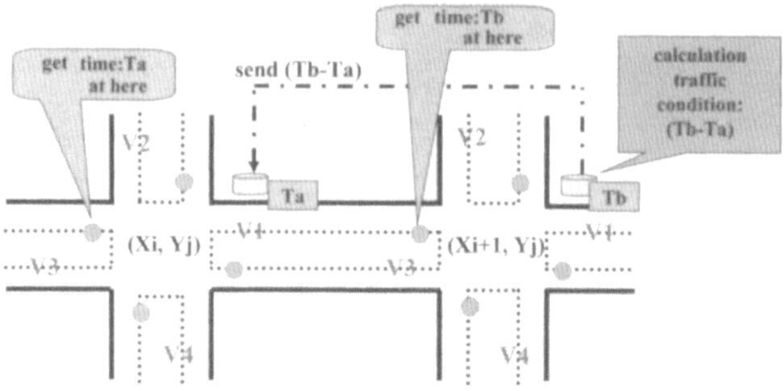

Fig. 10. Traffic Condition Calculation Scheme

$$
\begin{aligned}
L_{\alpha\beta} &= P_{dst} - P_{\alpha\beta} \\
&= \|X_{dst} - X_{\alpha}\| + \|Y_{dst} - Y_{\beta}\|
\end{aligned} \tag{1}
$$

Next, the minimum of $L_{\alpha\beta}$ is selected as the candidate of the path. As the natural result, there is possibility that plural candidate paths are selected. In the next discussion, we discuss that path selection method with considering the traffic condition.

4.2 Path Selection based on Traffic Condition

Using local stored information, the candidates of the path can be selected. Here, in order to realize efficient guidance, time which takes to get to the destination should be considered for the path selection. It means that traffic condition of near area should be referred. Therefore, we introduce a path selection method utilizing traffic condition information. Figure 10 shows the overview of our proposed scheme. . As Figure 10, the vehicle passes the intersection P_{ij} at time T_a and also reaches to the next intersection $P_{\alpha\beta}$ at time T_b. The vehicle sends T_a to IA at $P_{\alpha\beta}$ when IDC tags at $P_{\alpha\beta}$ is detected and IA calculates the time difference. Thus, IDC tags at the intersection are the sort of trigger for IA. Here, we set the traffic condition coefficient ${}^{P_{ij}}V_k$ which means the value for path V_k at the intersection P_{ij}.

$$
\begin{aligned}
{}^{P_{ij}}V_k &= \frac{T_b - T_a}{T^{reg}} \\
&= \frac{{}^{P_{\alpha\beta}}T - {}^{P_{ij}}T}{{}^{P_{ij}}T_k^{reg}}
\end{aligned} \tag{2}
$$

Here, ${}^{P_{ij}}T_k^{reg}$ indicates the regular time for passing through k path from the intersection P_{ij}, These traffic condition coefficients are supervised by each IA at the intersection. IA at $P_{\alpha\beta}$ sends the traffic condition coefficient to IA at P_{ij} The vehicle refers these value to select better path from the candidates. Figure 11 shows the path selection flow based on the distance value and the traffic condition.

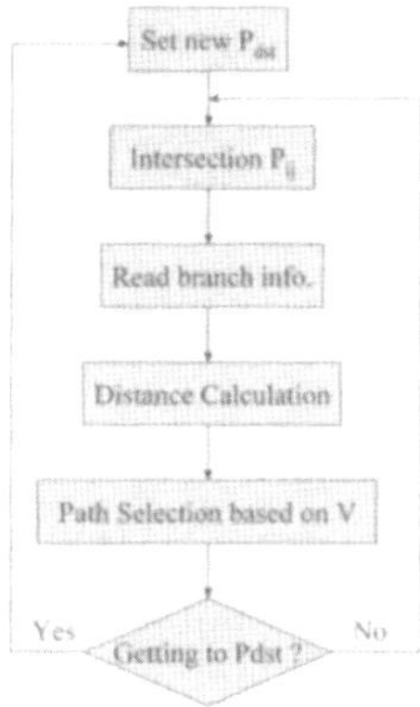

Fig. 11. Path Selection Flow

5 Computer Simulation

In this section, we confirm how our guidance method can work effectively. Here, we compare the result of utilizing only IDC tags and utilizing IDC tags and IA. Figure 12 (1) shows the simulation environment. In this environment, there is 22 vehicles and they set their initial position and destination at random. One of them is the target vehicle and its initial position and destination is set as [0,0] and [7,8](the unit in this map is 10[m]), respectively. Here, the vehicle moves at the speed : 1.8[km/h](the vehicle can pass the unit length:1.0[km] at 20[sec]) and takes 15[sec] at the intersection for local traffic management. Figure 12 (2) and Table. 2 show the simulation result. In figure 12, black bars in the graph indicate navigation time by using proposed method. Gray bars show navigation time by using only IDC tags. We can confirm that the advantage of our method for effective guidance based on local information management(approximately 22[%] improved).

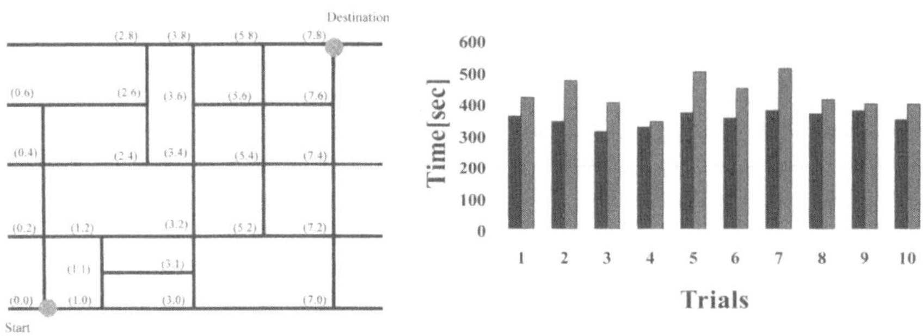

Fig. 12. (1)Simulation Environment (2) Simulation Result

Table 2. Average Time

condition	average time[sec]
only IDC tag	351
IDC tag and Assistant	429

6 Conclusion

The purpose of this research is to apply AGV technologies to flexible new traffic system for human transportation. In this paper, we developed a vehicle guidance system using local information assistants. The system consists of magnetic guidance line and distributed information devices on the running course. We had some experiments to confirm that developed system Also, we proposed a simple algorithm for utilizing our development guidance system. In our future work, we examine and discuss more effective guidance scheme based on a distributed manner with real-time local information, some part of global broadcasted information and so on.

References

1. http://www.agvp.com/
2. Oomichi, T., Kawauchi, N., Fuke. F.,(1999). Hierarchy control system for vehicle navigation based on information of sensor fusion perception depending on measuring distance layer, Proceedings of the International Conference on Field and Service Robotics, pp.197-201
3. Chiba, R., Ota, J. and Arai, T.(2002). Integrated Design with Classification of Transporter Routing for AGV Systems, Proc. 2002 IEEE/RSJ Int. Conf. Intell. Robots and Systems (IROS2002), pp.1820-1825.
4. Higashi,. T., Sekiyama, K., Fukuda., T., (2000) . "Self-organizing Control of Carrier Sequence in AGV Transportation System", Proc. of IECON2000, pp.706-711
5. http://www.yamahagolfcar.com/
6. Weiser, M. (1991). The Computer for the Twenty-First Century, Scientific American, pp. 94-104.
7. Gibson, J. J. (1979). The Ecological Approach to Visual Perception, Boston, MA: Houghton Mifflin.
8. Fujii, T., Asama, H., von Numers, T., Fujita, T., Kaetsu, H., Endo I. (1996). Co-evolution of a Multiple Autonomous Robot System and its Working Environment via Intelligent Local Information Storage, Robotics and Autonomous Systems, vol. 19, pp. 1-13.
9. Suzuki, T., Uehara, T., Kawabata, K., Kurabayashi, D., Paromtchik, I. E., Asama, H., (2003) : "Indoor Navigation for Mobile Robot by using Environment-embedded Local Information Management Device and Optical Pointer", Preprints of the 4th International Conference on Field and Service Robotics, 23-28.

Part III

Multi-Robot Perception

Topological Map Merging

Wesley H. Huang and Kristopher R. Beevers

Rensselaer Polytechnic Institute, Department of Computer Science
110 8th Street, Troy, New York 12180, U.S.A.
{whuang,beevek}@cs.rpi.edu

Summary. A key capability for teams of mobile robots is to cooperatively explore and map an environment. Maps created by one robot must be merged with those from another robot — a difficult problem when the robots do not have a common reference frame. This problem is greatly simplified when topological maps are used because they provide a concise description of the navigability of a space. In this paper, we formulate an algorithm for merging two topological maps that uses aspects of maximal subgraph matching and image registration methods. Simulated and real-world experiments demonstrate the efficacy of our algorithm.

1 Introduction

Systems of multiple mobile robots must be able to cooperatively explore and map an environment for applications such as urban reconnaissance, search & rescue operations, security monitoring, and even house cleaning. In order to create a map quickly, each robot can only explore part of the environment, and the robots' individual maps must be merged to form a complete map.

In this paper, we address the problem of map merging for topological maps. Topological maps use a graph to represent possibilities for navigation through an environment; vertices represent certain "places" in the environment and edges represent paths (or classes of paths) between these places. Often vertices and edges are annotated with certain metric information, such as path length (for edges) or relative orientations of incident paths (for vertices). Typically, vertices are junctions of hallways, and edges represent a path down a hallway from one junction to another. Topological maps provide a concise description of the environment specifically geared towards navigation. Our focus is on indoor environments, so the use of topological maps is appropriate.

Two robots that have explored overlapping regions of an environment should have topological maps that have common subgraphs with identical structure. Solving the map merging problem is thus analogous to identifying a matching between the two graphs. In general, we would expect exactly known attributes of matched vertices, such as the degree of the vertices in a static world, to match perfectly. However, attributes of vertices and edges that are subject to measurement error, such as the length of an edge or the angles between edges leaving a vertex, must only match closely.

If a robot's topological map contains geometric information about edges (e.g. path shape and the orientations of edges at vertices), there is enough information to estimate vertex locations with respect to the robot's world frame. This suggests that the map merging problem could also be solved using image registration techniques. The most widely used algorithms for image registration are iterative closest point (ICP) algorithms. An initial matching between feature points must be provided; the algorithm first estimates a transformation between the two feature sets by minimizing the (weighted) squared error between corresponding features. Next, the feature matching is expanded by finding features that are close together under this transformation, and the transformation is re-estimated. This process continues until the change in the transformation estimate between iterations is small.

In this paper, we describe an algorithm that uses the structural aspect of subgraph matching and the geometric aspect of image registration. We first create hypothesized matches by pairing compatible vertices in the two maps, and then locally grow each match using only the graph structure and vertex and edge annotations. Many hypotheses can be eliminated in this phase because of incompatibilities in the structure of the maps, or because of incompatible annotations (such as edge lengths that are too different). After growing the consistent matches, we estimate geometric transformations for each hypothesis and perform clustering in the transformation space. The best cluster (based on size and error) is returned as the algorithm's result, and the transformation from that cluster is used to merge the maps.

1.1 Related work

In recent years, there has been increased interest in map-making with multiple robots. Most multi-robot mapping work has focused on the creation of occupancy maps, e.g. [13, 10], though some work has been done on multi-robot feature-based or topological mapping [7, 6, 4]. In particular, Konolige *et al.* [11] have shown the benefits of feature-based approaches to map merging, as opposed to matching the raw sensor data of occupancy maps.

Much of the multi-robot map making and merging work assumes that all robots in the team begin the mapping process with a common reference frame — an assumption we do not make in this paper. One notable exception is the approach taken by Ko *et al.* [10] in which robots exchange occupancy maps and attempt to localize themselves in each others' maps.

The work most related to ours is that of Dedeoglu and Sukhatme [4], who presented a method for merging landmark-based maps without a common reference frame. They estimate a transformation between two maps using a single-vertex match found with simple heuristics, and match other vertices using this transformation. In contrast, we estimate geometric transformations between the two maps, but rather than use these transformations to generate vertex correspondences, we compute the transformations *using* correspondences that we find from the structure of the maps. Our use of the maps' structure results in quicker and more effective discovery of potential matches. Additionally, our transformations are computed from multiple-vertex matches using well-known image registration techniques [2], rather than from single-vertex matches.

Our work draws on ideas from the graph matching literature. In particular, the topological matching problem can be viewed as an instance of the maximal common subgraph problem [3], and is closely related to the problems of structural matching and error-tolerant subgraph isomorphism [14].

1.2 Assumptions

The maps to be merged must be consistent — no two vertices may represent the same "place." This means that the robots can either recognize or infer when they have revisited a place and thus are able to "close the loop." However, we do not assume that the vertices have "distinguishing" attributes, which would make the map merging problem significantly simpler.

The robots must record enough information that map vertices can be embedded in a metric space; path shapes and the angles between edges leaving vertices are sufficient. We assume there is a known error model for measurements and that, errors notwithstanding, only translation and rotation (but no scaling) are needed to merge the maps. Our examples are from rectilinear worlds, but the algorithm we develop will work with any topological maps.

2 Hypothesis building

The first phase of our algorithm creates hypotheses by locally growing single-vertex matches. A hypothesis is a list of vertex and edge correspondences between two maps; finding them is, in essence, the problem of finding all maximal common connected subgraph matchings between the two maps.

2.1 Vertex matching

We start with a pair of compatible vertices, one from each map. Vertices are tested for compatibility by examining their attributes: exactly known attributes (e.g. vertex type) must match perfectly; inexactly known attributes (e.g. edge lengths or orientations) must be compared with a similarity test.

It often makes sense to assume that the robots will know the degree of vertices exactly; robots with sufficiently powerful sensing should easily be able to determine the number of paths leading from a vertex. In dynamic worlds, where the degree of vertices may change — e.g., due to opened or closed doors — vertex degree cannot be treated as an exact attribute.

Relative edge orientations at a vertex are typically known with some uncertainty, so they must be compared using a similarity test which determines how similar two vertices must be in order to be paired.

2.2 Growing matches

We now grow the match by testing corresponding pairs of edges leaving the paired vertices. If the edges are compatible and the vertices at the ends are also compatible, then they are added to the match. If the edges or vertices are incompatible, the entire match is rejected.

The vertices are tested with the same criteria and similarity tests used to form the initial pair. Edges may also have both exactly and inexactly known attributes. Typically they have a path length, compared with a similarity test.

Our initial hypotheses are the unique matches that survive the growing process. We avoid generating duplicate hypotheses by keeping a table of vertex pairings. When vertices are paired during the growth phase, the corresponding entry is marked in the table. This entry is then ineligible as an initial pairing of vertices. A subgraph in one map can be matched to multiple subgraphs in the other (under separate hypotheses), but a pair of matched vertices (with a given edge correspondence) can appear in only one hypothesis. The matching/growing process is repeated until all valid vertex pairings are examined.

2.3 An example

For an example of hypothesis generation, we use two maps from a rectilinear world, shown in Figure 1. The rectilinear world assumption implies that we know exact orientations of paths leaving vertices (relative to the robot's coordinate frame). In the example, we also assume a static world, so vertex degrees must match exactly. Degree two and three vertices can match only for a single edge pairing; degree four vertices can match for four edge pairings.

To construct all unique hypotheses, we create tables as described previously. The matches are grown by adding compatible incident edges and vertices. When a match is found to be inconsistent (usually due to vertex degree mismatch), we still grow it as much as possible so that those vertex pairs can all be marked as incompatible. This avoids regrowing the same match from a different initial vertex pair.

3 Transform estimation & clustering

Our hypotheses now consist of maximal matched connected subgraphs. These matchings represent a single overlapping area, but the two maps may have several (separate) overlapping areas. In this phase of the algorithm, we consider the geometric relationships of hypotheses.

	b2	b3	b6
a3	×	×	H_1
a6	×	×	×
a7	H_2	×	H_3

Degree-2 Vertices

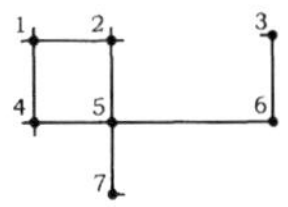

Map $\mathcal{A}$

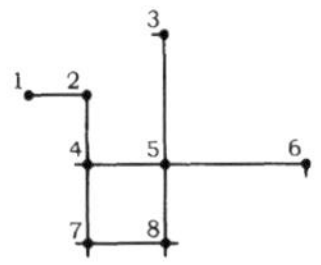

Map $\mathcal{B}$

Map $\mathcal{A}$ rotation 90° 180° + 0° 270°	b4	b5	b7	b8
a1	× × + H_4 H_{11}	× × + × H_{12}	× H_9 + H_5 H_{13}	× H_2 + H_6 H_{14}
a2	× × + × ×	× × + H_4 ×	× H_2 + H_7 H_{11}	H_8 × + H_5 H_{12}
a4	× H_9 + × H_{12}	× H_2 + × ×	× H_{10} + H_4 H_{14}	× × + × H_{15}
a5	× H_2 + × ×	× × + × ×	× × + × H_{12}	× × + H_4 ×

Degree-4 Vertices

Fig. 1. Hypothesis generation example for two maps from a rectilinear world. All possible single vertex matches are represented in the two tables; an "×" indicates that there is no valid hypothesis with that vertex match (and orientation). For example, hypothesis H_1 is generated from a matching with no rotation between vertex 3 in Map $\mathcal{A}$ and vertex 6 in Map $\mathcal{B}$ (a3–b6). There are no common edges from these vertices, so it cannot be grown further. For a 180° rotation of Map $\mathcal{A}$, hypothesis H_2 matches a1–b8, a2–b7, a4–b5, a5–b4, and a7–b2; this is the extent to which it can be grown. For a 90° rotation of Map $\mathcal{A}$, we can match a1–b7, a2–b4, a5–b5, a4–b8, and a6–b3. Though a7–b6 should also match, the edges a5–a7 and b5–b6 have significantly different lengths, so all these matches are marked invalid. There are 15 hypotheses that result from these two maps.

3.1 Transform estimation

In order to estimate a geometric transform, we must first embed the vertices of the maps in the plane. The problem of generating a consistent geometric map from the local distance and angular measurements added to a topological map has been addressed by several researchers, including Duckett *et al.* [5], Lu and Milios [12], and Golfarelli *et al.* [9]. Any of these methods would suffice; the reference frame for the vertices can be placed arbitrarily.

We can now estimate a transformation (translation and rotation) for each hypothesis using the vertex correspondences of the hypothesis. In image registration, this would typically be done using iteratively reweighted least-squares, where the weights help make the method robust to outliers. It is appropriate for us to use an unweighted least squares estimation because corresponding vertices should be close together; large error indicates that a hypothesis is geometrically bad (despite being structurally good).

3.2 Clustering

We group hypotheses into clusters to determine which hypotheses are consistent with each other. The clustering is done in the hypothesis transformation space, for which an appropriate distance function must be used.

Clustering requires some threshold distance to determine when two transforms are close enough to be compatible. This distance could be defined in terms of the map, e.g., a fraction of the minimum edge length in the hypothesis, or in terms of some aspect of the hypotheses themselves, such as metric error. Hypotheses within the threshold distance in transformation space are clustered together. There are a variety of techniques for clustering; we used a simple agglomerative clustering method in our implementation.

Once we have a set of hypothesis clusters, we order these clusters in terms of their "quality." The quality of a hypothesis is not straightforward to determine without information about the size, complexity, and self-similarity of the environment, which could be used in assessing the distinctiveness of the matched portions of the maps. Absent such information, we suggest the following heuristics and methods:

- Total number of vertices is a good primary indicator of cluster quality. Clusters containing only one or two pairs of vertices — particularly, clusters with only a small number of single-match hypotheses — are generally not significant unless the vertices are somehow unique.
- The amount of error (i.e., total squared error under the cluster transform) is a good secondary indicator of quality.
- The number or size of hypotheses in a cluster can be used as a secondary quality indicator. For example, a single large hypothesis is preferable to a cluster of small hypotheses.
- There is always some tradeoff between size and quality of a cluster: a single matched pair of vertices has no error, but generally does not constitute a significant match.

3.3 Example continued

For rectilinear worlds, the errors in orthogonal directions are decoupled. Embedding a map in the plane is thus reduced to two one-dimensional problems. We find the maximum likelihood embedding using a simple spring model.

We estimate a transform for each hypothesis using a two-dimensional version of the point-based rigid registration algorithm described by Fitzpatrick *et al.* [8]. This involves computing the singular value decomposition of a two-by-two covariance matrix.

Figure 2 shows the resulting clusters from our example. Because the example is in a rectilinear world, the transform space consists of a two-dimensional translation space for each of the four possible rotations. The quality of clusters was determined first by number of vertices and then by total error between vertex pairings under the cluster transform.

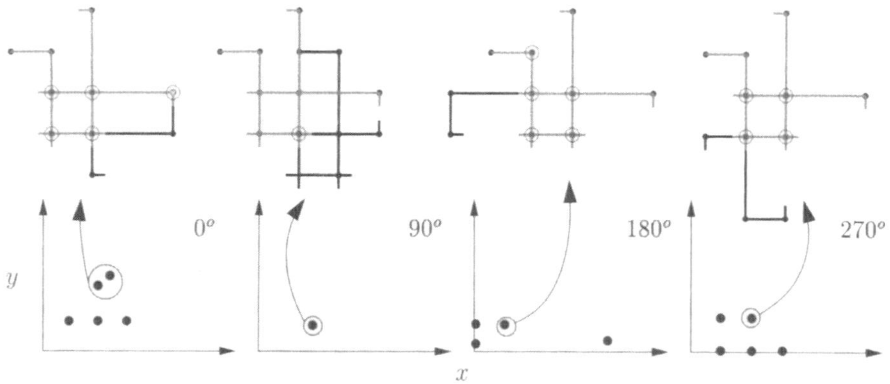

Fig. 2. Transformation spaces and example clusters for our example. The 180°, 5-vertex cluster (from a single hypothesis) is the best match; the 0°, 5-vertex cluster consisting of two hypotheses is also a good match. Map $\mathcal{A}$ is shown in black, map $\mathcal{B}$ is shown in gray, and matched vertices are circled.

4 Implementation issues

Once a hypothesis cluster has been deemed correct, it is fairly straightforward to merge the two maps. The estimates of path lengths can be updated by combining the measurements from the two maps for corresponding edges. The edge orientations at the corresponding vertices can be similarly merged. Portions of one map not present in the other should be added. Appropriate strategies for map storage and variations on the merging algorithm can simplify the implementation and improve its efficiency.

4.1 Map storage

Even the best cluster choice may later turn out to be incorrect. For example, early in the process of exploring a self-similar environment, the robots might seem to be exploring the same area when in fact they are exploring similar but distinct areas. We must consider how to merge and store maps so that incorrect hypotheses can be removed without discarding the whole map. Also, we would like to be able to merge the maps of several robots, not just two.

We can think of a robot's map as being represented in several layers. Layer 0 should be a robot's own map, recording only the measurements that robot has taken. Layer 1 is used to store other robots' Layer 0 maps that have been matched. Layer 2 contains maps that have been matched to Layer 1 maps (for which the matches have been computed either by another robot or locally using another robot's data), and so on.

This approach yields a "dependency" structure that makes it straightforward to discard hypotheses that are later determined to be incorrect, along with all other hypotheses that depend on them. Also, it does not require much extra storage; if necessary, upper layers can be compressed into a single layer.

4.2 Computational issues

When used online, robots may have additional information that can be used to merge maps more quickly. For example, if two robots exchange maps only when they are within communication range, we can start the hypothesis formation by pairing vertices near both robots.

After merging maps once, two robots may later merge their maps again. Here, an "incremental" map update can save a substantial amount of computation; this is possible with minimal bookkeeping effort. The other robot's map, stored separately, can be updated with new or changed vertices and edges. Updates can be used to verify and expand (or eliminate) hypotheses from previous mergings, and to form new hypotheses. All surviving hypotheses undergo the remainder of the merging process.

Although a map merging can be computed relatively quickly, it may be desirable to distribute the computation between robots. The hypothesis building and transform estimation steps can easily be split. The transformation clustering, however, is done most straightforwardly on a single robot.

5 Results

We have implemented the map merging algorithm for rectilinear worlds, and have tested this implementation in simulation with randomly-generated maps, and with maps generated from real-world data. Our results indicate that the algorithm is very effective, even for large maps with small overlap.

In our implementation, similarity of edge lengths was tested based on an odometry error model: when one edge was within a 95% confidence bound of the other, the edges were deemed acceptable matches. Thresholds on intra-cluster distance in transformation space were set to be equal to 3 times the largest mean squared (translational) error among the individual hypotheses; in rectilinear worlds, rotational error need not be considered.

For large maps (greater than 100 vertices), there were several thousand single-vertex correspondences; after the topological growth process, less than 100 hypotheses remained. Typically, there were only one or two large hypotheses (more than three vertices). Even for large maps, the matching process was quick, usually under one second on a 650 MHz Pentium 3 processor.

Clustering on the hypothesized matches worked well, but occasionally very small (correct) matches yielded transformations that were significantly different from the true transformation because of vertex positioning error; as such, the matches were not added to otherwise correct clusters of hypotheses. A potential solution is to take an "iterative" clustering approach, similar to the iterative closest point methods used in image registration, in which new hypotheses may be added to a cluster based on metric error under the transformation of the cluster, rather than on distance in transformation space.

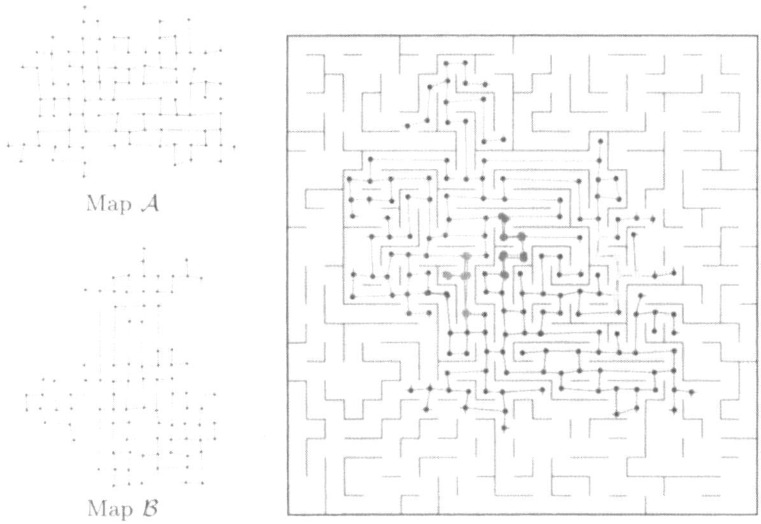

Fig. 3. Results of merging simulated rectilinear maps. Matched vertices are shown with large dots. The best matching occurs when map $\mathcal{B}$ is rotated $-90°$.

Figure 3 shows the matching found for two randomly generated maps in a maze-like world. The result is a cluster of three consistent hypotheses. In computing the matching, there were 3918 initial vertex pairs; after the topological growth process, there were 72 hypotheses, which were placed into 69 clusters. The entire process took 0.04 seconds. Notice that just to the left of the leftmost (magenta) hypothesis are two vertices that should be matched and are not, an example of the clustering problem with small hypotheses.

The algorithm was also tested with maps of a real-world indoor environment (an academic building at RPI). Though these maps were of a reasonably large environment (approximately 12 m $\times$ 30 m), they were relatively small, particularly in terms of complexity, when compared to our simulated tests. With the real-world data, the robot found the correct matchings between partial maps in all cases. For further details and additional experiments, see [1].

6 Conclusions

In this paper, we have presented an algorithm for merging two topological maps. The algorithm uses the structure of the maps to find a set of hypothesized matchings, and then uses the geometric transformations of hypotheses to group them into consistent clusters. In addition, we have proposed approaches to map storage that are effective for our hypothesis-based merging, and that facilitate merging maps from multiple robots. Finally, we have discussed ways to reduce the computational cost when robots re-merge updated maps.

We have demonstrated our algorithm on simulated and real-world maps. In our experiments, even maps with minimal overlap are often merged correctly; for those that are not, our algorithm returns a set of consistent mergings. Maps with meaningful overlap were always merged correctly.

Acknowledgement. We would like to thank Charlene Tsai and Chuck Stewart for discussions on the relationship between image registration and topological map merging. This work was supported by the NSF through grant IIS–9983642.

References

1. K. R. Beevers. Topological mapping and map merging with sensing-limited robots. Master's thesis, Rensselaer Polytechnic Institute, Troy, NY, April 2004.
2. P. Besl and N. McKay. A method for registration of 3-D shapes. *IEEE Trans. on Pattern Analysis and Machine Intelligence*, 14(2):239–256, February 1992.
3. H. Bunke. On a relation between graph edit distance and maximum common subgraph. *Pattern Recognition Letters*, 18(8):689–694, 1997.
4. G. Dedeoglu and G. S. Sukhatme. Landmark-based matching algorithm for cooperative mapping by autonomous robots. In L. E. Parker, G. W. Bekey, and J. Barhen, editors, *Distributed Autonomous Robotic Systems 4*, pages 251–260. Springer-Verlag, 2000.
5. T. Duckett, S. Marsland, and J. Shapiro. Learning globally consistent maps by relaxation. In *IEEE Intl. Conf. on Robotics & Automation*, pages 3841–3846, 2000.
6. G. Dudek, M. Jenkin, E. Milos, and D. Wilkes. Topological exploration with multiple robots. In *7th Intl. Symp. on Robotics with Applications*, 1998.
7. J. Fenwick, P. Newman, and J. Leonard. Cooperative concurrent mapping and localization. In *IEEE Intl. Conf. on Robotics & Automation*, 2002.
8. J. M. Fitzpatrick, D. L. Hill, and C. R. Maurer, Jr. Image registration. In M. Sonka and J. M. Fitzpatrick, editors, *Handbook of Medical Imaging, volume 2: Medical Image Processing and Analysis*, chapter 8. SPIE, 2000.
9. M. Golfarelli, D. Maio, and S. Rizzi. Elastic correction of dead-reckoning errors in map building. In *Intl. Conf. on Intelligent Robots and Systems*, pages 905–911, 1998.
10. J. Ko, B. Stewart, D. Fox, K. Konolige, and B. Limketkai. A practical, decision-theoretic approach to multi-robot mapping and exploration. In *Intl. Conf. on Intelligent Robots and Systems*, 2003.
11. K. Konolige, D. Fox, B. Limketkai, J. Ko, and B. Stewart. Map merging for distributed robot navigation. In *Intl. Conf. on Intelligent Robots and Systems*, pages 212–217, 2003.
12. F. Lu and E. Milios. Globally consistent range scan alignment for environment mapping. *Autonomous Robots*, 4(4):333–349, 1997.
13. S. Thrun. A probabilistic online mapping algorithm for teams of mobile robots. *Intl. Journal of Robotics Research*, 20(5):335–363, 2001.
14. R. Wilson and E. Hancock. Structural matching by discrete relaxation. *IEEE Transactions on Pattern Analysis and Machine Intelligence*, 19(6), June 1997.

An approach to active sensing using the Viterbi algorithm

Frank E. Schneider, Andreas Kräußling and Dennis Wildermuth

Research Establishment for Applied Sciences (FGAN)
Wachtberg, Germany
E-mail: {frank.schneider | a.kraeussling | dennis}@fgan.de

Summary. Surveillance is a typical task in the field of multi robot systems that operate as a security system. This paper is concerned with the special problem of observing expanded objects in such settings. A solution in form of a Viterbi based tracking algorithm is presented. Thus a Maximum–a–posteriori (MAP) filtering technique is applied to perform the tracking process. The mathematical background of the algorithm is proposed. The method uses the robot sensors in form of laser range finders and a motion and observation model of the objects being tracked. The special features of the Viterbi based algorithm can be used to support active sensing. The tracking information will facilitate the robots to enhance the perceptual processing via dexterous sensor positioning.

1 Introduction

One of the most relevant questions, when proceeding from single to multiple robots, is co-operative perception or sensing. A typical application for a MRS is for example reconnaissance and surveillance, which has its use in the area of security systems. Active sensing or perception can be described as the problem of navigational strategies in order to improve the data acquisition process (e.g. sensor data input). Projected on robotics this means a robot can use its motor control to enhance the perceptual processing via sensor positioning. When tracking people and other expanded objects in densely populated surroundings with devices like laser scanners the situation is usually as follows: there are a lot of readings from other objects like walls and only some readings from the object itself. Therefore every tracking–algorithm needs to use a gate which separates the signals belonging to the object from other signals. A second feature of all algorithms for tracking vast objects should be the description of the shape and the extension of the objects. Finally, an estimate for the centre of the object has to be computed. For this purpose two antithetic methods can be used. First, all points that have passed the gate successfully can be used for the calculation of the estimate in terms of a weighted mean. Second, only the point that fits best to the previous data is used for the further calculations. In this paper we introduce the Viterbi algorithm as an example for the latter of these two methods. The Viterbi algorithm has been introduced in [24], a good description of it is given in [6]. It has already been recommended for tracking punctiform targets in clutter [15]. The paper will show how the features of the Viterbi algorithm can be used to support the process of active sensing. We will present a concept for a novel approach to the problem of active sensing with multiple mobile robots. The following scenario will build the basis for the work: a security system has identified an object of interest. The object moves through the environment in a non-specific neutral way while a group of robots tracks and observes the object to acquire more information.

2 Related Work

The field of active sensing has been studied, mostly for single robots, by a relatively small community in robotics. The combination of even some of the issues like tracking and sensor positioning in conjunction with multiple robot systems is hard to find in the literature. The approach of Howard [10] is more related to the topic of monitoring a static environment. The same holds for the work of Clouqueur [3] who is looking on how to place sensors in a given infrastructure to cover most of the terrain. The work of Grocholsky et al. [9] is somewhat related to the presented work. However, we focus on the issues of tracking a moving object while navigating a number of robots/sensors to enhance the sensor information process.

Tracking people, e.g. for surveillance, is a well studied problem in machine vision [19]. Another approach for tracking objects uses laser range finders. Earlier work uses occupancy grids and linear extrapolation of occupancy maps to estimate trajectories [4]. Newer methods involve the use of advanced trajectory estimation algorithms [5]. Laser sensing differs significantly from vision in ways that can be exploited for tracking. In vision, variables like colour, intensity, and depth are available. In contrast, lasers are restricted to one plane of the observable space. Most of the useful information for tracking is in just one parameter – the range to the nearest obstacle over an arc. The range measurements are, however, of high accuracy. Thus, lasers have rapidly gained popularity for mobile robotic applications such as collision avoidance, navigation, localization, and map building [21] [22].

The problem of estimating the position of moving objects is an important problem in mobile robotics. The ability of estimating the position of moving objects allows a robot to adapt its velocity to the speed of the objects in the environment. This improves its collision avoidance behaviour in situations in which the trajectory of the robot crosses the path of a moving object [19]. Very notable in this context is the work of Schulz [18]. He combined the ideas of Joint Probabilistic Data Association Filtering (JPDAF) [2] [7] with Particle Filtering [8] [14] and called his method Sample–based Joint

Probabilistic Data Association Filtering (SJPDAF). Thereby, he is able to assign the measurements to the objects to be tracked and to reproduce multi-modal densities, a major improvement for example when handling obstacles.

3 Mathematical Background

3.1 The Model

The dynamics of the object to be observed and the observation process itself are modeled as usual by

$$x_k = A \cdot x_{k-1} + w_{k-1} \qquad \text{and} \qquad z_k = B \cdot x_k + v_k. \tag{1}$$

Thereby x_k is the object state vector at time k, A is the state transition matrix, z_k is the observation vector at time k and B is the observation matrix. Furthermore, w_k and v_k are supposed to be zero mean white Gaussian noises with $E(w_i(w_j)^\top) = Q \cdot \delta_{ij}$, $E(v_i(v_j)^\top) = R \cdot \delta_{ij}$ and $E(w_i(v_j)^\top) = 0$, which means the measurement noise and the process noise are uncorrelated. In these equations the object is supposed to be punctiform. Nevertheless, the model will also be helpful for the description of an expanded target. In this case x_k will get the state vector of the centre of the target at time k. Since the motion of a target has to be described a kinematic model is used. It is

$$x_k = \begin{pmatrix} x_{k1} & x_{k2} & \dot{x}_{k1} & \dot{x}_{k2} \end{pmatrix}^\top, \qquad A = \begin{pmatrix} 1 & 0 & \Delta T & 0 \\ 0 & 1 & 0 & \Delta T \\ 0 & 0 & e^{-\Delta T/\Theta} & 0 \\ 0 & 0 & 0 & e^{-\Delta T/\Theta} \end{pmatrix}, \tag{2}$$

$$Q = \rho^2 e^{-2\Delta T/\Theta} \begin{pmatrix} 0 & 0 & 0 & 0 \\ 0 & 0 & 0 & 0 \\ 0 & 0 & 1 & 0 \\ 0 & 0 & 0 & 1 \end{pmatrix}, \qquad B = \begin{pmatrix} 1 & 0 & 0 & 0 \\ 0 & 1 & 0 & 0 \end{pmatrix}, \qquad \text{and } R = \delta \begin{pmatrix} 1 & 0 \\ 0 & 1 \end{pmatrix}. \tag{3}$$

Thereby x_{k1} and x_{k2} are the Cartesian coordinates of the target in the plane at time k and $\dot{x}_{k1}$ and $\dot{x}_{k2}$ are the corresponding velocities. ρ and Θ are parameters that are modeling the acceleration of the target. A small Θ and a large ρ correspond to a large acceleration. ΔT is the temporal interval between two consecutive measurements. In the calculations $\Theta = 3$, $\rho = 40$ and $\delta = 1000$ have been used.

3.2 The Gate

The gate is realised by use of the Kalman filter [11] [13] [20], which makes a prediction $y(k+1|k)$ for the measurement of the object based on the last estimate $x(k|k)$ for its position via the formulae

$$x(k+1|k) = A \cdot x(k|k) \quad \text{and} \quad y(k+1|k) = B \cdot x(k+1|k). \tag{4}$$

Then the Mahalanobis distance

$$\lambda = (z_i(k+1) - y(k+1|k))^{\top}[S(k+1)]^{-1}(z_i(k+1) - y(k+1|k)) \tag{5}$$

with the innovations covariance $S(k+1)$ from the Kalman filter is computed for every sensor reading $z_i(k+1)$ at time $k+1$. The Mahalanobis distance has been introduced in [12]. In contrast to the Euclidian distance it incorporates the different uncertainnesses in the varying directions. To illustrate this one can think of $S(k+1)$ as a diagonal matrix. A greater uncertainty in a direction then results in a greater associated diagonal entry in $S(k+1)$ and thus in a smaller associated diagonal entry in $[S(k+1)]^{-1}$. Consequently, deviations from $y(k+1|k)$ in this direction will be multiplicated with a smaller factor and thus will be weighted less in the calculation of the Mahalanobis distance. More precisely, the points with the same Mahalanobis distance are lying on the surface of an ellipsoid with the semi–major axis in the direction of the greater uncertainty.

Deviant from the Kalman filter the Matrix $S(k+1)$ is computed as

$$S(k+1) = BP(k+1|k)B^{\top} + R + E(k) \tag{6}$$

with the matrix $E(k)$, that describes the expansion of the object in order to account for its shape and expansion. A detailed description of the matrix $E(k)$ follows in the next paragraph. The recursion formula for $P(k+1|k)$ is

$$P(k+1|k) = AP(k|k)A^{\top} + Q. \tag{7}$$

λ is χ^2 distributed with two degrees of freedom, so a χ^2 test is used. All sensor readings with a λ lower than a given threshold are passed through the gate.

3.3 The expansion of the Object

To describe the expansion of the tracked object a positive definite matrix can be used, because for a positiv definite matrix P and a given $\epsilon > 0$ all points y with $y^T P y = \epsilon$ are lying on the surface of an ellipsoid with the origin as the centre [23]. So shape, orientation, and expansion of the objects are approximated by an ellipsoid. The appropriate positive definite matrix is computed as follows. At first it is assumed that all points that have passed the gate successfully are originated from the target, so that it is reasonable to use all these points for the computation of the matrix $E(k)$. Hence the covariance of these points which is positive definite, is computed and used as the matrix $E(k)$.

3.4 The Viterbi–Algorithm

The Viterbi algorithm [24] is a recursive algorithm. Thus, as an initialisation $x(0|0)$, $P(0|0)$ and $E(0)$, that describe the object at the beginning, are used. Now a description of a single step of the recursion is given. The Viterbi algorithm uses a directed graph to determine the optimal sequence of measurements for a given set of measurements Z^T. Thereby it is $Z^T = \{Z_k\}_{k=1}^T$ and $Z_k = \{z_{k,j}\}$ is the set of selected measurements $z_{k,j}$ at time k. The selected measurements correspond to the nodes in the graph. Given the selected measurements Z_k at time k, the set of selected measurements at time $k+1$ is computed as follows:

For every measurement $z_{k,j}$ the gate is applied to the measurements at time $k+1$. That results in the sets $Z_{k+1,j}$ of measurements which have passed the particular gate for the measurement $z_{k,j}$ successfully. The set $Z_{k+1} = \{z_{k+1,i}\}_i$ of selected measurements $z_{k+1,i}$ at time $k+1$ is then just the union of these sets, i.e. it is

$$Z_{k+1} = \cup_j Z_{k+1,j}. \tag{8}$$

In the next step for every selected measurement $z_{k+1,i}$ its predecessor is identified. For this purpose only those selected measurements $z_{k,j}$ whose gates have been passed by $z_{k+1,i}$ successfully are considered. For each of these measurements the length $d_{k+1,j,i}$ of the path from $x(0|0)$ to $z_{k+1,i}$ through $z_{k,j}$ is calculated.

$$d_{k+1,j,i} = d_{k,j} + a_{k+1,j,i} \tag{9}$$

is used. Thereby $d_{k,j}$ is the length of the path that ends in the node $z_{k,j}$. $a_{k+1,j,i}$ is the distance between the nodes $z_{k,j}$ and $z_{k+1,i}$. For $a_{k+1,j,i}$

$$a_{k+1,j,i} = \frac{1}{2}\nu_{k+1,j,i}^{\top}[S_{k+1,j}]^{-1}\nu_{k+1,j,i} + \ln\left(\sqrt{|2\pi S_{k+1,j}|}\right) \tag{10}$$

is used. Thereby $\nu_{k+1,j,i}$ is the innovation defined as

$$\nu_{k+1,j,i} = z_{k+1,i} - y_j(k). \tag{11}$$

In doing so $S_{k+1,j}$ and $y_j(k)$ are the innovations covariance respectively the prediction evaluated by the Kalman filter based on the nodes or the measurements $\{z_{l,i(l,j)}\}_{l=1}^k$ belonging to the path ending in $z_{k,j}$. The sequence

$$Z_{k,j} = \{z_{l,i(l,j)}\}_{l=1}^k \tag{12}$$

of these nodes is called the tracking history belonging to the node $z_{k,j}$. This proceeding causes

$$p(z_{k+1,i}|Z_{k,j}) = \exp\left(-a_{k+1,j,i}\right) \tag{13}$$

and therefore with Bayes' rule as well

$$p(Z_{k,j}|Z_0) = \exp\left(-d_{k,j}\right) \tag{14}$$

with $Z_0 = x(0|0), P(0|0), E(0)$. Finding the tracking history with the minimal distance corresponds therefrom strictly to determining the tracking history with the maximal likelihood. The predecessor of $z_{k+1,i}$ is now just the node

$z_{k,j}$, by which $d_{k+1,j,i}$ is kept at a minimum. For this node it is written $z_{k,i(k,j)}$. With the appendant predictions $x(k+1|k)_i$, $P(k+1|k)_i$ and $y(k+1|k)_i$ then the Kalman filter is used to get the estimates $x(k+1|k+1)_i$ and $P(k+1|k+1)_i$ by the well known equations

$$x(k+1|k+1)_i = x(k+1|k)_i + W_{k+1,i}(z_{k+1,i} - y(k+1|k)_i) \quad (15)$$

and

$$P(k+1|k+1)_i = [I - W_{k+1,i}B]P(k+1|k)_i \quad (16)$$

with the Kalman gain

$$W_{k+1,i} = P(k+1|k)_i B^\top S_i^{-1}(k+1). \quad (17)$$

Thus the description of one step of the recursion is finished.

4 Active sensing

The following paragraph will explain how the results of the Viterbi algorithm can be used for sensor positioning. Figure 1 shows five different measurements for an U-shaped object in this case. The need for active sensing is obvious since each of the measurements will give only fragmentary information about the target. The sensors have to be placed in a dexterous way around the object to yield a maximum on information while tracking it. In order to do so, it is first essential to acquire some knowledge about the shape and the extension of the object. The results from the Viterbi algorithm can be used to derive the necessary information.

The situation at time k after conducting the recursion up to k is as follows. For every selected measurement $z_{k,i}$ we have an estimate $x(k|k)_i$ of the state of the object and the length $d_{k,i}$ of the appropriate path. Like in the case of a punctiform target in clutter we can now select the estimate with the minimal length of the path. But since each of the measurements $z_{k,i}$ stems from the interaction of the surface of the object for instance with the laser beam of the measurement device, accordingly, we can say that every estimate $x(k|k)_i$ pertains to a point on the surface of the object. As a result, depending on our prior knowledge about the target, it is possible to extract further information from the estimates $x(k|k)_i$. In the case when there is no knowledge at all at least a weighted mean $x(k|k)$ as

$$x(k|k) = \sum_i \mu_{k,i} \cdot x(k|k)_i \quad (18)$$

with the weights

$$\mu_{k,i} = \frac{e^{-d_{k,i}}}{\sum_i e^{-d_{k,i}}} \quad (19)$$

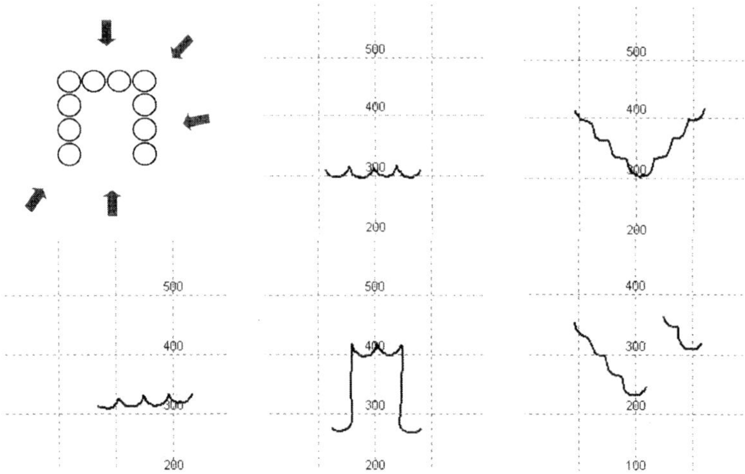

Fig. 1. Five laser measurements on U-shaped object. From top left to bottom right: Sensor positions, back view, 45-degree back view, side view, front view, 45-degree front view.

can be calculated. Then the matrix $E(k)$ can be calculated as described above. In practice, as our research has shown, the values $d_{k,i}$ differ only slightly for a given k, so that its sufficient to use $\mu_i = 1/N_k$ with N_k the number of gated measurements at time k. Alongside, when thinking of a symmetric target, the centre of the connecting line of the two estimates which have the greatest Euclidian distance from each other can be used as an estimate for the virtual centre of the target. This relies on the fact that in the case of a symmetric target at least half of the target is in the shadow of the other. After this the other estimates are mirrored at this estimate or the connecting line. Finally we should calculate the matrix $E(k)$ with all the estimates $x(k|k)_i$ and the mirrored points as described above. This is a reasonable approach for example in tracking the legs of a person, for which an ellipse is a meaningful approximation. It works the better the farer the measurement device is away from the target.

As an example a circular object with known radius r in the plane is considered, since this case can be treated analytically. A pair of different estimates $x(k|k)_i$ and $x(k|k)_j$, which belong to points on the surface of the object, is considered. Figure 2 shows how an estimate $c(k|k)_{i,j}$ for the position $c(k)$ of the centre of the target can be calculated from these estimates (see figure 2).

At first, it is

$$v = \begin{pmatrix} v_x \\ v_y \end{pmatrix} = \frac{x(k|k)_j - x(k|k)_i}{2}. \tag{20}$$

Next it is $r^2 = v^2 + w^2$ and thus

$$|w| = \sqrt{r^2 - v^2}. \tag{21}$$

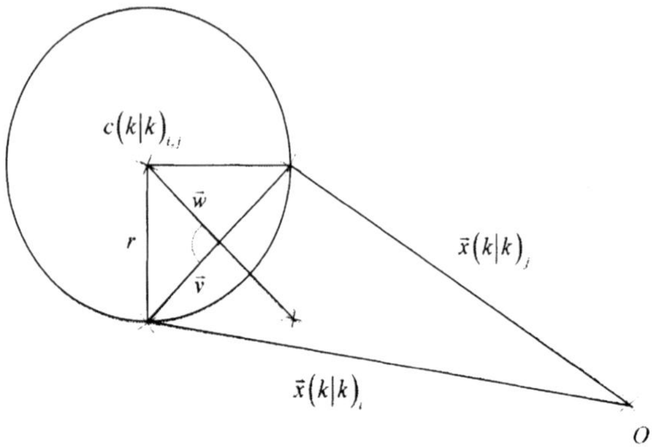

Fig. 2. Construction of $c(k|k)_{i,j}$

As w is orthogonal to v, there are for w the two possibilities

$$w_1 = \frac{|w|}{|v|} \begin{pmatrix} -v_y \\ v_x \end{pmatrix} \tag{22}$$

and $w_2 = -w_1$. So for $c(k|k)_{i,j}$ there are the two estimates

$$c(k|k)^1_{i,j} = x(k|k)_i + v + w_1 = \frac{x(k|k)_i + x(k|k)_j}{2} + w_1 \tag{23}$$

and

$$c(k|k)^2_{i,j} = x(k|k)_i + v + w_2 = \frac{x(k|k)_i + x(k|k)_j}{2} - w_1. \tag{24}$$

Obviously, the correct estimate is the estimate with the greater Euclidian distance from the measurement device and that one is chosen for the further calculations. If the number of estimates $x(k|k)_i$ is N_k, this calculation can be done for the $\frac{N_k(N_k-1)}{2}$ pairs of estimates $x(k|k)_i$ and then the weighted mean $c(k|k)$ is calculated as

$$c(k|k) = \sum_{i \neq j} \mu_{k,i,j} \cdot c(k|k)_{i,j} \tag{25}$$

with the weights

$$\mu_{k,i,j} = \frac{e^{-d_{k,i}} e^{-d_{k,j}}}{\sum_{i \neq j} e^{-d_{k,i}} e^{-d_{k,j}}}. \tag{26}$$

Again, for practical purposes $\mu_{k,i,j} = \left(\frac{N_k(N_k-1)}{2}\right)^{-1}$ can be used.

If the radius r is not known the following method can be used for an estimate: First, the centre $c(k)$ is computed by taking the estimates (points on the surface) of all robots into account that are involved in the tracking process. Second, the largest euclidian distance between two points of this set of surface points is computed. This distance and an additional scaling value are used to generate the radius r for a circle with centre $c(k)$. Now the available robots are distributed evenly around the tracked object on this circle by

$$angle\_stepping = (360/\#robots). \tag{27}$$

This input is used to establish and maintain a formation of robots around the tracked object, like shown in [16] [17]. Depending on the environment, task, sensors and robots it can be necessary to choose a larger scaling value.

5 Conclusion and Future work

In this paper a new method for tracking expanded objects by the application of the Viterbi algorithm has been introduced. It has been shown that the special features of the method also can enhance the perceptual processing. With a simple extension the results of the tracking process in conjunction with moving robots in formations deliver dexterous sensor positioning for expanded moving objects. Future work will concentrate on experiments and on the ability to deal with obstacles, occlusions, and crossing targets. Furthermore, we will compare our algorithm to an EM based method. The EM algorithm is a good example for a method that estimates the centre of the target as a weighted mean.

References

1. Yaakov Bar–Shalom and Thomas E. Fortmann, *Tracking and Data Association*, Academic Press, 1988.
2. Yaakov Bar–Shalom and Edison Tse, *Tracking in a Cluttered Environment With Probabilistic Data Association* Automatica, Volume 11, pp. 451–460, Pergamon Press, 1975.
3. Thomas Clouqueur, Veradej Phipatanasuphorn, Parameswaran Ramanathan and Kewal K. Saluja, *Sensor deployment strategy for target detection*, MOBICOM 2002: 42-48
4. A. Elfes, E. Prassler and J. Scholz, *Tracking people in a railway station during rush hour*, Proc. Computer Vision Systems, First International Conference, ICVS'99, 162–179, Las Palmas, Spain, 13–15 Jan 1999, Springer Verlag.
5. Ajo Fod, Andrew Howard and Maja J. Mataric, *Laser–Based People Tracking*, Computer Science Department, University of Southern California, Los Angeles, USA.
6. G. David Forney, Jr, *The Viterbi Algorithm*, Proceedings of the IEEE, Volume 61, Number 3, March 1973, 268–278.
7. Thomas E. Fortmann, Yaakov Bar–Shalom and Molly Scheffe, *Sonar Tracking of Multiple Targets Using Joint Probabilistic Data Association*, IEEE Journal Of Oceanic Engineering, Volume OE–8, Number 3, July 1983.

8. N. Gordon, D. Salmond and A. Smith, *A novel approach to nonlinear/non-Gaussian Bayesian state estimation*, IEEE Proceedings F, 140(2), 107–113, 1993.
9. B. Grocholsky, A. Makarenko and H. Durrant-Whyte, *Information-Theoretic Co-ordinated Control of Multiple Sensor Platforms*, IEEE International Conference on Robotics and Automation (ICRA), Taipeh, Taiwan, September 14–19, 2003.
10. A. Howard, M. Mataric and G. Sukhatme, *Mobile Sensor Network Deployment using Potential Fields: A Distributed, Scalable Solution to the Area Coverage Problem*, 6th International Symposium on Distributed Autonomous Robotics Systems (DARS02), Fukuoka, Japan, June 25-27, 2002.
11. R. E. Kalman, *A New Approach to Linear Filtering and Prediction Problems*, Trans. ASME, J. Basic Engineering, Volume 82, March 1960, 34–45.
12. P. C. Mahalanobis, *On the Generalized Distance in Statistics*, Proc. Natl. Inst. Science, Volume 12, Calcutta, 1936, 49–55.
13. Peter S. Maybeck, *Stochastic Models, Estimation, and Control*. Academic Press, New York, San Francisco, London, 1979.
14. M. Pitt and N. Shephard, *Filtering via simulation: auxiliary particle filters*, Journal of the American Statistical Association, 994(446), 1999.
15. T. Quach and M. Farrooq, *Maximum Likelihood Track Formation with the Viterbi Algorithm*, Proceedings of the 33rd Conference on Decision and Control, Lake Buena Vista, FL, December 1994.
16. F. E. Schneider and D. Wildermuth, *A potential field based approach to multi robot formation navigation*, IEEE International Conference on Robotics, Intelligent Systems and Signal Processing (RISSP), Changsha, Hunan, China, October 8–13, 2003.
17. F. E. Schneider and D. Wildermuth, *Motion co-ordination for formations of multiple mobile robots*, Conference on Field and Servive Robotics, Helsinki, Finnland, June 11–13, 2001.
18. Dirk Schulz, Wolfram Burgard, Dieter Fox and Armin B. Cremers, *Tracking Multiple Moving Objects with a Mobile Robot*, Proceedings of the 2001 IEEE Computer Society Conference on Computer Vision and Pattern Recognition (CVPR 2001), Kauai, Hawaii, 2001.
19. S. Sclaroff and R. Rosales, *Improved tracking of multiple humans with trajectory prediction and occlusion modeling*, IEEE Conference on Computer Vision and Pattern Recognition, Workshop on the interpretation of Visual Motion, Santa Barbara, CA, 1998.
20. Robert H. Shumway and David S. Stoffer, *Time Series Analysis and Its Applications*, Springer, 2000.
21. Sebastian Thrun, *Learning metric topological maps for indoor mobile robot navigation*, Artificial Intelligence, 1, 21–71, 1999.
22. Sebastian Thrun, Dieter Fox and Wolfram Burgard, *Markov localization for mobile robots in dynamic environments*, Artificial Intelligence, 11, 391–427, 1999.
23. Don J. Torrieri, *Statistical Theory of Passive Location Systems*, IEEE Transactions on Aerospace and Electronic Systems, Volume AES–20, Number 2, March 1984.
24. Andrew J. Viterbi, *Error Bounds for Convolutional Codes and an Asymptotically Optimum Decoding Algorithm*, IEEE Transactions On Information Theory, Volume IT–13, Number 2, April 1967.

Using Group Knowledge for Multitarget Terrain-Based State Estimation

Edward Sobiesk[1], Maria Gini[2], and John A. Marin[3]

[1] Dept of Electrical Engineering and Computer Science, United States Military Academy, West Point, NY 10996 edward.sobiesk@us.army.mil
[2] Dept of Computer Science and Engineering, University of Minnesota, 200 Union St. S.E., Minneapolis, MN 55455-0159 gini@cs.umn.edu
[3] BBN Technologies, Columbia, MD 21046 jamarin@bbn.com

Summary. Multitarget terrain-based tracking is a cyclic process that combines sensor information with state estimation and data association techniques to maintain an estimate of the state of an environment in which ground-based vehicles are operating. When the ground-based vehicles are military vehicles moving across terrain, most of them will being moving in groups instead of autonomously. This work presents a methodology that has been demonstrated to improve the estimation aspect of the tracking process for this military domain. A clustering algorithm identifies groups within a vehicular data set. Group characteristics are extracted and then a new, innovative technique is utilized to integrate these into the individual vehicles' state estimation process. A series of experiments shows that the proposed methodology significantly improves the performance of three classic estimation algorithms for multitarget terrain-based tracking.

1 Introduction

Multitarget tracking is essential for any surveillance system utilizing sensors to interpret an environment. A multitarget *terrain-based* tracking system attempts to maintain an accurate description of an environment in which ground-based vehicles are operating. This cyclic process involves state estimation and data association techniques as well as the fusion of intermittent sensor reports with *a priori* information [6]. This information includes both characteristics of the environment, such as detailed terrain information, and characteristics of the entities operating in the environment, such as the capabilities and movement techniques of the vehicles involved.

In this work, we present a methodology for improving the estimation process for the multitarget terrain-based tracking domain. We introduce algorithms that identify groups of military vehicles from within a data set, extract some of their group characteristics, and then integrate these characteristics into the individual vehicle estimation process. We present the results of 126

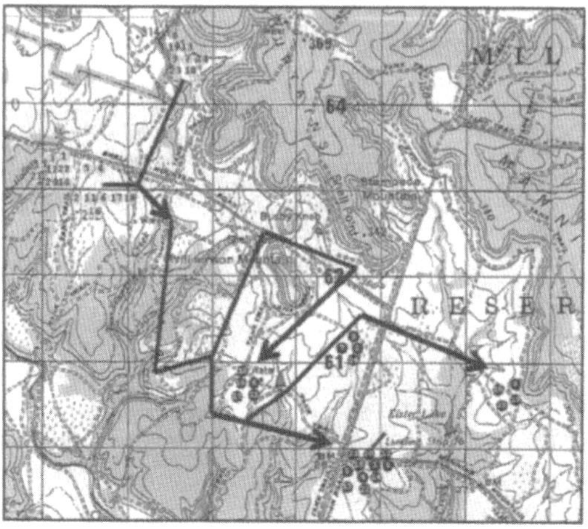

Fig. 1. Visualization of the movement routes taken by the 24 vehicles of Data Set One. The vehicles' start locations are marked by gray circles with black numbers, the finish locations by black circles with white numbers. The bold black arrows mark the general routes taken by the vehicles. Each grid square is 1000 meters long.

experiments conducted on large, real world data sets that test the methodology using three different state estimation models. In these experiments, an estimation model predicts the future locations of military vehicles as they move across terrain. The experiments compare the performance of the estimation models using group knowledge against their performance without it. Time intervals of 30, 60, and 90 seconds between sensor reports are tested. Two different ways of computing group knowledge and three different methods of integrating group knowledge are examined. The estimation models used include an α-β Filter, a Kalman Filter, and an iterative State-Space Model. The three different data sets are recorded from actual maneuvers by U.S. Army tanks in a training area at Fort Hood, Texas. Fig. 1 visualizes the movements of one of the vehicular data sets.

The major contribution of this work is demonstrating experimentally that group knowledge improves individual vehicle state estimation. Critical to the use of group knowledge is an algorithm for identifying groups and an innovative method of integrating group knowledge into state estimation models. It is not our intent to identify the best estimation model for terrain-based tracking. By using diverse models with different data sets and varying time intervals, we show the validity and robustness of our methodology.

2 Problem Background and Model Descriptions

The prediction phase of multitarget terrain-based tracking is the principal component of this research. Historically, the most common method used for prediction has been the Kalman Filter family of algorithms. Numerous mathematical models besides the Kalman Filter exist including the α-β, Bayes, and Least-Squares Filters. All of these are designed to predict the future location of an object based on past observations. These models are especially valuable for tracking aircraft or ships, where there is minimum change in velocity and direction. A major drawback to using these models for terrain-based tracking is that it is very difficult to integrate into the prediction process known terrain information for the areas involved. For instance, a basic Kalman Filter is incapable of integrating terrain constraints such as the observation that the entity is following a road or is headed toward an obstacle (a lake, a hill, etc.). A State-Space Model, which does include such terrain constraints, is therefore included as one of the models tested in this work.

For all the models, we assume the sensor reports for time k contain the reported x and y coordinates in meters, the past direction of movement, and the velocity in meters per time interval. Both the past direction of movement and the velocity are calculated based on the location data at times $k-1$ and k. To ensure consistency during testing, we assume that all sensor reports correlate correctly with their next reported location.

For the α-β Filter implementation [2], a value of 1 is used for both the α and β parameters because experimentation on a subset of the data found that the erratic movement and velocity of vehicles maneuvering across terrain made the long term target history of little value for near term estimation. The classic Kalman Filter was also implemented [3]. Based on previous work and for ease of implementation, both filters tracked a vehicle's x and y coordinate motion separately. The state of a vehicle at time k is defined by the estimated location and estimated velocity.

The State-Space Model (S-S) we use is based on the Reid and Bryson State-Space Model [8] and includes enhancements by both Nougues [6] and the authors of this work. This model removes the Gaussian assumption used in the Kalman Filter and accounts for the effects of on-road and off-road conditions on a vehicle's movement as well as terrain effects on a vehicle's speed and direction of movement. The terrain is represented as an eight-way connected grid of equally sized cells, which allows movement from a given cell to any one of the eight adjacent neighboring cells. Each grid cell represents a 20x20 meter square area of terrain. The probability distribution for the sensor report is initially assigned across the appropriate cells. The model's prediction algorithm then manipulates the probability distribution based on terrain information associated with each grid cell, the vehicle's velocity, and the vehicle's past direction of movement. The terrain information used includes data on vegetation, slope, hydrology, roads, and obstacles acquired from the U.S. National Imagery and Mapping Agency. The calculation and integration of the terrain factors utilizes the method [6] and data [5] created by Nougues.

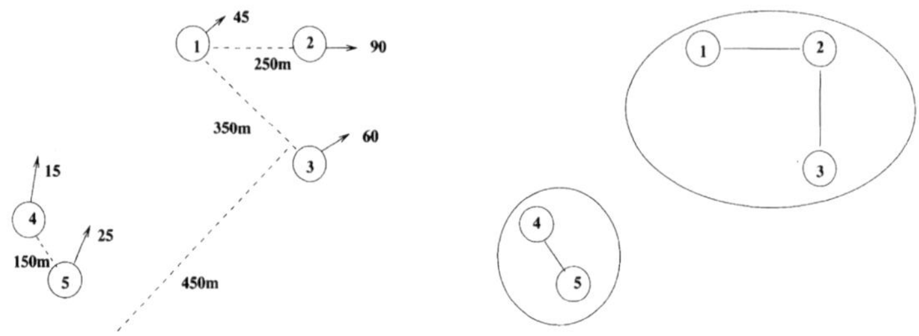

Fig. 2. Operation of the clustering algorithm. On the left is the relevant information about the vehicles derived from their sensor reports. On the right are the edges that result from applying the clustering criteria of δ = 300 meters and θ = 50 degrees to the vehicular data. Only three vehicular pairs meet the clustering criteria. Two groups of vehicles are formed by the connected components.

3 Group Knowledge

When multiple military vehicles move across terrain, most vehicles will be moving as part of a group instead of autonomously. This is evidenced by numerous U.S. Army manuals such as [1] that prescribe the different methods and formations to be used by groups of military vehicles. It is therefore logical to expect that information about the group would improve prediction.

For the sensor reports for time k, we define: *A set of vehicles is a group if, and only if, each member vehicle is located within δ meters and has a past direction of movement within θ degrees of at least one other vehicle in the set.* A simple example will illustrate this definition. Suppose vehicle 1 has a past direction of travel of 45 degrees, vehicle 2 of 90 degrees, and the vehicles are 200 meters apart. With $\delta = 100$ and $\theta = 50$, the vehicles *are not* a group. With $\delta = 300$ and $\theta = 25$, the vehicles *are not* a group. With $\delta = 300$ and $\theta = 50$, the vehicles *are* a group.

Identification of Groups. Groups are identified by a clustering algorithm that does not require prior specification of the number of clusters and utilizes the distance between elements and the past direction of movement as the criteria for clustering. The algorithm we designed is inspired by the line clustering algorithm of Yin and Chen [12]. Our algorithm creates a graph with the vehicles as vertices. The graph is initialized to have no edges. All possible vehicle pairs are considered, and an edge is placed between every vehicle pair that meets the user specified distance and direction criteria. The final graph is formed once all possible vehicle pairs have been considered. The different connected components of the final graph are the desired groups of vehicles. Fig. 2 presents a simple example of the results of the algorithm.

Extraction of Group Knowledge. Once groups are identified, the following group characteristics are extracted: the central direction of movement, the overall velocity, the x and y coordinate velocities, the rank of every vehicle in the group based on their location relative to the projected central direction

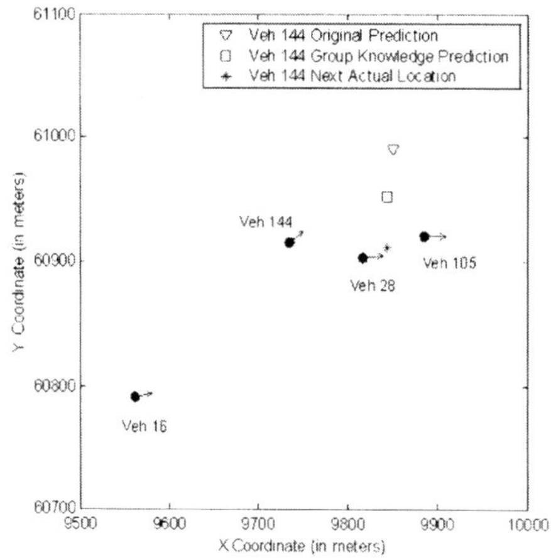

Fig. 3. Visualization of the application of 75% group knowledge adjusting the α-β Filter's prediction for Vehicle 144 closer to its next actual location.

of movement, and the number of vehicles in the group. The group velocities and the central direction of movement are calculated either as the mean, or as the median, of the individual vehicles' characteristics. The rank of a vehicle in the group is its position within the group relative to the group's projected central direction of movement. A rank of 1 is assigned to the vehicle whose x-y coordinate is furthest along the group's projected central direction of movement, a rank of 2 is assigned to the vehicle who is second furthest along, etc. The rank and the number of vehicles in the group are not integrated directly into any model, but are used together as part of a new method of determining the amount of group characteristic that is integrated into each vehicle's estimation.

Integration of Group Knowledge. We considered three methods for integrating group knowledge into the models:

1. the All-Group-Knowledge method uses only the calculated group characteristic and none of the individual vehicle characteristic;
2. the Half-Group-Knowledge method uses half of the group characteristic and half of the individual vehicle characteristic; and
3. the Varying-Group-Knowledge method uses different amounts of group characteristic for each vehicle in the group. This amount is calculated by dividing the vehicle's rank within the group by the number of vehicles in the group. This method assumes that lead vehicles in the central direction of movement are reflecting the future movement of the group more than the trail vehicles. Therefore, lead vehicles should use less of the group knowledge and more of their individual characteristic. Trail vehicles should use more of the group knowledge and less of their individual characteristic. Fig. 3 shows the results on Vehicle 144 whose rank is 3 out of 4.

4 Experimental Results

The three data sets used in the experiments are recorded from maneuvers by U.S. Army tanks at Fort Hood, Texas. The vehicular data consists of actual tank locations reported every 30 seconds. The locations were determined by mounting either a Global Positioning System or a Position Reporting and Recording System onto each vehicle. Both systems produce readings which are typically accurate to within a few meters. It is assumed that no goal destinations are known for the vehicles.

Data Set One[4] consists of 24 vehicles moving during daylight across wooded trails and open terrain in a variety of formations with dynamically changing groups and group sizes. There are a total of 2007 individual moving sensor reports during the 72 minute movement. The average speed is 107.5 meters per 30 second time interval with a standard deviation of 64.1.

Data Set Two[5] consists of 15 vehicles moving during darkness across wooded trails and open terrain almost exclusively in column and staggered column formations. Data Set Two remains a single group for most of the movement. There are a total of 1725 individual moving sensor reports during the 72 minute movement. The average speed is 78.5 meters per 30 second time interval with a standard deviation of 44.7.

Data Set Three consists of 12 vehicles moving during daylight across wooded trails and open terrain. There are a variety of formations with moderate changes in group composition and size. There are a total of 405 individual moving sensor reports during the 21 minute movement. The average speed is 136.1 meters per 30 second time interval with a standard deviation of 71.9.

Throughout their movements, the vehicles in the above three data sets did not conduct any evasive maneuvers based on contact with an enemy.

An experiment consisted of employing the original model without group knowledge and then employing the same model 12 more times with the exact same data using all combinations of δ values of 100, 300, and 500 meters and θ values of 25, 50, 75, and 180 degrees. Experiments were conducted with time intervals of 30, 60, and 90 seconds between sensor reports. Only moving sensor reports were tested.

The overall results are presented as the average Root Mean Square Error between a model's predicted location and the next reported location for all individual predictions made for the data set. To statistically compare the results of runs that used group knowledge against the ones that did not, paired one-tail t-tests were conducted using cutoff values of 5% and 1%.

---

[4]Battalion Size Tank Movement Conducted at Ft. Hood, vehicular digital data files from U.S. Army Texcom, Fort Hood, TX, 1994.

[5]Initial Operational Test and Evaluation of the M1A2 Tank, vehicular digital data files from U.S. Army Texcom, Fort Hood, TX, 1993. This data source produced Data Sets Two and Three.

4.1 Selecting the Group Knowledge Integration Method

81 experiments were conducted using the three state estimation models and the three data sets. Each experiment compared the use of no group knowledge against the use of the All-Group-Knowledge, Half-Group-Knowledge, or Varying-Group-Knowledge method. For all 81 experiments, group characteristics were calculated as the arithmetic mean. All-Group-Knowledge achieved at least a 5% statistical improvement over the original model without group knowledge 72.53% of the time, Half-Group-Knowledge 95.37%, and Varying-Group-Knowledge 96.30%.

All three group knowledge integration methods provided substantial statistical improvement. The performance of the Half-Group-Knowledge and Varying-Group-Knowledge methods was particularly stunning as statistical improvement was achieved with over 95% of the different parameter values. To determine which integration method performs most robustly, the average percentage of improvement by the different methods was calculated for each model and for each time interval. Fig. 4 shows the results of this comparison for all 81 experiments.

	30 seconds			60 seconds			90 seconds		
	α-β	Kalman	S-S	α-β	Kalman	S-S	α-β	Kalman	S-S
All-Group	1.54%	4.60%	-0.15%	7.64%	4.88%	1.45%	9.25%	5.85%	2.40%
Half-Group	6.48%	5.39%	1.29%	7.47%	4.43%	1.95%	7.48%	4.88%	2.21%
Varying-Group	6.62%	9.24%	1.23%	9.24%	7.08%	2.44%	9.78%	8.24%	2.79%

Fig. 4. Average percentages of improvement by the different integration methods over the original model employed without group knowledge.

Using the average percentage of improvement for comparison, Varying-Group-Knowledge is the most robust of the methods considered. The Varying-Group-Knowledge method performed better than the other two methods in eight out of the nine comparisons and was a very close second in the ninth. The Varying-Group-Knowledge method also demonstrated a consistency the other two methods lacked. The All-Group-Knowledge method had difficulty at the 30 second time interval, and the Half-Group-Knowledge method had difficulty at the 90 second time interval. Complete details on the models and full experimental and performance results are reported in [10].

4.2 Group Knowledge improves performance

Based on the above, we performed 27 additional experiments using the Varying-Group-Knowledge integration method and this time using the arithmetic median to calculate the group knowledge. The experiments included the different combinations of the data sets, models, time intervals, and δ and

	α-β mean	α-β median	Kalman mean	Kalman median	S-S mean	S-S median	Avg
30 sec, 1% impr	94.44%	83.33%	100.00%	100.00%	58.33%	58.33%	82.41%
30 sec, 5% impr	5.56%	8.33%	0.00%	0.00%	30.56%	25.00%	11.58%
30 sec, no impr	0.00%	8.34%	0.00%	0.00%	11.11%	16.67%	6.02%
60 sec, 1% impr	97.22%	97.22%	100.00%	97.22%	91.67%	91.67%	95.83%
60 sec, 5% impr	0.00%	0.00%	0.00%	2.78%	5.56%	0.00%	1.39%
60 sec, no impr	2.78%	2.78%	0.00%	0.00%	2.78%	8.33%	2.78%
90 sec, 1% impr	88.89%	88.89%	100.00%	97.22%	86.11%	80.56%	90.28%
90 sec, 5% impr	2.78%	2.78%	0.00%	0.00%	5.56%	8.33%	3.24%
90 sec, no impr	8.33%	8.33%	0.00%	2.78%	8.33%	11.11%	6.48%
Avg, 1% impr	93.52%	89.81%	100.00%	98.15%	78.70%	76.85%	89.51%
Avg, 5% impr	2.78%	3.70%	0.00%	0.93%	13.89%	11.11%	5.40%
Avg, no impr	3.70%	6.48%	0.00%	0.93%	7.41%	12.04%	5.09%

Fig. 5. Percentages of experimental runs, from all experiments that used Varying-Group-Knowledge, that achieved statistical improvement over the original model.

θ parameter values. In the 54 total experiments using the Varying-Group-Knowledge integration method, 89.51% achieved a statistical improvement over the original model at the 1% level and an additional 5.40% achieved a statistical improvement at the 5% level. The percentages of experimental runs that showed statistical improvement were calculated for each time interval and model. The results are in Fig. 5.

The single most important observation of Fig. 5 is the consistent statistical improvement across all models and time intervals. This work used diverse models that assume different amounts of error. It used multiple data sets, time intervals, and methods of calculating the group characteristic. Despite all of these variables, group knowledge continually improved state estimation.

It is also essential to note that the above results are the average of all the δ - θ parameter values, but not all parameter values performed the same. For the α-β Filter and the State-Space Model, which treated the sensor reports as more accurate than the Kalman Filter did, two interesting trends occur. First, the best δ value distance becomes larger as the amount of time between sensor reports becomes longer. For 30 seconds, the best δ value is 100 meters, for 60 seconds it is 300 meters, and for 90 seconds it is 500 meters. The second interesting trend is that 180 degrees is generally the best θ value. This implies that information on vehicles moving in any direction provides improvement for these models. For the Kalman Filter, the best δ value is 500 meters for all three time intervals. The best θ values are generally always in the 25 or 50 degree range. These optimum parameter values for the Kalman Filter indicate that it is more sensitive to variation in past direction of movement than to proximity.

4.3 A case where Group Knowledge does not help

An important aspect of any research is to establish the boundaries of applicability for the theory involved. This section presents the results of 18 experiments which demonstrate that in the special case when vehicles are conducting evasive maneuvers based on contact with an enemy, group knowledge does not generally improve the models' performance. Evasive maneuvers are actions taken by tanks when being fired upon by enemy vehicles. If evasive actions are conducted well, group knowledge will be of negligible assistance in prediction. To confirm this hypothesis, experiments were conducted using Data Set Four, which is a continuation of the movement of the 12 vehicles of Data Set Three as they conduct evasive maneuvers while in contact with an enemy. The 12 vehicles move during daylight across generally open terrain. There are a variety of formations with moderate changes in group composition and size. There are a total of 165 individual moving sensor reports during the 24 minute movement. The average speed is 107 meters per 30 second time period with a standard deviation of 75.8.

18 experiments were conducted with Data Set Four, using the Varying-Group-Knowledge method. Nine of the experiments calculated the group knowledge as the mean, and the other nine calculated it as the median. The experiments included the different combinations of the data sets, models, time intervals, and δ and θ parameter values. Only 19.4% of the experimental runs that used group knowledge showed statistical improvement over the original model. When in contact, tanks try to avoid being hit by enemy fire by taking such asymmetric actions as making radical turns, alternating speeds, and seeking cover and concealment in the terrain. These actions cause the characteristics of a group to be of little value in the estimation process.

5 Related Work

The iterative State-Space Model used in this work was designed by Reid and Bryson [8] and extended by Nougues [6], who demonstrated that it outperforms a Kalman Filter for terrain-based tracking of a single vehicle in a vegetative environment.

Pitman and Tenne [7] present a method of tracking ground-based vehicular convoys that combines convoy dynamics and topographical data to generate a probability density function that is used within a propagation algorithm. Their method also explores the influence a guide vehicle has on a trailing vehicle. Sidenbladh and Wirkander [9] argue that due to the variability of terrain, multitarget terrain-based tracking is a non-linear problem for which Kalman approaches are inappropriate. They instead propose a particle filter for tracking a varying or unknown number of vehicles in terrain and visualize their method using a simulated scenario with three vehicles moving in terrain with human observation reports and five second time steps. Particle filters have become popular in many signal processing applications [4], as well as in robotics [11], since they do not require any assumption about the probability distributions of the data.

6 Conclusions and Future Work

We presented a methodology that improves existing multitarget terrain-based state estimation models by adding group knowledge. Using a vehicle's location within its group to determine the amount of group knowledge to integrate into the estimation process produced the best results. Experiments were conducted with large data sets recorded from actual vehicle movements. The results demonstrate that group knowledge significantly improves state estimation except when the vehicles are conducting evasive maneuvers based on contact with the enemy. Important areas of future research include testing the methodology with shorter time periods between sensor reports, using group knowledge to improve the data association aspect of tracking, exploring additional heuristics for determining the optimal amount of group knowledge to use, and adapting and testing this methodology with other models, such as a particle filter.

References

1. Field manual 17-15: Tank platoon, 1996. U.S. Army Armor Center, ATTN: ATSB-SBB-D, Fort Knox, KY 40121.
2. E. Brookner. *Tracking and Kalman Filtering Made Easy.* Wiley, 1998.
3. R. Brown and P. Hwang. *Introduction to Random Signal Analysis and Kalman Filtering.* John Wiley & Sons, Inc., 1997.
4. A. Doucet, N. de Freitas, and N. Gordon, editors. *Sequential Monte Carlo Methods in Practice.* Springer Verlag, 2000.
5. P. Nougues. Terrain data files from Fort Hood, 1994. Terrain data files in raster format created at the Systems Engineering Dept, University of Virginia.
6. P. Nougues. *Tracking Intelligent Objects in Terrain.* PhD thesis, University of Virginia, 1996.
7. B. Pitman and D. Tenne. Tracking a convoy of ground vehicles. Technical report, State University of New York at Buffalo, January 2002.
8. D. Reid and R. Bryson. A non-Gaussian filter for tracking targets moving over terrain. In *12th Asilomar Conf. on Circuits, Systems, and Computers*, 1978.
9. H. Sidenbladh and S. Wirkander. Tracking random sets of vehicles in terrain. In *Proc. 2nd IEEE Workshop on Multi-Object Tracking*, 2003.
10. Edward J. Sobiesk. *Using Group Knowledge to track multiple vehicles moving across terrain.* PhD thesis, University of Minnesota, 2000.
11. S. Thrun, D. Fox, W. Burgard, and F. Dellaert. Robust Monte Carlo localization for mobile robots. *Artificial Intelligence*, 128(1–2):99–141, 2001.
12. P. Yin and L. Chen. A new non-iterative approach for clustering. *Pattern Recognition Letters*, 15:125–133, 1994.

Multi-AUVs for Visual Mapping Tasks

Botelho, S., Costa, R., Neves, R., Madsen, C., and Oliveira, V.

Fundação Universidade Federal do Rio Grande FURG Rua Eng. Alfredo Huch, 475 - Centro CEP: 96201-900 - Rio Grande/RS Brazil
{silviacb,costa,neves,madsen}@ee.furg.br, vinicius@ieee.org

Summary. The present paper describes a system for the construction of visual maps ("mosaics") and motion estimation for a set of AUVs (Autonomous Undersea Vehicle). Robots are equipped with down-looking camera which are used to estimate their motion with respect to the seafloor and to built real-time mosaic. As the mosaic increases in size, a systematic bias is introduced in its alignment, resulting in an erroneous output. The theoretical concepts associated with the use of an Augmented State Kalman Filter (ASKF) were applied to optimally estimate both visual map and the fleet position.

Key words: under water robotics, intelligent architectures, mobile robots

1 Introduction

The diversity of resources found at the bottom of the sea is of well-known importance. In this context, the domain of the technology to design and to develop unmanned Underwater Vehicles (UUVs) becomes a matter of strategy [3, 10]. The use of UUVs to create visual maps of the ocean floor becomes an important tool for underwater exploration [6, 8]. UUVs can be divided into two class: Remote Operated Vehicles (ROVs), that require a human pilot in the control loop; and Autonomous Underwater Vehicles (AUVs), that have the ability to perform high-level missions without user intervention. ROV operation is dependent on a skilled and experienced pilot to control the vehicle, meanwhile an autonomous navigation capability would significantly reduce this workload for pilots during many types of ROV exploration missions.

For visual-based underwater exploration, UUVs are equipped with down-looking cameras that produces images of the bottom of the sea, providing a visual map during the vehicle navigation. Every frame captured by the vehicle is used to compose the map. Consecutive frames are aligned and then a final map is generated, which is called mosaic [1]. Mosaics can also be used as reference maps in the navigation/location vehicle process [6]. This visual map can be used either in the surface for seabed exploration or for a visual-based AUV location/navigation. Thus, the vehicle is able to navigate itself using the real-time map it is generating. Given the vehicle altitude above the ocean floor (e.g. from an altimeter) and camera field of view, the actual area covered by the map (and individual frames) is known, the navigation in real world coordinates is possible.

Occasionally, the robot path may cross over itself, [6] propose a smooth adjust of the mosaic cumulative error detected by the crossover points in the map construction. [7] use the Augmented State Kalman Filter (ASKF) to estimate the correct position of both the vehicle and every frame of the mosaic, based on crossover and displacement measurements [5]. The ASKF strategies for the mosaic update and vehicle localization take into account simplified dynamic model of the AUV, as well as the detected crossover regions, which is very important to the accuracy of the system.

Two issues are considered in this paper:*i.* High costs and the complexity associated with more sophisticated and long missions unable the massive use of AUVs. On the other hand, simpler Vehicles are incapable to accomplish trivial tasks due to their low autonomy, poor number of sensors and other limitations.*ii.* The possibility of using its own explored image as initial information to locate the vehicle is an attractive and low cost operation. ASKF seems to be a good choice, specifically when the navigation provides a set of crossover regions.

In this context, we argue that a fleet of simple robots can be more efficient than a sophisticated AUV to seabed exploration tasks. These simple vehicles can explore a region in a fraction of the time needed by a single AUV. Mosaics can be more efficiently constructed and its visual information can be used to help the location and navigation of the fleet.

Thus, this paper presents the first results of the extension of the ASKF proposed by [7] for a set of Multi-AUVS. The fleet needs to explore a seabed region, providing its visual map. The mosaic is composed by a set of frames. These images are obtained by several simple and inexpensive robots associated with an undersea central station. The mosaic is computed by this central station. A distributed ASKF provides an image position estimation, as well as, each robot position.

An architecture to Multi-AUV Inspection Fleet

We have developed a generic architecture for multi-robot cooperation [2]. The proposed architecture deals with issues ranging from mission planning for several autonomous robots, to effective conflict free execution in a dynamic environment. Here, this generic architecture is applied to Multi-AUVs for a Visual Mapping Task, giving autonomy and communication capabilities for our AUVs. We suppose that the robots submerge inside a central station (CS). This CS is connected by a physical cable with the surface. The maps imaging by the robots are send to the surface through the CS. Besides physically connecting the AUVs with the surface, the CS does the decomposition of missions in a set of individual robots tasks. CS receives high level missions in a TCP/IP web link.

After a brief multi-robot context analysis, section 3 presents the theoretical extension of ASKF approach to Multi-AUV mosaicking. Next section details the implementation of the visual system, providing preliminary results and analysis of simulated tests. Finally a conclusion and future works are presented.

2 The context: Multi-AUVs and Visual Maps Models

Visual maps are an important tool to investigate seabed environments. An autonomous under water vehicle equipped with a down-looking camera can image the bottom of the sea, sending the image to the surface. A visual map is constructed with these images, and it can be used also to the location of the robot.

The cumulative error associated with the mosaic construction (frames location) decreases the performance of the system. Several approaches treat this problem [6, 7]. For instance, strategies based on the crossover detection realigns all frames of the mosaic according to the crossover position information. [7] proposes a crossover based system using an Augmented State Kalman Filter. This filter estimates both the state of the AUV and the state of each image of the map. Thus, each time a new state, associated with a new image added, needs to be estimated, resulting in an ASKF. This paper extends the theory developed by [7] to Multi-robots context.

Consider a set of M robots in a visual seabed exploration mission. During the exploration mission, each robot send frames to a central station (CS) to the mosaic composition. This CS congregates all information about the mosaic, adding new frames and updating their location in real time.

Every time k only one robot v_{add} sends to CS a new captured frame f_k^{add}, $0 \leq add \leq (M-1)$. The mosaic F is composed by a set of frames. This mosaic is used to estimate the future state of each vehicle v_r and the future state of each frame f_i^r [1].

At every time k, a robot v_r is described by a vector state x_r^v:

$$\mathbf{x}_r^v = \left[x \; y \; z \; \Psi \; \dot{x} \; \dot{y} \; \dot{z} \; \dot{\Psi} \right]^T , \tag{1}$$

where x, y are relative to a mosaic-fixed coordinate system [2]. z is relative to an inertial fixed coordinate system, Ψ (yaw) is the heading of the robot associated with a fixed coordinate system.

[1]We use r and i to describe r^{th} and i^{th} generic robot and frame, respectively, captured by robot v_r.

[2]We suppose that the robots have a known inertial referential, associated with the first frame of the mosaic.

Each robot can have a different dynamic model A_r^v, see [9] for a set of different dynamic models to AUVs. In this paper, we have chosen a simple kinematic model to demonstrate the strategy proposed:

$$\mathbf{A}_r^v(k) = \begin{bmatrix} \mathbf{I}_{4\times 4} & dt.\mathbf{I}_{4\times 4} \\ \mathbf{0}_{4\times 4} & \mathbf{I}_{4\times 4} \end{bmatrix} \tag{2}$$

where **I** is a 4-dimension identity matrix and dt is the sampling period between states at discrete-time instants.

A process noise, $\mathbf{Q}_r^v$, associated with each v_r, can be defined as:

$$\mathbf{Q}_r^v(k) = \begin{bmatrix} \frac{1}{4}dt^4\sigma_r^{v^2} & \frac{1}{2}dt^3\sigma_r^{v^2} \\ \frac{1}{2}dt^3\sigma_r^{v^2} & dt^2\sigma_r^{v^2} \end{bmatrix} \tag{3}$$

where $\sigma_r^{v^2}$ is a diagonal 4-dimension matrix of process noise variance in all coordinates (x,y,z,Ψ).

The Mosaic Construction Model

As described by [7], every frame has a state vector that contains the information required to pose the associated image in the map. In the multi-robot context, a frame f_i^r, captured by vehicle v_r, has the following state vector relative to a mosaic-fixed coordinate: $x_i^{f^r} = [x\ y\ z\ \Psi]^T$.

The Observation Model

Kalman Filters are based on the measurement of successive information of the system behavior. In this approach two vectors of measures are used: *i.* $\mathbf{z}_{adj}(k)$: this measurement is provided directly by each robot v_{add}, which is adding a new image to the mosaic map; *ii.* $\mathbf{z}_{cross}(k)$: it measures the displacement associated with the mosaic area, where the crossover has been detected. To provide this information we detect a crossover trajectory, analyzing the current image captured and the mosaic region using texture matching algorithm [1].

These vectors can be described by:$z_{adj,cross} = \left[\Delta x\ \Delta y\ z\ \Delta\Psi\right]^T$, where the subindex *adj* is associated with the coordinated relatives $(\Delta x, \Delta y, z, \Delta\Psi)$ between the present image k and the previous image of the same robot v_r. Similarly,the subindex *cross* is related to the displacement between the crossover area with respect to the closer node of the mosaic image (node j). We suppose that z can be obtained from a sonar altimeter, being the absolute measure of the altitude of the vehicle at time k.

Two measured sub-matrices need to be defined:

$$\mathbf{H}_r^v(k) = \left[\mathbf{I}_{4\times 4}\ \mathbf{0}_{4\times 4}\right] \tag{4}$$

that describes the vehicle measurement, and the image measurement sub-matrix:

$$\mathbf{H}^{f^r}_{(k-1)}(k) = \mathbf{H}^{f^s}_j(k) = diag\{1, 1, 0, 1\}, \tag{5}$$

which describes the image associated with *adjacent*(captured by a generic robot v_r) and *crossover* (captured by another some robot v_s) situations. One should observer that a component related to z coordinate is provided directly by the altimeter sensor.

If there is no crossover, the measurement covariance matrix is $\mathbf{R}(k) = \sigma^{add^2}_{adj}(k)$, what means that it is associated only with the covariance of the adjacent image addition measurements done by v_{add}. However, if it is present a crossover measurement, $\mathbf{R}(k)$ becomes:

$$\mathbf{R}(k) = \begin{bmatrix} \sigma^{add^2}_{adj}(k) & \mathbf{O}_{4\times4} \\ \mathbf{O}_{4\times4} & \sigma^2_{cross}(k) \end{bmatrix} \tag{6}$$

with:

$$\sigma^{add^2}_{adj}(k) = diag\{\sigma^{add^2}_x(k), \sigma^{add^2}_y(k), \sigma^{add^2}_z(k), \sigma^{add^2}_\Psi(k)\}, \tag{7}$$

where $\sigma^{add^2}_x(k), \sigma^{add^2}_y(k)$ and $\sigma^{add^2}_\Psi(k)$ are the measurement variances of v_{add} added images correlation in the mosaic, and $\sigma^2_z(k)$ the variance of the sonar altimeter of this robot.

Similarly,

$$\sigma^2_{cross}(k) = diag\{\sigma^2_x(k), \sigma^2_y(k), \sigma^2_z(k), \sigma^2_\Psi(k)\}, \tag{8}$$

with $\sigma^2_x(k), \sigma^2_y(k)$ and $\sigma^2_\Psi(k)$ are the measurement variances of all images correlation in the mosaic, and $\sigma^2_z(k)$ the variance of the sonar altimeter.

3 ASKF for Multi-AUV Mosaicking

Kalman Filter uses two sets of equations to predict values of the variable state. The *Time Update Equations* are responsible for `predicting` the current state and covariance matrix, used in future time to predict the previous state. The *Measurement Update Equations* are responsible for `correcting` the errors in the Time Update equations. In a sense, it is backpropagating to get new value for the prior state to improve the guess for the next state. The equations for our ASKF for Multi-AUV and mosaic localization are presented.

3.1 The Prediction Stage

From the kinematic model of the system (vehicles and mosaic), ASKF can propagate the following state estimative:

$$\hat{\mathbf{x}}(k) = \begin{bmatrix} \hat{\mathbf{x}}^v_0 \ldots \hat{\mathbf{x}}^v_r \ldots \hat{\mathbf{x}}^v_{M-1} \; \hat{\mathbf{x}}^{f^r}_{k-1} \ldots \hat{\mathbf{x}}^{f^r}_0 \end{bmatrix} \tag{9}$$

which means the estimated position of each robot v_r $(r = 0..(M-1))$ and each frame estimated position (from frame 0 to $(k-1)$). The covariance $\mathbf{P}(k)$ associated with this estimative is also propagated.

When a new mosaic frame is added by v_{add}, new predictions and covariance (for $(k+1)$ time) are obtained, according to *time update* equations:

$$\hat{\mathbf{x}}_{aug}(k+1) = \mathbf{A}_{aug}(k)\hat{\mathbf{x}}_{aug}(k) + \mathbf{B}_{aug}(k)\hat{\mathbf{u}}_{aug}(k), \tag{10}$$

$$\mathbf{P}^{-}_{aug}(k+1) = \mathbf{A}_{aug}(k)\mathbf{P}_{aug}(k)\mathbf{A}^{T}_{aug}(k) + \mathbf{B}_{aug}(k)\mathbf{Q}_{aug}(k)\mathbf{B}^{T}_{aug}(k). \tag{11}$$

Notice that $\mathbf{x}_{aug}(k+1)$ is the state of $\mathbf{x}$ augmented of a new image added state, $\hat{\mathbf{x}}^{f^{add}}_{k}(k+1)$, added by v_{add}, with

$$\hat{\mathbf{x}}^{f^{add}}_{k}(k+1) = \left[\mathbf{I}_{4\times 4}\ \mathbf{0}_{4\times 4}\right]\hat{\mathbf{x}}^{v}_{add}(k+1), \tag{12}$$

similarly,

$$\mathbf{P}_{k,(0,\ldots,r,\ldots,(M-1),k,k-1,\ldots,0)}(k+1) = \left[\mathbf{I}_{4\times 4}\ \mathbf{0}_{4\times 4}\right]\mathbf{P}_{v,(0,\ldots,r,\ldots,(M-1),r,k-1,\ldots,0)}(k+1)), \tag{13}$$

where equation 13 selects the information from the row and column associated with the vehicle position v_{add} which captured this new frame $f^{add}_{(k+1)}$.

As the position of images does not vary as a function of time, the system dynamics $\mathbf{A}_{aug}(k)$ and the noise covariance $\mathbf{Q}_{aug}(k)$ can be described by:

$$\mathbf{A}_{aug}(k) = diag\left[\mathbf{A}^{v}_{0}(k)\ \ldots\ \mathbf{A}^{v}_{r}(k)\ \ldots\ \mathbf{A}^{v}_{(M-1)}(k)\ \mathbf{I}\right] \tag{14}$$

$$\mathbf{Q}_{aug}(k) = diag\left[\mathbf{Q}^{v}_{0}(k)\ \ldots\ \mathbf{Q}^{v}_{r}(k)\ \ldots\ \mathbf{Q}^{v}_{(M-1)(k)}\ \mathbf{0}\right] \tag{15}$$

where the identity matrix $\mathbf{I}$ has a size $k.\dim(x^{f}_{i})$. Since the system does not have any input, $\mathbf{u}(k) = 0$ and $\mathbf{B}(k) = \mathbf{I}$ [7].

3.2 The Correction Stage

For each time step k, a robot v_{add} adds a new image to the visual map. The vehicle finds the observation measurement between two consecutive (adjacent) captured frames. Notice that the *adjacent* concept is associated with two consecutive images (i.e. f^{add}_{3}, f^{add}_{2}) of the same robot v_{add}. Two frames can be consecutive to the robot v_{add}, but not necessarily consecutive to the mosaic system, for instance, in the capture interval between f^{add}_{3}, f^{add}_{2}, one another robot $r_{s\neq add}$ can add an intermediate image to mosaic system. In this case, for example, the final sequence of the mosaic becomes: $f^{add}_{k}, f^{s}_{(k-1)}, f^{add}_{(k-2)}$. Thus, two mosaic frames $f^{r}_{k}, f^{r}_{(k-p)}$ are adjacent if they are captured in a successive order by the same robot v_r.

A new measured $\mathbf{z}_{adj}(k)$ is obtained at every time step by robot v_{add}. The value $\mathbf{z}(k)$ measures the position of the kth image (which corresponds to the position of the v_{add}) with respect to the $(k-p)^{th}$ previous frame of this robot in the mosaic, so that:

$$\mathbf{z}(k) = \mathbf{z}_{adj}(k) \tag{16}$$

$$\mathbf{H}_{aug}(k) = \mathbf{H}_{adj}(k) = \left[\mathbf{H}^{v}_{adj}(k)\ \mathbf{H}^{f}_{adj}(k)\right] \tag{17}$$

where adjacent measurement sub-matrix, see equations 4 and 5 associated with the vehicles, and the images are:

$$\mathbf{H}^{v}_{adj}(k) = \left[\mathbf{0}\ \ldots\ \mathbf{0}\ \mathbf{H}^{v}_{add}(k)\ \mathbf{0}\ \ldots\ \mathbf{0}\right] \tag{18}$$

$$\mathbf{H}^{f}_{adj}(k) = \left[\mathbf{0}\ \ldots\ \mathbf{0}\ -\mathbf{H}^{f^{add}}_{(k-p)}(k)\ \mathbf{0}\ \ldots\ \mathbf{0}\right] \tag{19}$$

However, when a crossover is detected, the current image k^{th} also intersects with the previous mosaic image. Then, the measurement vector $z(k)$ becomes:

$$\mathbf{z}(k) = \left[\mathbf{z}^{T}_{adj}(k)\ \mathbf{z}^{T}_{cross}(k)\right] \tag{20}$$

in this case we have two measurement: one regarding to the previous image of robot v_{add}, $\mathbf{z}_{adj}(k)$, and the other with respect to the area where the crossover has been detected $\mathbf{z}_{cross}(k)$. Notice that the crossover region could be captured by some another robot v_{cross}, $0 \leq cros \leq (M-1)$. If we suppose that the crossover corresponds to an image f^{cross}_{j}, the measurement matrix $H_{aug}(k)$ incorporates a measurement in column j, becoming:

$$\mathbf{H}_{aug}(k) = \left[\mathbf{H}_{adj}(k)\ \mathbf{H}_{cross}(k)\right]^{T} \tag{21}$$

$$\mathbf{H}_{cross}(k) = \left[\mathbf{H}^{v}_{cross}(k)\ \mathbf{H}^{f}_{cross}(k)\right] \tag{22}$$

with vehicle and image measurement sub-matrix defined as:

$$\mathbf{H}^{v}_{cross}(k) = \left[\mathbf{0}\ \ldots\ \mathbf{0}\ \mathbf{H}^{v}_{add}(k)\ \mathbf{0}\ \ldots\ \mathbf{0}\right] \tag{23}$$

$$\mathbf{H}^{f}_{cross}(k) = \left[\mathbf{0}\ \ldots\ \mathbf{0}\ -\mathbf{H}^{f^{cross}}_{j}(k)\ \mathbf{0}\ \ldots\ \mathbf{0}\right] \tag{24}$$

Innovation is the difference between the measurement $z(k)$ and the previous *a priory* estimation and according to [7] it is given by:

$$r(k) = \mathbf{z}_{aug}(k) - \mathbf{H}_{aug}(k)\hat{\mathbf{x}}_{aug}(k), \tag{25}$$

and its covariance $S(k)$ is defined as:

$$\mathbf{S}(k) = \mathbf{H}_{aug}(k)\mathbf{P}^{-}_{aug}(k)\mathbf{H}^{T}_{aug}(k) + \mathbf{R}(k), \tag{26}$$

where $R(k)$ is the measurement error covariance, see 6.

The adjacent and crossover measurements allow the correction of the estimated state (of the robots and frames) and its associated covariance are corrected according to the KF measurement update equations. So, the filter gain can be expressed as:

$$\mathbf{K}(k) = \mathbf{P}^{-}_{aug}(k)\mathbf{H}^{T}_{aug}(k)\mathbf{S}^{-1}(k). \tag{27}$$

Once the KF gain is computed, the estimate state can be obtained:

$$\hat{\mathbf{x}}_{aug}(k) = \hat{\mathbf{x}}_{aug}^{-}(k) + \mathbf{K}(k)(\mathbf{z}(k) - \mathbf{H}_{aug}(k)\hat{\mathbf{x}}_{aug}^{-}(k)) \tag{28}$$

and its corrected error covariance:

$$\begin{aligned}\mathbf{P}_{aug}(k) = (\mathbf{I} - \mathbf{K}(k)\mathbf{H}_{aug}(k))\mathbf{P}_{aug}^{-}(k)(\mathbf{I} - \mathbf{K}(k)\mathbf{H}_{aug}(k))^{T} + \\ + \mathbf{K}(k)\mathbf{R}(k)\mathbf{K}(k)^{T}.\end{aligned} \tag{29}$$

Once the stages of estimation and correction have been completed, the state vector and the covariance matrices are augmented to add the positioning of the new k^{th} image, captured by robot v_{add}. The mosaic final is composed by the set of frames f_i^r.

4 The implementation of Visual Mapping

We have a test environment where simple undersea robots can accomplish inspection tasks [4]. Starting from the proposed architecture, a CS can be accessed via TCP/IP web connection. Users can specify a set of missions for the robots: navigate, locate and inspect a specific region. The image processing services, called NAVision, is composed by two different modules: *i.* an individual robot module responsible for capturing and pre-processing the frames through the down-looking camera, and *ii* a central module which provides a real time mosaic and pattern recognition. The former can operate in two different ways: on board or in the simulator mode.

Figure 1 shows a final simulated result of three robots in a exploration task. R1 begins at (-20:75) coordinates, R2 begins at (-100:25) position, and R3 begins at (0,-70) coordinates, see figure (a). Each vehicle has a different dynamical model. Their perception system have different noise features. Circular points represents the true trajectory of each robot. In addition to real trajectory, the simulator gives the estimated trajectory provides by perception system without crossover detection (observed). We can see that we have a cumulative error associated with the image processing observation (star points). Cross points show the smoothed trajectory obtained with our approach. We can see that, when R3 cross a old mosaic area imaging by R1 (near (20,25) coordinates), the ASKF provides a new estimation motion to R3, reseting its cumulative error. There is a second crossover, providing a second correction in the final mosaic. It happens between R3 and R1. Since that R1 has a lower cumulative error, it is used as setpoint. Thus this robot holds the same old trajectory (and localization). However,R2 and R3 have an enhancement of their cumulative error (both robots have old intersection region - crossover 1).

The error between true and estimated trajectory is showed in (b), (c) and (d) figures. With our approach, the mean errors of robot R2 and R3 have fallen by near 80 per cent. As expected, R1 mean error has not changed.

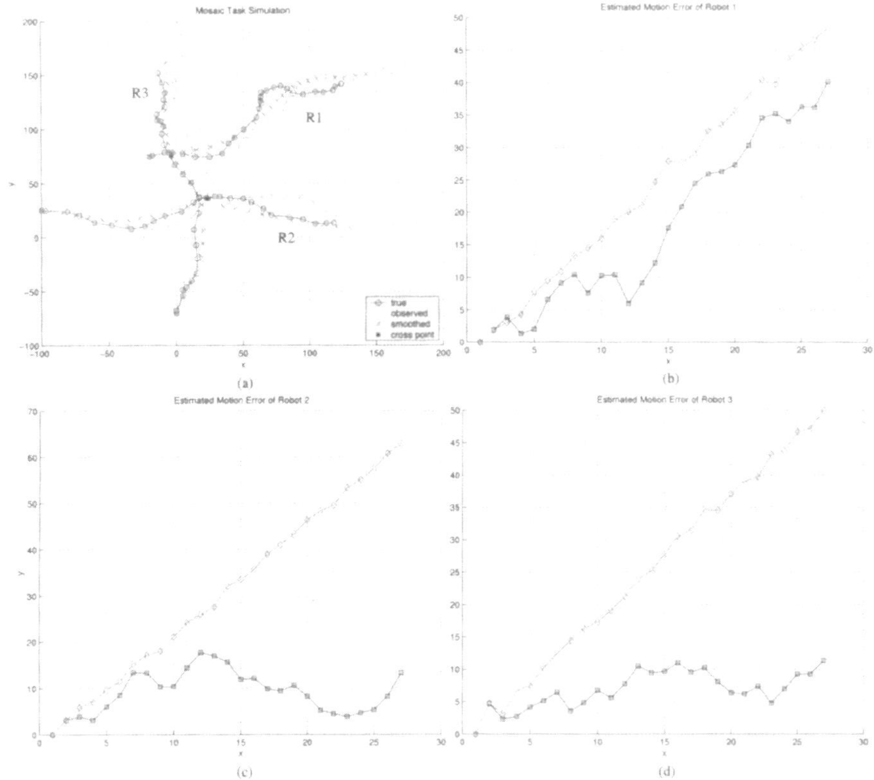

Fig. 1. The simulated multi-AUV mosaicking with ASKF filter

5 Conclusion

We have proposed and discussed a theoretical scheme for cooperative multi-AUVs Mosaicking. A set of robots can explore the seabed in a more efficient and fast way that single vehicle. We have built a generic architecture for multi-robot cooperation. Its interest stems from its ability to provide a framework for cooperative decisional processes at different levels: mission decomposition and high level plan synthesis, task allocation and task achievement.

For Visual Mapping, our approach is an extension of the [7] to the multi-vehicle context. ASFK is used for both estimating the state of each robot of the mission, and the position of each mosaic image. This estimation changes with the error observation based on adjacent and crossover measurements obtained by the fleet. The proposed methodology treats the mosaic as a centralized map, where different robots add frames and provide informations of the observation. We intend to simulate our approach through a number of significant different dynamical models and parameters, taking into account more distributed workloads among the robots.

References

1. Vargas A., Madsen C.A., and S. S. C. Botelho. Navision - sistema de visão subaqüático para navegação e montagem de mosaicos em auvs. In *Seminário e Workshop em Engenharia Oceânica*, 2004.
2. R. Alami and S. S. C. Botelho. Plan-based multi-robot cooperation. In Michael Beetz, Jachim Hertzberg, Malik Ghallab, and Martha E. Pollack, editors, *Advances in Plan-Based Control of Robotic Agents*, volume 2466 of *Lecture Notes in Computer Science*. Springer, 2002.
3. D Blidberg. The development of autonomous underwater vehicles (auvs); a brief summary. In *IEEE ICRA*, 2001.
4. S. S. C. Botelho, R. Mendes, L. Tadei, and M. Teixeira. Lambdari um robô subaqüático autônomo. In *Simpósio Brasileiro De Automação Inteligente - VI SBAI*, 2003.
5. R. Deaves. Covariance bounds for augmented state kalman filter application. *IEEE Electronics Letters.*, 35(23):2062–2063, 1999.
6. S. Fleischer. *Bounded-error vision-based of autonomous underwater vehicles.* PhD thesis, Stanford University, 2000.
7. R. Garcia, J. Puig, P. Ridao, and X. Cufi. Augmented state kalman filtering for auv navigation. In *IEEE ICRA*, 2002.
8. M.G. González, P. Holifield, and M. Varley. Improved video mosaic construction by accumulated alignment error distribution. In *British Machine Vision Conference*, pages 377–387, 1998.
9. A. Tavares. Um estudo sobre a modelagem e o controle de veículos subaquáticos não tripulados. Master's thesis, Engenharia Oceânica,Fundação Universidade Federal do Rio Grande, 2003.
10. R. Wernli. Auv's - a technology whose time has come. In *IEEE International Conference in Underwater Tecnologies 2002*, 2002.

Part IV

Reconfigurable Robots II

Cellular Robots Forming a Mechanical Structure
(Evaluation of structural formation and hardware design of "CHOBIE II")

Michihiko KOSEKI, Kengo MINAMI, and Norio INOU

Department of Mechanical and Control Engineering, Tokyo Institute of Technology, 2-12-1, O-okayama, Meguro-ku, Tokyo, 152-8552 JAPAN
koseki@mech.titech.ac.jp

Summary. This paper deals with group robots called CHOBIE that cooperatively transform a mechanical structure. The CHOBIE have slide motion mechanisms with some mechanical constraints for large stiffness even in movement. First of all, a way of structural transformation including the mechanical constraints is discussed. Second, dissipative energy in the structural transformation based on experimental data of the CHOBIE is estimated. Third, for autonomy of the robots, CHOBIE II is developed and the performance test is demonstrated.
Key words: Modular robotic system, Self-reconfiguration, Mechanical structure, Slide motion mechanism

1 Introduction

Reconfigurable group robots have potential to fulfill various missions such as cooperative transportation, collection and construction [1]. To realize the group robots, variety of mechanisms have been developed [2]–[7]. However, there are few robots designed for supporting large outer forces. The reason why is that the almost developed robots are putting more emphasis on a mobile function than a supporting function.

Our study focuses on group robots forming a mechanical structure. The group robots consist of cellular robots. Each cellular robot communicates with adjacent robots and determines the behavior where it should be positioned. They form the structure by successive cooperative movements. We call the cellular robot "CHOBIE" (Cooperative Hexahedral Objects for Building with Intelligent Enhancement).

Figure 1 shows a concept of our study that CHOBIE cooperatively construct a mechanical structure. There are many cellular robots in the working space where the robots provide arrangements for constructing a structure. The construction of a structure is performed within the constructing area.

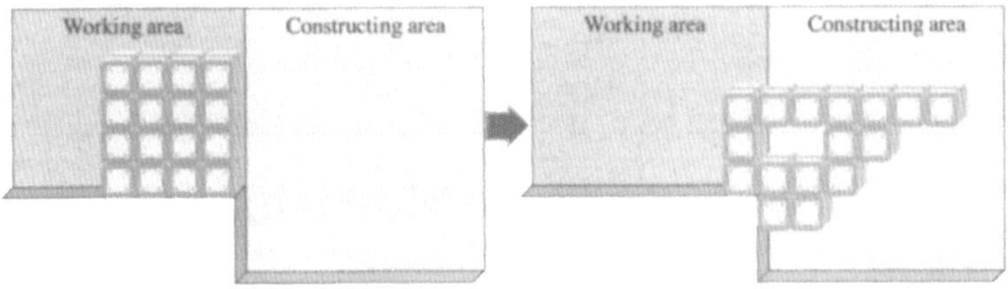

Fig. 1. Idea of cellular robots "CHOBIE" forming a mechanical structure

In our previous studies, we proposed a motion mechanism that cellular robots move in the rectangular directions keeping large stiffness. The cellular robots (we call them "CHOBIE I") demonstrated that they adaptively changed the structure to support a large outer force [8].

This paper discusses three subjects to improve performance of the robots. The first subject is to discuss a way of structural formation of the proposed cellular robots because they have some constraints in the structural formation. The second one is to estimate energy dissipation of the cellular robots for the structural formation. The third one is to develop revised robots CHOBIE II with autonomous functions since the previous robots CHOBIE I were demonstrated by a preprogrammed control without local communication between the neighboring robots.

2 Structure of a cellular robot

Figure 2(a) shows the proposed slide motion mechanism in our previous paper [8]. It consists of two lateral boards and a central board. The central board is sandwiched by the two lateral boards and all the boards are tightly connected.

The two lateral boards include symmetrical motion mechanisms that consist of two sets of wheels as shown in Fig. 2(b). They are allocated in vertical and horizontal directions, which enable the two directional motions of cellular robots. The only one DC motor is embedded in each lateral board, and jointly drives 4 wheels that are placed on the same plane through a drive shaft in the central board.

The central board has grooves as sliding guides, which maintain high rigidity even in transformation as shown in Fig. 2(c). For this motion mechanism, cellular robots successfully connect to other robots. The central board can easily change the depth. A controller, sensors and batteries for autonomic functions of cellular robots are embedded in the board as described later.

3 Structural formation of cellular robots

The proposed motion mechanism has some constraints in the transformation of the structure. Figure 3 illustrates basic structural transformation of three

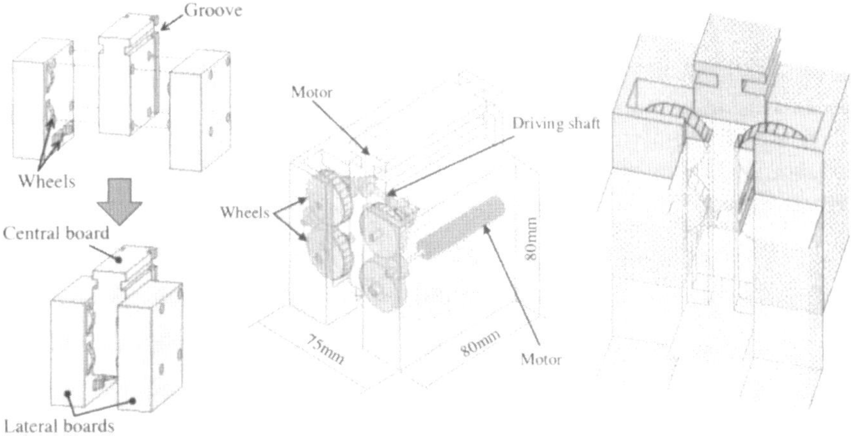

Fig. 2. Slide motion mechanism of the cellular robot

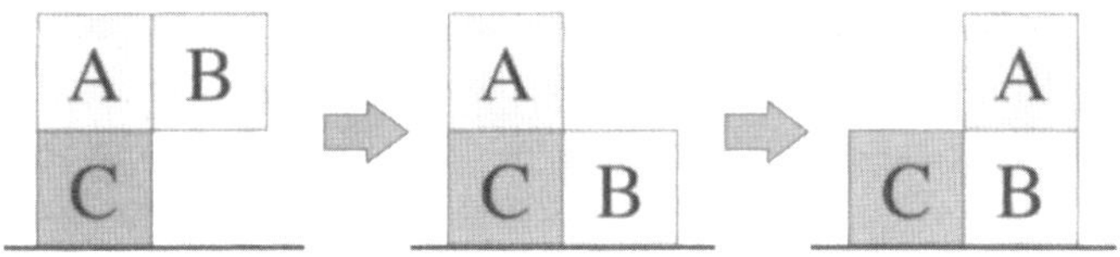

Fig. 3. Transformation of cellular robots by the proposed slide motion mechanism

group robots with the proposed motion mechanism. First, the robot B slides down the faces of robots A and C. After the robot B reached the bottom, the robot A horizontally slides on the faces of the robots C and B. Since the robot has slide motion mechanisms, it can neither separate nor connect each other. In spite of the mechanical constraints, it is possible to build up various structures. We will show the way of the transformation as follows.

We start from an initial structure with a long straight arrangement as shown in Fig. 4. If the initial structure is plainly straight, it cannot change to other structure. The initial structure has a cellular robot called "seed" at the right angle to the trunk in the working area. The seed robot plays an important role in the structural transformation as it breeds another seed called "sub-seed" in the working area. Figure 5 shows a breeding way of the sub-seed. Plural sub-seeds are easily bred in the same manner.

The sub-seed makes a structure in the constructing area by moving them to the constructing area. Figure 6 shows the way of the construction that sliding movement of a sub-seed robot produces a cellular robot called "sprout" in the constructing area. The continuous production of sprout extends the structure like a trunk as in Figure 7. Using the ways of transformation, various structures can be built up because it is possible to transform a two dimensional structure by recurrent productions of the trunk structure. A topological structure including holes is also produced in the same manner. When the robots in the working area start from a rectangular arrangement as shown in Fig. 1, they can also form a various structure in the constructing area because the straight arrangement with a seed is possible to transform the rectangular arrangement.

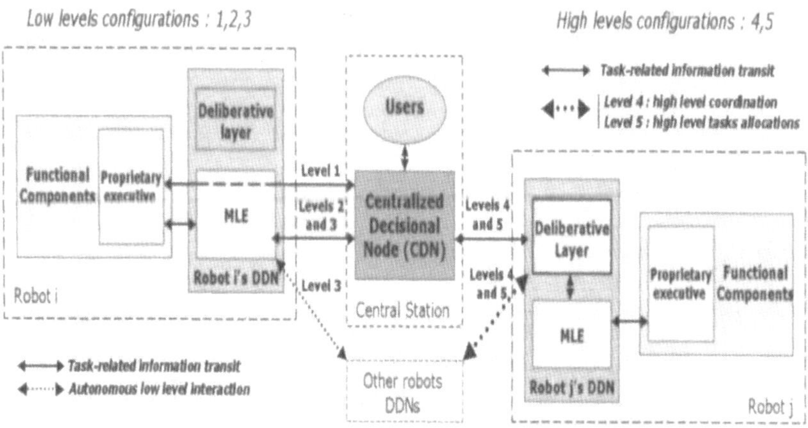

Fig. 4. Initial straight structure with a "seed"

Fig. 6. Construction of a structure called "sprout"

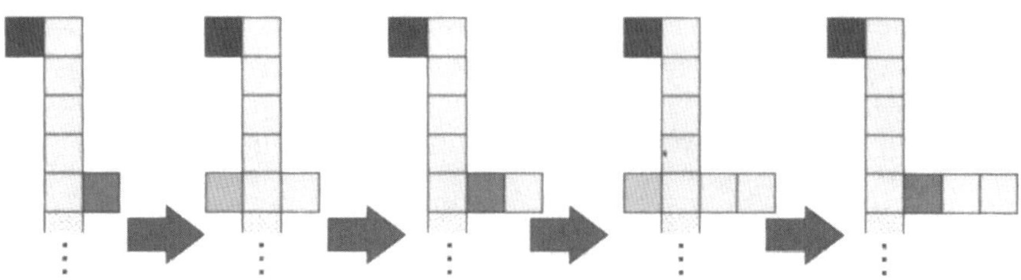

Fig. 7. Extension of a structure like a trunk

4 Estimation of dissipative energy

For effective structural formation, estimation of dissipative energy in the transformation is important. In the previous study, we reported a method to find the shortest routes in structural transformation when an initial configuration of robots and a final one are given. This paper describes a way to find a preferable route considering energy consumption in each step.

To evaluate the dissipative energy for each transformation step, we made formulation of energy consumption based on the experimental results for typical movements. As the required electric power depends on configuration of robots, we measured them under various conditions as shown in Fig. 8. These data are available to estimate total energy that group robots form a different configuration. Based on these data, the formulation is classified into three equations considering the gravitational effect as follows:

$$\begin{aligned} &\text{movement} \\ \text{Horizontal}: E &= 0.45(F_1 + F_2)L + 0.5 \\ \text{Upward}: E &= 0.45(F_1 + F_2)L + 2.0G \\ \text{Downward}: E &= 0.45(F_1 + F_2)L + 0.5 - 0.3G \end{aligned} \tag{1}$$

where F_1, F_2 = reaction forces produced at the moving face as a foundation [N], L = side length of the robot [m], G = the number of robots whose gravitational position is changed

The above equations are formulated on the assumption that dissipated energy is basically represented by multiplication of reaction forces produced at the sliding part and the moving distance of the robots. The reaction forces F_1, F_2 are derived using the cantilever model as shown in Fig. 9.

The simulator calculates the dissipated energy along structural formation using the above equations considering driving power of the robots. That is, the simulator judges possibility of transformation of the robots comparing the required energy for the movement with the maximum possible power of the robots ($E_{\max}$). If the required power is greater than $E_{\max}$, the simulator passes the calculation of the movement because the transformation is impossible. $E_{\max}$ is calculated as follows:

$$\begin{aligned} E_{\max} &= U \times I_p \times T \times R_s \times n \\ &= 6.0 \times 0.17 \times 10 \times 0.5 \times n \\ &= 5.1n \end{aligned} \tag{2}$$

where U = nominal voltage [V], I_p = maximum continuous current [A], T = required time to move unit length [sec], R_s = safety ratio and n = the number of driving motors.

Figure 10 shows the energy consumption for each step computed by the above method when an initial and a final configuration are given. There are six possible transformations. We easily find the most effective transformation that needs the minimum dissipated energy. This transformation includes a transformation step that six robots are moving at the same time. The simulation result shows that we should consider not only the number of steps of transformation but also the amount of the energy consumption.

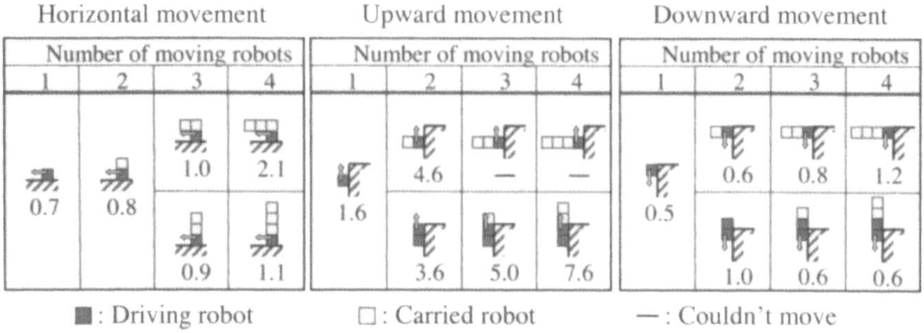

Fig. 8. Dissipative energy for the typical movements (unit of energy: joule)

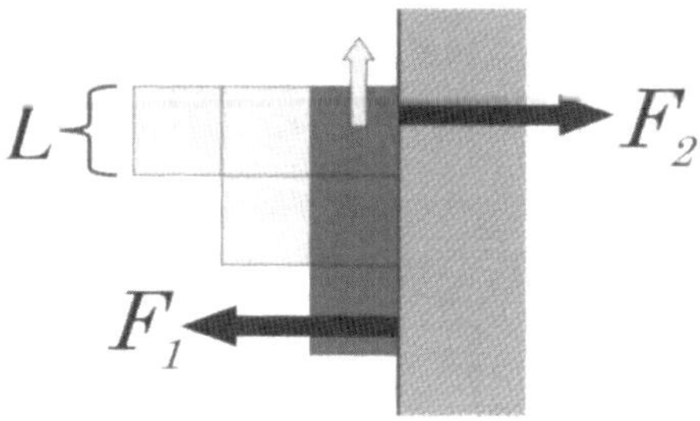

Fig. 9. Parameters to calculate energy consumption of the cellular robots

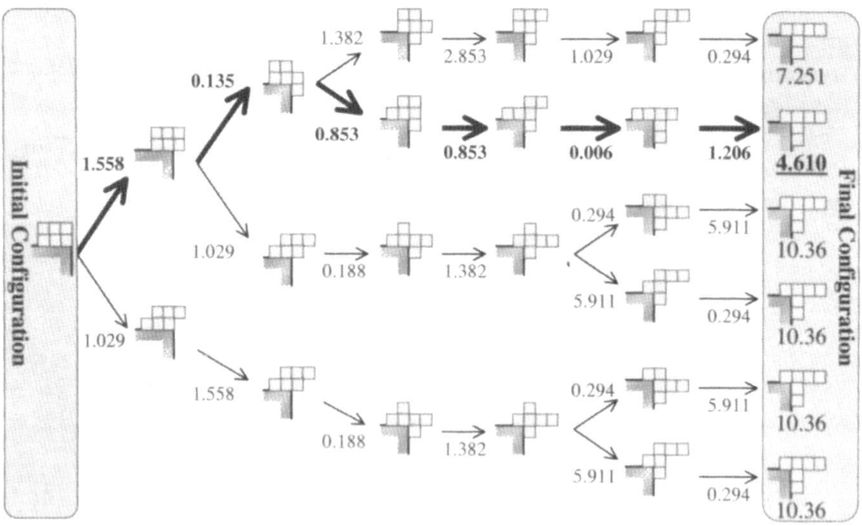

Fig. 10. Total energy for structural transformation (unit of energy: joule)

5 Development of devices for autonomous cellular robots

Our study aims to develop the autonomous cellular robots adaptively forming a mechanical structure. The robot needs the following functions: locomotive function, connecting and separating functions, sensing function for stressed states and information function. We challenge to boost up autonomy of the robots adding required devices to them.

5.1 Connecting and separating functions

For realizing cellular robots forming a mechanical structure they must have a connecting function to endure outer forces and their own weight. The proposed cellular robots provide slide motion mechanism that enable to move the cellular robot maintaining a connecting state at any time. The mechanism has large stiffness for normal direction of the sliding face. However, it is liable to be misaligned for sliding direction. We introduced a locking mechanism for CHOBIE II so that they hold a precise position with large stiffness in the sliding direction.

The locking mechanism is composed of a rod and a hole as shown in Fig. 11. The rod passively protrudes with a spring force and also actively retreats with a wired tension. There are two rods on the sliding sides for each robot and they are placed at the bottom side and the left side of the robot. On the other hand, the hole is placed at the top and right side. When the rod comes to a hole that belongs to other cellular robot, it automatically protrudes to the hole with a spring force. The protrusion is detected by a photo interrupter and firmly fixes in the sliding direction. A tether is used for release of the lock. The rod is retreated by winding up the tether with DC motor.

5.2 Integration of information functions

The cellular robots must communicate the information signals. To endow the robot with autonomy, we must integrate several devices into each robot: sensors, an electric controller and electric battery. To put these devises into the robot we increased the width of the central board from 25mm to 50mm.

Photo sensors that communicate with neighboring robots are embedded on the surface of the frame and force sensors are attached at the corner of a portion that produces large strain by outer forces as shown in Fig. 12. We use PIC (Peripheral Interface Controller) as the electric controller. This devise (PIC16F84) is a microcomputer chip and has a programmable function with I/O ports. Lithium batteries are also embedded in the central board. Figure 13 shows the control circuits embedded in the central board of CHOBIE II.

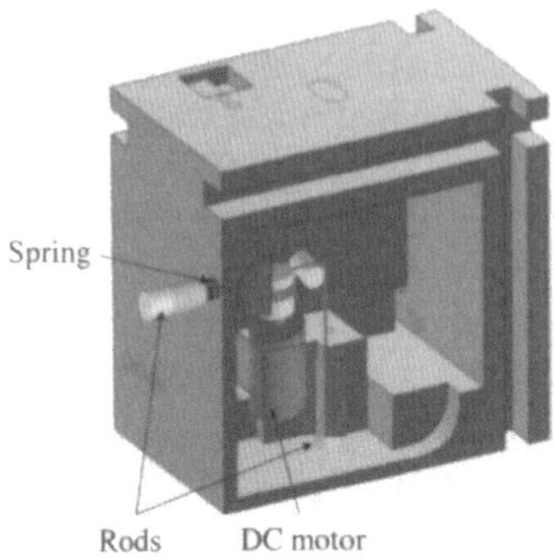

Fig. 11. Lock and unlock mechanism

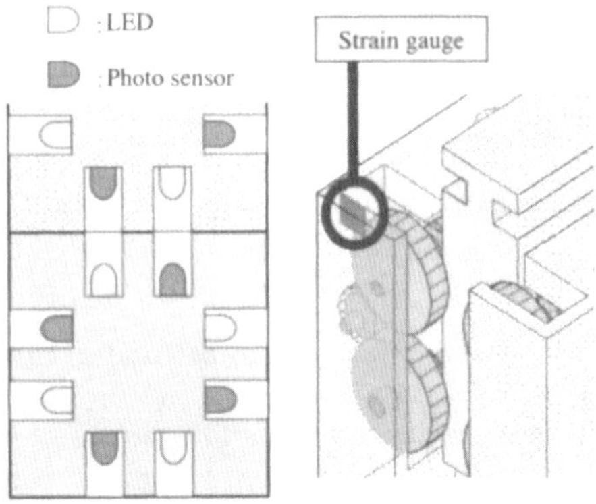

Fig. 12. Photo sensor and force sensor

Fig. 13. Control circuits and central board of CHOBIE II

6 Demonstration of cellular robot

We fabricated three cellular robots including the autonomous information functions as stated above. It is difficult for the present group robots to take a transformation path with the minimum dissipative energy because they do not still have enough intelligence. We demonstrate autonomy of the robots using several behavioral rules this time. Each robot follows the rules in all steps of structural transformation. It is expected that an initial configuration of robots is transformed into a stable configuration by iterative steps. The behavioral rules are set up as the following program code.

(1) Each robot inspects the neighboring existence by use of the photo sensors.
(2) IF the left side of a robot is vacant and the robot is connected to neighboring robots both in the right side and the under side,
THEN the heading direction of the robot is "left".
GOTO (5).
(3) ELSE,
IF the under side of a robot is vacant and the robot is connected to neighboring robots both in the right side and the upper side,
THEN the heading direction of the robot is "downward".
GOTO (5).
(4) ELSE,
The robot will not move actively to any direction and waits for messages from other robots; EXIT.
(5) The robot send the heading direction to other robots.
(6) The robots slide with the other robots that are positioned in-line simultaneously.
(7) GOTO (1).

Figure 14 shows the sequential motions based on the above criterion. As the embedded battery was not enough power to move DC motors smoothly, we added an additional battery at an external surface of the robot. An autonomy of CHOBIE II was successfully demonstrated by the addition. As the installed program code in the PIC is very simple and signals by force sensors are not used in this experiment, sophisticated transformation is not yet realized. However it is possible to construct a more complicated structure considering outer forces by revising the program code.

7 Conclusions

The cellular group robots "CHOBIE" forming a mechanical structure were discussed. It is possible to transform various structures although the proposed robots have mechanical constraints. We proposed a method to estimate energy consumption along the transformation route. This method is useful to perform effective structural transformation. To realize autonomy of the robots we developed CHOBIE II. Experimental demonstration showed the robots changed the structure communicating neighboring robots.

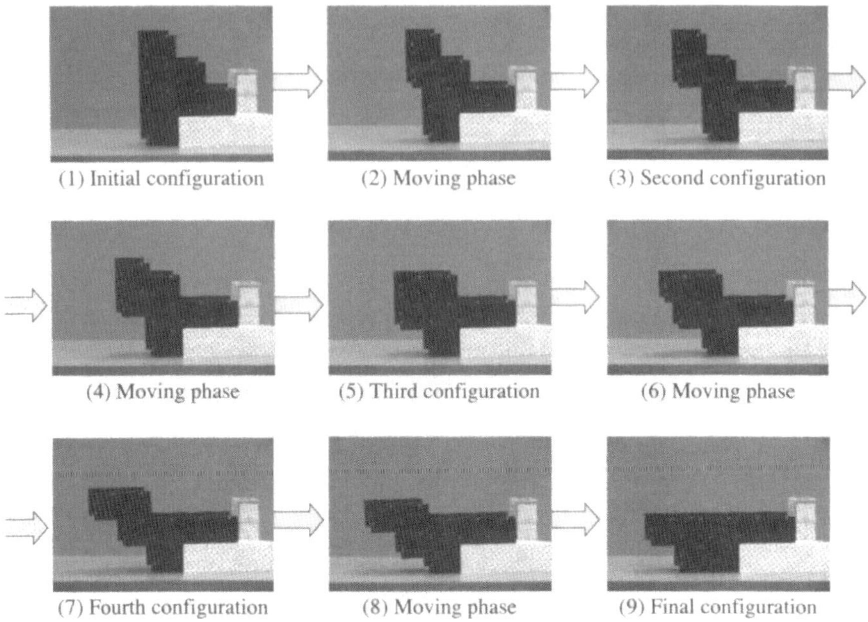

Fig. 14. Experimental demonstration of autonomous transformation

References

1. Cao YU, Fukunaga AS, Kahng AB (1997) Cooperative Mobile Robotics: Antecedents and Directions. Autonomous Robots 4: 7–27
2. Fukuda T, Nagasawa S, Kawauchi Y, Buss M (1989) Structure Decision Method for Self Organizing Robots Based on Cell Structures-CEBOT. Proc IEEE Int Conf on Robotics and Automation: 698–700
3. Murata S, Kurokawa H, Kokaji S (1994) Self-Assembling Machine. Proc IEEE Int Conf on Robotics and Automation: 441–448
4. Chirikjian G, Pamecha A (1996) Bounds for Self-Reconfiguration of Metamorphic Robots. Proc IEEE Int Conf on Robotics and Automation: 1452–1457
5. Yoshida E, Kokaji S, Murata S, Tomita K, Kurokawa H (1999) Miniaturized Self-reconfigurable System using Shape Memory Alloy. Proc IEEE/RSJ Int Conf on Intelligent Robots and Systems: 1579–1585
6. Yim M, Duff D, Roufas K, Kissner L (2000) Plybot: demonstrations of modular reconfigurable robot. Video Proc IEEE Int Conf Robotics and Automation
7. Østergaard EH, Lund HH (2003) Evolving Control for Modular Robotic Units. Proc IEEE Int Symp on Computational Intelligence in Robotics and Automaton: 886–892
8. Inou N, Minami K, Koseki M (2003) Group Robots Forming a Mechanical Structure (Development of slide motion mechanism and estimation of energy consumption of the structural formation): Proc IEEE Int Symp on Computational Intelligence in Robotics and Automation: 874–879

Planning Behaviors of Modular Robots with Coherent Structure using Randomized Method

Eiichi Yoshida[1], Haruhisa Kurokawa[1], Akiya Kamimura[1], Satoshi Murata[2], Kohji Tomita[1], and Shigeru Kokaji[1]

[1] Intelligent Systems Institute, National Institute of Advanced Industrial Science and Technology (AIST), 1-2 Namiki, Tsukuba, Ibaraki 305-8564 Japan
{e.yoshida, kurokawa-h, kamimura.a, k.tomita, s.kokaji}@aist.go.jp

[2] Interdisciplinary Graduate School of Science and Engineering, Tokyo Institute of Technology, 4259 Nagatsuta-cho, Midori-ku, Yokohama, Kanagawa 226-8502 Japan
murata@titech.ac.jp

Summary. A behavior planning method is presented for reconfigurable modular robots with coherent structure using a randomized planning. Coherent structure is introduced to cope with difficulty in planning of many degrees of freedom, in terms of control system and robot configuration. This is realized by a phase synchronization mechanism together with symmetric robot configuration, which enables the robot to generate various coherent dynamic motions. The parameters of control systems are explored using a randomized planning method called rapidly exploring random trees (RRTs). The RRT planner has an advantage of simple implementation as well as possibility of integrating differential constraints. The dynamic robot motion is thus planned and preliminary simulation results are shown to demonstrate the proposed planning scheme can generate appropriate behaviors according to environments.

1 Introduction

Self-reconfigurable modular robots are recently investigated intensively starting from earlier work [1, 2], since their flexibility, versatility, and fault-tolerance are considered to be suitable for wide range of tasks [3, 4, 5]. They are especially expected to be used as robots that can operate in unknown or unstructured environments, sometimes hostile to humans, through their adaptability. Those applications include a robot that tries to find survivors in corrupted buildings, planetary exploring vehicles, or inspection robots in nuclear plants. Many hardware systems have been developed as well as software. As to the hardware, many types of three-dimensional (3-D) self-reconfigurable robots have been developed and recently their reliability and self-containedness made a remarkable progress [6].

In this paper we focus on a behavior planning of such modular robots, to decide how those robots with many degrees of freedom act according to the environments. There are mainly two contexts of research on the software of self-reconfigurable robots. One is discrete reconfiguration planning that gives how the robot should make a sequence of configuration changes so that it can transform itself from one configuration to another [7]–[12]. The other is continuous aspect about how to generate dynamic behavior to allow the robot make useful and meaningful motions for the application [6, 13].

Although the reconfiguration planning has been addressed intensively, dynamic behavior planning remains less exploited except some work on CPG network [6]. In view of unifying the above two research contexts, we propose a method for planning behaviors of a modular robot with coherent structure. Here the "behaviors" corresponds to various different dynamic motions. With "coherent" structures in robot configuration and control system such as synchronized motions or structure symmetry, various behaviors of the robot with many degrees of freedom (DOFs) can be controlled with reduced numbers of parameters.

As control structure, we adopt a simple phase synchronization mechanism where such parameters as phase difference determine the robots' behavior. However, search space is still very large even if those parameters are not numerous. We use a randomized planning method called rapidly-exploring random tree (RRT) [15] to explore this search space for the behavior.

In the next section, a simple phase synchronization mechanism is introduced, and then the planning with randomized method is presented in Section 3. Some simulation results using modular robot M-TRAN [14] are shown in Section 4 before concluding the paper.

2 Using Phase Synchronization for Dynamic Motion

2.1 Simple Phase Synchronization Mechanism

In order a robot with many DOFs such as modular robot to generate useful motions, coherency is necessary for its control system as well as its configuration. There are several ways to realize coherent control system. One of most popular methods is using a neural network that generates oscillatory signals, like central pattern generator (CPG) [16, 17].

In this paper, to give a clear perspective of the problem, we adopt a simple phase synchronization mechanism. We begin with a simple illustrative example composed of two connected elements 1, 2. The value θ_i is a phase variable that describes the internal state of each element. In this example, both elements try to have constant velocity ω so that θ_1 of element 1 is always greater than θ_2 of element 2 by phase difference ϕ. The differential equation can be written as

$$\begin{aligned}\dot{\theta_1} &= \omega - k(\theta_1 - \theta_2 - \phi)\\ \dot{\theta_2} &= \omega - k(\theta_2 - \theta_1 + \phi)\end{aligned} \tag{1}$$

where k is a gain coefficient. By solving this equation, we have

$$\begin{aligned}\theta_1(t) &= \omega t + \frac{1}{2}(\theta_1^0 + \theta_2^0 + \phi) + e^{-2kt}(\theta_1^0 - \theta_2^0 - \phi)\\ \theta_2(t) &= \omega t + \frac{1}{2}(\theta_1^0 + \theta_2^0 - \phi) + e^{-2kt}(-\theta_1^0 + \theta_2^0 + \phi)\end{aligned} \tag{2}$$

where θ_1^0, θ_2^0 are the initial values at t=0. The difference converges to ϕ.

This mechanism can be extended to the case of n elements:

$$\dot{\theta_i} = \omega - k\sum_{j=0}^{n}(\theta_i - \theta_j - \phi_{ij}) \tag{3}$$

If there are not loops with the element connection, the phase difference between element i and j converges to the given value ϕ_{ij} [18]. In this paper we only address the cases without loops for simplicity, although some methods are proposed to cope with cases of existence of loops [19]. This mechanism has also an advantage that distributed implementation into modular robots.

We will apply this simple mechanism to generate various dynamic motions of modular robots. Since the output of (3) is linearly increasing along the time, we use a sinusoidal function to generate an oscillatory movement like

$$a_i(t) = \beta_i + \alpha_i \sin\theta_i(t) \tag{4}$$

where α_i and β_i is the amplification and offset of oscillation.

By assigning this synchronization element to each actuator, the modular robot is expected to generate diverse coordinated actuator outputs for dynamic motion. However, it is difficult for the robot to generate motions if the control system does not have a good accordance to the robot structure.

For this reason, another coherency is introduced in the robot structure, namely symmetrical configurations in our case. Even if the control system has coherency, a robot that has some irregular structure can hardly be controlled. Looking at the nature, animals indeed have symmetric form to make efficient motions.

Fortunately, self-reconfigurable modular robots can have a variety of configurations. Symmetric configurations can be used to apply the phase synchronization control mechanism to realize the dynamic motions such as gait patterns. Moreover, not only changing the dynamic motion, but they can choose different configurations according to the application, sometimes a four-legged robot or a snake-like robot. This is one of the major advantages of modular robots as mentioned earlier.

2.2 Self-Reconfigurable Modular Robot M-TRAN

To fully exploit these advantages, we adopt a symmetric structure that can realize different dynamic motions as well as configuration using a modular robot platform M-TRAN.

M-TRAN (Modular TRANSformer) has been developed in AIST that can realize both self-reconfiguration and dynamic motions in three dimensions. The module has a simple bipartite structure. Each part rotates about an parallel axis by geared motors and has three magnetic connecting faces as shown in Fig. 1. Each module is a self-contained with embedded a controller circuit board and a battery. Figure 2 shows the newest hardware model "M-TRAN II." For details on hardware, please see other references [6, 14].

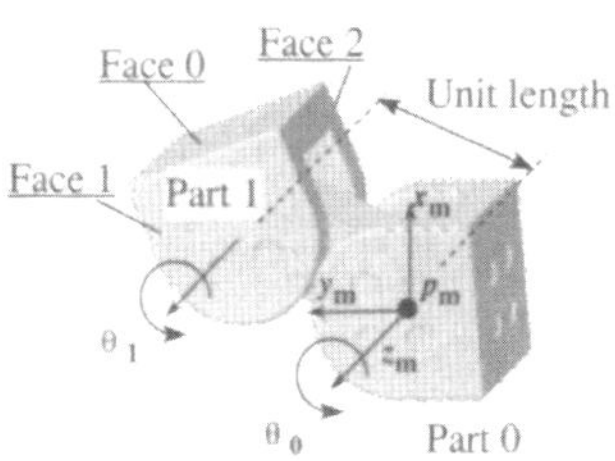

Fig. 1. A module of M-TRAN.

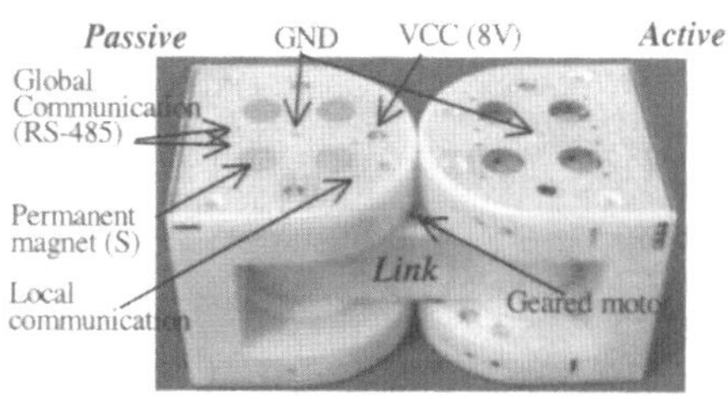

Fig. 2. A hardware module of M-TRAN II.

2.3 Applying the Synchronization to M-TRAN Modular Robot

In this paper, we deal with a configuration composed of nine modules as shown in Fig. 3. By assigning the phase synchronization mechanism to each module's actuator appropriately, generated oscillatory output enables the robot to make efficient locomotion.

Owing to the symmetry of the robot structure, all the parameters do not have to be controlled individually. Instead, the same parameters can be used repeatedly to symmetrically corresponding actuators and synchronization connections. In the case of configuration in Fig. 3, the parameters can be considerably reduced since there are two symmetry axes as shown Fig. 4. In this figure, circles and dotted lines denote oscillatory elements and connection for synchronization respectively. The arrow is defined the direction in such a way that ϕ is the phase difference from outgoing element to incoming one.

Exception of the symmetrical parameter assignment exists regarding Module 1 that is in the center of the structure. Its amplification and offset of the oscillation α_i, β_i and phase difference ϕ_5 are need to be controlled separately. The value ϕ_4 to Module 1 is also applied in different direction.

Based on this parameter assignment, the modular robot with this configuration can realize different structures for locomotion. We assume that the actuators are controlled by velocity.

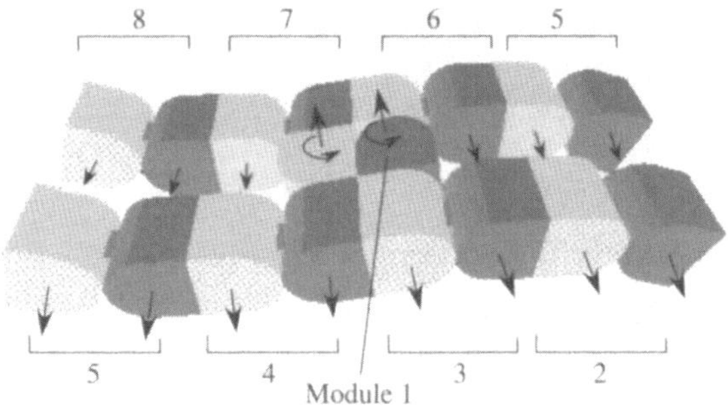

Fig. 3. A Coherent configuration of M-TRAN modules

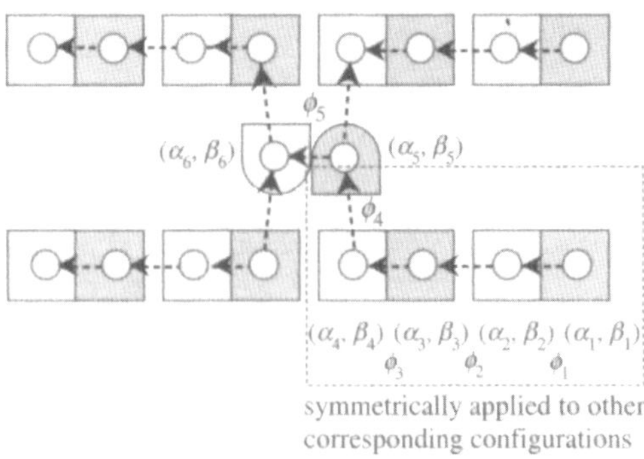

Fig. 4. Assignment of phase synchronization parameters to symmetric robot structure

One is a flat configuration whose corresponding motion snake-like wave motion (Fig. 5). Another motion is four-leg configuration that requires certain gait pattern to move, using parameters illustrated in Fig. 6. Rotational motion can be realized by setting non-zero values to α_5, α_6 for Module 1. In general, the opposite direction of motion can be generated by reversing ϕ_i values.

It is noteworthy that a single synchronization scheme can realize those very different locomotion modes are by changing parameters and synchronizing connection. In the next section, we will describe how those behaviors can be planned using a randomized method.

3 Behavior Planning using a Randomized Method

There are several ways to derive behaviors determined a number of parameters. A heuristic planning method for static motion is proposed to enable a legged robot to move rough terrain [20]. Støy et al. proposed a control model for chain-type modular robots using coupled oscillators [22]. As a more general scheme, gradient method has been proposed to optimize those phase synchronization mechanism [21].

Kamimura et. al use a genetic algorithm (GA) to obtain CPG parameters for locomotion [6]. This pattern generation method is more dedicated to real-time control and adaptation according to external stimuli, where many parameters of CPG model should be regulated. However, in the planning phase, a simplified control model is more preferable than direct usage of rather complex model of CPG since it is important to reduce the number of parameters to explore.

Fig. 5. A snake-like wave motion: with parameters $\alpha_1 \sim \alpha_4 = 10°, \alpha_5, \alpha_6 = 0°, \beta_1 \sim \beta_4 = 0°, \beta_5 = -90°, \beta_6 = 90°, \phi_1 \sim \phi_3, \phi_5 = 30°, \phi_4 = 0°, \omega = 180°$

Fig. 6. Legged locomotion: with parameters $\alpha_1 \sim \alpha_4 = 5°$, $\alpha_5, \alpha_6 = 0°$, $\beta_1 = 0°, \beta_2 = 10°, \beta_3 = 20°, \beta_4 = 40°$, $\beta_5 = -90°$, $\beta_6 = 90°$, $\phi_1, \phi_3, \phi_4 = 90°$, $\phi_2, \phi_5 = -90°$, $\omega = 180°$

In this paper, focusing on planning dynamic behavior using a simple control model based on coherent structure, a randomized planning method called rapidly-exploring random trees (RRTs) is introduced. Using this method incremental and reactive behavior planning can be implemented in a simple manner in terms of both planning and control mechanism. The CPG pattern generator [6] can then be used for real-time adaptive control to execute the planned motion. To apply this method, we assume that the robot has knowledge about local environment and its goal through external sensor capacity.

3.1 Rapidly-Exploring Random Trees (RRTs)

RRTs have been proposed by LaValle as a randomized motion planning method [15]. The idea is to incrementally construct search trees that attempt to rapidly and uniformly explore the state space. It has been proven that this method is probabilistic complete, namely desired path will be found eventually as the number of vertex becomes infinity. Since it has such advantages as simplicity of implementation for exploring many DOFs and possibility of including differential constraints, we adopt this method for dynamic behavior planning.

The basic algorithm is shown in Fig. 7 to explore configuration q. The tree $\mathcal{T}$ rapidly explores through biased search of large unexplored region of the state space. At each step, after generating a random configuration node q_{rand}, the function EXTEND($\mathcal{T}$, q_{rand}) is called to extend the tree. As illustrated in Fig. 7(b), this function first selects a node q_{near} nearest to q based on given metric, and then generates a new node q_{new} that advances to q and adds it to the tree $\mathcal{T}$. Several RRT-based motion planners have been proposed according to the problem. For example, RRT-Connect [23] is a bidirectional planner that explores RRTs from initial and goal position in environments, possibly with obstacles. The search can be accelerated using bidirectional planning.

3.2 Applying RRTs to Modular Robot's Behavior Planning

Now RRT is applied to modular robot's behavior planning based on phase synchronization mechanism. As mentioned in 2.3, synchronization mechanism can be described using parameters ϕ_{ij}, ω, α and β for the configuration shown in Fig. 3.

In our case, the robot generates motions according those parameters, then it causes oscillatory motion that brings to the robot to a different position. This is a differential constraint where the relationship between control input and resulting configuration must be specified, since simple interpolation between configurations does not apply. In the algorithm Fig. 7, the configuration q should be replaced by the state x that describes robot's current state [24]. Also,

```
BUILD_RRT(q_init)
T.init(q_init)
for k=1 to K do
    q_rand ← RAND_CONFIG( );
    EXTEND(T, q_rand);
Return T;
```

(a) Building a RRT

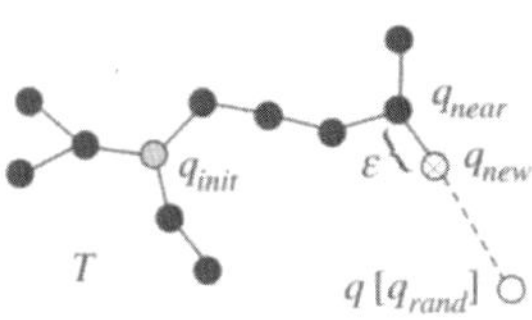

(b) function EXTEND($\mathcal{T}$, q)

Fig. 7. Basic Algorithm of RRT

NEW_CONFIG(q, q_{near}, q_{new}) is replaced by GEN_STATE(x, x_{near}, x_{new}, Δt, *Inputs*) that generates next state x_{new} advancing to given state x from its nearest neighbor x_{near} in the tree, by selecting control input from the possible set *Inputs*.

In our example, the state x includes the position and orientation of the whole robot represented by a reference frame fixed in the Module 1 as well as each module's the output angles, position, orientation, and velocity. To compute the distance between the states x, we adopt the distance between these representative positions and orientations as the metric.

The RRT planner explores in the space of those synchronization parameters as the above *Inputs*. Then the next state x is computed as a result of dynamic motion described by the control system based on the phase synchronization mechanism. Usually, those inputs are selected in such a way that it determines directly the state of robot, like position or velocity of each actuator. However, direct search of those input results in meaningless motions in most cases. In contrast, by applying RRT to those "indirect" parameters through coherency of control system and robot structure, we expect our goal of dynamic behavior planning can be achieved. On the other hand, this causes a disadvantage of heavy computation of next state in GEN_STATE(). Currently we must calculate the next state using a dynamics simulator that is computationally expensive. To accelerate planning, this needs to be substituted simple and fast solver for dynamics and collision detection.

4 Simulation Results

Based on the behavior planning scheme presented in the previous sections, we have conducted several preliminary simulations. This section shows its results.

In the simulations, the state of the robots are described using its representative position x, y on the plane and orientation Θ of Module 1. Given its goal state, the modular robot tries to find a sequence of synchronization parameters. In this simulation, the input parameters are selected from following sets shown in Table 1. Although small numbers of values are used to reduce the search space, this simple combination turned out to be sufficient to generate various motions.

The simulation is implemented using Vortex dynamics simulator [25] and MSL library [26] for RRT planner. Collision detection is implemented in both libraries. According to each input, the next state (x, y, Θ) after time Δt = 2 (sec) are calculated by the dynamics simulator. We use relatively large Δt to allow the robot to make oscillatory motion for a certain period. Then RRT planners explore this state space to reach the goal. Here, goal-biased RRT planner is used as the planner.

Table 1. Input parameter sets for phase synchronization

amplification	$\alpha_1 \sim \alpha_4$	$5, 10, 20$
	α_5, α_6	$0, 5$
offset	$\{\beta_1, \beta_2 \beta_3, \beta_4\}$	{0,0,0,0}, {0,-5,-5,-5}, {0,10,20,40}
	$\{\beta_5, \beta_6\}$	{-90, 90}, {-80, 80}
phase difference	$\phi_1 \sim \phi_3, \phi_5$	$\pm30, \pm60, \pm90$
	ϕ_4	$0, \pm20$
	$\{\phi_1, \phi_2, \phi_3, \phi_4, \phi_5\}$	$\{\mp90, \pm90, \mp90, \mp90, \pm90\}$
angular velocity	ω	140, 180, 220

In the initial state, the robot is at (0, 0, 0) and goal state is (18, 0, 0) on a plane, where the unit length in Fig. 1. Two simulations are conducted where there are obstacles with different shape at different positions.

The simulation results are shown in Figs. 8 and 9. In Fig. 8, the robot moves around the obstacle by changing its moving direction to reach the goal position. In the next simulation, first the robot uses the four-leg locomotion to advance in free space. It could go around the obstacle, but goal-biased RRT planner found the motion to go under the obstacle with snake-like locomotion before finally restoring the four-leg locomotion. Those preliminary results demonstrated the effectiveness of the proposed method.

Figure 10 shows the explored tree in the state space projected to x-y plane in the simulations, where the thick lines are the path from the initial position to the goal. Note that the states are actually more smoothly connected than it appears because they are plotted only at every Δt.

The computation time took several minutes in both cases using Pentium M processor with 1.4GHz. In future development, improvements will be addressed using alternative fast solver for dynamics and collision detection.

Fig. 8. Simulation results (1): avoiding obstacle in front

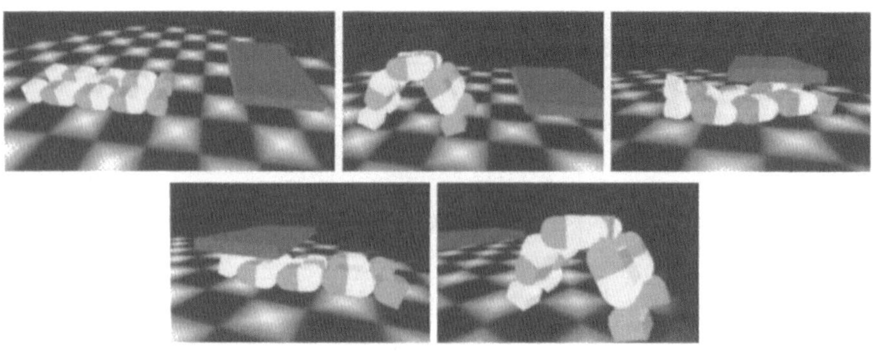

Fig. 9. Simulation results (2): going under the obstacle

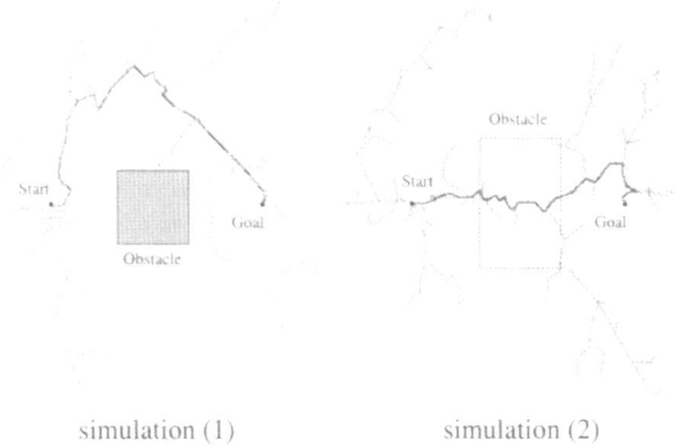

Fig. 10. Explored RRT projected on x-y plane for simulations

5 Conclusions and Future Work

In this paper we presented a behavior planning method for modular robot with coherent structure using a randomized planning method. Coherency is discussed in two aspects, in control system and robot configuration. We have introduced a simple phase synchronization mechanism for the control system and symmetry for the robot configuration. This coherency can reduce the control parameters of various dynamic motion of modular robot. A randomized method called rapidly-exploring random tree (RRT) was adopted to plan the robot's dynamic behavior. This method allows to plan a dynamic system with differential constraints based on simple implementation. The proposed method was implemented for a coherent structure of a modular robot platform M-TRAN. The preliminary simulation results showed the feasibility of the proposed method.

Future work includes such issues as integrating self-reconfiguration process, selection of coherent structures and control parameters, and more efficient implementation. Since the RRT planning scheme can include both discrete planning and differential constraints, the first issue is important in the next stage of development. This is related to the second issue, as several coherent structures can be possible according to the application. Concerning the control parameter, the control input sets are currently defined empirically. To improve the applicability of the method, improvement toward automatic acquisition of those inputs, through learning or evolutionary computation like in [6], will be addressed in the future development. Efficient implementation is also to be addressed so that to reduce the planning time. Likewise, self-learning of dynamic property through interaction with environments is also a challenging issue for implementation in real robots.

References

1. T. Fukuda and S. Nakagawa (1998) Approach to the dynamically reconfigurable robotic system, *Journal of Intelligent and Robot Systems*, **1**, 55/72.
2. S. Murata, et al (1994) Self-assembling machine, *Proc. 1994 IEEE Int. Conf. on Robotics and Automation*, 441/448.
3. *Autonomous Robots* (2001) *Special issue on self-reconfigurable robots*, **10**-1.

4. *IEEE/ASME Trans. on Mechatronics* (2002) *Special issue on reconfigurable robots*, **7**-4.
5. *Science* (2003) Shape Shifters Thread a Daunting Path Toward Reality, 301, 754/756.
6. A. Kamimura, et al (2003) Automatic Locomotion Pattern Generation for Modular Robots, *Proc. 2003 IEEE Int. Conf. on Robotics and Automation*, 2003.
7. K. Kotay and D. Rus (1998) "Motion Synthesis for the Self-Reconfigurable Molecule," *Proc. 1998 IEEE/RSJ Int. Conf. on Intelligent Robots and Systems*, 843/851.
8. E. Yoshida, et al (1999) A Distributed Method for Reconfiguration of 3-D homogeneous structure, *Advanced Robotics*, **13**-4, 363/380.
9. C. Ünsal, et al.: "A modular self-reconfigurable bipartite robotic system: Implementation and Motion Planning," *Autonomous Robots*, **10**-1, 23/40.
10. M. Yim et al (2001) "Distributed Control for 3D Metamorphosis," Autonomous Robots **10**-1, 41/56.
11. D. Rus and M. Vona (2001) Crystalline Robots: Self-reconfiguration with Compressible Unit Modules, *Autonomous Robots*, **10**-1, 107/124.
12. E. Yoshida, et al (2002) A Self-Reconfigurable Modular Robot: Reconfiguration Planning and Experiments, *Int. J. Robotics Research*, **21**-10, 903/916.
13. Y. Zhang et al (2003) Scalable and Reconfigurable Configurations and Locomotion Gaits for Chain-type Modular Reconfigurable Robots *Proc. 2003 IEEE Int. Conf. on Computational Intelligence in Robotics and Automation (CIRA2003).*
14. S. Murata, et al (2002) M-TRAN: Self-reconfigurable Modular Robotic System, *IEEE/ASME Transactions on Mechatronics*, **7**-4, 431/441.
15. S. LaValle and J. Kuffner (2001) Rapidly-Exploring Random Trees: Progress and Prospects, InB. R. Donald, K. M. Lynch, and D. Rus, eds., Algorithmic and Computational Robotics: New Directions, 293/308, A K Peters, Wellesley.
16. G. Taga (1995) A model of the Neuro-Musculo-Skeletal System for Human Locomotion II / Real-Time Adaptability under Various Constraints," *Biolog. Cybern.*, **73**, 113/121.
17. H. Kimura, et al (1999) Realization of dynamic walking and running of the quadruped using neural oscillator, Autonomous Robots, **7**-3, 247/258.
18. Hiroaki Yamaguchi, et al (2001) A distributed control scheme for multiple robotic vehicles to make group formations, Robotics and Autonomous Systems, **36**, 125/147.
19. S. Kokaji, et al (1996) Clock synchronization algorithm for a distributed Autonomous System, *J. Robotics and Mechatronics*, **8**-5, 317/328.
20. C. Eldershaw and M. Yim (2001) Motion planning of legged vehicles in an unstructured environment, *Proc. 2001 Int. Conf. on Robotics and Automation*, 3383/3389.
21. H. Yuasa and M. Ito (1990) Coordination of Many Oscillators and Generation of Locomotory Patterns, *Biol. Cybern.*, **63**, 177/184
22. H. Støy et al (2002) Using Role Based Control to Produce Locomotion in Chain-Type Self-Reconfigurable Robots *IEEE Trans. on Mechatronics*,
23. J. Kuffner and S. LaValle (2000) RRT-Connect: an dfficient approach to single-query path planning, *Proc. 2000 IEEE Int. Conf. on Roboics and Automation*, 995/1001.
24. S. LaValle and J. Kuffner (1999) Randomized kinodynamic planning, *Proc. 1999 IEEE Int. Conf. on Roboics and Automation* 473/479.
25. http://www.cm-labs.com
26. http://msl.cs.uiuc.edu/msl

In-Place Distributed Heterogeneous Reconfiguration Planning

Robert Fitch[1], Zack Butler[1], and Daniela Rus[1,2*]

[1] Department of Computer Science, Dartmouth College, Hanover, NH 03755, USA {rfitch,zackb,rus}@cs.dartmouth.edu
[2] Computer Science and Artificial Intelligence Laboratory, MIT, Cambridge, MA, USA rus@csail.mit.edu

Summary. This paper describes a distributed planner that allows a self-reconfiguring robot consisting of heterogeneous modules to change configuration without using extra space. This work builds on previous work on reconfiguration planning for homogeneous robots and our recent work on heterogeneous reconfiguration planning that requires temporary working space to execute. Motivated by a specific algorithm for the distributed Crystal robot, we describe and analyze an $O(n^2)$-time reconfiguration algorithm for systems of modules with assigned types that uses a constrained amount of free space.

1 Introduction

In self-reconfiguring (SR) robotics, the reconfiguration problem is most often studied in the context of systems of identical units. Hardware prototypes of such robots have been developed by a number of groups [4, 5, 9], and reconfiguration algorithms have been studied by [6, 7, 8, 10, 11]. It is desirable, however, to consider robots with functionally specialized units such as sensor modules, battery modules, or modules of different shapes. Unfortunately, reconfiguration algorithms for homogeneous systems cannot guarantee the absolute position of any given unit and are insufficient for heterogeneous systems where module placement is important. Examples include placing sensors at the front of the robot or maintaining heavy battery modules at the robot's base. We have previously developed simple centralized and distributed algorithms for heterogeneous reconfiguration that require a large amount of temporary working space to execute [3]. In the present paper we improve this work with a demonstration of decentralized heterogeneous reconfiguration in hardware, and with a reconfiguration algorithm optimized for temporary working space.

*Support for this work was provided by NSF awards EIA-9901589, IIS-9818299, IIS-9912193, EIA-0202789 and 0225446, and ONR award N00014-01-1-0675.

We start with the hardware experiment, in which the algorithm is tightly coupled to a particular hardware system, and then present a generic algorithm that can be instantiated to many different kinds of hardware.

The *heterogeneous reconfiguration planning* problem for lattice-based systems is a variant of standard reconfiguration planning that specifies exact positions of particular module types in the goal configuration [3]. Given two configurations with module type labels, we are trying to find a series of module motion primitives, known as a *reconfiguration plan*, that transform the first configuration into the second. Reconfiguration algorithms can be classified as *in-place* or *out-of-place* based on the quantity of required free space [3, 10]. In our previous work, we likened heterogeneous reconfiguration to the *Warehouse* problem, where free space is in contention, and showed that out-of-place heterogeneous reconfiguration is asymptotically no harder than homogeneous reconfiguration. In this paper we develop an in-place solution, retaining the asymptotically optimal time bound [6]. Benefits are a more efficient solution and clear opportunities for parallelism.

2 Experiments with a Heterogeneous SR Robot

The Crystal robot [1] is an example of a distributed SR robot system that uses expansion and contraction for actuation. The robot system is computationally decentralized. Each module has its own processor and a thin software infrastructure based on message passing that enables algorithm implementation. In our previous work [7], we developed algorithms that treat the robot as a homogeneous structure. In this section, we view the robot as a heterogeneous system, using color to differentiate the modules. More generally, we can encode arbitrary types in each unit. We demonstrate that the modules can be sorted by color using the robot's specific degrees of freedom. Since many kinds of computation reduce to sorting, applications that rely on this basic operation will be possible with this robot. In this section we describe the distributed sorting algorithm and present hardware implementation results. The lessons learned from this implementation lead us to a more generic algorithm for heterogeneous reconfiguration planning, which we present in Section 3.

2.1 Sorting Modules by Color

To sort the heterogeneous Crystal robot we instantiate the generic sorting step of our previous algorithm for heterogeneous planning [3] to the Crystal actuation method. Given a configuration of modules with color labels, the objective is to reconfigure the structure such that modules are arranged according to a given sequence of colors.

The algorithm will sort a row of Crystal modules using a technique similar to the familiar selection sort. Because of the actuation requirements in a unit-compressible system, this operation needs surrounding structure to succeed, in the form of two additional rows. This allows any single module to retract fully from the current row. The procedure to reposition a module consists of three steps. First, the module is removed from its row by contraction of its upper neighbors. Then, the remaining modules move back and forth to align the module with its new position. Finally, the contracted modules expand to place the module back into its original row but in the new order.

A top level protocol synchronizes the sorting such that modules are placed in their correct order beginning at the left of the configuration and progressing successively to the right. The order is determined by the color sequence of the middle row. The algorithm is illustrated in Figure 1. Step (a) shows the initial configuration. The first operation is to move the module in the last column to the first column, which happens in steps (b) through (f). Next, the correct module for column two is located and repositioned in steps (g) and (h). The final operation exchanges modules in columns three and four and results in the sorted configuration shown in (i). Pseudocode for the decentralized implementation is listed in Algorithm 1.

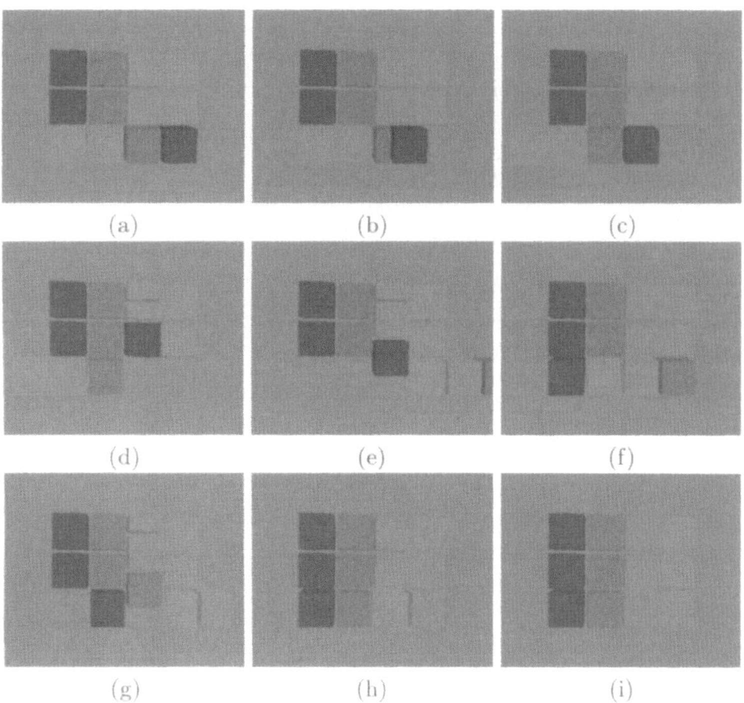

Fig. 1. Screenshots from CrystalSort implementation in SRSim simulator. Bottom row is in reverse order in (a), and is sorted to match column type. The first swap (4-1), is completed in (f), the second (4-2) is completed in (h), and the final swap (4-3) completes the sorting in (i).

Algorithm 1 Decentralized algorithm CrystalSort. Assumes three rows of modules, and begins with *sortToken* message sent to leftmost module (column one) in middle row. Bottom row will be sorted to match middle row.

State:
my_color, my color label
my_column, my current column

Messages:
sortToken, sent to synchronize sorting
Action: Request color from bottom neighbor. If neighbor color matches *my_color*, send *sortToken* to right neighbor. Else search for matching color by sending *sortQuery(my_color, my_column)* to bottom neighbor.
sortQuery(requested_color, column), searches for matching color
Action: If *my_color* does not match *requested_color*, send sortQuery message to right neighbor. Else, execute moveToPosition(*column*). When done, pass *sortToken* back to new upper neighbor.

Procedures:
moveToPosition(column)
1. initiate locomotion to align myself with center of structure
2. send command to upper neighbors to contract
3. command bottom row to reconnect
4. send locomotion command to lower neighbor to produce hole beneath me at desired column
5. command upper neighbors to expand
6. initiate locomotion to realign row with rest of structure

Our experimental setup consisted of a 12-module robot placed on plexiglass (see Figure 2). The possible configurations consist of all permutations of a row of four types. We ran the algorithm from a number of initial configurations, summarized in Table 1.

2.2 Results and Discussion

The robot successfully completed sorting from numerous initial configurations. In general, the longer trials requiring more motions were more prone to failure. Failures were due exclusively to mechanical limitations of the hardware; the software and algorithms performed correctly. The main hardware errors were caused by IR communication failures and by connector failures. Better hardware prototypes are needed to address the mechanical concerns, although performance can also be enhanced algorithmically by reducing the number of module motions. Friction encountered during expansion and contraction through free space leads to misalignment and hence connection failure. Manual intervention was required during the experiment to ensure connections. Another way to address this issue is by spatially constraining the motion of

Fig. 2. CrystalSort implementation in hardware using the Crystal robot. Steps correspond to simulation in Figure 1.

the modules. This constraint can be accomplished in part by reducing the amount of free space used by the system, allowing stationary modules to become alignment tracks.

Table 1. Hardware Results: CrystalSort Algorithm on Crystal Robot from various initial configurations. Goal configuration was 1-2-3-4.

Initial Config	Number of Swaps	Swap Sequence	Successful Attempts	Total Attempts	Manual Interventions
2-1-3-4	1	2↔1	1	1	12
2-3-1-4	1	3↔1	1	1	12
2-4-3-1	1	4↔1	1	1	16
1-3-2-4	1	3↔2	1	1	10
1-3-4-2	1	4↔2	1	1	12
1-2-4-3	1	4↔3	2	4	11
2-1-4-3	2	2↔1; 4↔3	2	3	23
3-2-1-4	2	3↔1; 3↔2	1	1	22
3-4-1-2	2	3↔1; 4↔2	1	2	24
4-3-2-1	3	4↔1; 4↔2; 4↔3	1	10-15 (est.)	35
Totals:			**12**	**25-30**	

3 Heterogeneous Reconfiguration Planning with Free-Space Constraints

Inspired by the hardware experiments, we wish to develop an algorithm that reduces the number of free space actions in the system. The algorithm presented here uses only the union of the start and goal configurations plus empty lattice positions immediately adjacent to modules on the surface of this union. We call this extra space the *crust* since it can be thought of as growing the union by one unit. This method has the added benefit of enabling reconfigurations in confined spaces such as among obstacles.

Heterogeneous reconfiguration can be divided into two phases: the first forms the goal shape regardless of type, and the second adjusts the goal shape to ensure type consistency. Decomposing the problem in this way is helpful since the first phase is purely homogeneous and can therefore be solved using existing algorithms such as those described by Yim [10] or Kotay [4]. The challenge lies in the second phase, where the modules must be sorted by type. We present a solution, *TunnelSort*, in this section, first in centralized form for clarity as Algorithm 2, then as a decentralized implementation in Algorithm 3. Both versions have been implemented in the SRSim simulator [3].

3.1 Algorithm: TunnelSort

Our approach is to repeatedly swap modules until all type requirements in the goal configuration are satisfied. The challenge is to "unlock" modules from the structure while both maintaining global connectivity and avoiding backtracking caused by displacing correctly positioned modules. For example, to access a module buried deeply in the structure, it might be necessary to temporarily displace a large number of other modules. Or, consider rearranging a sparse shape such as a line. Removing a module in the interior clearly divides the line. We address these challenges as follows. To swap modules, we will unlock each by creating a path, or tunnel, to the crust. Displaced modules are temporarily stored in the crust, while the two desired modules exchange positions. Then the remaining modules reverse their motions. See Figure 3 for a simple example.

(a) (b) (c) (d)

Fig. 3. Simulation of TunnelSort algorithm. Some modules not shown. Initial configuration with two incorrectly placed modules is shown in (a). Modules are unlocked (b) and swapped (c), resulting in the final configuration (d).

Algorithm 2 Centralized TunnelSort.

1: Reconfigure homogeneously to match goal shape
2: Label crust
3: **while** Reconfiguration is not complete **do**
4: Choose modules m_1, m_2 where m_2's type matches goal type at m_1's position
5: Find minimum-length straight-line path from m_1 to crust
6: **for** each segment s from m_1 to crust **do**
7: **if** s is tunnelable **then**
8: Find path around s or create bridge
9: **if** s was bridged **then**
10: Move all modules in s into adjacent free space
11: Repeat with m_2
12: Find path between m_1 and m_2 and swap
13: Replace all other modules

We present the algorithm in centralized fashion as Algorithm 2 in order to simply describe the algorithmic idea. Line 1 uses existing algorithms as noted earlier. This constructs the correct shape, so we can label surface modules to localize the crust and begin the main loop. The choice of modules to swap can be done in any linear-time search. Unlocking occurs in lines 6-10, and repeats in line 11. Line 13 prevents backtracking and prepares for the next swap.

Most of the complexity lies in the unlocking procedure, which we now detail. In line 5, we find a minimum-length path by searching out from m_1 in a straight line in all dimensions until reaching the crust. The resulting path consists of either a single segment of connected modules or alternating segments of modules and holes. The possible cases are described in Figure 4. If modules along the path can be removed without disconnecting the structure, then they can be temporarily stored in the crust or in an adjacent hole (Figure 4a and b). Otherwise, there may exist a path around the segment (Figure 4c). If not (Figure 4d) then it is possible to "bridge" the segment by moving some free module from the perimeter of the hole into a position that prevents disconnection of the structure (Figure 4e). Such a position much exist since it was not possible to circumvent the segment entirely. A path from m_1 to the crust is thus created. Note that since the modules follow one another along the tunnel, these motions can be executed in parallel for improved efficiency.

3.2 Decentralized Algorithm

The decentralized version of TunnelSort is listed as Algorithm 3. Our general approach is to dynamically choose modules as controllers over local operations, and to use message passing for global synchronization. Most often, message passing is implemented such that a message traverses modules sequentially. For example, with procedure DFS-send(), the message arrives at modules in the robot in the same order in which depth-first search (DFS) would visit nodes in the module connectivity graph.

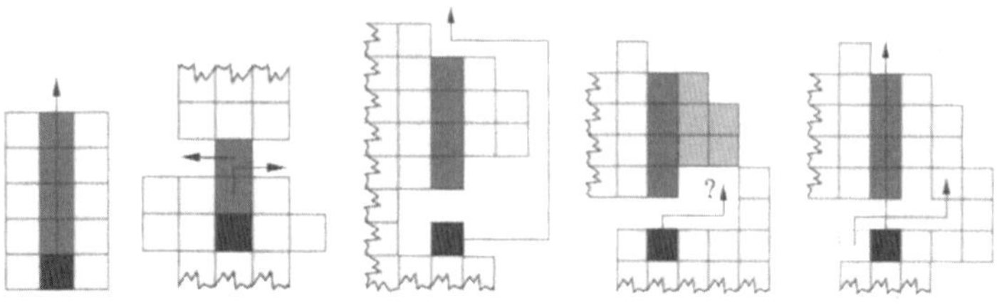

Fig. 4. Tunneling through a segment. Module to be unlocked is shown in black; shading indicates the original tunnel path. The case with no holes is shown in (a). Shaded modules will be stored in the crust. If the path does contain holes, as in (b), shaded modules can be stored there. In (c), the tunnel segment cannot be removed but an alternate path exists. A similar case is shown in (d), but no such path exists. Rather, a bridging module allows the segment to be safely removed by connecting the lightly shaded modules to the rest of the structure, resulting in (e).

The algorithm begins as message *type_check* is sent to any module. This initial message can originate from outside the system or from another module using this algorithm as a subroutine. When a module receives *type_check*, it initiates a swap procedure if necessary and then resends *type_check*. The algorithm terminates after all modules have received *type_check*. This implements the outer loop of the centralized algorithm. Control over the swap procedure is shared between the two modules, m_1 and m_2, that exchange positions. Module m_1 unlocks itself by locally creating a tunnel in procedure unlock(), and then searches for m_2 by sending the *swap* message. Module m_2 similarly unlocks itself, moves into position, and signals m_1. Then m_1 is free to move to its final position. All remaining modules return to their original positions via the *return* message and the swap is complete.

3.3 Analysis

Theorem 1 (Correctness). *Algorithm 2 produces a reconfiguration plan that transforms any given start configuration into any given goal configuration.*

Proof. When the algorithm begins, the current configuration matches the goal shape. For each color c, the number of c-colored modules equals the number of c-colored goal positions. Therefore, suitable m_1, m_2 always exist in Line 4.

The unlocking procedure is correct based on case descriptions in Section 3.1. There is sufficient space in the crust for temporarily storing modules since the size of the surface is greater than the length of any path. A path between m_1 and m_2 through the crust always exists since temporary modules never form a self-cycle. After swapping, the original configuration is restored except for the position reversal of m_1 and m_2. Each iteration therefore correctly positions at least one module and the loop eventually terminates with the robot in the correct final configuration.

Algorithm 3 Decentralized TunnelSort. Algorithm begins with message *type_check* sent to any module. Each module executes identical code.

```
State:
type, my type label
goal_type, type in goal configuration at my current position

Messages:
type_check, sent to search for modules whose type conflicts with goal type
  Action: If type = goal_type, DFS-send(type_check). Else execute unlock(), send
  swap(goal_type, my position).
swap(t, position), sent to search for a module of requested type t
  Action: If type = t, handleSwap(position). Else DFS-send(swap(t, position)).
swap_done(position), sent to synchronize swap
  Action: Find path to position, follow path, DFS-send return, send type_check.
tunnel, sent to move modules out of the way
  Action: Send tunnel in same direction. Move to side of path, recording motions.
return, sent to return displaced modules to original positions
  Action: While not in original position, wait for next position to be free and
  move there. Return true.

Procedures:
handleSwap(p)
  unlock()
  Find path and move to position p
  send swap_done(my_original_position) to m1
unlock()
  search for minimum length straight line path
  while I am not in crust do
    if current segment not tunnelable then
      if path exists around segment then
        follow path and continue
      else
        request mobile modules to bridge sides of tunnel to rest of structure
    send tunnel to segment
    move through tunnel to next segment
DFS-send(message)
  send message to first child, wait for response
  repeat for all children and compute result
  send result in return message to parent
```

It remains to show the time bound of $O(n^2)$. First, the time spent in unlocking a given module m is equal to the sum of distances traveled by modules along the path, by m itself, and by modules used in creating bridges. The number of moves required to create bridges plus the path length of m is $O(n) + O(n) = O(n)$. The work done in moving all other modules is proportional to the squared length of the tunnel. Since we always choose the straight line path of minimum length, we can amortize the cost of all tunnels to $O(n^2)$.

The total work done in unlocking is therefore $n * O(n) + O(n^2) = O(n^2)$. Total time overall is n iterations of $O(n)$ time for searching and swapping plus $O(n^2)$ amortized time for unlocking, or $O(n^2)$ overall.

4 Discussion and Future Work

In our earlier work, we asked whether an in-place heterogeneous reconfiguration solution was possible in polynomial time. The algorithm presented here answers that question with an asymptotically optimal $O(n^2)$-time solution. Important theoretical questions that remain include approximating the optimal number of moves for a given configuration, and generating an approximate goal shape using the optimal number of moves. Another interesting direction is to consider relative placement of modules based on function, without specifying a goal shape *a priori*. For example, a camera module could be constrained to the front of the robot during reconfiguration. We are currently investigating these types of relative position constraints [2].

References

1. Z. Butler, R. Fitch, and D. Rus. Distributed control for unit-compressible robots: Goal-recognition, locomotion and splitting. *IEEE/ASME Trans. on Mechatronics*, 7(4):418–30, Dec. 2002.
2. R. Fitch. *Heterogeneous Self-Reconfiguring Robotics.* PhD thesis, Dartmouth College, Expected August, 2004.
3. R. Fitch, Z. Butler, and D. Rus. Reconfiguration planning for heterogeneous self-reconfiguring robots. In *Proc. of IROS*, 2003.
4. K. Kotay. *Self-Reconfiguring Robots: Designs, Algorithms, and Applications.* PhD thesis, Dartmouth College, Computer Science Department, 2003.
5. S. Murata, E. Yoshida, A. Kamimura, H. Kurokawa, K. Tomita, and S. Kokaji. M-TRAN: Self-reconfigurable modular robotic system. *IEEE/ASME Trans. on Mechatronics*, 7(4):431–41, 2002.
6. A. Pamecha, I. Ebert-Uphoff, and G. Chirikjian. Useful metrics for modular robot motion planning. *IEEE Trans. on Robotics and Automation*, 13(4):531–45, 1997.
7. D. Rus and M. Vona. Crystalline robots: Self-reconfiguration with unit-compressible modules. *Autonomous Robots*, 10(1):107–24, 2001.
8. W.-M. Shen, B. Salemi, and P. Will. Hormone-inspired adaptive communication and distributed control for conro self-reconfigurable robots. *IEEE Transactions on Robotics and Automation*, October 2002.
9. J. Suh, S. Homans, and M. Yim. *Telecubes*: Mechanical design of a module for self-reconfiguring robotics. In *Proc of IEEE ICRA*, 2002.
10. S. Vassilvitskii, M. Yim, and J. Suh. A complete, local and parallel reconfiguration algorithm for cube style modular robots. In *Proc. of IEEE ICRA*, 2002.
11. J. Walter, J. Welch, and N. Amato. Concurrent metamorphosis of hexagonal robot chains into simple connected configurations. *IEEE Transactions on Robotics and Automation*, 18(6):945–956, 2002.

Distributed Metamorphosis of Regular M-TRAN Structures

Esben H. Ostergaard[1], Kohji Tomita[2], and Haruhisa Kurokawa[2]

[1] The Maersk Mc-Kinney Moller Institute for Production Technology, University of Southern Denmark esben@mip.sdu.dk
[2] National Institute of Advanced Industrial Science and Technology (AIST), Japan, (k.tomita,kurokawa-h)@aist.go.jp

Abstract

A key issue in controlling the morphing process of a self-reconfigurable robot is how to deal with the complexity introduced by limitations in the mechanical capabilities of an actual physical system. In this paper, we explore how structural regularity can be exploited to reduce this complexity, by considering three specific structures that consist of many M-TRAN modules. Two control approaches are presented and compared for an example 2-dimensional flow motion. One approach considers programming using subroutines and local variables, and the other considers a direct mapping from the local physical state of a module to the modules' action space. Also, we discuss the concept of sprouting, a process in which structures grow substructures.

1 Introduction

Self-reconfigurable robotics is still a relatively young field of research and still presents many great challenges. A general programming method that permits shape change so as to better fulfill a given task while utilizing the inherent robustness and versatility of a self-reconfigurable modular robotic system, still remains elusive for researchers. However, rapid progress is being made toward subproblems of this challenge.

A central problem of controlling self-reconfigurable robots seems to be the complexity involved in doing so. A number of factors contribute to this complexity; the sheer number of actuators and sensors in a large cluster of modules, the constraints on motion capabilities of the modules causing complicated motion heuristics, the difficulties of coordinating the efforts of a large distributed system, and the requirement that the system should be fault tolerant and have self-repair capabilities.

A system that deals with all the above described complexities while at the same time performing some useful function, might become a reality in the future. At present time, research is progressing by various kinds of simplifications of the involved complexity.

In this work, we consider the metamorphosis of several regular structures consisting of a large number of robotic modules. Since centralized approaches are not suitable for dealing with large number of modules, we present two distributed control methods, *program-based* and *rule-based*, and discuss their advantages and disadvantages.

1.1 Related Work

A number of approaches have so far been tried for controlling the self-reconfiguration process of a modular robotic system, both centralized and distributed. We can divide the centralized approach into off-line centralized approaches, where the behaviour sequence is precomputed [18], and on-line centralized approaches, where a central computer determines the actions based on the given state of the entire robot.

For the distributed approaches we can divide existing work into hormone message based approaches[13], rule based approaches [1], finite state machine approaches [14] and stochastic approaches [3].

Also, the tasks considered seems to fall into one of two categories. In one kind, the goal is to have the robot morph into some predefined target shape, either using a local [8, 15] or a global [14] description. In another kind, the goal is to produce collective behaviour that makes the robot interact with its environment [2]. Approaches to these problems differ by the amount of global information used, as well as on the availability of local and global communication.

Murata et al [8] and Butler et al [1, 2] have described how local rules can produce flow-type motion, and Stoy [14] has described how a gradient attraction scheme can be used to construct an arbitrary-given 3D shape. So far there has been good progress in terms of controlling modules with omnipotent motion capabilities [2, 14]. However, due to mechanical reasons, omnipotent motion capabilities for a robotic module are not feasible, so more research is needed on dealing with the complexities that arise when considering the constraints of an actual robotic system.

2 Self-reconfiguration, the M-TRAN System

The M-TRAN modules, shown in figure 1, is a unit-modular self-reconfigurable robot system. A module consists of a passive and an active part connected by a link. Each part can rotate to ±90° on the link. Since the axis of rotation is the same for both parts, cooperation between two or more modules are required to make a module change its plane of operation. Several experiments have been published on self-reconfiguration on the real M-TRAN hardware[9, 4]. While performing self-reconfiguration, care must be taken to avoid collisions, violating the joints' angle and torque limits and separating the structure.

The complexity of the reconfiguration problem can be reduced by introducing macro-scale regularity in the structure. A "meta-module" is one example, which is made of modules connected with each other and works as a larger module [11, 12, 3].

In this paper, we explore a different approach to complexity reduction by regularity. We designed three regular structure families, called Type-1, -2, and -3 hereafter, shown in figure 3, 6 and 7 respectively. Each of them is assembled by identical building blocks made of multiple modules, similar to a meta-module structure. Though each block does not work as a meta-module by itself, it can change its position by the help of other modules and keep the regularity after a sequence of motions. Such a sequence is used as a subroutine, and as long as it is used general problems of self-reconfiguration such as connectivity loss and module collisions are avoided.

So far, only offline centralized planning approaches for control has been reported for the type-1 and -2 structures. As for the type-3 structure, it was only recently suggested [5] and no work on controlling self-reconfiguration in the structure has yet been reported. This paper presents and discusses decentralized control algorithms for two of these regular structures, namely the type-1 and -3 structure. Videos from simulation can be found at
`http://unit.aist.go.jp/is/dsysd/mtran/English/clusterflowE.htm`

In this work, we consider the M-TRAN module with obstacle detectors on five sides of each part and local communication capabilities. The physical state of an M-TRAN module is described in figure 2, as well as the action capabilities for each module.

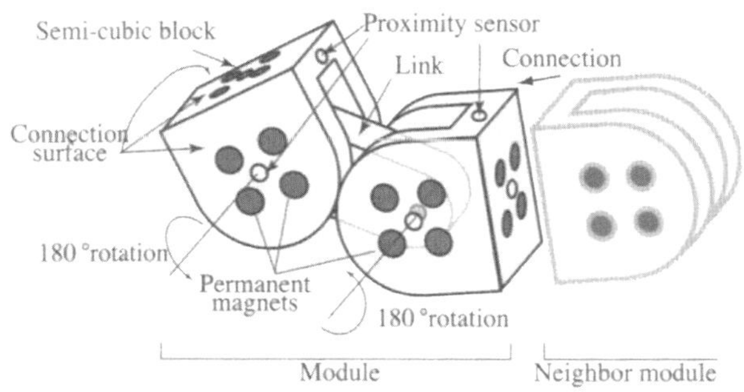

Fig. 1. The M-TRAN module. The passive part has three female connectors and the active part has three male connectors. Each part has five obstacle detectors. Due to the geometrical properties of the module, connector gender polarity problems do not arise as long as joint angles are integer multiples of 90°. We use this property and simple local message passing to simulate unisex connectors.

State Information	Values
Joint act. j, ×2	$j \in \{\emptyset, -90..90\}$
Connectors c, ×6	$c \in \{\emptyset, \texttt{DISCONNECTED}, 0..3\}$
Communication x, ×6	$x \in \{\emptyset, \texttt{NOSIGNAL}, \texttt{OK}\}$
Proximity sensor o, ×10	$o \in \{\emptyset, \texttt{CLEAR}, \texttt{OBSTACLE}\}$

Action Space	Parameter Values
setJointPos(j_p,j_a)	$j \in \{-90..90\}$
Connect(C)	$C \subset \{0..5\}$
Disconnect(C)	$C \subset \{0..5\}$

Fig. 2. The physical state information for one M-TRAN module. The $\emptyset$ symbol is used when the information is not available, for example during connection or joint actuation. Connector symmetry of the M-TRAN system makes it possible for modules to be connected in four different ways through the same connector pair. This information is represented in the connector status as the values $\{0..3\}$.

Fig. 3. A Type-1 structure climbing a step. The modules use their obstacle detector to find and navigate over the step.

3 Two Control Approaches for a Regular Structure

3.1 Linear flow motion of a Type-1 structure

The Type-1 structure is based on a parallel linear structure with two additional modules, called converters[16]. Without converters, this structure can move straight, up or down along the terrain, as shown in figure 3. We have implemented two instances of this behaviour. In the first instance, modules are controlled in a distributed manner using conventional programming. In the second instance, modules are controlled using simple local rules. We call these two instances a *program-based* and a *rule-based* approach. In both cases, we provide one bit of information from the global coordinate system that represents the direction of flow.

Two different strategies for cluster flow was implemented. In one, a pair of modules travels from the tail to the head, while in the other a void space travels from the head to the tail. In both cases, modules in one side remain passive and connected while nearby modules in the other side are moving, to preserve connectivity.

3.2 The program-based approach

In this approach, two modules at the tail walk as a pair along the line of modules, until they reach the head, using a simulator that can simulate concurrent processes [4]. For each step of walking, the next walking motion and the position of the next foot base are determined based on the shape of modules' chain within a distance of one to three modules, as illustrated in figure 4. Making a foot base is easy in the middle of a straight part, but requires coordination with neighbours near a corner (fig. 4 b & c). By this method, the structure makes a flow motion that can deal with flat terrain and steps larger than a single unit size. The method contains 18 rules, 12 walking strides and $\approx$ 11 bits of memory used in each module.

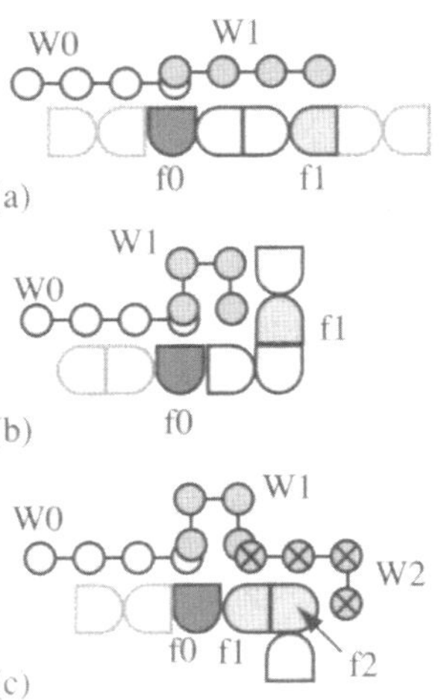


```
init(11,0)     //initialize all modules
rollmotion(11,0) //initiate walker
makeFootBase(11,0) //initiate foot base maker
end

proc rollmotion(id,pt)
waitbf(...); mov(); //wait for foot base complete
if(...) //decide next motion
  setWalkStride()
  makeFootBase(...)
else
  ...

proc makeFootBase(id,pt,inc) //concurrent process
waitbf(id,pt,0) //wait for flag change
mov(id0,~pt,aa,bb) //move joint angle
  ...
setFlag(..) //permit another proc
  ...
```

Fig. 4. Left: Three cases of walker motion. The four chained circles (`W0`) represent a walkers' current position. In each of the three cases, the next foot base (`f1`) for the walker is determined by joint angles of the modules at the right side of the walkers' current foot base (`f0`). **Right:** The algorithm for controlling the motion. Subroutines declared by `proc fname()` run concurrently, simulating a process running in each module.

3.3 The rule-based approach

For the sake of minimalism, we experimented with a memoryless (reactive) local controller [10] for dealing with the motion complexities of the M-TRAN self-reconfigurable robot. In this control method, each module simply react to its own physical state combined with the physical state of the neighbour modules. The physical state space of a module, S ($\approx$ 18bits), and the action space of a module, A, are described in figure 2. Since a single M-TRAN module can have up to six neighbours, a memoryless local controller is a mapping $S^7 \rightarrow A$ (a module's own physical state plus the physical state of the six neighbours, mapped to an action for the module).

In the memoryless local approach, we define a *rule* as a mapping from a point in S^7 to a point in A, constructed as a two-tuple $\{precondition, action\}$ where *precondition* $\in S^7$ and *action* $\in A$. A *rule set* is a set of such rules, and each module has an identical copy of the rule set. At regular intervals, a program running on each module searches the rule set for a precondition matching the modules' physical local state, and if a matching precondition is found, executes the associated action. An example rule is shown in figure 5.

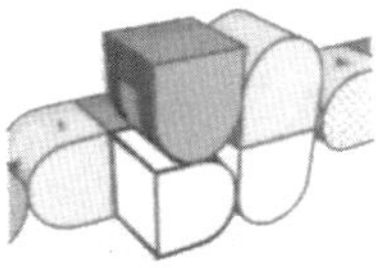

Precondition

	o_0	o_1	$c_0x_0o_2$	$c_1x_1o_3$	$c_2x_2o_4$	j_p	j_a	$c_3x_3o_5$	$c_4x_4o_6$	$c_5x_5o_7$	o_8	o_9
S_m	O	-	—	C0XO	-XO	90	0	C2XO	—	—	O	O
S_0						∅						
S_1	O	-	—	—	—	90	90	—	C0XO	—	O	-
S_2	O	-	C2XO	C0XO	—	0	0	—	C0XO	-XO	-	O
S_3	O	-	C2XO	C0XO	—	0	0	—	C0XO	-XO	-	O
S_4						∅						
S_5						∅						

Action: `setJointPos(0,0)`

Fig. 5. An example rule. The precondition is the state of the module S_m and the state of all connected modules S_0 to S_5. The rule is applicable to the white module on the drawing to the left, causing it to set new joint angular position target values for the motor controllers of this module.

To test the memoryless local control method, we implemented a cluster walk gait for a configuration of M-TRAN modules, shown in figure 3. The gait method used is that of a void space traveling backward from the head to the tail.

In the implementation, the cluster travels forward until the step is encountered. The change of physical state that occurs when the front modules' proximity sensor detects the step, causes the entire system of modules to go into a different behavioural pattern, and they start climbing upward, as shown in figure 3, middle. Again, when the top module detects the top of the step, the modules start traveling horizontally again. Similarly for climbing down. This behaviour was realized by implementing 629 rules. During motion over the step 6532 actions were performed, averaging 10 executions of each rule.

3.4 Comparison of the program-based and rule-based approach

The main differences between the two approaches are;

1. In the program-based approach, a number of internal state variables are used in the modules, while the rule-based approach relies purely on the local physical state of a module to determine the modules' action.
2. In the rule-based approach, modules cannot synchronize their actions, while the program-based approach can maintain local synchronization based on message exchange among neighbours. This provides modules with the ability to synchronize actuation with its connected neighbours.

Each approach has a number of advantages and disadvantages. In terms of organization, the program-based method has better potential for hierarchical control of the cluster, while the rule-based must rely on a completely flat organization. The rule-based approach has natural limitations to its performance due to the lack of memory. However, this also gives the rule-based approach some robustness with respect to rearrangement or replacement of modules at run-time, since no information will be lost when doing so.

4 Further Study of Regular Structures

4.1 Flow motion of a Type-2 structure

In [17] Yoshida proposed a Type-2 structure and developed a centralized method for generating sequences of its flow motion. A Type-2 structure is a series of four-module blocks connected serially and two additional modules called converters, shown in figure 6 a). It can make flow motions including a straight motion and orthogonal turns [18]. An experiment was also made for a two block structure [9]. Two studies have considered decentralized algorithms to solve a simplified subset of this problem [6, 1]. A decentralized algorithm for the full problem needs a number of subroutines for elemental motion sequences, of which only some have been made. All the above described flow motions are of the forward moving walker type, where four modules moves from the tail along the structure to reach the head. Currently, no method for a backward moving void space type flow motion has been found.

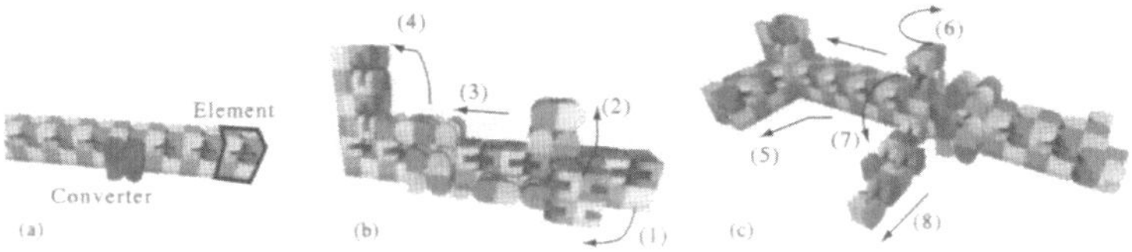

Fig. 6. a) & b) A Type-2 structure. **c)** A Type-2 structure sprouting a Type-1 structure.

4.2 Flow motion of a Type-3 structure

A Type-3 structure is made of numbers of four-module elements, as shown in figure 7. An element can be either a cross shape or a square, both of which are made of two sub elements. As we showed in [5], a cross element can walk by itself after separation from the structure.

The flow motion is made by 1) transferring a sub element upward on the flat structure, 2) sliding motion of the sub element across the structure with other motions to make foot bases, 3) transferring it downward after reaching the target position or the edge of the structure, illustrated in figure 7 (b). Two successive such sequences result in one element's transportation.

Both a void-type flow method and a walker-type flow method is possible. The process 1) includes five step motions, the process 2) needs two foot-base constructions and two step sliding motions to reach the same position of a neighbour element, and the process 3) is a reverse process of 1).

This type-3 structure cluster walk has been implemented using both the program-based approach and the rule-based approach, shown in figure 8. The rule-based implementation is using ≈ 150 rules for moving forwards, and an additional ≈ 150 rules for turning. Details of the implementation are omitted due to limited paper length. With an appropriate mechanism to suppress activation of neighbour modules' motions within the distance of two elements, this process can work spatially in parallel.

Since the above processes are primitives for the flow motion of the Type-3 structure, and since the structure is two dimensional different from the other type structures above, we need an upper level process of decision-making to determine a direction of the flow.

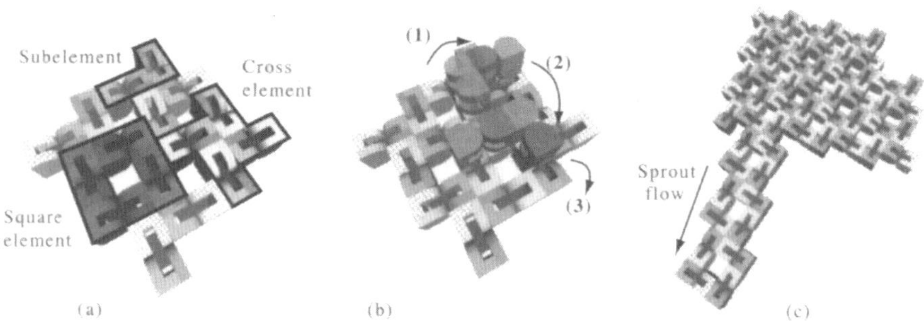

Fig. 7. The Type-3 Structure and suggested sub elements.

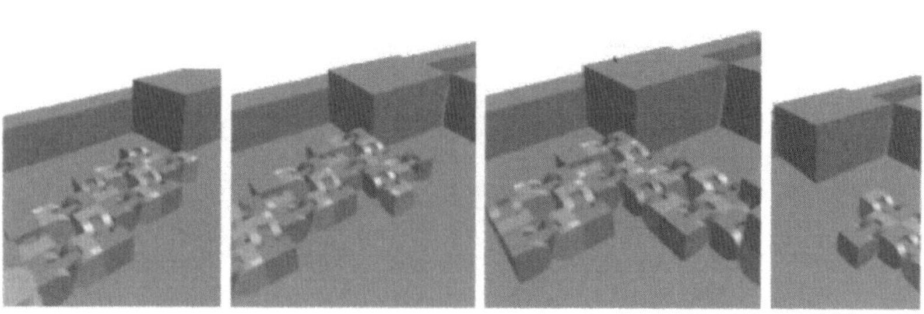

Fig. 8. A Type-3 structure turning to avoid an obstacle encountered during cluster flow. The obstacle activates a modules proximity sensor, causing a change in the flow of the structure. This is achieved by an appropriately designed rule set.

4.3 Sprouting

In [5], the separation of substructures from three types of structures was examined. Without separation, these structures can form substructures by using the above primitive processes, similar to sprouting of living organisms.

A Type-2 structure can grow sprouts in the form of a Type-1 structure (see figure 6, (c)). Primitive motions (1)-(3) are the same ones for the Type-2 flow motion, the motion (8) is the same as the Type-1 flow motion, and the motions (6) and (7) are necessary for sprouting. Figure 7 (c) shows sprouting of a Type-3 structure. For this, a mechanism to initiate and direct a flow is necessary, such as a message or a hormone [7].

5 Conclusion

In this study, we have examined a number of approaches for dealing with the complexity of controlling a self-reconfigurable robot, the M-TRAN, at the individual actuator level. We have found that regularity in the structure is a significant advantage when dealing with the constraints of module motion capabilities, not just in the form of meta modules, and that we, by exploiting these regularities, can construct both distributed controllers that have simple internal state, as well as memoryless local and asynchronous controllers. However, we also found that the memoryless local controllers bring an increased complexity in rule design. Where the program-based controllers have rules counted in tens, the rule-based controllers have rules counted in hundreds. Among the flow methods for the three types of structures, Type-3 seems to be the most versatile and also requires a simpler control process than the Type-2 structure. Future work will be on exploring upper level control methods for the Type-3 structure.

Acknowledgment

This work was partly funded by EU's Information Society Technologies Programme, IST-2001-33060, HYDRA, Future and Emerging Technologies.

References

1. Z. Butler, K. Kotay, D. Rus, and K. Tomita. Generic decentralized control for a class of self-reconfigurable robots. In *Proceedings, IEEE Int. Conf. on Robotics and Automation (ICRA'02)*, volume 1, pages 809–815, Washington, DC, USA, 2002.
2. Z. Butler and D. Rus. Distributed locomotion algorithms for self-reconfigurable robots operating on rough terrain. In *Proceedings of IEEE International Symposium on Computational Intelligence in Robotics and Automation (CIRA)*, pages 880–885, 2003.
3. D. J. Christensen, E. H. Ostergaard, and H. H. Lund. Metamodule control for the ATRON self-reconfigurable robotic system. In *Proceedings of the The 8th Conference on Intelligent Autonomous Systems (IAS-8)*, pages 685–692, Amsterdam, 2004.

4. H. Kurokawa, A. Kamimura, E. Yoshida, K. Tomita, S. Kokaji, and S. Murata. M-TRAN II: Metamorphosis from a four-legged walker to a caterpillar. In *Proceedings of IEEE/RSJ International Conference on Intelligent Robots and Systems*, pages 2454–2459, 2003.
5. H. Kurokawa, E. Yoshida, K. Tomita, A. Kamimura, S. Murata, and S. Kokaji. Deformable multi M-TRAN structure works as walker generator. In *Proceedings of the The 8th Conference on Intelligent Autonomous Systems (IAS-8)*, pages 746–753, Amsterdam, 2004.
6. H. H. Lund, R. L. Larsen, and E. H. Østergaard. Distributed control in self-reconfigurable robots. In *Proceedings of ICES, The 5th International Conference on Evolvable Systems: From Biology to Hardware*, Trondheim, Norway, March 2003. Springer-Verlag.
7. H. Meinhardt. A model for pattern formation of hypostome, tentacles, and foot in hydra: how to form structures close to each other, how to form them at a distance. *Developmental Biology*, 157(2):321–333, June 1993.
8. S. Murata, H. Kurokawa, and S. Kokaji. Self-assembling machine. In *Proceedings, IEEE Int. Conf. on Robotics & Automation (ICRA'94)*, pages 441–448, San Diego, California, USA, 1994.
9. S. Murata, E. Yoshida, A. Kamimura, H. Kurokawa, K. Tomita, and S. Kokaji. M-TRAN: Self-reconfigurable modular robotic system. *IEEE/ASME Transactions on Mechatronics*, 7(4):431–441, 2002.
10. E. H. Ostergaard and H. H. Lund. Distributed cluster walk for the ATRON self-reconfigurable robot. In *Proceedings of the The 8th Conference on Intelligent Autonomous Systems (IAS-8)*, pages 291–298, Mar. 2004.
11. K.C. Prevas, C. Unsal, M.O. Efe, and P.K. Khosla. A hierarchical motion planning strategy for a uniform self-reconfigurable modular robotic system. In *Proceedings, IEEE International Conference on Robotics and Automation (ICRA'02)*, volume 1, pages 787–792, Washington, DC, USA, 2002.
12. D. Rus and M. Vona. Crystalline robots: Self-reconfiguration with compressible unit modules. *Autonomous Robots*, 10(1):107–124, 2001.
13. W.-M. Shen, B. Salemi, and P. Will. Hormone-inspired adaptive communication and distributed control for conro self-reconfigurable robots. *IEEE Transactions on Robotics and Automation*, 18(5):700–712, Oct. 2002.
14. K. Stoy. Controlling self-reconfiguration using cellular automata and gradients. In *Proceedings of the The 8th Conference on Intelligent Autonomous Systems (IAS-8)*, pages 693–702, Amsterdam, 2004.
15. K. Tomita, S. Murata, H. Kurokawa, E. Yoshida, and S. Kokaji. Self-assembly and self-repair method for a distributed mechanical system. *IEEE Transactions on Robotics and Automation*, 15(6):1035–1045, 1999.
16. K. Tomita, S. Murata, E. Yoshida, H. Kurokawa, A. Kamimura, and S. Kokaji. Development of a self-reconfigurable modular robotic system. In *Sensor Fusion and Decentralized Control in Robotic Systems III, Proceedings of SPIE, Vol. 4196*, pages 469–476, 2000.
17. E. Yoshida, S. Murata, A. Kamimura, K. Tomita, H. Kurokawa, and S. Kokaji. A motion planning method for a self-reconfigurable modular robot. In *Proceedings, IEEE/RSJ International Conference on Intelligent Robots and Systems (IROS'01)*, pages 590–597, Maui, Hawaii, USA, 2001.
18. E. Yoshida, S. Murata, A. Kamimura, K. Tomita, H. Kurokawa, and S. Kokaji. A self-reconfigurable modular robot : Reconfiguration planning and experiments. *The International Journal of Robotics Research*, 21(10):903–916, 2002.

Part V

Task Allocation — Multi-Robot Cooperation

Multi-Robot Task Allocation Method for Heterogeneous Tasks with Priorities

José Guerrero and Gabriel Oliver

Mathematics and Computer Science Department,
Universitat de les Illes Balears (UIB), Cra. de Valldemossa, km 7.5,
07122, Palma de Mallorca (SPAIN)
{jose.guerrero, goliver}@uib.es

Summary. Task allocation is a complex and open problem for multi-robot systems and very especially if a priority is associated to each task. In this paper, we present a method to allocate tasks with priorities in a team of heterogeneous robots. The system is partially inspired on auction and thresholds-based methods and tries to determine the optimum number of robots that are needed to solve specific tasks taking into account their priorities and characteristics. Thus, we can minimize the interference effect between robots and increase the system performance. The method has been extensively tested for a modification of the well-known foraging task, using different kinds of robots. Experimental results are presented to show the benefits of the proposed method.

1 Introduction

Multi-robot systems can provide several advantages over single-robot systems: robustness, flexibility and efficiency among others. To benefit from these potential aspects the robots must cooperate to carry out a common mission. It is well known that several problems have to be solved to achieve this aim. Among these problems, we focused on the task allocation aspects, that is, selecting the best robot or robots to carry out a task. As it has been demonstrated in different studies [1, 12], the number of robots has an important impact on the system performance, among other factors, due to the interference effect. Interference is the result of competition for the shared resources, especially the physical space. That is, two or more robots need to reach the same point at the same time. Besides, if a certain task captures the attention of an excessive number of robots, other tasks can be forsaken. This effect can increase if a priority is associated to each task. Therefore, a good task allocation mechanism must decide on the 'optimal' number of robots needed to carry out each task.

In this paper, we extend our decentralized method of task allocation for groups of heterogeneous robots. We mainly focus on deciding the optimal number of robots to execute each task when priority is associated to each one. Our method is inspired in both swarm systems, and, very especially auction-like methods. In most cases multi-robot researchers studied the interference effect only using methods based on the swarm intelligence paradigm. In these systems each robot decides the task to execute using only its own information. As will be shown later, the pure swarm based systems has some limitation. Our task allocation method mitigates some of these problems.

To test our system we use a foraging like task, where the robots must find a set of objects and carry them to a delivery point. A priority and a weight are associated to each object and each robot has a load capacity. Unlike the classical foraging task, multiple robots can cooperate to transport the same object. In this case we have to decide how many robots and which ones do we need to transport each object according to its priority and weight. This is a new task that has not been tested from the interference point of view.

The performance of a task allocation mechanism is closely connected to the diversity level of the team of robots, as will be shown later. To measure the heterogeneity of the group collectivity we use, among others, the social entropy proposed by T. Blach [3]. We will also study the relation between this metric and some of our architecture parameters. Finally, two different strategies or variations of our method are tested: 'preemption' and 'no preemption'. The preemption ability is the capacity of changing the task assigned to a robot. As will be exposed, among other advantages, if we use a 'preemption' strategy, when a robot finds a high priority task it can ask for help other robots with a lower priority task.

The rest of this paper is organized as follows: section 2 presents some relevant work in the field of multi-robot task allocation; section 3 describes our methods and their implementation; section 4 shows the experiments carried out to validate the different approaches; finally, section 5 exposes some conclusions and future work is stated.

2 Related work

The computer engineering community has done a lot of research to solve the task allocation problem. In recent years, some studies on multi-robot systems have used some similar ideas to solve the problem of how robot teams can distribute their individual work capacity to efficiently achieve a common task. This section shortly relates some of those researches that have inspired us.

Dias and Stenz [7] have proposed cooperation mechanisms based on explicit coordination between robots in the so-called market-based mechanisms. In the same line, Gerkey and Matari'c are working on auction-based mechanisms [8]. In this kind of systems, the robots act as self-interest agents and they bid for tasks. The robot with the highest bid wins the auction process and gets the task. The bids are adjusted to the robots' interest (capacity) to carry out the goal. Thus, the best robot for a specific task can be chosen, but they need communication mechanisms between robots.

Other papers proposed swarm intelligence inspired solutions. To implement these systems some authors make use of the response thresholds systems [4, 5, 11]. In these systems, each robot has a stimuli associated with each task to execute. When the level of the stimuli exceeds a threshold, the robot starts its execution. The pure threshold-based systems don't require any kind of communication mechanisms.

Nonetheless, a disadvantage of these systems is the absence of knowledge about the other robots. Thus, a robot can decide by itself to execute a task when other option could be better.

Many authors [1, 12] try to study and solve the interference problem using swarm methods. For example, K. Lerman [12] studies mathematically the interference effect using this kind of systems. This work shows how system performance decreases as number of robots is incremented. Few systems have studied auction methods to solve the interference problem. These systems use the 'classical' foraging mission, where a single robot is assigned to each object and the set of robots and tasks are to be supposed homogeneous and without priorities. Other authors [6] use an auction like system, similar to our method, but the number of robots assigned to each task is predefined.

Our approach is partially inspired in the auction mechanisms and, consequently the best for a specific task can be chosen. However, while previous work cannot determine the optimal number of robots to execute a task, our method allows deciding this number as a function of the amount of work required to complete the task, the priority of this task and the work capacity of the robots involved in the auction process.

3 Working group leading, formation and updating

The coordination mechanisms description, including the groups' formation, the membership policy, the task assignment and the preemption strategies is briefly described in the following paragraphs. In this paper we focus on the issues related to the priorities. A more detailed description of our method without priorities can be found in [9].

3.1 Single robot to leader negotiation

This section describes the initial negotiation that corresponds to a non preemption strategy, and therefore it doesn't allow the exchange of robots (work capacity) between tasks.

In an initial stage, each robot is looking for a task. When a robot finds a new task by itself, it will try to lead it. As there can only be one leader per task, the candidate will first check if this task is assigned to any other robot or not. If there are two or more robots requesting the leadership of the same task, it will be assigned to the 'best' robot. When a robot is promoted to leader of a task, it evaluates the work needed to carry it out. Then, it will create, if necessary, the work group; that is, the set of robots that will cooperate to execute this specific task. In that case, the leader must decide which the optimum group size is. We propose the leader to decide using the following equation:

$$TH_g = \frac{priority * taskWorkLoad}{\sum_{1 \leq i \leq N} workCapacity_i} < TH \qquad (1)$$

Where N is the number of robots of the group and $workCapacity_i$ is the individual work capacity of the ith robot. TH is the group threshold; this value is a parameter that will be used to compare the efficiency of the group formation policy.

taskWorkLoad is the amount of work required to finish the assigned task that is calculated by the leader. Finally, *priority* is the priority of the task. A high value of priority represents a task with a high priority. Using equation (1) the leader fixes the maximum ratio between work to do and available work capacity and, therefore, it fixes the maximum number of robots that will be part of the group. As it can be seen, a task with a high priority will require more robots and, therefore, more work capacity than other with lower priority.

To select the robots that will be part of its group the leader uses an auction process. Unlike other auction-based methods, we use the inequality (1) to select the robots. As it is described bellow, initially the leader informs to other robots that it needs to form a work group. Then, each robot without any assigned task sends to the leader its work capacity. Finally, the leader selects the robots with the highest work capacity provided inequality (1) is verified. If, after this process, the equation (1) is not fulfilled, the leader starts a new auction round. This new auction round will include a leader to leader negotiation as will be explained in section 3.2. When the task is finished, the working group is dissolved immediately.

Using only this kind of negotiation, the size of the group can vary during the execution of the task by means of two mechanisms. Group size can be reduced by a robot segregation process. During this process a non-leader robot finds a new task and leaves its group. On the other hand, the size of the group can increase thanks to a robot aggregation process. Using this process a new member can be accepted in a group if, after a certain time has passed, this robot hasn't any task to execute. In that way, inactive robots are avoided. In the next section an additional mechanism to modify the group size will be described thanks to the leader to leader negotiation.

3.2 Leader to leader negotiation: Preemption

The leader to leader negotiation allows the exchange of robots between groups. If, after the single robot to leader negotiation, the equation (1) is not fulfilled, the group's leader tries to contract the robots which are working in tasks with equal or lower priority. To select robots from another group, each leader bids for the task using both its robots load capacity (as in the previous negotiation) and its working group energy. Using the operating systems vocabulary, this kind of algorithm will be called strategies with preemption.

The working group energy is a measure to indicate the group's tendency to send its robots to other groups. A group with a high energy value is a potential sender of robots to other tasks and, on the other hand, a low energy value indicates that this group is a receptor of new robots. The following equation is one of the simplest to modelize the described behavior for the energy:

$$E_g = \frac{GroupSize}{TH_g} \tag{2}$$

Where *GroupSize* is the number of group robots and TH_g is obtained from equation (1). A high value of TH_g indicates that the group has a low value of work capacity compared to the task to be carried out. In this case the group needs more robots and, therefore, the energy is low. Moreover, the leader has to try to create a group with the minimum number of robots to reduce the interference effect between robots. Thus, a group with a lot of robots needs to reduce its number, and this is expressed by a high energy value. The energy concept is similar to the stimulus intensity value used by some threshold based algorithms [4].

To select robots from other groups the leader uses another auction process. The robots from other groups bid during the auction process not only using its work capacity, like in the single robot to leader negotiation, but using the following value:

$$B = workCapacity * E'_{g2} \tag{3}$$

Where *workCapacity* is the individual work capacity of the robot and E'_{g2} is the energy of the robot's group if this robot is selected. The robot with a higher bid, B, is selected.

The goal of this selection is to create a more stable system. By the way, a system where the groups have low energy will be more stable than a system made up of groups with high energy. A more appropriate definition of system stability is now under study. To get this objective, the leader only selects a robot if reassigning it to the new group significantly reduces the maximum value of the global energy; that is, if the following condition is verified:

$$MAX(E_{g1}, E_{g2}) > O * MAX(E'_{g1}, E'_{g2}) \tag{4}$$

Where $MAX(v1, v2)$ returns the maximum value of $v1$ and $v2$. E_{g1} is the energy of the acceptant group and E_{g2} is the energy of the requesting group before the transaction. E'_{g1} and E'_{g2} are new energies of the groups in the case that the transaction would take place. Finally, O is the percentual overhead produced by the group change. This factor avoids the robot interchange when the benefit obtained by the group is very low. Alternatively, the overhead factor O can be seen as a cost function to leave a task. That can be related, for example, to the complexity of the environment. Thus, in a simple or diaphanous environment a lower value of O should be used compared to a complex or a crowded one. During the experiments the overhead is equal to 50% and, therefore, the O value is equal to 1,5.

Having the above described rules in mind, the auction process finishes when the equation (1) is fulfilled or when no more robots validate the equation (4). If after this process the equation (1) is not fulfilled another auction process is started.

4 Experiments and validation

This section explains the experiments carried out to validate our approach and how the value of the group threshold, the social entropy of the group and the preemption strategy affect to the mission. We also evaluate the suitability of our system to carry out tasks with priorities.

4.1 Measure of the homogeneity of the collectivity

It has to be emphasized that each robot can have a different work capacity. Thus, we have focused part of our study on the measure of the performance of the collective while its homogeneity is changed. To characterize the homogeneity of the collective of robots we use two entropy measures. On the one hand, we use the simple entropy measure based on the Shannon's information entropy. On the other hand, we use the hierarchic social entropy as formulated by Balch [3]. This measure extends simple entropy to take into account the quantitative differences between groups. In our experiments, these differences are the work capacity of each robot; therefore, two robots belong to the same group if they have the same work capacity.

4.2 Platform and task description

We use as test bed a multi-robot simulator called RoboCoT (Robot Colonies Tool). RoboCoT is a software tool developed and used by the authors at the University of the Balearic Islands that allows testing the performance of individual or colonies of mobile robots. This simulator implements robots that work according to a control architecture based on behaviours, as they were introduced by Ronald C. Arkin [2]. A detailed discussion about the RoboCoT architecture can be found in [10].

The task to be carried out by the robots is described as follows: some randomly placed robots must locate objects, randomly placed too, and carry them to a common delivery point. To maintain the initial conditions, when an object is transported to the delivery point immediately appears, randomly placed, another one, with identical characteristics. Figure 1 shows a typical situation, where the squares represent the objects to collect, the delivery point is the big circle in the middle of the image and the robots are the little circles. Each object to gather has a weight and each robot has a load capacity. The robot load capacity is the amount of weight that it can carry. Thus, if a robot cannot carry the entire object at once, it takes a part of it, goes to the delivery point and comes back to the object for more bits. Moreover, a priority is associated to each task. This priority is an integer value between 1 and 5, where 1 is the lowest priority and 5 is the highest. This priority value is the priority value, the $taskWorkLoad$ value is the object weight and the $workCapacity$ of a robot is its load capacity.

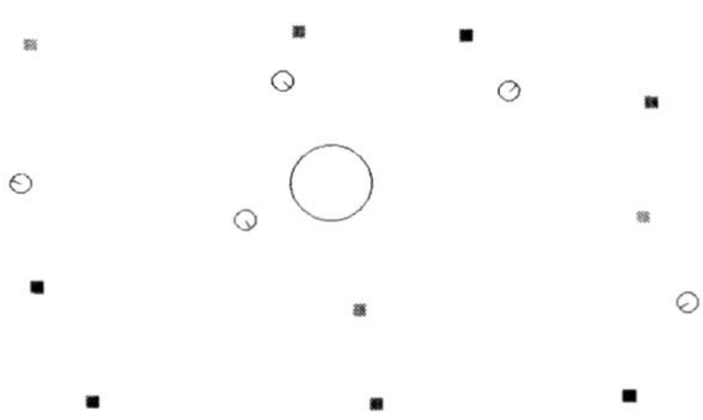

Fig. 1. Example of initial situation of the experiments

In all the experiments presented here, we have used ten objects to load and five robots. The weight of the objects is 30 units in all cases. During the experiments with priorities, we use two object of each priority, that is, two object with priority 1, two objects with priority 2, etc. All the robots have the same sensorial, behavioral and communication capabilities, and they only differ in their load capacity. To study the impact of the robots' homogeneity we have used four different configurations or combinations for the load capabilities of the robots, as it can be seen in table 1. For each configuration, we have used as values for the group threshold (TH): 0, 2, 4, 6 and 8. In the case TH=0, equation (1) has not been used, and therefore the number

Table 1. Robots' load capacities used during the experiments. R1..R5 represent the robots' load capacities. The H represents the simple entropy of the configuration and SH the social entropy.

Configuration	R1	R2	R3	R4	R5	H	SH
1 (Homogeneous)	3	3	3	3	3	0	0
2	1	1	1	1	11	0,72	7,22
3	1	1	3	5	5	1,52	4,49
4	1	2	3	4	5	2,32	2,32

of robots per group is not limited. The robots carry out the mission during 35000 time units. After this period, we get the total weight transported, the average time required to finish each task, etc.

4.3 Results without priorities.

During the first set of experiments we evaluate our system using a set of homogeneous objects. These objects have no priority associated. Thus, we can study the impact of the threshold value on the system.

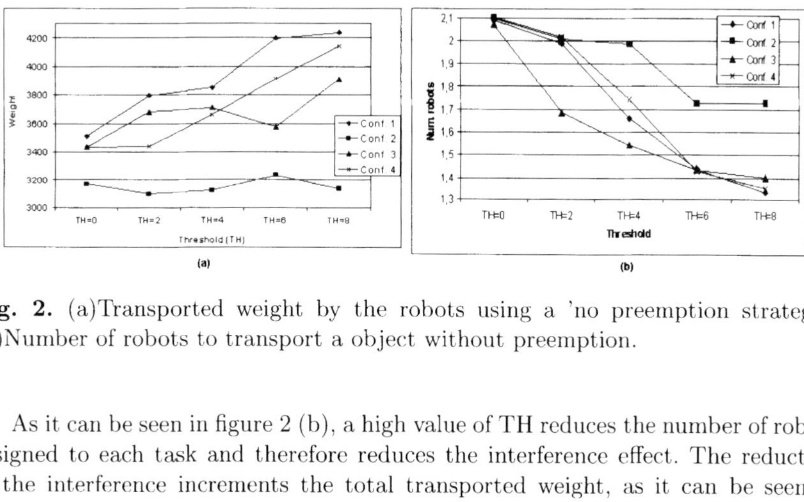

Fig. 2. (a)Transported weight by the robots using a 'no preemption strategy'. (b)Number of robots to transport a object without preemption.

As it can be seen in figure 2 (b), a high value of TH reduces the number of robots assigned to each task and therefore reduces the interference effect. The reduction of the interference increments the total transported weight, as it can be seen in figure 2 (a). It can also be seen that, in most cases, the configuration with less SH value presents the best results and the configuration with the highest SH is the worst. Moreover, concerning the weight transported, the more inhomogeneous configuration seems not be significantly affected by the variation of the TH value . This is due to a better distribution of robots efforts between tasks when the configuration is more homogeneous (SH low).

Figure 3 (a) shows the ratio between the transported weight with and without pre-emption. A ratio value grater than one means that the preemption solution is better than the non preemption strategy. As it can be seen, the results with and

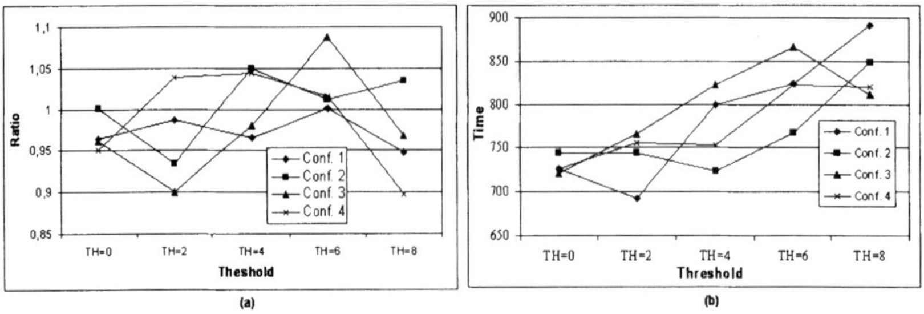

Fig. 3. (a)Ratio between transported weight with preemption/weight without preemption. (b)Average time required to fully transport an object without preemption.

without preemption are similar, in all cases the difference is lower than 10%. However, in general terms, when TH is very low or very high this benefit decreases. When the threshold is low the number of robots assigned to each task can be excessive, and therefore the system needs to interchange a lot of robots between groups. In this case, the time needed to stabilize the system can be greater than the time required to finish the task. When the threshold is very high, the number is too low to allow the interchange of robots between groups.

Finally, 3 (b) shows the average time required by the set of robots to gather single objects without pre-emption. This time is computed from the moment a robot finds an object to the time the object is fully transported. As it can be seen, the average time is increased as the threshold increases, because when the number of robots is lower also decreases the amount of load capacity assigned to a task.

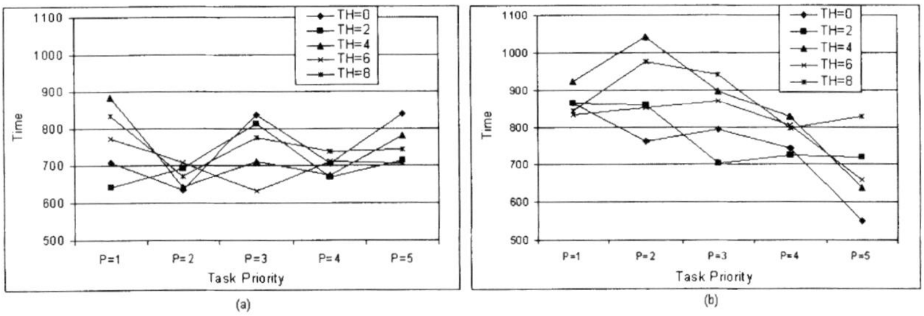

Fig. 4. Average time required to fully transport an object with priorities and configuration 3. (a)Results without preemption. (b)Results with preemption.

4.4 Results of tasks with priorities.

We evaluate here our method when a priority is associated to each task, as described in section 4.2. Figure 4 (a) shows the average time required to completely gather an object as a function of its priority for different values of threshold without preemption when we used configuration 3. As it can be seen, there is no correlation between priority and execution time, because this time strongly depends on the placement of

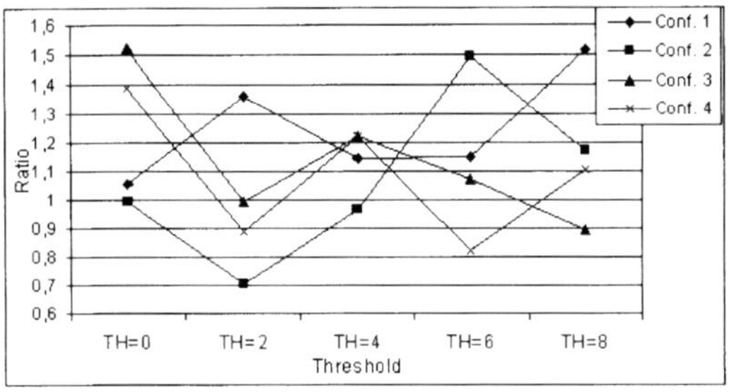

Fig. 5. Ratio between time to finish a task with priority 5 without preemption/time to finish with preemption.

the robots when the task is found. If the same system uses preemption, the robots can be transferred from a lower priority task to high priority one and thus, high priority tasks can be carried out faster, as it is a shown in figure 4 (b). Finally, figure 5, show the ratio between time required to finish a task with the highest priority (priority=5) without and with preemption. As it can be seen, in most cases this ratio is higher than 1, that is, the preemption strategy is better than non preemption system. This benefit increases as the SH value of the configuration decreases. For example, configuration 1 (SH=0) presents a ratio greater than 1 in all cases. On the other hand, the configuration with the highest SH (configuration 2) only presents a benefit when TH=6 or TH=8.

5 Conclusion and future work

This paper presents a simple and efficient way to solve the task allocation problem and, more specifically, to decide how many robots are needed to execute a specific task. Our algorithm adapts the classical auction process using some threshold-based systems concepts to find the optimal number of robots when a priority is associated to each task. In addition, our method allows both changing the robots assigned to a task as new objectives are found and interchanging robots between working groups. Thus, we have provided a faster and flexible way to regulate the optimal number of robots as a function of the kind of task, the priority of the task and the available robots. The execution results of the foraging task prove that our mechanism increments the amount of objects transported during a foraging-like mission, very specially if a non preemption strategy is used. On the other hand, these results show that our system can reduce the average time required to transport the objects with a high priority, only if the preemption strategy is used.

The work presented is in progress and has some challenging aspects to add and to improve. For the time being we are focused on a deep analysis of the data available, obtained from a huge set of experiments that should take us to a precise understanding of the relations between the parameters of our architecture and the heterogeneity level of the robots. We are also working on developing a better definition of system stability, including task constraints. Finally, another aspect of the systems that should be improved is the use of a non fixed threshold like [5].

Acknowledgement. This work has been partially supported by project CICYT-DPI2001-2311-C03-02 and FEDER fundings.

References

1. Agassounon W., Martinoli A. (2002). Efficiency and Robustness of Threshold-Based Distributed Allocation Algorithms in Multi-Agent Systems. 1st Int. Joint Conf. on Autonomous Agents and Multi-Agents Systems, Bologna (Italy).
2. Arkin R.C. (1998), Behaviour-Based Robotics, The MIT Press.
3. Balch T. (1997) Social Entropy: A New Metric for Learning Multi-Robot Teams, 10th international Florida Artificial Intelligent Research Society Conference, CA:AAAI Press, Daytona Beach, FL (USA)
4. Bonabeau E., Sobkowski A., Theraulaz G. and Deneubourg J. L. (1997) Adaptive Task Allocation Inspired by a Model of Division of Labour in Social Insects, Bio Computation and Emergent Computing.
5. Campos M., Bonabeau E. and Thraulaz G., Deneubourg J. L. (2001) Dynamic Scheduling and Division of Labour in Social Insects, Adaptive Behaviour Vol. 8-2, Singapore.
6. Chamowicz L., Campos M., Kumar C. (2002) Dynamic Role Assignment for Cooperative Robots. 2002 IEEE International Conference on Robotics and Automation, Washington - DC (USA).
7. Dias M.B. and Stentz A. (2000) A Free Market Architecture for Distributed Control of a Multirobot System, 6th International Conference on Intelligent Autonomous Systems, Venice (Italy).
8. Gerkey B. P. and Mataric M. J. (2002) Sold!: Auction methods for multi-robot coordination, IEEE Transactions on Robotics and Automation, Special Issue on Multi-robot Systems, Vol. 18 No. 5
9. Guerrero J. and Oliver G. (2003) Multi-robot Task Allocation Strategies Using Auction-Like Mechanisms, Artificial Intelligence Research and Development in Frontiers in Artificial Intelligence and Application, Vol. 100, IOS Press
10. Guerrero J., Oliver G. and Ortiz A. (2001) On Simulating Behaviour-based Robot Colonies in the Classroom, First EURON Workshop on Robotics Education and Training, Weingarden (Germany)
11. Krieger M. J., Billeter J. G. and Keller L. (2000) Ant-like task allocation and recruitment in cooperative robots, Nature 406.
12. Lerman K., Galstyan A. (2002) Mathematical Model of Foraging in a Group of Robots: Effect of Interference, Autonomous Robots 13 (2)

Decentralized Markov Decision Processes for Handling Temporal and Resource constraints in a Multiple Robot System

Aurélie Beynier[1], Abdel-Illah Mouaddib[1]

GREYC-CNRS, Bd Marechal Juin, Campus II, BP5186, 14032 Caen Cedex, France abeynier,mouaddib@info.unicaen.fr

Summary. We consider in this paper a multi-robot planning system where robots realize a common mission with the following characteristics : the mission is an acyclic graph of tasks with dependencies and temporal window validity. Tasks are distributed among robots which have uncertain durations and resource consumptions to achieve tasks. This class of problems can be solved by using decision-theoretic planning techniques that are able to handle local temporal constraints and dependencies between robots allowing them to synchronize their processing. A specific decision model and a value function allow robots to coordinate their actions at runtime to maximize the overall value of the mission realization. For that, we design in this paper a cooperative multi-robot planning system using distributed Markov Decision Processes (MDPs) without communicating. Robots take uncertainty on temporal intervals and dependencies into consideration and use a distributed value function to coordinate the actions of robots.

1.1 Introduction

Although a substantial progress with formal models for decision process of individual robots using Markov Decision Process (MDP), extensions of MDP to multiple robots is lacking. Recent attempts identify different classes of multi-robot decision process that include Multi-Agent Markov Decision Process (MMDP) proposed by Boutilier [Bou99], the Partial Observable Identical Payoff Stochastic Game (POIPSG) proposed by Peshkin et al. [PKMK00], the multi-agent decision process by Xuan and Lesser [XLZ00], the Communicative Multiagent Team Decision Problem (COM-MTDP) proposed by Pynadath and Tambe [NPY$^+$03], the Decentralized Markov Decision Process (DEC-POMDP and DEC-MDP) by Bernstein et al. [BZI00], DEC-MDP with a central coordination proposed by Hanna and Mouaddib [HM02], the DEC-POMDP with communication proposed by Goldman and Zilberstein [GZ03] and Transition Independent DEC-MDP proposed by Becker et al. [BZLG03]. Bererton et al. [BGT03] present an extension of MDPs and apply auction mechanisms to coordinate multiple robots. Therefore, they reduce communication. Nevertheless, agents communicate during the executing.

Our approach is a specific structured DEC-MDPs to control multiple robots realizing a common mission with temporal and dependency constraints. Indeed, the formal model we use can be seen as an extension of the structured DEC-MDP, proposed by Becker et al. [BZLG03], to increase the expressiveness of the model to handle temporal constraints. The formal model we develop is based on an augmented MDP per robot using an augmented reward function that represents the reward a robot gains when it achieves a task and the opportunity cost of violating temporal constraints. As described in [NPY+03], our system builds the policy in a centralized way but the execution is decentralized. Nonetheless, in our approach, the robots don't have to communicate during their execution. This model is motivated by a more general scenario than the one introduced in [CMZW01] to control the operations of a rover. The extended scenario, considered, is a multiple robot scenario in which each robot has a mission to achieve (a set of tasks) similar to the planetary rovers [BDM+02] where each one has to visit sites and to take pictures, conduct experiments and collect data (see the example in [CMZW01]).

These scenarios are mainly characterized by

1. Single rover activities have an associated temporal window : the examples involve measurements of the environment – a "gravity wave" experiment that needs to be done "preferably in the morning", and atmospheric measurements at sunrise, sunset, and noon (look at the sun through the atmosphere).
2. There is uncertainty on time realization tasks and the temporal interval activity of rovers.
3. The temporal constraints are, in general, soft.
4. Precedence dependencies among tasks exist.
5. Interacting tasks could not be assigned to the same rover because of its limited resource : one popular scenario is preparing a site with a set of robotic bulldozers (pushing dirt and small stones). In this paper, we don't take into account tasks achieved by more than one robot.

We present in the next sections the multi-robot decision process system and how the decision process is distributed and coordinated among rovers.

1.2 A Formal Description of Decentralized MDP

One approach to this problem is to represent the system as a large Markov Decision Process (MDP) [SB98] where the "macro-action" is a joint action of all of the robots and the reward is the total reward of all of the robots. The problem is the large action space that is for n robots and m actions for each one, we have m^n macro-actions [GDP01]. Some approaches use Factored MDPs while other approaches are designed in such a way that most of them are based on distributed robots with partial information. Such approaches are known to be intractable [BZI00]. In our approach, we view the entire multi-robot system as distributed Markov Decision Process robots without communication but some information on the effect of the local robot's actions, on the plans of the other robots are assumed to be available. Differently speaking, the system is composed on robots each of which constructs a local MDP and derives a local policy, taking into account the temporal dependencies. Indeed, in the local MDP we introduce an opportunity cost due to the current activity of the robot which delays the successor activities of the other robots. It measures the loss in value when starting the activity with a delay δt. This cost is the difference between the value when we start at time and the value when we start with a delay δt. Consequently, each robot develops a local MDP independently of the local policies of the other robots but, introducing in its expected value the opportunity cost due to the possible delay of his activity. The construction of the local MDP is based, first, on the construction of the state space, second, on the computation of the opportunity cost at each state and then, the computation of the value of each state to construct the local optimal policy. To define the state space, we need to define what is the state for the decision process. To do that, let us recall the characteristics of the problem.

The mission of the robot, as described previously, is a graph of tasks where each task is characterized by the execution time window, and the uncertainty on the execution time and resource consumption. In the rest of the paper, we assume that the mission graph is given and each task has a set of possible execution times and their probabilities, and a set of possible amounts of resource and their probabilities. It means that the representation of execution time and resource consumption are discrete. Given these information, the problem is to choose the best decision about which task to execute and when to execute it. This decision is based on the available resources and the temporal constraints. The respect of the temporal constraints requires to know the interval of time during which the current task has been executed. From that, the decision process constructs its decision given the current state of the last executed task, the remaining resources and the interval during which this task has been executed. The state of this decision process is then, $[l_i$, r, I] that corresponds to the last executed task l_i, the remaining resource r and the interval of time. Given the uncertainty on the execution time, there exist many possible intervals of time during which a task could be executed. In order to develop the state space of the decision process, we need to know for each task in the graph the set of its possible execution intervals of time. To do that, we develop an algorithm that computes for each task in the graph all the possible intervals of time by propagating different execution times.

The algorithm of developing a local MDP is divided into 5 major steps :

- Propagating the temporal constraints and computing the set of execution intervals of time for each task (node in the graph) among the graph (section 1.4).
- Computing the probability for each interval of time (section 1.5).
- Constructing the state space of the Markov Decision Process using the transition model (section 1.6).
- Computing the opportunity cost at each state (section 1.6).
- Using the value iteration algorithm to solve the MDP.

In the rest of the paper, we describe each step of the algorithm and we give the formal description of the local MDP and their interactions.

1.3 Preliminaries

In the previous section, we describe the overall basis of the model we develop and give its main lines. For that we consider a distribution probability on execution time and resource, and probabilities on start and end time of tasks.

1.3.1 Uncertain computation time

The uncertainty on execution time has been considered in several approaches developed in [CMZW01]. All those approaches ignore the uncertainty on the start time. We show in this paper how extensions can be considered in those approaches taking different temporal constraints into account.

Definition 1 *A probabilistic execution time distribution,* $Pr(\delta_c = t_c) = P_c(t_c)$ *is the probability that the activity takes* t_c *time units for its execution.*

The representation adopted of this distribution is discrete. We use a set of couples (t_c,p) where each couple means that there is a probability p that the execution will take t_c time units.

1.3.2 Uncertain resource consumption

The consumptions of resources (energy, memory, etc ...) are uncertain. We assume a probability distribution on the resource consumptions of a rover when performing an activity.

Definition 2 *A probabilistic resource consumption is a probability distribution,* $Pr(\Delta r)$ *of resource consumption measuring the probability that an activity consumes* Δr *units of resources.*

The representation adopted of this distribution is discrete. We use a set of couples (r,p) where each couple means that there is a probability p that the execution will consume r units.

1.3.3 Temporal window of Tasks

Each task is assigned a temporal window [EST,LET] during which is should be executed. EST is the earliest start time and LET is the latest end time. The temporal execution interval of the activity (start time and the end time) should be included in this interval.

1.3.4 Rewards

Each robot, i, receives a reward R presumably based on the quality of the solution and the remaining resources. For all states $[l_i, r, I]$, the reward function is assumed given $R([[l_i, r, I])$. However, we assume that all failure states have a zero reward.

1.4 Temporal interval propagation

1.4.1 A simple temporal interval propagation algorithm

Given the possible transition times of different tasks, we determine the set of temporal intervals during which a task can be realized. Firstly, the possible *start times* is a one of instants $EST, EST + 1, \ldots, LET - \min \delta_i$ (Last start time that we node LST). However, a robot needs to know when its predecessor terminates its processing to validate some of those start times. For that, we compute off-line all the possible end times of all its predecessors and compute its possible start times consequently. The possible intervals I of a robot are determined with a simple temporal propagation constraints in the graph. This propagation organizes the graph into levels such that : l_0 is the root of the graph, l_1 contains all the successors of the root (successors(root)), ..., l_i contains the successors of all nodes at level l_{i-1}. For each node in given level l, we compute all its possible intervals of time from its predecessors.

• $level_0$: the start time and the end times of the root node (the first task of the mission) are computed as follows : $st(root) = EST(root)$ and $ET(root) = \{st(root) + \delta_c^{root}, \forall \delta_c^{root}\}$ where δ_c^{root} is the execution time of the first activity (task) of the mission.

• $level_i$: for each node in level i, it starts its execution when all its predecessors end their own activities. The set of the possible end times of the node is then given thanks to the start times and the task's durations : $ET(node) = \bigcup_{\forall \delta_c^{node}, st}\{st + \delta_c^{node}\}$ where δ_c^{node} is the execution time of the activity (task) at node $node$ and $st \in ST(node)$. We recall also here that there is a probability that some end times can violate the deadline LET_{node}.

1.5 A Probability propagation algorithm

After computing off-line all the possible execution intervals of each activity (or node), we describe, in this section, how we can weight each of those intervals with a probability. This probabilistic weight allows us to know the probability that an activity can be executed during an interval of time. For that, a probability propagation algorithm among the graph of activities is described using the execution time probability and the temporal constraints EST, LST, and LET. This algorithm has to take the precedence-constraint that affects the start time of each node, and the uncertainty of execution time that affects the end time of the node. In the following, we describe how the conditional start time probability (DP) is computed and the probability of an execution interval P_w using DP and the probability of execution time P_c.

Conditional probability on start time

The computation of conditional start time has to consider the **precedence-constraint** that expresses the fact that an activity cannot start before its predecessors finish. The start time probability of an activity should express the uncertainty on the precedence-constraint dependency. This constraint expresses the fact that the activity starts its execution when all activities of its predecessors have been finished. Consequently, the probability $DP(t)$ that a robot starts an activity at t is the product of the probability that all predecessor robots terminate their activities before time t and there is, at least, one of them that finish at time t. More formally speaking :

•for the root : $DP_s(i) = 1, i \in [EST_{root}, LET_{root}]$

•for the other nodes : $DP_s(\delta_s = t) = \Pi_{a \in predecessors(c)} Pr^a(\delta_e \leq t) - \sum_{t'<t} DP(t')$

Where a is an activity of a predecessor robot of the robot performing the considered activity c and $Pr^a(\delta_e \leq t)$ is the probability that predecessor a finishes before time t. This probability is the sum of probabilities that the predecessor a executes its task in an interval I with an end time $et(I)$ less than t. More formally speaking, $Pr^a(\delta_e < t) = \sum_{t_i+t_j=t} DP_s(t_i).P_c(t_j)$ such that $[t_i, t]$ is one of intervals $interval(a)$ computed for the activity of robot a. This probability can be rewritten as follows :

$Pr^a(\delta_e < t) = \sum_{I \in intervals(a), et(I) < t} Pr(I)$

In the following, we show how we compute the probability that an execution occurs in an interval I.

Probability on a temporal interval of an activity

Given the probability on start time and end time, we can compute the probability that the execution of an activity occurs during the interval I where *st(I)* is the start time and *et(I)* is the end time.

Definition 3 *A probabilistic execution interval I is the probability $P_w(I)$ of the interval during which an activity can be executed. This probability measures the probability that an activity starts at $st(I)$ and it ends at $et(I)$.*

$$P_w(I) = DP_s(st(I)).P_c(et(I) - st(I))$$

An activity l_{i+1} of an agent is executed during an interval I' when the agent finishes its activity l_i and that all predecessor agents finish their activities. To compute the probability of the interval I', we need to add the fact that we know the end time of activity l_i. For that, we compute the probability $P_w(I'|et_{l_i}(I))$ such that :

$$P_w(I'|et_{l_i}(I)) = DP_s(st(I')|et_{l_i}(I)).P_c(et(I')_{l_{i+1}} - st(I)_{l_{i+1}})$$

And the probability $DP_s(st(I')|et_{l_i}(I))$ is computed as follows :

$$DP_s(st(I')|et_{l_i}(I)) = \prod_{a \in predecessors(l_{i+1}) - l_i} P_r^a(\delta_e \leq st(I')|et_{l_i}(I)) - \sum_{t_1 < st(I')} DP_s(t_1|et_{l_i}(I))$$

$$P_r^a(\delta_e \leq t|et(I)_{l_i}) = \sum_{I_1 \in, et(I_1) = et(I)_{l_i}, et(I_1) \leq t} P_w(I_1)$$

This equation expresses the fact that we know activity l_i has finished at $et(I)$, and allows us to consider only the probability that the other predecessor activities finish.

1.6 A Decision Model for Multi-robot Planning system with temporal dependencies

As mentioned above, we formalize this problem with Markov Decision Processes. To do that, we need to define what is the state space, the transition model and the value function for each robot.

Each rover develops its local policy using the off-line temporal interval propagation. Robots don't need to communicate since all information needed for each to make a decision are available. The consequence of representing all the intervals and all the remaining resources, is that the state space becomes fully observable and the decision process can start its maximization action selection using the modified Bellman equation defined bellow. However, the maximization action selection uses an uncertain start time that is computed from an uncertain end time of the predecessors.

The start time selected by the policy can be earlier or later than the end time of the predecessors. When the start time is later than the end time of the predecessors, we need to handle the situation where the start time is later than $LET - \min \delta_i$. The other case is when the policy selects a start time earlier than the end time of the predecessors, in such case, the action selected won't succeed since the precedence constraint is not respected. In such case, we assume that the new state is a partial failure and a penalty should be paid. For example, rover A assumes that at time t_1, rover B (bulldozer) finishes its processing and it can moves toward its destination. When rover B finishes later than t_1, the moving action of rover A fails. In the rest of this section we formalize this decision process.

State Representation, Transition model and Values

Each robot, i, observes its resource levels and the progress made in achieving its tasks which represent the state of the robot. The state is then a triplet $[l, r, I]$ where l is the last task, r is the available resource and I is the interval during which the task has been executed.

We assume that the actions of one robot is enough to achieve a task. The robot should make a decision on which task to execute and when to start its execution. However, the actions to perform consist of *Execute the next task* l_i at time st ($\mathsf{E(st)}$) that is the action to achieve task i at time st when the task $i-1$ has been executed . This action is probabilistic since the processing time of the task is uncertain. This action allows the Decision process to move from state $[l_i, r, I]$ to state $[l_{i+1}, r', I']$. When a robot starts before its predecessors terminate, this transition leads to a failure state that we recover by a state $[l_i, r - \Delta r, [st, unknown]]$ where Δr is the consumed resources. This recovery allows us to represent the situation where the robot acts with no results, except the fact that further resource has been consumed. Finally, the execution can lead to a failure state that we represent with $[failure, 0, [st, +\infty]]$ when the remaining resources are not enough to realize a task. It can also lead to another failure state when the execution starts too late ($st > LET - \min \delta_i$). We use ∞ or *unknown* in order to indicate to the policy that those states needs a special consideration by considering a special value that we explain in the next section. Let us just give a meaning to $+\infty$: $+\infty$ means that the robot is never be able to achieve the task while *unknown* means that the robot tries and fails but there is a chance to succeed another time by starting later. The transitions are formalized as follows :

- *Successful transition :* The action allows the policy to transition to a state $[l_{i+1}, r', I']$ where task l_{i+1} has been achieved during the interval I' respecting the EST and LET time of this task and that r' is the remaining resource for the rest of the plan. The expected value to move to the state $[l_{i+1}, r', I']$ is : $V1 = \sum_{\Delta_r \le r} \sum_{et(I') \le LET} P_r(\Delta_r).P_w(I'|et(I)).V([l_{i+1}, r', I']$
- *Too late start time Transition :* The action starts too late and the execution meets the deadline LET. In such case, the action allows the policy to move to a $[failure, r, [st, +\infty]]$. The expected value to move to this state is : $V2 = Pr(st > LET - min\delta_i).V([failure, r, [st, +\infty]]) =$

$$\prod_{a \in pred(l_{i+1}) - \{l_i\}} \sum_{I_a} P_w(I_a) - \prod_{a \in pred(l_{i+1}) - l_i} \sum_{I_a : et(I_a) \le LET - min\delta_i} P_w(I_a).V([failure, r, [st, +\infty]])$$

- *Deadline met Transition :* The action starts an execution at time st but the duration δ is so long that the deadline is met. This transition moves to the state $[failure, r, [st, +\infty]]$. The expected value to move to this state is : $V3 = \sum_{\Delta r \leq r} \sum_{st+t_c > LET} P_r(\Delta r).DP(st|et(I)_{l_i}).P_c(t_c).V([failure, r, [st, +\infty]])$
- *Insufficient resource Transition :* The execution action requires more resources than available. This transition moves to the state $[failure, 0, [st, +\infty]]$. The expected value to move to this state is : $V4 = \sum_{\Delta_r > r} P_r(\Delta_r).V([failure, 0, [st, +\infty]])$
- *Too early start time Transition :* The action starts too early before one of the predecessor robots has finished its tasks. In such case, the action allows the policy to move to $[failure, r, [st, st+1]]$. Conceptually, we proposed that the end time should be unknown, but in our model we formalize it by the fact that when a robot starts before its predecessor robots finish, it realizes it immediately. This means that the robot, at $st+1$, realizes that it fails. This state is a non-permanent failure because the robot can retry later. The robot should pay a penalty k visiting such states. The expected value to move to this state is :

$$V5 = \sum_{\Delta r \leq r} \Big(\big(\prod_{a \in predecessors(l_{i+1}) - \{l_i\}} \sum_{I_a : et(I_a) > LET - min\delta_i} P_w(I_a) \big)$$

$$- \sum_{s \leq st} DP(s|et(I)_{l_i}) \Big) .P_r(\Delta r).V([failure, r, [st, st+1]])$$

Given these different transitions, we adapt our former to Bellman equation as follows :

$$V[l_i, r, I] = \overbrace{R([l_i, r, I])}^{\text{immediat gain}} - \overbrace{\sum_{k \in \text{sucessors}} OC_k(et(I) - et(I_{first}))}^{\text{Opportunity Cost}}$$

$$+ \overbrace{max_{E(l_{i+1}, st), st > current\_time}(V1 + V2 + V3 + V4 + V5)}^{\text{Expected value}}$$

This equation means that robot rewards $R([l_i, r, I])$ are reduced by an opportunity cost due to the delay caused in the successor robots.

Opportunity cost is the lost in value when a robot starts with a delay. This means it is a difference between V^0 when we start with no delay and $V^{\Delta t}$ when we start with a delay Δt : $OC_k(\Delta t) = V^0 - V^{\Delta t}$ with : i) $\forall t \leq 0, OC(t) = 0$, ii) $OC(unknown) = 0$, iii) $\forall t > LET - \min_i \delta_i$

$$OC([failure(l_i), r, [+\infty, +\infty]]) = R(l_i) + \sum_{a \in AllSucc(l_i)} R(a)$$

The opportunity cost has to be computed off-line for all the delays from 0 to latest start time (LET - $\min \delta_i$) and for each task. We store all these costs in library that allows to each robot to know the opportunity cost of its delay in the values of its successors. Each robot has, for each successor i, the corresponding opportunity cost OC_i function for each task.

1.7 Implementation

This multi-robot decision system has been implemented and run on a scenario involving two robots : R1 and R2. Each one must exit a warehouse, move to an object, and catch it. Then, it must bring back the object to another warehouse. This problem implies constraints :

- *Temporal constraints* : warehouses have opening and closing hours. They open from 1 to 4 and from 25 to 32.
- *Precedence constraints* : the object O1 must be moved in order R2 to bring O2 back to warehouse 1.

The robots' mission can be represented by the graph shown in the figure 1.1. A fictitious task is added at the begin of the mission in order to have only one root. The intervals stand for the temporal constraints. For each task, the graph specifies the robot that must complete it. We have tested our decision system on this scenario. The results are promising. The robots perform the whole mission and act in an optimal way : for each step, each robot chooses the action that leads to the highest expected value.

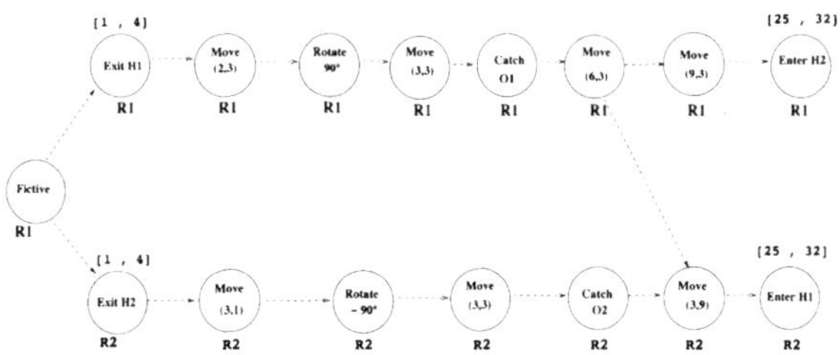

Fig. 1.1. Graph of the mission

More complex scenarios, involving 3 or 4 robots, are under development. They deal with planetary rovers or crisis scenario. For instance, a set of rovers must operate on the surface of Mars, they must complete different observations or experiments : take pictures, complete atmospheric measurements,... Another scenario deal with crisis control : an earth-quake arises and the firemen, the policemen and the ambulance men have several tasks to complete in order to rescue the inhabitants.

Currently, we are developing different experiments by modifying three parameters : *number of tasks*, *number of constraints* and *number of robots*. The first results show the robustness of our approach and its ability to support a large problem (200 tasks, 3 robots and 50 constraints) where the state space is almost 300000 states. Furthermore, given a mission, it can be shown that when we increase the number of agents, the state space size of each local MDP decreases. Indeed, each agent has less tasks to complete and the number of triplet $[l_i, r, I]$ decreases. If we increase the number of tasks, the state space size increases. When we increase the number of precedence constraints or we tighten the temporal constraints (smaller temporal windows $[EST, LET]$), the number of possible execution intervals goes down. Thus, the state space size diminishes. Also, the initial resource rate has effects on the space state size. Formalization of these fluctuations is under development.

1.8 Conclusion

In this paper we have deal with a multi-robot system under temporal and complex constraints. In this multi-robot system, robots are with limited and uncertain resources. This problem can be seen as a multi-robot planning under uncertainty with temporal and complex dependencies and limited resources. To address this problem, we proposed a decentralized MDPs framework. In this framework, MDPs are with no communication and they don't have a complete observation about the states of the other robots. This framework is based on the notion of opportunity cost to derive an optimal joint policy. Each robot constructs its optimal policy to achieve all its tasks taking for each one the dependency with the other tasks of the other robots. This policy respects the local temporal constraints (EST, LST and LET) and the temporal dependency between robots (precedence constraint).

Future work will concerns many techniques for computing exact or approximate opportunity cost.

References

[BDM+02] J. Bresina, R. Dearden, N. Meuleau, S. Ramakrishnan, D. Smith, and R. Washington. Planning under continuous time and resource uncertainty : A challenge for ai. In *UAI*, 2002.

[BGT03] C. Bererton, G. Gordon, and S. Thrun. Auction mechanism design for multi-robot coordination. In S. Thrun, L. Saul, and B. Schölkopf, editors, *Proceedings of Conference on Neural Information Processing Systems (NIPS)*. MIT Press, 2003.

[Bou99] Graig Boutilier. Sequential optimality and coordination in multiagents systems. In *IJCAI*, 1999.

[BZI00] D. Bernstein, S. Zilberstein, and N. Immerman. The complexity of decentralized control of mdps. In *UAI*, 2000.

[BZLG03] R. Becker, S. Zilberstein, V. Lesser, and C. Goldman. Transition-independent decentralized markov decision processes. In *AAMAS*, 2003.

[CMZW01] S. Cardon, AI. Mouaddib, S. Zilberstein, and R. Washington. Adaptive control of acyclic progressive processing task structures. In *IJCAI*, pages 701–706, 2001.

[GDP01] C. Guestrin, D.Koller, and R. Parr. Multiagent planning with factored mdps. In *NIPS*, 2001.

[GZ03] C. Goldman and S. Zilberstein. Optimizing information exchange in cooperative multiagent systems. In *AAMAS*, 2003.

[HM02] H. Hanna and AI Mouaddib. Task selection as decision making in multiagent system. In *AAMAS*, pages 616–623, 2002.

[NPY+03] R. Nair, D. Pynadath, M. Yokoo, M. Tambe, and S. Marsella. Taming decentralized pomdps: Towards efficient policy computation for multiagent settings. In *Proceedings of the Eighteenth International Joint Conference on Artificial Intelligence*, 2003.

[PKMK00] L. Peshkin, K.E. Kim, N. Meuleu, and L.P. Kaelbling. Learning to cooperate via policy search. In *UAI*, pages 489–496, 2000.

[SB98] R.S. Sutton and A.G. Barto. Reinforcement learning : An introduction. *MIT press, Cambrige, MA*, 1998.

[XLZ00] P. Xuan, V. Lesser, and S. Zilberstein. Communication decisions in multiagent cooperation. In *Autonomous Agents*, pages 616–623, 2000.

Emergent Robot Differentiation for Distributed Multi-Robot Task Allocation

Torbjørn S. Dahl[1], Maja J Matarić[2], and Gaurav S. Sukhatme[2]

[1] Norwegian Defence Research Establishment (FFI), Kjeller, Norway
Torbjorn-Semb.Dahl@ffi.no
[2] Center for Robotics and Embedded Systems
University of Southern California, Los Angeles, California
mataric|gaurav@usc.edu

Summary. We present a distributed mechanism for automatically allocating tasks to robots in a manner sensitive to each robot's performance level without hand-coding these levels in advance. This mechanism is an important part of improving multi-robot task allocation (MRTA) in systems where communication is restricted or where the complexity of the group dynamics makes it necessary to make allocation decisions locally. The general mechanism is demonstrated as an improvement on our previously published task allocation through vacancy chains (TAVC) algorithm for distributed MRTA. The TAVC algorithm uses individual reinforcement learning of task utilities and relies on the specializing abilities of the members of the group to produce dedicated optimal allocations. Through experiments with realistic simulator we evaluate the improved algorithm by comparing it to random allocation. We conclude that using softmax action selection functions on task utility values makes algorithms responsive to different performance levels in a group of heterogeneous robots.

1 Introduction and Motivation

Multi-robot task allocation (MRTA) algorithms for heterogeneous groups of robots have to be able to differentiate between robots based on their performance in order to optimize allocation. Existing MRTA algorithms [12, 8] generally do this based on hand-coded information about the task utilities relative to each robot. Using hand-coded task utilities, however, these algorithms are typically not sensitive to the effects of group dynamics, such as interference and synergy. These effects typically have to be estimated at runtime as they are difficult to model due to their volatility and complexity.

We have previously [6] presented a model of distributed MRTA as task distribution through *vacancy chains*, a distribution process common in animal and human societies [4]. We have also presented an algorithm based on task allocation through vacancy chains (TAVC) that validated our TAVC model and demonstrated improvements in performance over existing MRTA algorithms in dynamic domains. The TAVC algorithm is a partial solution to the problem of modeling the effects of group dynamics. The original TAVC algorithm, however, only optimized allocation for homogeneous groups of robots and was not designed to differentiate between robots based on individual performance.

In this paper we show how softmax action selection [13], on an inter-robot level, has the effect of reliably allocating high-performance robots to high-value tasks. This provides us with a general mechanism for producing performance sensitive task allocations in a distributed manner. We demonstrate this mechanism by using it to extend our original TAVC algorithm. We also present experimental evidence that the resulting allocation leads to an improved group performance.

2 The Prioritized Transportation Problem

Cooperative transportation is a multi-robot MRTA problem where group dynamics can have a critical impact on performance. In the basic transportation problem, a group of robots traverse a given environment in order to transport items between the sources and the sinks. We call the time taken to traverse the environment once from a sink via a source and back to a sink the *traversal time*. To perform optimally on this task the robots must minimize the total traversal time. The basic transportation problem is one of the sub-problems of *foraging* [2]. If the locations of sources and sinks are given, the foraging problem is reduced to a problem of transportation. The prioritized transportation problem extends the basic transportation problem by dividing the sinks into sets of different priority.

When sources and sinks are spatially distributed into distinct pairs or *circuits*, the optimal allocation will have to dedicate each robot to one of these circuits in order to avoid the increased processing implied by crossing between circuits. We call transportation tasks in such environments *spatially classifiable* [6]. To optimize its performance on these transportation problems, a group of robots must strike the correct balance between the different values of tasks within a class and the different traversal times as defined by the current levels of interference and synergy on each circuit.

3 Task Allocation through Vacancy Chains

The inspiration for our TAVC algorithm is the vacancy chain process through which resources are distributed to consumers. The typical example of resource distribution through a vacancy chain is a bureaucracy where the retirement of a senior employee creates a vacancy that is filled by a less senior employee. This promotion, in turn, creates a second vacancy to be filled, and so on, resulting in a chain of vacancies linked by the promotions. The resources that are distributed in this example are the positions, and the consumers are the employees.

3.1 The TAVC Model

Here we generalize our previous TAVC model to include groups of *heterogeneous* robots. According to our model, any number of robots can be assigned to tasks from a given *class*. Two tasks are in the same class if doing one affects the efficiency with which it is possible to do the other. We restrict ourselves to problems where these classes are disjunct.

When a robot, j, is assigned to a task from a class, i, currently being serviced by a set of robots J, we say that *service-slot*, (i, J, j) is filled. A particular set of robots, J, servicing the same task, i, will have a mean task processing frequency, $c_{i,J}$, dependent on the degree to which the robots are able to work concurrently without interfering with each other. The difference in mean task processing frequency together with the task value, w_i, define the contribution of the last robot added or the last service-slot filled. We call this contribution, which can be negative, the *slot-value*, $s_{i,J,j}$. The formal definition is given in Equation 1. When assigning an additional robot to a task leads to a decrease in the task processing frequency, the slot-value correspondingly becomes negative.

$$s_{i,J,j} = w_i(c_{i,J\cup\{j\}} - c_{i,J}) \tag{1}$$

In a scenario where the service-slots are allocated optimally, i.e., where the total value of the filled service slots are maximized, a failure in a robot servicing a high-value task will result in an empty high-value service-slot that must be re-allocated to reestablish optimality. Expressed in the vacancy chain framework, a vacant, high-value service-slot is a resource to be distributed between the robots.

The TAVC model formalizes the system level performance-related effects of group dynamics in terms of individual time measurements. Hence, it can be used as a model for optimizing MRTA problems that satisfy the *greedy property* [5] such as homogeneous MRTA. To optimize problems that do not satisfy this property, such as heterogeneous MRTA, it is necessary to introduce elaborations and additional allocation mechanisms such as the emergent performance-based robot differentiation mechanism presented here.

3.2 The TAVC Algorithm

In the TAVC algorithm, each robot keeps a local estimate of task utilities and choses its next task according to an action selection function. The TAVC algorithm uses Q-learning for task utility estimation. As Q-learning is time insensitive, it is necessary to make the temporal aspect of performance explicit in the reward function in order to use Q-learning to improve performance over time. We used a reward function based on the last task processing time, t, and task value, w_i as presented in Equation2

$$r = \frac{w_i}{t} \tag{2}$$

This reward function promotes the tasks with the highest value because these will on average provide a higher reward. However, if a robot consistently occupies a service-slot that is sub-optimal due to too much interference, the increased average traversal time will reduce the average reward for that slot below the average reward of the optimal service-slots. This change in average reward will attract the robot back to an optimal slot.

3.3 Softmax Action Selection as a Performance Differentiator

When robots keep a set of task related utilities, different functions can be used to select the next task to undertake based on these utilities [13]. With a Greedy-ϵ function, all the tasks, apart from the one with the highest utility, have equal probability, ϵ, of being explored. With a softmax function however, the probability of trying a suboptimal task is correlated with the relative estimated utility of that task. Our contribution in this paper is to demonstrate that the use of a Boltzmann softmax function on a task-selection level, has reliable effects on an inter-robot level, where it functions as a mechanism for allocating high-value tasks to high-performance robots. A robot that on average can service tasks in time $\overline{p}$ will have a difference in estimated task utility that correlates with the expression $\frac{w_h - w_l}{\overline{p}}$ where w_h and w_l are the values of high- and low-value tasks respectively. A fast robot with a lower average service time $\overline{p}$ will have a correspondingly higher difference between the estimated utility of high- and low-value tasks. Using a softmax action selection function, this greater difference in estimated utility theoretically translates into a greater probability of servicing high-value tasks. Such persistence will lead to the fast robots servicing high-value tasks and may, depending on the group dynamics and the task values, lead to the slow robots servicing low-value tasks.

4 Controller Architecture

All the robots in the experiments presented here used the same adaptive, behavior-based controller [11]. However, our TAVC algorithm and the performance sensitive robot differentiation mechanism we present here are independent of the underlying control architecture, being defined purely in terms of task utilities and the action selection function.

The robots were divided into two groups in order to make them heterogeneous. The first group was made to operate at a default speed of 300 mm/sec. The second group had a default speed of 200 mm/sec. The speeds were chosen to be equidistant from the speed used in our homogeneous robot experiments [6] in order to preserve the distribution where three robots served the high-value circuit and three robots serving the low-value circuit. This general difference in task processing speed encompasses more specific differences such as differences in robot morphology and task competence. The results here therefore generalize to all heterogeneous multi-robot systems where the differences between the participating robots can be expressed in terms of differences in task processing speed.

The input- or state-space reflected which circuit the robot used for its last traversal. The action space corresponded to the available tasks. For two tasks and six robots this resulted in six two-by-two Q-tables or 24 Q-values. The robots used temporal difference Q-learning [13] to associate the different input states with one of the high level approach behaviors. The Q-tables were initialized with random values between -0.1 and 0.1, the learning rate, α, was set to 0.1, and the discount factor, γ, was set to 0.95. For action selection we used either a Greedy-ϵ or a Boltzmann softmax function. For the Greedy-ϵ function, ϵ was set empirically to 0.1. For the Boltzmann softmax function, the temperature parameter τ was set empirically to 0.005. With these values the experimentation rate was significant without being overwhelming. Because we wanted the system to remain adaptive to changes in the environment we did not decrease ϵ or τ over time, as is common.

5 Experimental Design

Having previously performed experiments to demonstrate how the TAVC algorithm with a Greedy-ϵ action selection function allocates tasks in a group of homogeneous robots [7], the work focused on testing our hypothesis that using a softmax action selection function would make a MRTA algorithm sensitive to differences in robot skill levels.

The experiments were done in simulation on the *Player/Stage* software platform. From experience, controllers written for the Stage simulator work with little or no modification on real Pioneers. The experiment used simulated Pioneer 2DX robots with SICK laser range-finders and PTZ color cameras and took place in a twelve by eight meter environment with two sets of sources and sinks. Figure 1 shows a graphical rendering of the simulated environment, with the sources and sinks labeled.

The sources and sinks were simulated bar-codes made from high-reflection material and recognizable by the laser range finder. We call the circuit with the highest related reward the *high-value* circuit and correspondingly, the circuit with the lowest related reward is called the *low-value* circuit.

In previous experiments with homogeneous robots [7], in order to produce a task allocation where three robots serviced one circuit and three robots serviced the other circuit, it was necessary that it was less attractive to the robots to be one of four robots servicing the high-value circuit than to be one

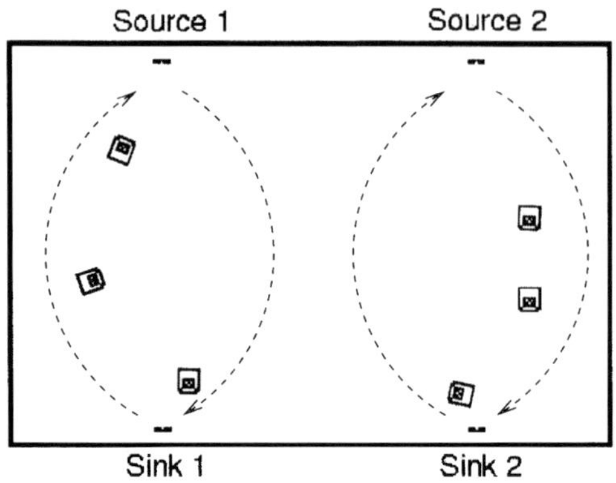

Fig. 1. The Simulated Environment with Circuits Indicated

of three servicing the low-value circuit. We empirically estimated the relevant average traversal times. To satisfy the given constraints we chose the circuit values as used in the reward-function given in Equation 2 to be $w_1 = 2200$ and $w_2 = 2000$. Assuming that the allocation with three robots on each circuit persists in the case of heterogeneous robots, the optimal allocation in terms of the TAVC model is to have the three fast robots on the high-value circuit and the three slow robots on the low-value circuit.

We had two aims for the heterogeneous robot experiment. First, to test whether the modified TAVC algorithm would produce the allocation predicted to be optimal by the TAVC model, where the three fast robots serviced the high-value circuit and the three slow robots serviced the low-value circuit. Second, to demonstrate that this allocation improved the performance of the system to a level significantly above the performance level of a group where tasks were allocated randomly.

6 Results

We defined a convergence period of 15 hours based on the stability of the system performance. The current allocations was identified by looking at which of the robots visited the high-value circuit last. We used three fast and three slow robots, yielding fifteen possible system states. We refer to each state using the notation $f : s$, where f is the number of fast robots whose last target was on the high-value circuit. Correspondingly, s is the number of slow robots whose last target was on the high-value circuit. The columns labeled $\hat{\mu}_a$ and s_a in Table 1 show the mean and standard deviation of the time the system spent in each of the states while running the modified TAVC algorithm. The values are percentages of the total stable period. The columns labeled $\hat{\mu}_r$ and s_r give the mean and standard deviation of the time the system spent in each of the states for a set of 15 trials using a group of robots that randomly chose between tasks. The column labeled T lists the combinatorial probability of

$f:s$	$\hat{\mu}_a$	s_a	$\hat{\mu}_r$	s_r	T	$\hat{\mu}_a - \hat{\mu}_r$	$\frac{\hat{\mu}_a - \hat{\mu}_r}{\hat{\mu}_r}$
0:0	0.2	0.2	1.4	0.5	1.5	-1.3	-0.88
0:1	1.5	1.2	3.4	1.7	4.7	-3.2	-0.67
0:2	2.0	2.6	5.0	1.4	4.7	-2.7	-0.58
0:3	0.5	0.7	2.0	1.1	1.5	-1.1	-0.70
1:0	2.9	1.6	3.6	1.4	4.7	-1.8	-0.38
1:1	12.1	4.9	13.5	3.3	14.1	-2.0	-0.14
1:2	14.4	6.1	14.4	2.4	14.1	0.3	0.02
1:3	4.1	2.7	4.0	1.4	4.7	-0.6	-0.13
2:0	7.0	5.0	5.1	1.2	4.7	2.3	0.48
2:1	19.4	7.4	15.7	2.3	14.1	5.30	0.37
2:2	18.5	5.0	14.7	3.0	14.1	4.4	0.3
2:3	5.7	3.6	4.4	2.4	4.7	1.0	0.20
3:0	2.7	4.1	1.7	0.6	1.5	1.2	0.78
3:1	5.2	4.6	5.2	1.6	4.7	0.5	0.10
3:2	3.3	2.6	4.2	1.2	4.7	-1.4	-0.29
3:3	0.7	0.7	1.5	0.8	1.5	-0.9	-0.56

Table 1. State-Time Dist. for Heterogeneous Robots

choosing a sample of size f from a population of $g = 3$ fast robots as well as choosing a sample of size s from a population of $h = 3$ slow robots. This probability is given in Equation 3. It is worth noticing that the time distribution produced by random allocation is closely aligned with the theoretical estimate, though the differences are statistically significant.

$$T = \frac{100g!h!}{f!(g-f)!s!(h-s)!2^g 2^h} \tag{3}$$

The difference between the state-time distribution produced by the modified TAVC algorithm and the distribution produced by random allocation is presented in the column labeled $\hat{\mu}_a - \hat{\mu}_r$. This difference is presented as a percentage of the mean times from the random distribution, $\hat{\mu}_r$, for each state in the last column, labeled $\frac{\hat{\mu}_a - \hat{\mu}_r}{\hat{\mu}_r}$. The difference between the distributions produced by the adaptive controllers and the random controllers, i.e., the last two columns, are also presented graphically by the two histograms in Figure 2.

Over these 15 experiments, the increase in time spent in state $0:3$ is statistically significant. In the second histogram in Figure 2 the optimal state, $0:3$, stands out as the state with the highest relative increase in time. This confirms that the group's set of Q-tables have converged to promote the state defined as optimal according to the TAVC model. The performance data also show that the performance of a group of robots controlled by the TAVC algorithm, 0.081 of target value per 10 seconds, is significantly higher than the

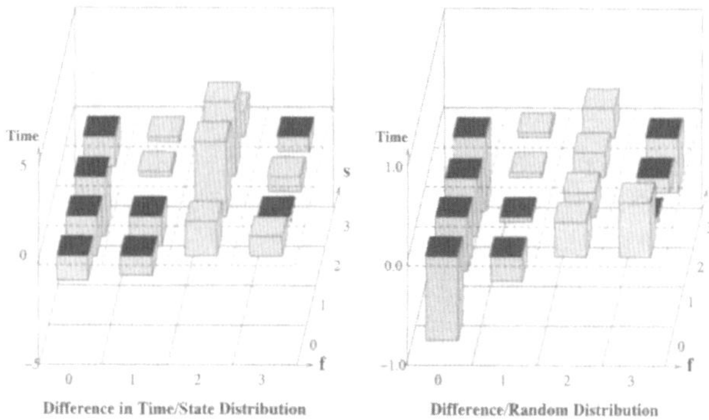

Fig. 2. Difference in Distributions

performance of random allocation, 0.078 of target value per 10 seconds. Together, the state-time distribution data and the performance data show that the modified TAVC algorithm improve the group's performance by adopting a dedicated service structure that conforms to the predictions of the TAVC model. The learned Q-tables show that, as predicted, the fast robots, on average, have higher estimated utilities for high-value tasks and higher average differences between the estimated utilities of high- and low-value tasks.

7 Related Work

In addition to existing MRTA algorithms that rely on hand-coded information about performance levels, task utility and group dynamics [12, 8], there is a large body of work on modeling group dynamics.

Goldberg and Matarić [9] developed Augmented Markov Models, or transition probability matrices with additional temporal information, to learn statistical models of interaction in a space of abstract behaviors. Yan and Matarić [14] have used multi-level modeling of group behaviors from spatial data to describe both human and robot activity. Lerman *et al.* [10] have studied three different types of models of cooperation and interference in groups of robots: sensor-based simulations, microscopic numerical models, and macroscopic numerical models. Currently there are no models of the effects of group dynamics with the speed, generality, and predictive accuracy necessary to specify the effects of interaction on task processing times in task allocation problems when the aim is to constructing optimal schedules on the fly. Simulation-based models are in general too slow while macroscopic mathematical models make too many simplifying assumptions in order to be of predictive use.

In the absence of models, learning has been used to increase the applicability of both centralized and distributed MRTA and related scheduling algorithms. In the L-ALLIANCE work by Parker [12], each robot explicitly estimates its own performance and the performance of other robots on selected tasks and uses these values to reallocate tasks by taking them over or acquiescing. The L-ALLIANCE algorithm uses local utility estimates to make local allocation decision, but needs hand-coded estimation procedures that reduce the general applicability of this algorithm. Brauer and Weiss [3] use a distributed RL mechanism for Multi-Machine Scheduling (MMS) where each machine estimates locally the optimal receiver of the material it has processed. This approach, like ours, uses local action selection and utility estimates. The MMS problem however does not contain the complex group dynamics of the transportation problem. Balch [1] studied performance-related reward functions for robot using Q-learning. Our results build on Balch's work and further explore the use of local performance based reward functions for optimizing group performance.

8 Conclusions

Our experiments show that when using a Boltzmann softmax function, the modified TAVC algorithm was sensitive to the different operating speeds of the robots and promoted the optimal state as defined by the TAVC model. As a percentage of the total system time, the increase in time the system spent in the optimal state is too small to create significant improvements in the performance of cooperative multi-robot systems. Also, a comparison with random allocation does not demonstrate the performance of the modified TAVC algorithm relative to existing MRTA algorithms. For such purposes a basic greedy allocation algorithm would have provided a better basis for comparison. Finally, we do not know how the TAVC algorithm would scale up to problems that are too complex for each robot to evaluate every circuit, but it seems feasible that satisfactory global solutions could emerge through migration based on local interactions. In spite of these limitations, our results are important as they show that the theoretical probability of using a softmax action selection function to make distributed MRTA sensitive to different robot abilities does translate into observable effects in realistic multi-robot simulations.

Acknowledgments

This work is supported in part by a Department of Energy (DOE) Robotics and Intelligent Machines (RIM) grant DE-FG03-01ER45905 and in part by an Office of Naval Research (ONR) Defense University Research Instrumentation Program (DURIP) grant 00014-00-1-0638.

References

1. T. R. Balch. Reward and Diversity in Multirobot Foraging. In *Proceedings of the 16th International Joint Conference on Artificial Intelligence (IJCAI'99)*

Workshop: Learning About, From and With other Agents, Stockholm, Sweden, July 31 - August 6 1999.

2. Tucker R. Balch. The impact of diversity on performance in multi-robot foraging. In Oren Etzioni, Jörg P. Müller, and Jeffrey M. Bradshaw, editors, *The proceedings of the Third International Conference on Autonomous Agents (Agents'99)*, pages 92–99, Seattle, Washington, May 1 - 5 1999. ACM Press.
3. Wilfried Brauer and Gerhard Weiß. Multi-machine scheduling - a multi-agent learning approach. In *Proceedings of the 3rd International Conference on Multi-Agent Systems (ICMAS'98)*, pages 42–48, Paris, France, July 4 -7 1998. IEEE Press.
4. Ivan D. Chase, Marc Weissburg, and Theodore H. Dewitt. The vacancy chain process: a new mechanism of resource distribution in animals with application to hermit crabs. *Animal Behavior*, 36:1265–1274, 1988.
5. Thomas H. Cormen, Charles E. Leiserson, Ronald L. Rivest, and Clifford Stein. *Introduction to algorithms.* MIT Press, Cambridge, Massachusetts, second edition, 2001.
6. Torbjørn S. Dahl, Maja J Matarić, and Gaurav S. Sukhatme. Scheduling with group dynamics: A multi-robot task allocation algorithm based on vacancy chains. Technical Report CRES-002-07, Center for Robotics and Embedded Systems, University of Southern California, Los angeles, CA, 2002.
7. Torbjørn S. Dahl, Maja J. Matarić, and Gaurav S. Sukhatme. Multi-robot task-allocation through vacancy chains. In *Proceedings of the 2003 IEEE International Conference on Robotics and Automation (ICRA'03)*, pages 2293–2298, Taipei, Taiwan, September 9 - 14 2003. IEEE Press.
8. Brian P. Gerkey and Maja J Matarić. Sold!: Auction methods for multi-robot coordination. *IEEE Transactions on Robotics and Automation*, 18(5):758–768, October 2002.
9. Dani Goldberg and Maja J Matarić. Learning multiple models for reward maximization. In Pat Langley, editor, *Proceedings of the 17th International Conference on Machine Learning (ICML'00)*, pages 319–326, Stanford, California, June 29 - July 2 2000. Morgan Kaufmann.
10. Kristina Lerman, Asram. Galstyan, Alcherio Martinoli, and Auke J. Ijspeert. A macroscopic analytical model of collaboration in distributed robotic systems. *Artificial Life*, 7(4):375–393, 2001.
11. Maja J. Matarić. Behavior-based control: Examples from navigation, learning, and group behavior. *Journal of Experimental and Theoretical Artificial Intelligence, special issue on Software Architectures for Physical Agents*, 9(2-3):323–336, 1997.
12. Lynne E. Parker. L-ALLIANCE: Task-Oriented Multi-Robot Learning in Behaviour-Based Systems. *Advanced Robotics, Special Issue on Selected Papers from IROS'96*, 11(4):305–322, 1997.
13. Richard S. Sutton and Andrew G. Barto. *Reinforcement learning: an introduction.* MIT Press, Cambridge, Massachusetts, 1998.
14. Helen Yan and Maja J Matarić. General spatial features for analysis of multi-robot and human activities from raw position data. In *Proceedings of the 2002 IEEE/RSJ International Conference on Intelligent Robots and Systems (IROS'02)*, pages 2770–2775, Lausanne, Switzerland, September 30 - October 4 2002. IEEE Press.

Cooperative Control Method Using Evaluation Information on Objective Achievement

Hikari Fujii[*1], Daiki Sakai[*1] and Kazuo Yoshida[*2]

[*1]Graduate School of Science and Technology, Keio University

[*2]Faculty of Science and Technology, Keio University

Abstract. This paper presents a cooperative control method for a multi-agent system. The fundamental concept of this method is the control based on the evaluation on objective achievement of multi-agent systems. Each robot communicates not directly by sensory data, but by qualitative evaluation of achievement level. Each robot calculates global evaluation on the achievement of the team's objective by agents' evaluations. This method enables a robot to perform flexible cooperation based on the global evaluation on achievement of objectives. As an example, the method is applied to the EIGEN team robots for the Middle Size League of RoboCup which is international soccer robot project, since it is necessary for the soccer robots to cooperate each other under dynamic environment. As a result its effectiveness was demonstrated.

1 Introduction

The efficiency of achieving task is improved by cooperation among multi-agent systems. Some tasks that are difficult to be accomplished by one agent are executed by a multi-agent system. Therefore, many researchers have studied about cooperative action of multi-agent systems [1] and [2]. Recently it has been expected to realize robots that are symbiotic with human in open environment. These robots are required to act cooperatively with human, other robots and artifacts in complicated environment. To realize this action, it is necessary to develop a more flexible cooperative control method.

This paper presents a cooperative control method of multi-agent systems using the evaluation on objective achievement. In this method, each agent possesses the objective of the multi-agent system and the evaluation about achievement. The cooperation is carried out by making robots share the abstracted evaluation information on achievement of the global objective, so that the multi-agent system executes the objective. The RoboCup is chosen as a test-bed.

RoboCup Middle Size League is a soccer game executed by autonomous mobile robots. Robots are forbidden to use a global sensor. Since there are many robots in a field, it is difficult to construct a global model and the wireless LAN used in communication is not stable. The cooperation at the RoboCup Middle Size League is required to adapt to the dynamic environment and possess the flexibility, which is a good test bed of a multi-agent system. In this study, the proposed method is applied to the control of the team EIGEN for the RoboCup Middle Size Robots as shown in Fig.1

Cooperative behavior is one of important subjects in the RoboCup Middle Size League and many researchers have studied about it. There are representative examples of Dynamic Task Assignment [3], Dynamic Role Assignment [4-6] and so on. Dynamic Role Assignment is realized for the efficient cooperation among the agents. It assigns a role without confliction. However, the method described in [6] needs accurate data of the agents' position. Dynamic Task Assignment requires the agents to have the same action modules.

In this study, the flexible objective selection is realized by using the method of the qualitative information about own achievement level of the objective and the evaluation about team's achievement level of the objective calculated based on the sum of the respective self-evaluations of each robot. This method enables robots to change appropriately the multi-agent system's behavior with keeping the high autonomy of each agent. The agents are able to select the same role before achieving the desired state to accomplish the global objective. They are also able to cooperate without sharing accurate position data and the same action modules.

Fig. 1. Robots of team EIGEN

2 Cooperative control method

2.1 Concept

In this study, the cooperative behavior among agents is realized by calculating the evaluation of degree of achieving an objective and by sharing the evaluation among them. In this method, each agent possesses the information about the objective of multi-agent system called global objective. They can estimate the current state from the environmental information and the internal state, such as selecting role and action. With these information, Each agent calculates two evaluation information, own degree of achieving objective which is called Self-evaluation and degree of achieving the global objective from the point of view of each agent which is called System Satisfaction. In this study, the function for calculate Self-evaluation is defined based on the knowledge of designer. Each evaluation is represented by a simple integral number. Each agent knows the desired state for achieving the objective which is called System Objective. This value is also represented by an integral number. These values of evaluation can be considered as qualitative abstracted data of degree of achieving the objective.

The quantitative information such as sensory data and the physical quantities has been directly used in many studies, where agents needs other agents' physical quantities to cooperate with other agents and a lot of exchanges of information are necessary. However, a method using qualitative information might be more useful than the method using quantitative information in open environment which varies from moment to moment. The effectiveness of using the qualitative information for the control was shown in the studies [7] and [8]. Therefore, the qualitative information is used in this study.

To calculate the System Satisfaction, each agent compares the value of Self-evaluation and summarizes Self-evaluations which have higher value than own value of it. With this operation, each agent has a deferent value of System Satisfaction according to its state. In the case that an agent satisfies the global objective, the others' action to achieve the objective is inhibited according to the high value of System Satisfaction.

In some situation, the priority exists among agents. In order to realize the effective cooperation, the priority agent should be selected according to the agent's role. Special consideration is paid to the evaluation of the priority agent. This kind of agent should have weighted big evaluation or negative evaluation, so that it has powerful influence on the evaluation of the whole system. The priority agent induces an action according to the own evaluation using negative evaluation. The evaluation of the priority agent is always considered by other agents when they calculate the System Satisfaction.

The outline of the proposed method is as followings.

STEP1: Each agent evaluates its state about the respective objective of a robot, and the evaluated information is shared among the agents.

STEP2: Each agent calculates the System Satisfaction based on the sum of the agents which have higher evaluation than itself.

STEP3: The agent selects the role according to the System Satisfaction and the System Objective. The concept and the flow of this method are shown in Fig.2.

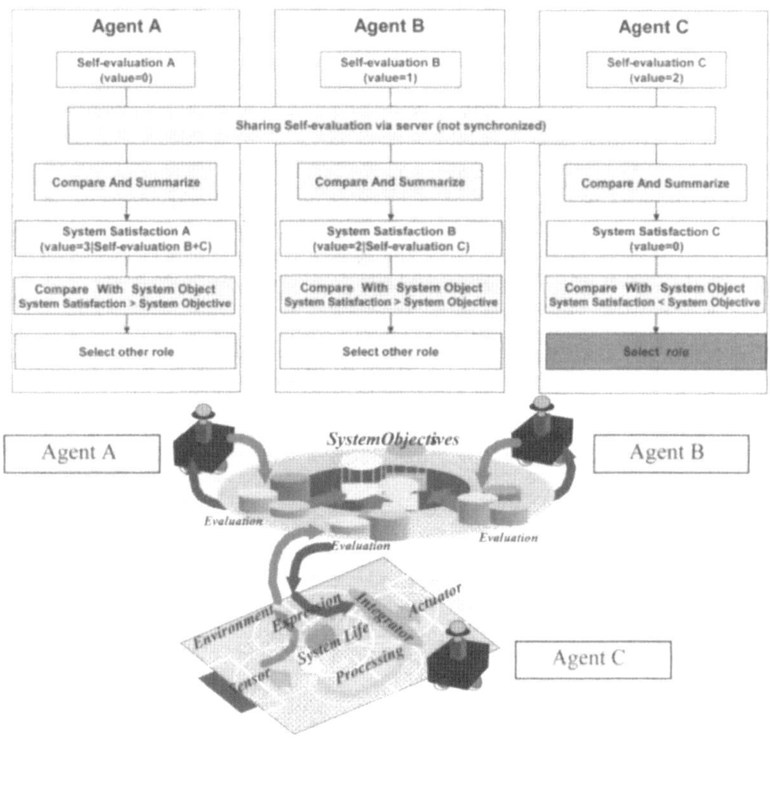

Fig. 2. The concept and the flow of the proposed method

2.2 Formulation for proposed method

According to the above-mentioned concept, the formulas for the proposed method are defined.

These variables are defined as followings:

- *System Objective $_i$* : Desired state of the Objective (i). Each agent has the same value.
- *System Satisfaction $_i$* : Satisfaction and evaluation index of the Objective (i) through the position of *Agent $_{own}$*. Each agent has a different value.
- *$Agent^k{}_{priority}$* : Priority agent.
- *Evaluate $_i$ (Agent $_j$)* : Value of Self-evaluation about Objective (i) from *Agent $_j$*.
- *E_i (Agent $_j$)* : Value of the evaluation about the Objective (i) from *Agent $_j$* considered by *Agent $_{own}$*
- *m* : Number of the priority agent through the position of *Agent $_{own}$*.
- *n* : Number of the non-priority agent through the position of *Agent $_{own}$*.

Agents always take into account the evaluation of priority agent. The evaluation about Objective (i) is the sum of these evaluations. The formulation of the evaluation about Objective (i) is as the following equation:

$$SystemSatisfaction_i = \sum_{k=1}^{m} Evaluate_i\left(Agent^k{}_{priority}\right) + \sum_{j=1}^{n} E_i\left(Agent_j\right) \tag{1}$$

where

$$E_i\left(Agent_j\right) = \begin{cases} Evaluate_i\left(Agent_j\right) & (Evaluate_i\left(Agent_j\right) > Evaluate_i\left(Agent_{own}\right)) \\ 0 & (Evaluate_i\left(Agent_j\right) \le Evaluate_i\left(Agent_{own}\right)) \end{cases} \tag{2}$$

This *System Satisfaction $_i$* is different in each agent. When the *System Satisfaction $_i$* is bigger than the *System Objective $_i$*, the agent thinks that the Objective (i) is achieved and inhibits the action for achieving Objective (i). When the *System Satisfaction $_i$* does not reach the *System Objective $_i$*, the agent thinks that the Objective (i) is not achieved and selects the action for achieving Objective (i). As a result, agents behave cooperatively.

2.3 Expansion to the multi-objective system and objective selection method

The effectiveness of this method is shown in a single objective system [9]. A complex task is required in a multi-objective system. This method can be applied to a multi-objective system by evaluating multiple objectives of the system. To select the objective, we define the v_i as the following equation:

$$v_i = SystemObjective_i - SystemSatisfaction_i \tag{3}$$

The v_i is the difference between the objective state and the evaluation of the system which is obtained through the position of the agent. The scale of the v_i is determined by considering the desire level of achieving that objective. The objective is achieved when the v_i become zero.

The methods of selecting objective are shown as followings:

1. The method setting priority on objectives.
   First, the agent checks the v_i of the objective which has the highest priority. Second, if the v_i is bigger than zero, the agent selects that objective. If the v_i is not over zero, the agent checks other objective and repeats these steps.
2. The method selecting the objective has the biggest v_i value.
3. The method using selecting function which is defined according to knowledge of a designer.
4. The combination of the methods 1, 2 and 3.
   The agent uses the method 1 basically. When it is difficult to set the priority, the agent uses the method 2 or 3.
   The proposed method is expected to be applicable to various systems.

3 Application to the RoboCup Middle Size League

3.1 Objective and self-evaluation

The proposed method is applied to RoboCup Middle Size League.

To keep a ball is one of most basic actions of the soccer playing robots. However, if all robots approach to a ball simultaneously, it is not suitable for team play. It is not necessary for all agents to approach a ball. It is required for one member of the team to keep a ball. Furthermore, defensive action and combination of robots are also important. Therefore, three kinds of objectives; offence, support and defense, are defined in the proposed method.

The variables are defined as following:

- $e_{offence}$: Quantitative Self-evaluation of the offence
- $e_{support}$: Quantitative Self-evaluation of the support
- $e_{defense}$: Quantitative Self-evaluation of the defense
- $Field_{length}$: Length of the field. Field length = 9000[mm]
- $Field_{width}$: Width of the field. Field width = 5000[mm]
- d_{diag} : Diagonal distance of the Field.

$$d_{diag} = \sqrt{Field_{length}^2 + Field_{width}^2}[mm] \quad (4)$$

- θ_{ball} : Angle between the directions of the front of agent and the ball
- d_{ball} : Distance between the agent and ball
- θ_{goal} : Angle between the directions of the front of agent and the goal
- d_{goal} : Distance between the agent and goal
- $\theta_{support}$: Angle to the direction of the support's position calculated by the dynamic potential.
- $d_{support}$: Distance to the support's position calculated by the dynamic potential.
- $\theta_{defense}$: Angle to the direction of the defense's position calculated by the dynamic potential.
- $d_{defense}$: Distance to the support's position calculated by the dynamic potential.

The coordinate system is shown in Fig. 3.

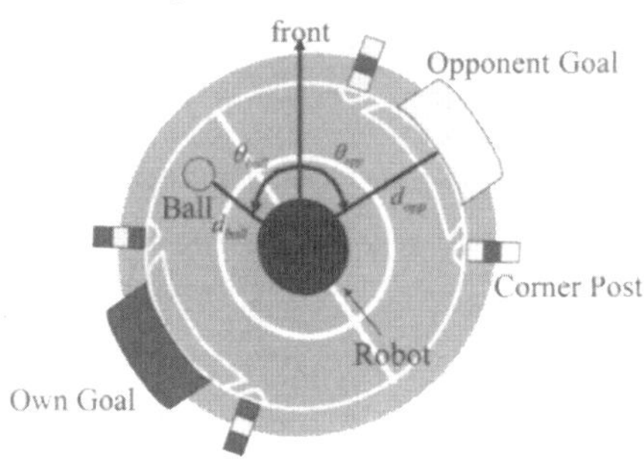

Fig. 3. Coordinate system based on omni directional camera.

With these variables, the evaluation of the offence is formulated as the following equation:

$$e_{offense} = a_1 \exp\left(-\sqrt{\left(\frac{\theta_{ball}^{\ 2}}{\alpha_{11}^{\ 2}\pi^2} + \frac{d_{ball}^{\ 2}}{\alpha_{12}^{\ 2}d_{diag}^{\ 2}} + \frac{\theta_{opp}^{\ 2}}{\alpha_{13}^{\ 2}\pi^2} + \frac{d_{opp}^{\ 2}}{\alpha_{14}^{\ 2}d_{diag}^{\ 2}} + \frac{(\theta_{ball}-\theta_{opp})^2}{\alpha_{15}^{\ 2}\pi^2}\right)}\right) \tag{5}$$

$$+a_2 \exp\left(-\sqrt{\left(\frac{\theta_{ball}^{\ 2}}{\alpha_{21}^{\ 2}\pi^2} + \frac{d_{ball}^{\ 2}}{\alpha_{22}^{\ 2}d_{diag}^{\ 2}} + \frac{\theta_{opp}^{\ 2}}{\alpha_{23}^{\ 2}\pi^2} + \frac{d_{opp}^{\ 2}}{\alpha_{24}^{\ 2}d_{diag}^{\ 2}} + \frac{(\theta_{ball}-\theta_{opp})^2}{\alpha_{25}^{\ 2}\pi^2}\right)}\right)$$

The wrap-around about the ball is expressed as the first term. The approaching to a ball is expressed as the second term. The example of the offence behavior is shown in Fig. 4 and the parameters are shown in Table 1.

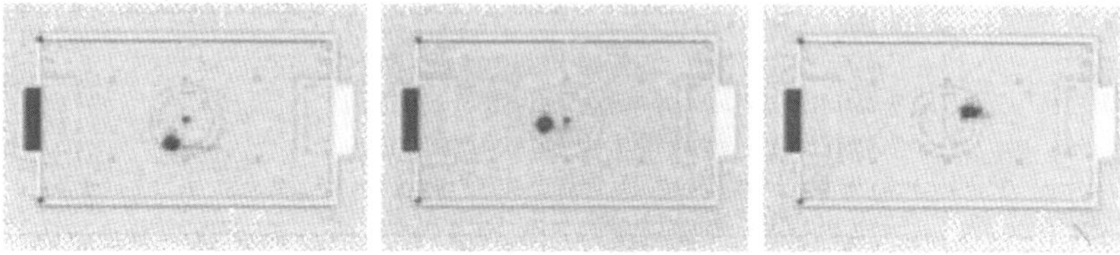

(a) $e_{offence} = 0.35$ (b) $e_{offence} = 0.63$ (c) $e_{offence} = 0.77$

Fig.4. Typical situation in offence action

Table 1. Coordinates of the representative points

$\theta_{ball} = 1.92[rad]$	$\theta_{ball} = 0.42[rad]$	$\theta_{ball} = 0.06[rad]$
$d_{ball} = 500.71[mm]$	$d_{ball} = 294.71[mm]$	$d_{ball} = 0[mm]$
$\theta_{goal} = 2.76[rad]$	$\theta_{goal} = 0.58[rad]$	$\theta_{goal} = 0.26[rad]$
$d_{goal} = 4716.25[mm]$	$d_{goal} = 5185.43[mm]$	$d_{goal} = 3865.32[mm]$
$Evaluate_{offense} = 0.35$	$Evaluate_{offense} = 0.63$	$Evaluate_{offense} = 0.77$

When the objective is achieved, each θ and d is zero and the evaluation $e_{offence}$ is 1.0. The evaluation of the situation in Fig.4 (a) is decided as 0.35 by designer. According to this value, the constant parameter shown in equation (5) is defined as $a_1 = 0.51$, $a_2 = 0.49$ .The value of evaluation is abstracted according to the following equation.

$$Evaluate_{offense}\left(Agent_{own}\right) = \begin{cases} 2 & \left(0.5 < e_{offense}\right) \\ 1 & \left(0.2 < e_{offense} \le 0.5\right) \\ 0 & \left(e_{offense} \le 0.2\right) \end{cases} \tag{6}$$

In this case, *System Objective*$_{offence}$ is set to 2. If any agents do not keep a ball, two agents are going to approach to the ball.

The self-evaluation of the support is designed as the following equation. The role of this objective is to support the offence agent. The evaluation is calculated with the objective position.

$$e_{support} = \exp\left(-\sqrt{\left(\frac{\theta_{support}^{\ 2}}{\beta_1^2\pi^2} + \frac{d_{support}^{\ 2}}{\beta_2^2 d_{diag}^{\ 2}}\right)}\right) \tag{7}$$

The $e_{support}$ is abstracted according to the following equation:

$$Evaluate_{support}\left(Agent_{own}\right) = \begin{cases} 1 & \left(0.6 < e_{support}\right) \\ 0 & \left(e_{support} \le 0.6\right) \end{cases} \tag{8}$$

System Objective$_{support}$ is set to 1.

The self-evaluation of the defense is designed as the following equation. The evaluation is calculated with the objective position.

$$e_{defense} = \exp\left(-\sqrt{\left(\frac{\theta_{defense}^{2}}{\gamma_1^{2}\pi^{2}} + \frac{d_{defense}^{2}}{\gamma_2^{2}d_{diag}^{2}}\right)}\right) \quad (9)$$

The e $_{defense}$ is abstracted according to the following equation

$$Evaluate_{defense}\left(Agent_{own}\right) = \begin{cases} 1 & \left(0.5 < e_{defense}\right) \\ 0 & \left(e_{defense} \le 0.5\right) \end{cases} \quad (10)$$

System Objective$_{defense}$ is set to 1.

The parameters of each evaluator are shown in Table 2.

Table 2. Parameters of each evaluator

$\alpha_{11} = 10.0$ $\alpha_{12} = 0.3$ $\alpha_{13} = 10.0$ $\alpha_{14} = 10.0$ $\alpha_{15} = 0.5$	$\alpha_{21} = 0.3$ $\alpha_{22} = 0.05$ $\alpha_{23} = 0.7$ $\alpha_{24} = 1.0$ $\alpha_{25} = 0.6$	$\beta_1 = 1.0$ $\beta_2 = 0.2$ $\gamma_1 = 1.0$ $\gamma_2 = 0.2$

When an agent is stuck, the evaluation changes into zero. The actions based on each objective and parameters of each evaluator are designed with the knowledge of the designer.

3.2 Design of sending information by priority agent

In this study, the goal keeper robot and the dribbring agent are selected as the priority agents. The evaluations on the goal keeper are as the following equation:

$$Evaluate_{offense}\left(Agent_{priority}^{goalie}\right) = \begin{cases} 1 & (defense) \\ 0 & (normal) \end{cases} \quad (11)$$

$$Evaluate_{support}\left(Agent_{priority}^{goalie}\right) = \begin{cases} 0 & (normal) \\ -1 & (kick\ action) \end{cases} \quad (12)$$

$$Evaluate_{defense}\left(Agent_{priority}^{goalie}\right) = \begin{cases} 1 & (normal) \\ 0 & (stack) \end{cases} \quad (13)$$

By using the above equations the stick between the goal keeper and the other agent is avoided. The evaluation of the dribbring agent is as the following equation:

$$Evaluate_{support}\left(Agent_{priority}^{dribble}\right) = -1 \quad (14)$$

This evaluation makes the other agents behave as a supporter.

3.3 Selecting objective

To apply the proposed method to the RoboCup Middle Size League robots, the objective selection method 3 is utilized. The objective selecting function is defined as the following equation:

$$V_i = v_i\left(1 + \kappa_i e_i\right) \quad (15)$$

The objective having the biggest V_i is selected. In this study, the parameters are determined as : $\kappa_{offense} = 0.5$, $\kappa_{support} = 0.1$, $\kappa_{defense} = 0.3$.

The processing flow of the agent is shown as followings.

STEP1: Quantitative Self-evaluation e_i is calculated according to the equations (5), (7) and (9) with sensory data.

STEP2: Self-evaluation $Evaluate_i(Agent_{own})$ is calculated according to equations (6), (8) and (10).

STEP3: $SystemSatisfaction_i$ is calculated according to equations (1) and (2).

STEP4: V_i is calculated according to equation (15).

STEP5: Objective(i) with the biggest V_i is selected.

STEP6: Action module is selected according to the objective(i)

STEP7: Output is determined according to the action module and fuzzy potential method [10].

4 Experimental results

4.1 Simulation results and considerations

The 4 on 4 soccer game is simulated by a computor. Figure 5 shows the time history of the self-evaluation and the selected objective as the results of the charactaristics of this mehod.
The time histories of the selected objective show the flexible changing of the role of robots according to the situation.

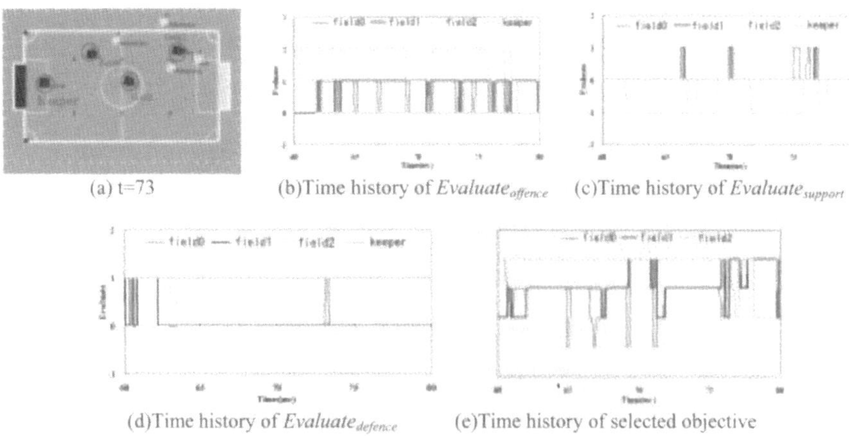

(a) t=73 (b)Time history of $Evaluate_{offence}$ (c)Time history of $Evaluate_{support}$

(d)Time history of $Evaluate_{defence}$ (e)Time history of selected objective

Fig.5. The simulation result (case 1)

CASE 1: In Fig. 5(a), the field player 2 is holding the ball. The other agents are inhibitting the approaching action. The field player 2 becomes the priority agent. As a result, the field player 1 selects the role of the support.
CASE 2: The field player 1 sticks during selecting the offence. Accordint to this situation, the other robots select the offence action. After that, the field player 2 satisfies the offence objective of the system and the field player 1 behaves as a support action as shown in Fig. 7(e).
CASE 3: When the goal keeper having the priority takes the defense action, other agents' ball approaching action is inhibitted. As a result, the collision between the goalie and the agent is reduced. After that, the ball exists behind the goal keeper, the goal keeper decreases the

evaluation information of the defense. According to this change in the common information, the defense action of the field player is brought about.

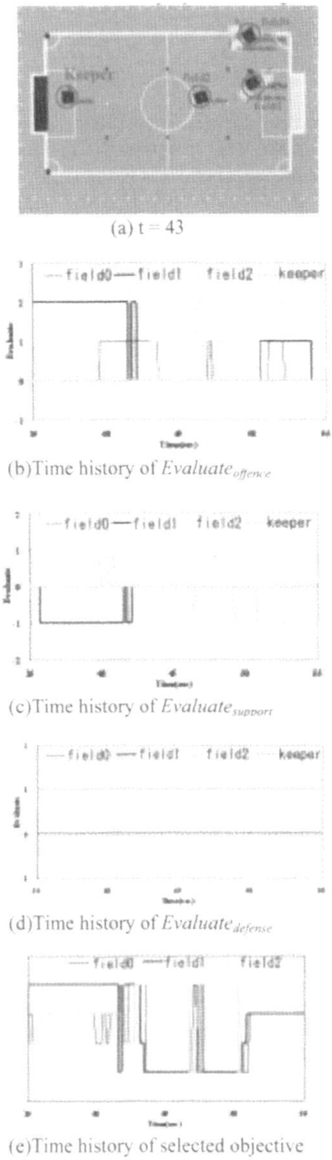

(a) t = 43

(b)Time history of $Evaluate_{offence}$

(c)Time history of $Evaluate_{support}$

(d)Time history of $Evaluate_{defense}$

(e)Time history of selected objective

Fig.6. The simulation result (Case2)

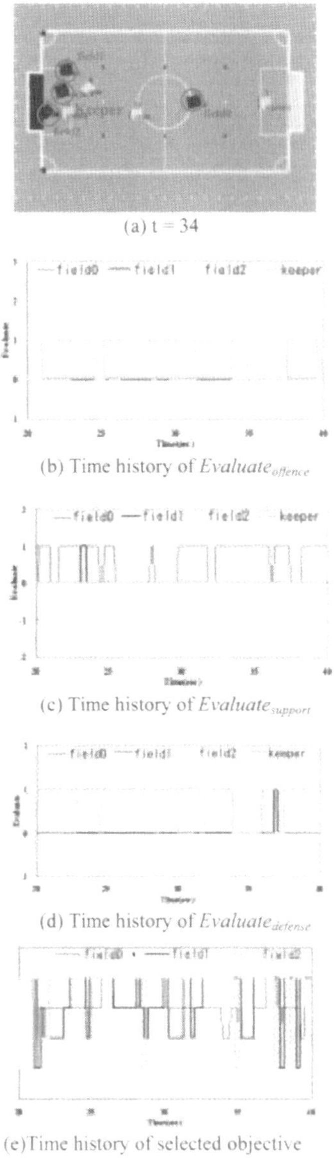

(a) t = 34

(b) Time history of $Evaluate_{offence}$

(c) Time history of $Evaluate_{support}$

(d) Time history of $Evaluate_{defense}$

(e)Time history of selected objective

Fig.7. The simulation result (Case3)

In most of the conventional methods, the cooperative behavior as shown in the simulation result has been realized by the complex adjustment among the agents at a restricted situation. In the proposed method, two features of the existance of the priority agents and the utilization of the qualitative information of the self-evaluation make a multi-agent system to act cooperatively.

5 Conclusion

In this study, the cooperative control method using the evaluation information on objective achievement was proposed. In this method the qualitative information is used and the satisfaction degree of the system is caluculated from the information communicated from each agent. And the cooperative control was applied to the robots of RoboCup Middle Size League, where each autonomous system has only local sensors. The effectiveness of the proposed method were demonstrated in the computor simulation. The future subject of this study will be the construction of an unified control method from the action selection level to the objective selection level.

Acknowledgments

This work is supported in part by Grant in Aid for the 21st century center of excellence for "System Design: Paradigm Shift from Intelligence to Life" from Ministry of Education, Culture, Sport, and Technology in Japan and the foundation of Keio university.

References

[1] L. Parker, ALLIANCE: An Architecture for Fault Tolerant Multirobot Cooperation, IEEE Transaction on Robotics and Automation, vol.14, no.2, pp.220-240, 1998.

[2] K. Ozaki, H. Asama, Y. Ishida, A. Matsumoto, K. Yokota, H. Kaetsu and I. Endo, Synchronized Motion by Multiple Mobile Robots using Communication, Proceedings of the 1993 IEEE/RSJ International Conference on Intelligent Robots and Systems, pp.1164-1170, 1993

[3] E. Uchibe, T. Kato, M. Asada and K. Hosoda, Dynamic Task Assignment in a Multiagent/Multitask Environment based on Module Conflict Resolution, Proceedings of IEEE International Conference on Robotics & Automation , pp.3987-3992,2001

[4] R. Emery, K. Sikorski and T. Balch, Protocols for Collaboration, Coordination and Dynamic Role Assignment in a Robot Team, Proceedings of IEEE International Conference on Robotics & Automation, pp3008-3015,2002

[5] L. Chaimowicz, M.F.M. Campos and V. Kumar, Dynamic Role Assignment for Cooperative Robots Proceedings of IEEE International Conference on Robotics & Automation, pp3008-3015,2002.

[6] Thilo Weigel, Jens-Steffen Gutmann, Markus Dietl, Alexander Kleiner, and Bernhard Nobel, CS Freiburg: Coordination Robots for Successful Soccer Playing, IEEE Transactions on Robotics and Automation, Vol.18, No.5, 2002

[7] H. Kidohshi, K. Yoshida and M. Kamiya, Intelligent control method using cubic neural network with multi-levels of information abstraction, Proceedings of the IEEE International Conference on Neural Networks, Vol.5, pp.2326-2331, 1995.

[8] M. Takahashi and K. Yoshida, Intelligent Failure-Proof Control using Cubic Neural Network - Application to Swinging up and Stabilizing Double Pendulum -, Proceedings of the 6th International Conference on Motion and Vibration Control (MOVIC2003), Vol.1, No.02-201, pp.265-270, 2002.

[9] Sakai, R. Hayashi, K. Yoshida, Cooperative Control Method Using Information on Purpose and Evaluation, the 20th Annual Conference of the Robotics Society of Japan (RSJ2002), CD-ROM(1B33), 2002.

[10]R. Tsuzaki, K. Yoshida, Motion Control Based on Fuzzy Potential Method for Autonomous Moblie Robot with Omnidirectional Vision, Journal of the Robotics Society of Japan, Vol.21, No.6, pp.656-662, 2003

Multiple UAV cooperative searching operation using polygon area decomposition and efficient coverage algorithms

Ivan Maza and Anibal Ollero

{imaza,aollero}@cartuja.us.es
Grupo de Robotica, Vision y Control
Escuela Superior de Ingenieros. University of Seville
Camino de los Descubrimientos, s/n
41092 Seville, SPAIN

This paper focuses on the problem of cooperatively searching a given area to detect objects of interest, using a team of heterogenous unmanned air vehicles (UAVs). The paper presents algorithms to divide the whole area taking into account UAV's relative capabilities and initial locations. Resulting areas are assigned among the UAVs, who could cover them using a zigzag pattern. Each UAV has to compute the sweep direction which minimizes the number of turns needed along a zigzag pattern. Algorithms are developed considering their computational complexity in order to allow near-real time operation. Results demonstrating the feasibility of the cooperative search in a scenario of the COMETS multi-UAV project are presented.

1 Introduction

This paper addresses cooperative search problems for UAVs. Cooperative coverage of *a priori* unknown rectilinear environments using mobile robots is discussed in [5]. Ref. [10], uses neural networks to direct robots for complete coverage in complex domains with dynamically moving obstacles. For execution and coordinated control of a large fleet of autonomous mobile robots, Alami et al. propose a Plan Merging Paradigm [1] . The robots incrementally merge their plans into a set of already coordinated plans, through exchange of information about their current state and their future actions.

In recent years, there has been a great deal of work on cooperative control for UAVs. The cooperative control problem that has received the most attention is formation flying [8, 12]. In formation flight, the UAV trajectories are dynamically coupled through the physics of close flight. By exploiting the physical structure of the problem, path planning for formation flying applications can be reduced to path planning algorithms for single vehicles [13].

Unfortunately, there are many other cooperative control problems that do not admit solutions that are extensions of single vehicle solutions. These include cooperative rendezvous [11], coordinated target assignment and intercept [3], multiple task allocation [4], and ISR scenarios [6].

The full solution to many of these cooperative control problems are NP-hard. While formation flight problems can be solved efficiently using numerical methods, there is a need to identify others classes of cooperative control problems that can also be solved efficiently.

Research presented in this paper has been carried out in the framework of the COMETS Project (Real-time coordination and control of multiple heterogeneous unmanned aerial vehicles). In this EU Project, several missions have been considered: detection, aerial mapping, alarm confirmation, fire monitoring, object/person tracking, communications relay, etc. In the mission considered in this paper, a team of heterogeneous UAVs has to cooperatively search an area to detect objects of interest (fire, cars, etc). The problem has been decomposed into the subproblems of (1) determine relative capabilities of each UAV, (2) cooperative area assignment, and (3) efficient area coverage.

The paper is organized as follows: In Section 2 an algorithm based on a divide-and-conquer, sweep-line approach is applied to solve the area partition problem. In Section 3 we introduce the sensing capabilities considered on board the UAVs and the implications with respect to the following sections. A discussion about the covering algorithm that each UAV should use is presented in Section 4. In Section 5 the flexibility in case of re-planning and the complexity of the method outlined is analyzed. Simulations are presented in Section 6 and finally conclusions are given in Section 7.

2 Area decomposition for UAV workspace division

In [9] it was presented a polygon decomposition problem, the *anchored area partition problem*, which has applications to our multiple-UAV terrain-covering mission. This problem concerns dividing a given polygon $\mathcal{P}$ into n polygonal pieces, each of a specified area and each containing a certain point (site) on its boundary. In our case, there are n UAVs U_i, $i = 1, \ldots, n$, each placed at a distinct starting point S_i on the boundary of the polygonal region $\mathcal{P}$ (see Figure 1). The team of UAVs has the mission of completely covering the given region, and to do this most efficiently, the region $\mathcal{P}$ should be divided among the UAVs accordingly with their relative capabilities. Within its assigned region, each vehicle will execute a covering algorithm which is discussed in Section 4.

The algorithm applied in this paper solves the case when $\mathcal{P}$ is convex and contains no holes (no obstacles), which is a preliminar scenario considered in COMETS. A generalized version that handles nonconvex and nonsimply connected polygons is also presented in [9], but computational complexity increases in this case.

Fig. 1. Initial scenario considered.

2.1 Relative capabilities of the UAVs

The low cost UAVs currently involved in the COMETS system are strongly constrained in flying endurance and range. Then, in a first approximation, maximum range of the UAVs seems to be a good measure of their capabilities to perform the mission considered. As UAVs are heterogeneous, range information should be scaled taking into account factors like flight speed and altitude required for the mission, sensitivity to wind conditions, sensing width (due to different camera's fields of view), etc.

Based on the relative capabilities of the vehicles, it is determined what proportion of the area of the region $\mathcal{P}$ should be assigned to each of them. These proportions are represented by a set of values c_i, $i = 1, \ldots, n$, with $0 < c_i < 1$ and $\sum_{i=1}^{n} c_i = 1$. Therefore, the problem considered is as follows: Given a polygon $\mathcal{P}$ and n points (sites) $S_1, \ldots, S_n$ on the polygon, divide the polygon into n nonoverlapping polygons $\mathcal{P}_1, \ldots, \mathcal{P}_n$ such that $\text{Area}(\mathcal{P}_i) = c_i \text{Area}(\mathcal{P})$ and S_i is on $\mathcal{P}_i$.

2.2 Algorithm

Let $S_1, \ldots, S_n$ be a set of sites (start positions of the UAVs), each of them with an *area requirement*, denoted $\text{AreaRequired}(S_i)$, which specifies the desired area of each polygon $\mathcal{P}_i$.

A polygon $\mathcal{P}$ which contains q sites is called a *q-site polygon*, and is called *area-complete* if $\text{AreaRequired}(S(\mathcal{P})) = \text{Area}(\mathcal{P})$ where $\text{AreaRequired}(S(\mathcal{P}))$ is the sum of the required areas by the sites in $\mathcal{P}$.

As it has been stated before, it is assumed a polygon $\mathcal{P}$ convex and with no holes (no obstacles). In this case, it has been shown (see Ref. [9]) that the desired area partition can be achieved using $n-1$ line segments, each of which divides a given q-site ($q > 1$) area-complete polygon $\mathcal{P}$, into two smaller convex polygons — a q_1-site area-complete polygon and a q_2-site area-complete polygon with $q_1 + q_2 = q$ and $q_1, q_2 > 0$. The computation of each segment can be done using an algorithm based on a divide-and-conquer, sweep-line approach presented in [9]. This procedure should be called exactly $n-1$ times to partition a convex, n-site area-complete polygon into n convex, 1-site area-complete polygons.

3 Sensing capabilities

A team of UAVs has to perform a cooperative search operation over an area to detect objects of interest. Consider a Base Coordinate System (BCS), fixed in the environment (x-axis towards north, y-axis west, z-axis up) and UAVs equipped with sensors and cameras. Sensors allow the vehicles to determine their own coordinates relative to BCS and those of any point detected in its sensing region.

The UAVs are assumed to have cameras without orientation devices. In fact, light UAVs have strong payload constraints that may preclude the use of ginbals and other devices to change the orientation of the on-board cameras.

Each UAV has associated an UCS (UAV Coordinate System) that changes its point of origin and its orientation with the movement of the vehicle (x-axis forward, y-axis left, z-axis up). On board cameras are fixed, oriented in the x-z plane of the UCS and defined by the angle $(-\alpha)$ with respect to the x-axis.

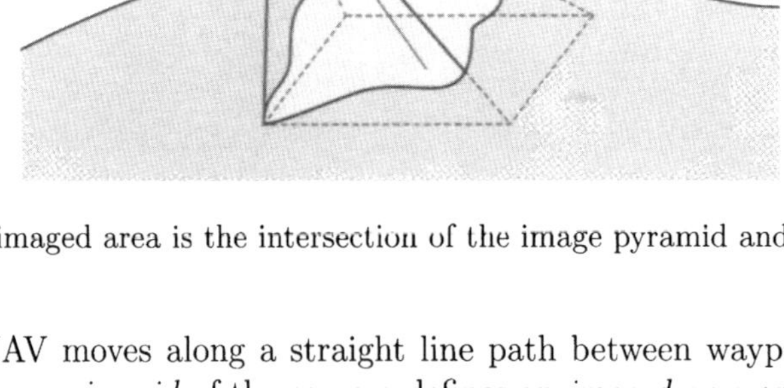

Fig. 2. The imaged area is the intersection of the image pyramid and the terrain.

As the UAV moves along a straight line path between waypoints taking shots, the *image piramid* of the camera defines an *imaged area* on the terrain (see Figure 2). Considering a plain terrain, it can be shown that the sensing width of an UAV moving in the x-z plane of the UCS is given by:

$$w = 2z_{BCS}\tan\gamma\left[\sin\alpha + \cos\alpha\tan\left(\frac{\pi}{2} - \alpha - \beta\right)\right] \tag{1}$$

where z_{BCS} is the altitude of the UAV, and angles β and γ determine the field of view of the camera.

As the planar algorithm for covering a given area is based on a zigzag pattern, the spacing of the parallel lines will be determined in first approximation by equation (1). To be able to generalize this planar algorithm directly to the three-dimensional environment considered, the nonplanar surface (area) to be covered must be a *vertically projectively planar surface.* That is, a vertical line passing through any point on the surface intersects it at only one point.

4 Individual areas coverage algorithm

Once each UAV has an area assigned (corresponding to a convex polygon $\mathcal{P}_i$), an algorithm is needed to cover this area searching for objects of interest. Those convex areas can be easily and efficiently covered by back and forth motion along rows perpendicular to the sweep direction (simulations have shown that in general this pattern is faster than the spiral pattern). The time to cover an area in this manner consists of the time to travel along the rows plus the time to turn around at the end of the rows. Covering an area for a different sweep direction results in rows of approximately the same total length; however, there can be a large difference in the number of turns required as illustrated in Figure 3. In the COMETS Project, autonomous helicopters are included in the heterogeneous team of UAVs. Helicopter turns take a significant amount of time: the helicopter must slow down, stay in hovering, make the turn, and then accelerate.

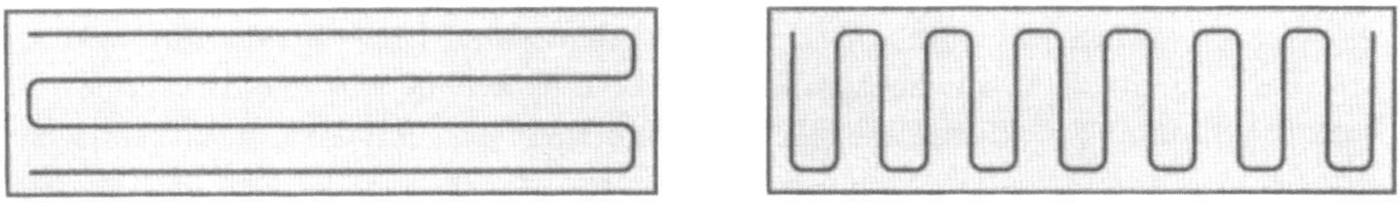

Fig. 3. The number of turns is the main factor in the cost difference of covering a region along different sweep directions.

We therefore wish to minimize the number of turns in an area, and this is proportional to the altitude of the polygon measured along the sweep direction. The *altitude* of a polygon is just its height. We can use the *diameter function* $d(\theta)$ to describe the altitude of a polygon along the sweep direction. For a given angle θ, the diameter of a polygon is determined by rotating the polygon by $-\theta$ and measuring the height difference between its highest and lowest point. The altitude of a polygon $\mathcal{P}_i$ for a sweep direction at an orientation of α is $d_{\mathcal{P}_i}\left(\alpha - \frac{\pi}{2}\right)$.

The shape of a diameter function can be understood by considering the height of the polygon as it rolls along a flat surface (Figure 4). Starting with one edge resting on the surface, we can draw a segment from the pivot vertex to another vertex of the polygon, and the height of the polygon will be determined by this vertex. Whenever the polygon has rolled on to the next side

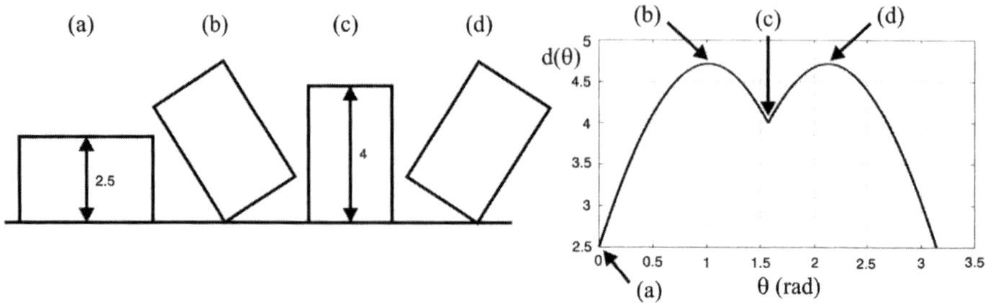

Fig. 4. Example of a simple diameter function.

or when an edge at the top of the polygon becomes parallel to the surface, we will change to a different segment (from a different pivot vertex or to a different top vertex). Therefore, a diameter function has the following form for an n sided convex polygon:

$$d(\theta) = \begin{cases} k_1 \sin(\theta + \phi_1) & \theta \in [\theta_0, \theta_1) \\ k_2 \sin(\theta + \phi_2) & \theta \in [\theta_1, \theta_2) \\ \vdots \\ k_{2n} \sin(\theta + \phi_{2n}) & \theta \in [\theta_{(2n-1)}, \theta_{2n}) \end{cases} \tag{2}$$

where $\theta_0 = 0$ and $\theta_{2n} = 2\pi$. The diameter function is piecewise sinusoidal; its "breakpoints" θ_i occur when an edge of the rotated polygon is parallel to the horizontal. The minimum of the diameter function must lie either at a critical point ($d'_{\mathcal{P}_i} = 0$) or at a breakpoint. However, for any critical point in between breakpoints $d''_{\mathcal{P}_i} < 0$, which means that it corresponds to a maximum. Therefore, the minimum must lie at a breakpoint and these breakpoints correspond to when the sweep direction is perpendicular to an edge of the perimeter. Testing each of these sweep directions, the minimum can be determined.

A similar approach can also be applied when obstacles are present inside the areas. In this case, the altitude to be minimized is the sum of the diameter function of the perimeter plus the diameter functions of the obstacles.

5 Complexity analysis and reconfiguration process

Special attention has been focused to this issue due to the real time operation required in the COMETS Project.

The computation of the full partition of a convex n-site polygon $\mathcal{P}$ with v vertices, requires $\mathcal{O}(n-1)(n+v)$ time in the worst case. Resulting polygons are assigned to the UAVs, and each of them has to compute the sweep direction which minimizes the number of turns needed along the zigzag pattern. It only implies that each UAV has to test a number of directions equal to the number of edges of its assigned polygon $\mathcal{P}_i$.

Therefore, the whole process has a low computational cost which could also be shared easily among a control centre and the UAVs:

- Each UAV computes its relative capabilities (simple algebraic expressions).
- The control centre computes the complete partition and assigns the resulting areas to each UAV ($\mathcal{O}(n-1)(n+v)$ time in the worst case).
- Each UAV determines its more efficient sweep direction (test a number of directions equal to the number of edges of its polygon).

The functionality of this control centre could also be performed by an UAV with enough computational capability.

If system reconfiguration is needed, the system can quickly adapt to the new scenario. For example, if an UAV is lost, remaining UAVs have to perform the detection mission properly. It implies that a new area partition process must be triggered. In that case, initial locations of the UAVs are not in the boundary of the given area and the algorithm described in Section 2.2 is not valid. A different algorithm described in [9] should be applied, but computational complexity remains bounded and low. In the next section, this re-planning has been handled with a minor modification of the algorithm described in Section 2.2, due to the low number of vehicles involved.

6 Implementation details and simulations results

Algorithms have been implemented in C++ using the CGAL library [7] for computational geometry support.

Table 1. Initial coordinates, camera angles, sensing width and relative capabilities of the UAVs.

	x_{BCS}(m)	y_{BCS}(m)	z_{BCS}(m)	α_i(rad)	β_i(rad)	γ_i(rad)	w_i(m)	c_i(%)
UAV1	190.00	0.00	29.00	$\pi/2$	$\pi/8$	$\pi/8$	24.02	24.92
UAV2	550.00	100.00	34.00	$\pi/3$	$\pi/7$	$\pi/8$	25.45	41.81
UAV3	225.38	412.69	20.00	$\pi/3$	$\pi/6$	$\pi/6$	20.00	33.27

In this simulation, three UAVs have to search an area defined by a convex polygon with seven edges. We assume different cameras on board the UAVs, each of them defined by different values of the angles α, β and γ. In Table 1, initial coordinates of the UAVs and their relative capabilities (c_i – see Section 2.1) are listed. Those values for c_i have been obtained via an estimation of the maximum range in function of parameters like remaining fuel, specific consumption, flight speed, etc. (see Ref. [2]) Using equation (1), and assuming constant altitudes during the mission, sensing width (w_i) of each UAV can be easily derived (see also Table 1).

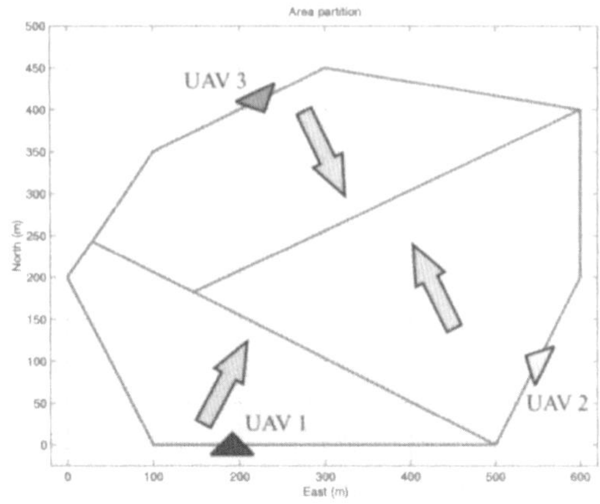

Fig. 5. Area partition simulation results. Optimal sweep directions have been represented by arrows.

Area partition has been computed using the algorithm presented in Section 2.2. The resulting assignment is shown in Figure 5. It can be seen that each UAV has been assigned an area (convex polygon) according with its relative capabilities.

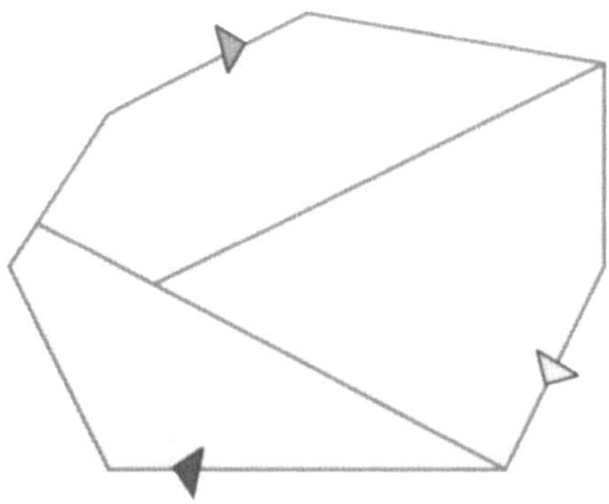

Fig. 6. Resulting zigzag patterns minimizing the number of turns required.

Each UAV has to find the optimal sweep direction which minimizes its assigned polygon's altitude. As it has been explained in Section 4, only the directions which are perpendicular to the edges of each polygon have to be tested. Resulting directions have been represented by arrows in Figure 5. Then, each UAV has to compute the waypoints needed to follow a zigzag pattern perpendicular to those directions (see Figure 6). Distance between parallel lines depends on the sensing width of the UAV.

Finally, a reconfiguration process has been simulated. When UAV3 is lost, remaining UAVs have to cover the whole area. A new area partition process has to be triggered and new sweep directions are followed (see Figure 7).

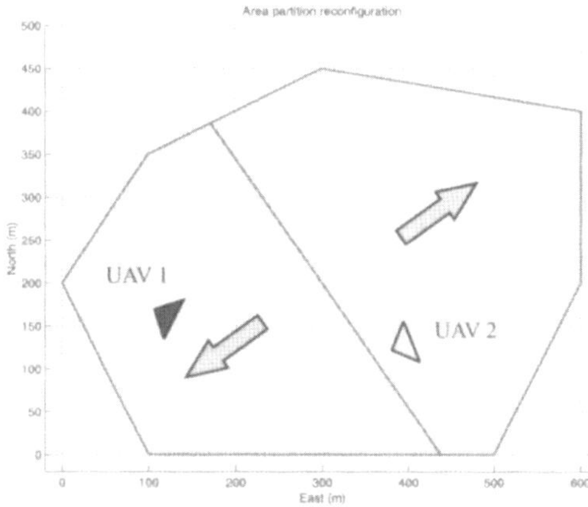

Fig. 7. UAV3 is lost and remaining UAVs have to reconfigure their flight plans to cover the whole area.

7 Conclusions and future research

The problem of cooperative searching a given area (convex polygon) by a team of UAVs taking into account their different sensing and range capabilities has been considered. The solution adopted in this paper is fully adapted to a simple scenario inspired by experiments developed in the COMETS Project (convex polygons and no obstacles), but all the algorithms could be extended to more complex problems with bounded (and relatively low) computational load. It provides a spectrum of solutions useful for real-time implementation (experiments are expected for next year).

It would be very interesting to modify the altitude of the UAVs during the mission execution to maximize the capabilities of their cameras (increase the altitude in sectors with low detection probability).

The methods presented in this paper can be easily extended to the cooperation of autonomous aerial and ground vehicles, which is being addressed in the framework of the CROMAT Spanish project.

Acknowledgements

This work was partially funded by European Union COMETS Project (IST-2001-34304). The authors would like to thank LAAS (Laboratory for Analysis and Architecture of Systems) for their technical reports in COMETS Project, which have been very useful to focus this research. Partial funding of the CROMAT Project of the Spanish Research and Development Program (DPI2002-04401-C03-03) is also acknowledged.

References

1. Alami R, Robert F, Ingrand F, Suzuki S (1995) Multi-robot cooperation through incremental plan-merging. In: Proceedings of the IEEE International Conference on Robotics and Automation, pp. 2573–2579. Nagoya, Japan
2. Barcala M, Rodriguez A (1998) Helicopteros. EUIT Aeronautica, Madrid
3. Beard R W, McLain T W, Goodrich M (2002) Coordinated target assignment and intercept for unmanned air vehicles. In: Proceedings of the IEEE International Conference on Robotics and Automation, pp. 2581–2586. Washington
4. Bellingham J, Tillerson M, Richards A, How J P (2001) Multi-task allocation and path planning for cooperating UAVs. In: Cooperative Control: Models, Applications and Algorithms, pp. 1–19, Conference on Coordination, Control and Optimization.
5. Butler Z J, Rizzi A A, Hollis R L (2000) Cooperative coverage of rectilinear environments. In: Proceedings of the IEEE International Conference on Robotics and Automation, pp. 2722–2727. San Francisco, CA
6. Chandler P R, Pachter M, Swaroop D, Fowler J M, Howlett J K, Rasmussen S, Schumacher C, Nygard K (2002) Complexity in UAV cooperative control. In: Proceedings of the American Control Conference. Anchorage, AK
7. Computational Geometry Algorithms Library (CGAL). Web address: `http://www.cgal.org/`
8. Giulietti F, Pollini L, Innocenti M (2000) Autonomous formation flight. IEEE Control Systems Magazine 20:34–44
9. Hert S, Lumelsky V (1998) Polygon area decomposition for multiple-robot workspace division. International Journal of Computational Geometry and Applications, 8(4):437-466.
10. Luo C, Yang S X, Stacey D A, Jofriet J C (2002) A solution to vicinity problem of obstacles in complete coverage path planning. In: Proceedings of the IEEE International Conference on Robotics and Automation, pp. 612–617. Washington DC
11. McLain T, Beard R (2000) Cooperative rendezvous of multiple unmanned air vehicles. In: Proceedings of the AIAA Guidance, Navigation and Control Conference, paper no. AIAA 2000–4369. Denver, CO
12. Pachter M, D'Azzo J J, Proud A W (2001) Tight formation flight control. AIAA Journal of Guidance, Control and Dynamics 24:246–254
13. Pledgie S T, Hao Y, Ferreira A M, Agrawal S K, Murphey R (2002) Groups of unmanned vehicles: Differential flatness, trajectory planning, and control. In: Proceedings of the IEEE International Conference on Robotics and Automation, pp. 3461–3466. Washington DC
14. Wesley H., Huang W H (2001) Optimal line-sweep-based decompositions for coverage algorithms. In: Proceedings of the 2001 IEEE International Conference on Robotics and Automation, 1:27–32. Seoul, Korea

Part VI

Control Architectures

A Distributed Architecture for Autonomous Unmanned Aerial Vehicle Experimentation

P. Doherty, P. Haslum, F. Heintz, T. Merz, P. Nyblom, T. Persson, and B. Wingman

Linköping University
Dept. of Computer and Information Science
SE-58183 Linköping, Sweden
patdo@ida.liu.se

Summary. The emerging area of intelligent unmanned aerial vehicle (UAV) research has shown rapid development in recent years and offers a great number of research challenges for distributed autonomous robotics systems. In this article, a prototype distributed architecture for autonomous unmanned aerial vehicle experimentation is presented which supports the development of intelligent capabilities and their integration in a robust, scalable, plug-and-play hardware/software architecture. The architecture itself uses CORBA to support its infrastructure and it is based on a reactive concentric software control philosophy. A research prototype UAV system has been built, is operational and is being tested in actual missions over urban environments.

1 Introduction

The emerging area of intelligent unmanned aerial vehicle (UAV) research has shown rapid development in recent years and offers a great number of research challenges in the area of distributed autonomous robotics systems. Much previous research has focused on low-level control capability with the goal of developing controllers which support the autonomous flight of a UAV from one way-point to another. The most common type of mission scenario involves placing sensor payloads in position for data collection tasks where the data is eventually processed off-line or in real-time by ground personnel. Use of UAVs and mission tasks such as these have become increasingly more important in recent conflict situations and are predicted to play increasingly more important roles in any future conflicts.

Intelligent UAVs will play an equally important role in civil applications. For both military and civil applications, there is a desire to develop more sophisticated UAV platforms where the emphasis is placed on development of intelligent capabilities and on abilities to interact with human operators and additional robotic platforms. Focus in research has moved from low-level control towards a combination of low-level and decision-level control integrated in sophisticated software architectures. These in turn, should also integrate well with larger network-centric based C^4I^2 systems. Such platforms are a prerequisite for supporting the capabilities required for the increasingly more complex mission tasks on the horizon and an ideal testbed for the development and integration of distributed AI technologies.

The WITAS[1] Unmanned Aerial Vehicle Project[2] [4] is a basic research project whose main objectives are the development of an integrated hardware/software VTOL (Vertical Take-Off and Landing) platform for fully-autonomous missions and its future deployment in applications such as traffic monitoring and surveillance, emergency services assistance, photogrammetry and surveying.

Basic and applied research in the project covers a wide range of topics which include the development of a distributed architecture for autonomous unmanned aerial vehicles. In developing the architecture, the larger goals of integration with human operators and other ground and aerial robotics systems in network centric C^4I^2 infrastructures has been taken into account and influenced the nature of the base architecture. In addition to the software architecture, many AI technologies have been developed such as path planners, chronicle recognition and situational awareness techniques. The architecture supports modular and distributed integration of these and any additional functionalities added in the future. The WITAS UAV hardware/software platform has been built and successfully used in a VTOL system capable of achieving a number of complex autonomous missions flown in an interesting *urban* environment populated with building and road structures. In one mission, the UAV autonomously tracked a moving vehicle for up to 20 minutes. In another, several building structures were chosen as survey targets and the UAV autonomously generated a plan to fly to each and take photos of each of its facades and then successfully executed the mission.

Fig. 1. Aerial photo over Revinge, Sweden

Figure 1 shows an aerial photo of our primary testing area located in Revinge, Sweden. An emergency services training school is located in this area and consists of a collection of buildings, roads and even makeshift car and train accidents. This provides an ideal test area for experimenting with traffic surveillance, photogrammetric and surveying scenarios, in addition to scenarios involving emergency services. We have also constructed an accurate 3D model for this area which has proven invaluable in simulation experiments and parts of which have been used in the onboard GIS.

Fig. 2. The WITAS RMAX Helicopter

In the remainder of the paper, we will focus primarily on a description of the engineered on-board system itself, parts of the distributed CORBA-based software architecture and interaction with the primary flight control system. There are a great many topics that will not be considered due to page limitations, particularly in the area of knowledge representation [5, 7, 8], symbol grounding [14, 13], and deliberative capabilities [24, 6], task-based planning [6], specific control modes [3, 16, 27] and their support, image processing [18]

[1]WITAS (pronounced *vee-tas*) is an acronym for the Wallenberg Information Technology and Autonomous Systems Laboratory at Linköping University, Sweden.

[2]This work and the project is funded by a grant from the Wallenberg Foundation.

and in technologies such as dialogue management for support of ground operation personnel [26].

2 The VTOL and Hardware Platform

The WITAS Project UAV platform we use is a slightly modified Yamaha RMAX (figure 2). It has a total length of 3.6 m (incl. main rotor), a maximum take-off weight of 95 kg, and is powered by a 21 hp two-stroke engine. Yamaha equipped the radio controlled RMAX with an attitude sensor (YAS) and an attitude control system (YACS). Figure 3 shows a high-level schematic of the hardware platform that we have built and integrated with the RMAX platform.

The hardware platform consists of three PC104 embedded computers (figure 3). The primary flight control (PFC) system consists of a PIII (700Mhz) processor, a wireless Ethernet bridge and the following sensors: a RTK GPS (serial), and a barometric altitude sensor (analog). It is connected to the YAS and YACS (serial), the image processing computer (serial) and the deliberative computer (Ethernet). The image processing (IP) system consists of a second PC104 embedded computer (PIII 700MHz), a color CCD camera (S-VIDEO, serial interface for control) mounted on a pan/tilt unit (serial), a video transmitter (composite video) and a recorder (miniDV). The deliberative/reactive (D/R) system runs on a third PC104 embedded computer (PIII 700MHz) which is connected to the PFC system with Ethernet using CORBA event channels. The D/R system is described in more detail in section 4.

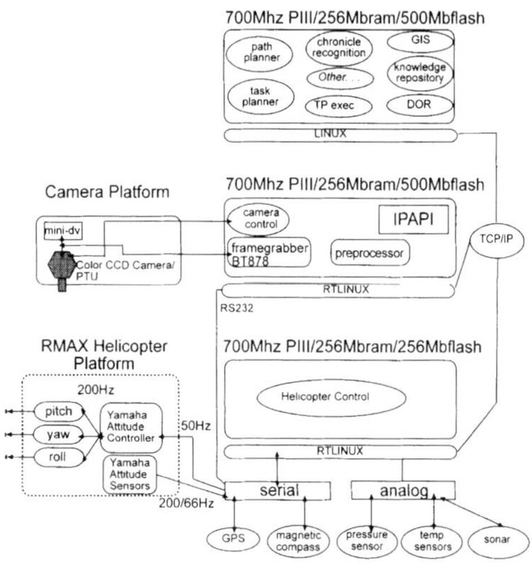

Fig. 3. On-Board Hardware Schematic

3 Control

A great deal of effort has gone into the development of a control system for the WITAS UAV which incorporates a number of different control modes and includes a high-level interface to the control system. This enables other parts of the architecture to call the appropriate control modes dynamically during the execution of a mission. The ability to switch modes contingently is a fundamental functionality in the architecture and can be programmed into the task procedures associated with the reactive component in the architecture. We developed and tested the following autonomous flight control modes:

- take-off (TO-Mode) and landing via visual navigation (L-Mode, see [16])
- hovering (H-Mode)
- dynamic path following (DPF-Mode, see [3])
- reactive flight modes for interception and tracking (RTF-Mode).

These modes and their combinations have been successfully demonstrated in a number of missions at the Revinge testflight area. The primary flight control system (bottom PC104 in Figure 3) can be described conceptually as consisting of a device, reactive, behavior and application layer as depicted in figure 4a.[3]Each layer consists of several functional units which, with the exception of the application layer, are executed periodically with comparable period and worst case execution times. The implementation is based on RTLinux (GPL), which runs an ordinary Linux distribution as a task with lower priority. The application layer is realized in user space as no hard real-time execution is required, while the other layers contain functional units running as kernel modules in hard real-time.

A CORBA interface is set up between the PFC system and the deliberative/reactive system of the software architecture (top PC104 in Figure 3). Network communication between the two is physically realized using Ethernet with CORBA event channels and CORBA method calls. Task procedures in the D/R system issue commands to the PFC system to initiate different flight modes and receive helicopter states and events from the PFC system which influence the activity of the task procedure.

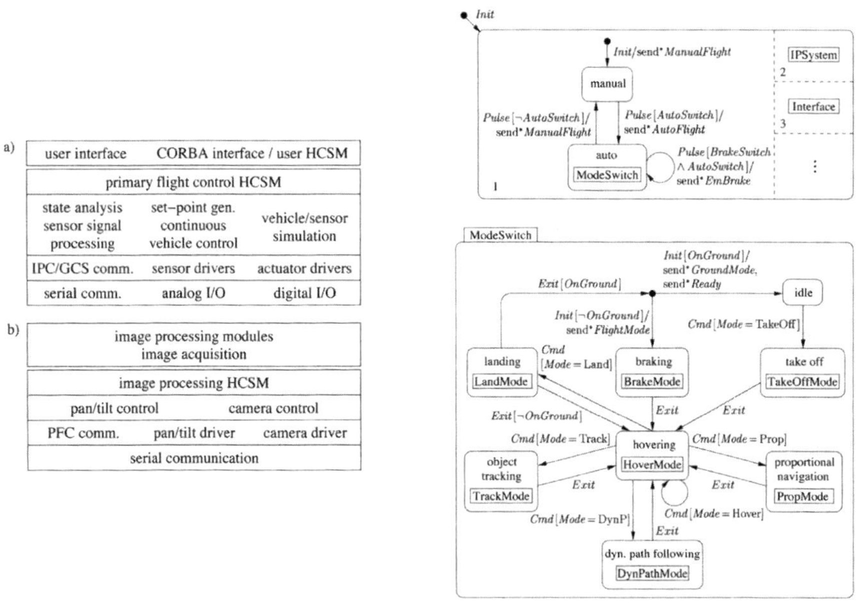

Fig. 4. PFC and IP System Schematic (left) and HCSMs for Flight Mode Switching (right)

Hierarchical concurrent state machines (HCSMs) are used to represent states of the PFC system. HCSMs are mixed Mealy and Moore machines with hierarchical and orthogonal decomposition. These are executed explicitly in the PFC system. We use a visual formalism similar to Harel's statecharts [12] to describe HCSMs.

Figure 4b shows that part of the flight control HCSM involving mode switching for different flight modes.[4]

[3]The image processing (IP) system (middle PC104 in Figure 3) has a similar structure, but will not be considered in this paper.

[4]Nested state machines are symbolized as rectangular boxes in a state node, *Pulse* is an event sent periodically, *Init* triggers a transition from an entry state (circular node) when condition holds, *Exit*

4 Software System

A great deal of effort in the artificial intelligence community has recently been directed towards deliberative/reactive control architectures for intelligent robotic systems. Two good sources of reference are [1] and [15]. Many of the architectures proposed can be viewed in a loose sense as layered architectures with a control, a reactive and a deliberative layer (see [9]). The software architecture we have developed does have deliberative, reactive and control components, but rather than viewing it from the perspective of a layered architecture, it is best viewed as a *reactive concentric* architecture which uses services provided by both the deliberative and control components in a highly distributed and concurrent manner. At any time during the execution of a mission, a number of interacting concurrent processes at various levels of abstraction, ranging from high-level services such as path planners to low-level services such as execution of control laws, are being executed with various latencies.

We have chosen CORBA[5] as a basis for the design and implementation of a loosely coupled distributed software architecture for our aerial robotic system. We believe this is a good choice which enables us to manage the complexity of a deliberative/reactive (D/R) software architecture with as much functionality as we require for our applications. It also ensures clean and flexible interfacing to the deliberative and control components in addition to the hardware platform via the use of IDL (Interface Definition Language).

In short, CORBA (Common Object Request Broker Architecture) is middleware that establishes client/server relationships between objects or components. A component can be a complex piece of software such as a path planner, or something less complex such as a task procedure which is used to interface to helicopter or camera control. Objects or components can make requests to, and receive replies from, other objects or components located locally in the same process, in different processes, or on different processors on the same or separate machines.

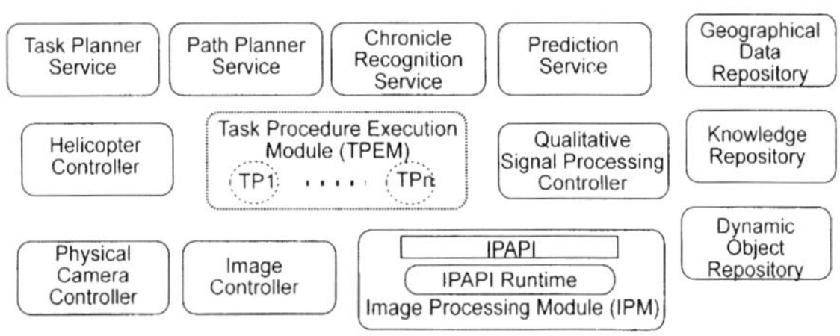

Fig. 5. Some deliberative, reactive and control services

Many of the functionalities which are part of the architecture can be viewed as clients or servers where the communication infrastructure is provided by CORBA, using services such as standard and real-time event channels. [6] Figure 5 depicts an (incomplete) high-level schematic of some of the software components used in the architecture. Each of these functionalities has been implemented and are being used and developed in our applications. This architectural choice provides us with an ideal development environment and versatile run-time system with built-in scalability, modularity, software relocatability on various hard-

is sent to the superstate when entering an exit state (square node), superstate transitions are executed prior to substate transitions.

[5]We are currently using TAO/ACE [20]. The **A**ce **O**rb is an open source implementation of CORBA 2.6.

[6]Event channels are specified in the CORBA Event Service standard (`http://www.omg.org/technology/document/corbaservices/_spec/_catalog.htm`) and allow decoupled passing of arbitrary data from one or more senders to one or more receivers.

ware configurations, performance (real-time event channels and schedulers), and support for plug-and-play software modules.

4.1 The Modular Task Architecture

The conceptual layers used to describe an architecture are less important than the actual control flow and interaction between processes invoked at the various layers. What is most important is the ability to use deliberative services in a reactive or contingent manner (which we call *hybrid deliberation*) and to use traditional control services in a reactive or contingent manner (which is commonly called *hybrid control*). Figure 6 depicts some of these ideas. Due to the reactive concentric nature of the architecture, *task procedures* and their execution are a central feature of the architecture.

The modular task architecture (MTA) is a reactive system design in the procedure-based paradigm developed for loosely coupled heterogeneous systems such as the WITAS aerial robotic system. A *task* is a behavior intended to achieve a goal in a limited set of circumstances. A *task procedure* (TP) is the computational mechanism that achieves this behavior. A TP is essentially event-driven; it may open its own (CORBA) event channels, and call its own services (both CORBA and application-oriented services such as path planners); it may fail to perform its task due to the limited set of circumstances in which it is specified to operate; it may be called, spawned or terminated by an outside agent, or by other TPs; and it may be executed concurrently.

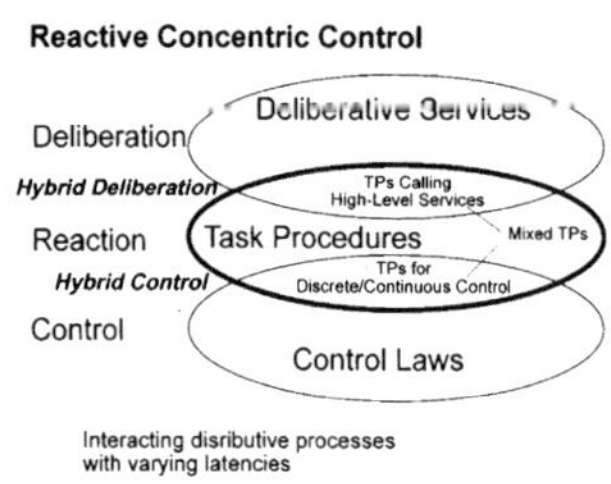

Fig. 6. Reactive Concentric Control

Formally, a TP is any CORBA object that implements the Witas::Task interface and adheres to the behavioral restrictions of the MTA specification.[7]

To increase the ease and flexibility of specifying and implementing individual TPs, a Task Specification Language (TSL) has been developed. TSL is an XML-based code markup language intended to simplify the implementation of TPs. The idea is that for any TP there is an application dependent or operative part which can be implemented in any host language supported by TSL [19]. Currently TSL supports C++ and it would be straightforward to extend it to support languages such as JAVA and C. The application independent part of the TP is set up automatically in a translation process from TSL to the host language. Figure 7 shows a schematic of a special type of TP specified in the TSL with some of the essential markup tags, including those for a finite state machine (fsm) block. Task procedures can be used for many different purposes. Figure 6 depicts several types of usage as hybrid deliberation TPs, hybrid control TPs or mixed TPs.

A good way to view and represent a hybrid controller is as an augmented automaton. We have provided structural and macro tags for this purpose which can be used in a TP. Figure 7b shows a TSL schematic for the finite state machine specification which would be included in the $\langle\mathbf{fsm}\rangle \ldots \langle/\mathbf{fsm}\rangle$ tag block in Figure 7a. Some additional tags not listed allow for specification of jumps to other states, exits on failure and the setting up of execution checkpoints.

An important property of D/R architectures is their ability to integrate and execute processes at the deliberative, reactive and control levels concurrently and to switch between

[7]The TPEM (Task Procedure Execution Module) in Figure 5 represents the set of TP CORBA objects used. There is no centralized execution mechanism.

```
⟨tp name = Taskname⟩
⟨declarations⟩
        ⟨parameter . . . /⟩ . . . ⟨parameter . . . /⟩
        // other declarations, e.g. local variables and constants
⟨/declarations⟩
⟨services⟩
        // CORBA server objects, event channels used, etc.
⟨/services⟩
⟨init⟩
        // Host code for task specific initialization
⟨/init⟩
⟨destroy⟩
        // Host code for task specific cleanup
        // CORBA cleanup handled automatically
⟨/destroy⟩
⟨function⟩ . . . ⟨/function⟩ . . . //more fns
⟨start⟩
        // Executed with call to TP start() method
        // Host code plus host code macros
        // Typically will perform some setup then
        // a ⟨jump⟩ to FSM state
⟨/start⟩
⟨fsm⟩
        // Main behavioral specification in form
        // of a finite state machine
⟨/fsm⟩
⟨/tp⟩
```

(a) TSL tags and partial schematic for a TP specification.

```
⟨fsm⟩
⟨statename =sname⟩
⟨action⟩
        // Executed whenever TP enters this state
⟨/action⟩
        // State specific reactions to events
⟨reaction event = "event_name"⟩ . . . ⟨/reaction⟩
        .
        .
        .
⟨reaction event = "event_name"⟩ . . . ⟨/reaction⟩
⟨/state⟩

//More state specifications . . .

        // Global reactions to events
⟨reaction⟩ . . . ⟨/reaction⟩
        .
        .
        .
⟨reaction⟩ . . . ⟨/reaction⟩
⟨/fsm⟩
```

(b) TSL tags and partial schematic for an fsm specification

Fig. 7. TP Specifications with TSL tags.

them seamlessly. Reactive components must be able to interface in a clean manner to helicopter and camera control and adapt activity to contingencies in the environment in a timely manner. Likewise, they must be able to interface to deliberative services in a clean manner. Consequently, use of TPs for hybrid control and hybrid deliberation is of fundamental importance. An example is provided in the following section.

4.2 Using Task Procedures for Flight Control

In this section, the interface between TPs for hybrid control and flight control modes, in addition to some limited hybrid deliberation, will be described. The main ingredients are TPs in the shape of finite state machines and the use of a declarative flight command language (FCL). Events from the flight controller (passed via event channels) create a partial view of the state of the UAV. Based on this, and its own state, a TP can dynamically generate appropriate flight commands, which are sent back to the control system.

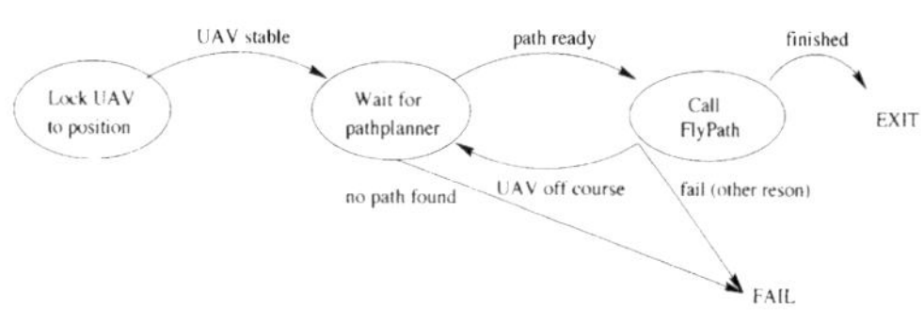

Fig. 8. The NavToPt automaton

Suppose the UAV has been given the task of finding and tracking a vehicle with a particular signature, believed to be in a certain region of Revinge. A TP for such a mission consists of several subtasks, each carried out by the top-level TP invoking other TPs: The first is NavToPt (depicted in Figure 8), which navigates the UAV safely to a waypoint in the region of interest (ROI).

To do so, NavToPt makes use of both deliberative and control services. First, it calls a path planning service [25] which provides a (segmented) path from the UAVs current position to the waypoint (provided such a path can be found). Then, it invokes a FlyPath TP which takes care of passing the path segments, target velocity and other parameters to the dynamic path following flight mode in the control system. During flight, FlyPath receives events from the control system (via event channels) and responds appropriately, for example issuing a command to enter hovering mode when the end of the last path segment is reached. It also generates events reporting on the progress of the flight, which are in turn monitored by NavToPt.

Once the UAV has arrived at the goal point, the parent TP configures and starts image processing services [18] to attempt identification of vehicles in the area. If a vehicle matching the initial signature is found, the parent TP starts a new TP FlyTo (see Figure 9) which uses proportional navigation mode [27], one of the existing RTF-modes, to position the UAV relative to the vehicle. Since the vehicle is probably moving, FlyTo instances will be continually terminated and restarted with new parameters. Each instance of FlyTo, while running, generates a stream of flight commands passed to the PFC system, where they are handled by the PN flight mode functional unit, and receives a stream of events from the flight control system and image processing system. Both automata described are in fact implemented using TPs with structure similar to that in the two TSL schemata.

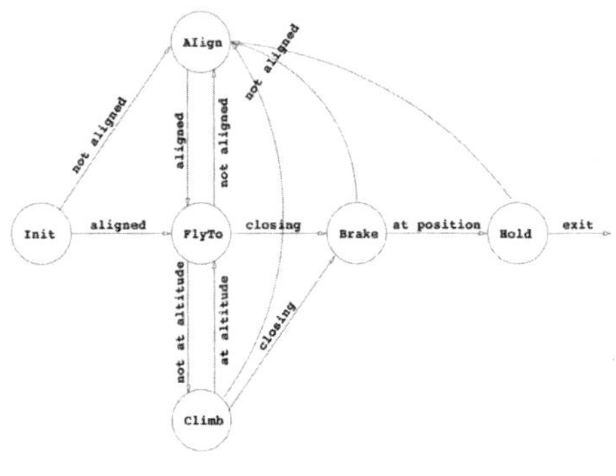

Fig. 9. The FlyTo automaton

The flight command language (FCL) is used to bridge the abstraction gap between task procedures in the reactive component and flight and camera modes in the control component. Figure 10 shows a number of representative flight commands used for the proportional navigation flight control mode.

In this case the PN mode is controlled by providing commands to XY, Z and Yaw channels, in addition to an administration channel. The administration channel is used to set various parameters. It is also used to inform the PN mode which object to fly to (FlyObject), this may be a waypoint or a moving object on the ground previously identified, and the object to look at with the camera (LookObject). Additional flight commands are provided for other flight control modes and camera control. In the case of dynamic path following, representations of parameterized curves are passed as arguments to flight commands, these arguments are generated via a task procedure call to the path planning service.

XY Channel	**Z Channel**
Cruise(), SlowDown()	Land(), FlyAtAltitude()
Slow(), Brake()	ClimbTo()
FlyToFlyObject(), FlyWithVelocity()	**Yaw Channel**
FlyTowards(), LockOnFlyObject()	FlyWithYaw(), FlyWithYawOfYourself()
LockOnPosition(), LockOnLookObject()	FlyCleanly()

Fig. 10. Representative Flight Commands for Proportional Navigation.

4.3 Benefits of the Approach

The use of TSL with XML markup tags and a host language translator provides a modular, extensible means of constructing and experimenting with TPs that hides the syntactic overhead associated with initialization of CORBA event channels, services and error handling code. Each TP is automatically generated as a standard CORBA object and can be spawned or terminated using the Witas::Task interface. We believe that the greatest flexibility is provided if the actual semantic content of a TP can be specified in a familiar language such as C++ or Java. The structuring of code in this manner also provides a clean way to analyze the behavior of a TP formally. The use of a flight command language provides a smooth transition from discrete to continuous control behavior between the reactive and control components.

New control modes can be added in a modular manner and interfaced by adding new flight commands. The flight command streams themselves can be generated and analyzed in many different ways. The use of event channels and filters by TPs also provides a flexible means of dynamically constructing partial state representations which drive the behaviors of TPs. In a similar manner, deliberative functionalities can be modularly added and packaged as CORBA objects. These may include legacy code. For example, the chronicle recognition software used is based on IXTET [11][8] and is wrapped directly as a CORBA object . These objects may physically be distributed between the UAV itself and various ground stations. This is useful if the computational resources required for a service exceed those provided by the UAV platform.

5 Related Work

There is, without a doubt, much activity with UAVs in the military sector, primarily with fixed-wing high altitude vehicles. Much of the focus is on design of physical platforms, low-level control, and sensing [21]. Less focus has been placed on the type of system described here. This also applies to the majority of commercial attempts at marketing UAVs, although here, there has been more activity with VTOL platforms. Academic research with UAVs is increasing at a rapid pace. Here again, the majority of projects have focused on low-level control, although there is more activity with software architectures and integration of some deliberative capabilities. The work closest to ours is that being pursued at Georgia Tech[10]. Rather than list publications, we refer the reader to the following (non-exhaustive) list of websites for information about interesting university and institute UAV activities: Georgia Tech[10], M.I.T[17], Carnegie-Mellon[2], U. of C., Berkeley[29], Stanford University[28], ONERA RESSAC [22].

References

1. R. C. Arkin. *Behavior-Based Robotics*. MIT Press, 1998.
2. Carnegie Mellon University. `http://www.ri.cmu.edu/projects/project_93.html`.
3. G. Conte, S. Duranti, and T. Merz. Dynamic 3D path following for an autonomous helicopter. In *Proceedings of the 5th IFAC Symposium on Intelligent Autonomous Vehicles*, 2004.

[8] In fact, we use a new implementation, CRS, developed by C. Dousson at France Telecom (`http://crs.elibel.tm.fr/en/`).

4. P. Doherty, G. Granlund, K. Kuchcinski, K. Nordberg, E. Sandewall, E. Skarman, and J. Wiklund. The WITAS unmanned aerial vehicle project. In *Proceedings of the 14th European Conference on Artificial Intelligence*, pages 747–755, 2000. `http://www.liu.se/ext/witas`.
5. P. Doherty, J. Kachniarz, and A. Szałas. Using contextually closed queries for local closed-world reasoning in rough knowledge databases. In *[23]*, 2003.
6. P. Doherty and J. Kvarnström. TALplanner: A temporal logic based planner. In *AI Magazine*, volume 22, pages 95–102. AAAI Press, 2001.
7. P. Doherty, W. Łukaszewicz, A. Skowron, and A. Szałas. Combining rough and crisp knowledge in deductive databases. In *[23]*, 2003.
8. P. Doherty, W. Łukaszewicz, and A. Szałas. Approximative query techniques for agents using heterogenous ontologies. In *Int'l Conference on Principles of Knowledge Representation and Reasoning (KR-04)*, 2004.
9. E. Gat. Three-layer architectures. In D. Kortenkamp, R. P. Bonassao, and R. Murphy, editors, *Artificial Intelligence and Mobile Robots*, chapter 8, pages 195–210. MIT Press, 1998.
10. Georgia Tech University. `http://controls.ae.gatech.edu/uavrf/`.
11. M. Ghallab. On chronicles: Representation, on-line recognition and learning. In *Proceedings of the International Conference on Knowledge Representation and Reasoning (KR-96)*, 1996.
12. D. Harel. Statecharts: A visual formalism for complex systems. *Science of Computer Programming*, 8(3):231–274, 1987.
13. F. Heintz and P. Doherty. Dyknow: An approach to middleware for knowledge processing. *Journal of Intelligent and Fuzzy Systems*, 2004. Accepted for publication.
14. F. Heintz and P. Doherty. Managing dynamic object structures usin hypothesis generation and validation. In *Proceedings of 2004 AAAI Workshop on Anchoring Symbols to Sensor Data*, 2004. National Conference on Artificial Intelligence.
15. D. Kortenkamp, R. P. Bonassao, and R. Murphy, editors. *Artificial Intelligence and Mobile Robots*. AAAI Press/MIT Press, 1998.
16. T. Merz, S. Duranti, and G. Conte. Autonomous landing of an unmanned helicopter based on vision and inertial sensing. In *Proc. of the 5th IFAC Symposium on Intelligent Autonomous Vehicles*, 2004.
17. M.I.T. `http://gewurtz.mit.edu/research.htm`.
18. K. Nordberg, P. Doherty, P-E. Forssen, J. Wiklund, and P. Andersson. A flexible runtime system for image processing in a distributed computational environment for an unmanned aerial vehicle. *International Journal of Pattern Recognition and Artificial Intelligence*, 2004. To appear.
19. P. Nyblom. A language translator for robotic task procedure specifications. Master's thesis, Linköping university, Sweden, 2003. LITH-IDA-EX-03/034-SE.
20. Object Computing, Inc. *TAO Developer's Guide, Version 1.1a*, 2000. See also `http://www.cs.wustl.edu/~schmidt/TAO.html`.
21. USA Office of the Secretary of Defence. Unmanned aerial vehicles roadmap, 2002-2025, 2001. `http://www.acq.osd.mil/uav_roadmap.pdf`.
22. ONERA RESSAC. `http://www.cert.fr/dcsd/RESSAC/`.
23. S.K. Pal, L. Polkowski, and A. Skowron, editors. *Rough-Neuro Computing: Techniques for Computing with Words*. Springer–Verlag, Heidelberg, 2003.
24. P-O. Pettersson and P. Doherty. Probabilistic roadmap based path planning for autonomous unmanned aerial vehicles. In *Proceedings of the Workshop on Connecting Planning and Theroy with Practice*, 2004. 14th Int'l Conf. on Automated Planning and Scheduling, ICAPS'2004.
25. Per-Olof Pettersson. Helicopter path planning using probabilistic roadmaps. Master's thesis, Linköping university, Sweden, 2003. LITH-IDA-EX-02-56.
26. E. Sandewall, P. Doherty, O. Lemon, and S. Peters. Words at the right time: Real-time dialogues with the WITAS unmanned aerial vehicle. In *Proc. of the 26th German Conference on Artificial Intelligence*, 2003.
27. E. Skarman. On proportional navigation. Technical note, 2003.
28. Stanford University. `http://sun-valley.stanford.edu/users/heli/`.
29. University of California, Berkeley. `http://robotics.eecs.berkeley.edu/bear/`.

Aerial Shepherds: Coordination among UAVs and Swarms of Robots

Luiz Chaimowicz and Vijay Kumar

GRASP Laboratory – University of Pennsylvania
3330 Walnut. St. - Levine Hall - Philadelphia, PA, 19014.
{chaimo, kumar}@grasp.cis.upenn.edu

Abstract - We address the problem of deploying groups of tens or hundreds of unmanned ground vehicles (UGVs) in urban environments where a group of aerial vehicles (UAVs) can be used to coordinate the ground vehicles. We envision a hierarchy in which UAVs with aerial cameras can be used to monitor and command a swarm of UGVs, controlling the splitting and merging of the swarm into groups and the shape (distribution) and motion of each group. We call these UAVs *Aerial Shepherds*. We show a probabilistic approach using the EM algorithm for the initial assignment of shepherds to groups and present behaviors that allow an efficient hierarchical decomposition. We illustrate the framework through simulation examples, with applications to deployment in an urban environment.

1 Introduction

The use of Unmanned Aerial Vehicles (UAVs) in concert with Unmanned Ground Vehicles (UGVs) affords a number of synergies. First, UAVs with cameras and other sensors can obtain views of the environment that are complementary to views that can be obtained by cameras on UGVs. Second, UAVs can fly over obstacles while keeping UGVs in their field of view, providing a global perspective and monitoring the positions of UGVs while keeping track of the goal target. This is especially advantageous in two and a half dimensions where UAVs can obtain global maps and the coordination of UAVs and UGVs can enable efficient solutions to the mapping problem. Third, if UAVs can see the UGVs and the UGVs can see UAVs, the resulting three-dimensional sensor network can be used to solve the simultaneous localization and mapping problem, while being robust to failures in sensors like GPS and to errors in dead reckoning. In addition to this, the use of air and ground robotic vehicles working in cooperation has received a lot of attention for defense applications because of the obvious tactical advantages in such military operations as scouting and reconnaissance.

We are interested in deploying teams of heterogeneous vehicles in urban environments to perform tasks such as searching for targets or mapping the environment. We are particularly interested in having several UAVs shepherding large groups of UGVs. We are motivated by our experience with deploying a team of aerial and ground vehicles at a Military Operations on Urban Terrain (MOUT) site at Fort Benning. The MOUT site is a replica of a small city consisting of 17 two and three store buildings, streets and access roads. It is engineered with cameras that allow a multiple view tracking of training missions. It also features a small airfield, being a suitable test ground for air-ground cooperation. Our multi-robot team [9] consists of a 30 foot unmanned blimp and ten ground vehicles. Figure 1 depicts our blimp and two of our ground robots performing a leader-follower task at the MOUT site.

Fig. 1. Blimp and two ground robots during the experiments at the MOUT site.

In this paper, we focus on a scalable approach for deploying tens and hundreds of ground vehicles in order to search the urban terrain for targets or simply to disperse and perform a coverage task. We propose a paradigm in which a group UAVs, equipped with cameras, can act as aerial shepherds, monitoring and commanding a swarm of UGVs. Differently from our previous work [5], where we needed a central planner for controlling the robots, in this paper we present behaviors that allow the UAVs to coordinate the splitting and merging of the swarm into groups and control the shape and motion of each group in a distributed manner. We also propose a probabilistic approach using the EM algorithm for the initial assignment of shepherds to groups and illustrate this architecture with simulation examples.

It is important to mention that in spite of the growing number of works in cooperative robotics, few tackle specifically the cooperation between UAVs and UGVs. A general application is to use the UAVs as an "extra sensor" for the UGVs. In this context, Stenz et al. [12] present an application where an autonomous helicopter helps the navigation of an UGV by exploring the terrain and informing the UGV about potential hazards (holes, obstacles, etc.) on its way. Other important applications include environmental monitoring, cooperative control, and cooperative localization. Elfes et al. [8] and Lacroix et al. [11], for example, present some preliminary ideas on the integration of blimps and ground robots for environmental surveillance and other tasks. Sukhatme et al. [13] propose an architecture for coordinating an autonomous helicopter and a group of ground vehicles using decentralized behavior based controllers and minimal top down planning. Sukkarieh et al. [14] present a decentralized architecture for data fusion and control of multiple UAVs and UGVs. In our previous work, we report on coordination between aerial and ground vehicles for robust localization in the presence of GPS errors [4].

2 Architecture

We developed a hierarchical architecture to allow a group of UAVs to coordinate and control swarms of ground robots. At the lowest level, there are individual robots i, $i = 1, \ldots, N_r$. They are assumed to be holonomic and the state of each robot is described by its cartesian coordinates:

$$\begin{aligned} X^i &= (x^i, y^i), \\ \dot{X}^i &= u^i = f(X^i, a^j). \end{aligned}$$

Robots are assembled together in a set of groups Γ. Each group $\gamma^j \in \Gamma$, $j = 1, \ldots, N_g$ is modeled by a double (g^j, ρ^j): g is an element of the Lie group $SE(2)$ while ρ is the shape of the group formation [7]. Thus each group γ^j can be represented by an abstraction a^j that comprises the group pose g^j and the group shape ρ^j [1]. More specifically:

$$\begin{aligned} a^j &= (g^j, \rho^j), \\ g^j &= (\mu_x^j, \mu_y^j, \theta^j), \\ \rho^j &= (s_1^j, s_2^j). \end{aligned}$$

This abstraction provides a scalable way of controlling groups of robots. It can be viewed as an *equipotential* or *concentration ellipse* centered in μ with orientation θ and principal axis given by $\sqrt{cs_1}$ and $\sqrt{cs_2}$. The number c is given by $c = -2\ln(1-p)$, where p is the percentage of a large number of normally distributed robots that lies inside the ellipse. The abstraction a^j of a certain group γ^j is computed using only the states of the robots that belong to this group, as explained in [1].

The third level of the hierarchy is composed by medium altitude UAVs (blimps) that hover over the groups. Every group must have a shepherd blimp but one blimp can escort one or more groups simultaneously (see Figure 2). Let's consider that each blimp k, $k = 1, \ldots, N_b$ is shepherding a subset of groups S^k, $S^k \in \mathcal{P}(\Gamma)$, where $\mathcal{P}(\Gamma)$ is the power set of Γ. In this paper, we adopt a simple kinematic model for the blimps that considers its position (τ, η, z) in the 3D space and its yaw angle (ϕ). More complex dynamic models of our blimp are derived in [10].

$$\begin{aligned} Z^k &= (\tau^k, \eta^k, z^k, \phi^k,), \\ \dot{Z}^k &= v^k = f(Z^k, S^k). \end{aligned}$$

Basically, after an initial assignment phase, the blimp controller tracks the shape and pose of the groups in S^k, trying to keep all the robots of these groups inside the blimp's visibility region. For now, we are considering that this visibility region is a circle with radius varying proportionally to the blimps' altitude z and that all the robots inside this circle can be localized by the blimp. The blimps have a superior limit for their altitude (z_{max}) and, consequently, the maximum visibility area is also limited.

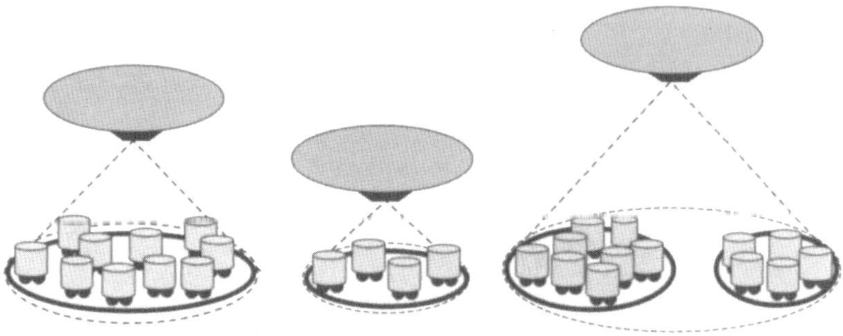

Fig. 2. Hierarchical architecture in which three blimps shepherd groups of UGVs.

The controllers used by each of these hierarchical levels are described in detail in [1] and [5]. Basically, the ground robots only require state feedback and information about the pose and shape of its own group. The group's pose and shape variables are controlled by the blimps based on the robots' positions using simple proportional equations. Finally, the blimp controller tracks the pose and shape of its groups to keep the robots inside its field of view.

One of the main advantages of this architecture is that the amount of information that must be exchanged among the different hierarchical levels is small. Blimps have to broadcast to the ground robots the pose and shape of their groups and the desired group velocities (basically, a^j and $\dot{a}^j$). Other than that, as will be explained in the next section, ground robots broadcast their position once for the initial distribution and respond sporadically to inquires from blimps regarding its shepherd during the merge process.

The communication within the hierarchical levels is also small. As mentioned, the ground robots only require state feedback and information about their own group. No communication is necessary internally among them. Blimps have to communicate with each other to coordinate the split and merge behaviors. But this is not frequent and the number of blimps is much smaller than the number of ground robots. Thus, the amount of information exchanged between the ground vehicles and the blimps increases linearly with the number of robots making the proposed architecture scalable. Moreover, the information being broadcast is sparse: only information about the groups' poses and shapes is being broadcast, regardless of how big the groups are.

3 Coordination

3.1 Initial Distribution

We consider that the initial position of the robots can be described by a *Mixture of Gaussians*. A mixture of gaussians is a distribution Θ composed by m components, each of which is a gaussian distribution in its own right. Each gaussian have its own mean and covariance (μ,Σ), and has a weight w in the mixture Θ. Ideally, each individual gaussian will represent one group γ^j of the hierarchy, and will be shepherded by one blimp. Initially, we do not know which robot belongs to each component and, consequently, the parameters of each gaussian are unknown. In the approach used here, these parameters are determined by the blimps through the *EM – Expectation Maximization* algorithm [6]. The EM algorithm was developed to compute the maximum likelihood estimate of the parameters of a distribution from a given data set when the data is incomplete or has missing values. But one of its main uses is when the optimization of the likelihood function is intractable, but the function can be simplified considering its data incomplete [2]. This is exact the case of mixtures of gaussians.

Thus, the initial distribution of robots into groups and the assignment of blimps to the groups is done as follows. Initially all robots broadcast their position once to the blimps. Each blimp apply the EM algorithm to determine the distribution parameters and allocate itself to one of the computed distributions. Since the blimps have the same data and run the same algorithm they will obtain the same results, and the assignment can be done in a sequential way: blimp b_1 to the first distribution, blimp b_2 to the second, etc. Each distribution determined by the algorithm will become a group in the hierarchy.

The only drawback of this method is that the number of distributions (components of the mixture) in which to divide the robots should be given a priori to the algorithm. We consider that this number is equal to the number of available blimps (N_b). The problem is that this number can be different of the number of "real distributions" (N_d). For example, if we have the robots distributed in three gaussians ($N_d = 3$) but we have four blimps ($N_b = 4$), the algorithm will divide one of the real distributions in two, resulting in four groups instead of three. So, the ideal situation will be when the number of blimps is equal the number of distributions ($N_b = N_d$). If $N_b < N_d$ there will be one blimp allocated to more than one distribution and if $N_b > N_d$ there will be two or more blimps allocated to a single distribution. In these two situations, the groups will probably have to split or merge, as will be explained in the next section. Figure 3 exemplify these three scenarios. In this figure, the small circles represent the robots, the dashed circles represent the blimps and the large solid circles represent the visibility area of each blimp. There are basically 3 robot distributions and 2, 3 and 4 blimps respectively in the pictures.

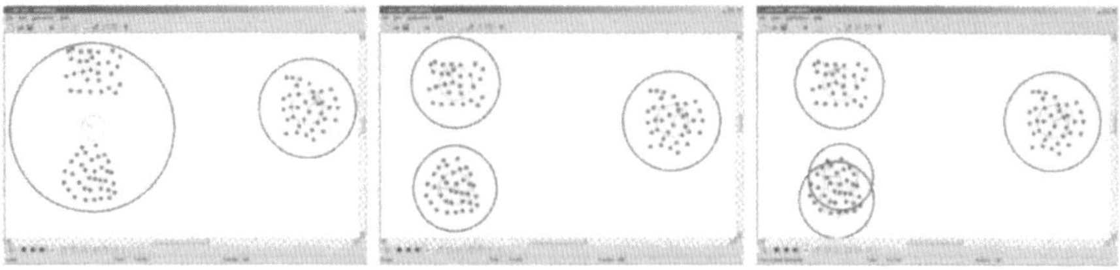

Fig. 3. Initial distribution when: a) $N_b < N_d$, b) $N_b = N_d$, c) $N_b > N_d$.

3.2 Merge and Split

The ability of splitting and merging groups is a desired behavior in several different scenarios. Basically, one group may divide into two or more groups and different groups can merge into a single one. For example, a robot group may have to split to avoid obstacles or to cover different regions as shown in Figure 4. Also, split and merge behaviors can be used to fix an unbalanced initial distribution as mentioned in the previous section.

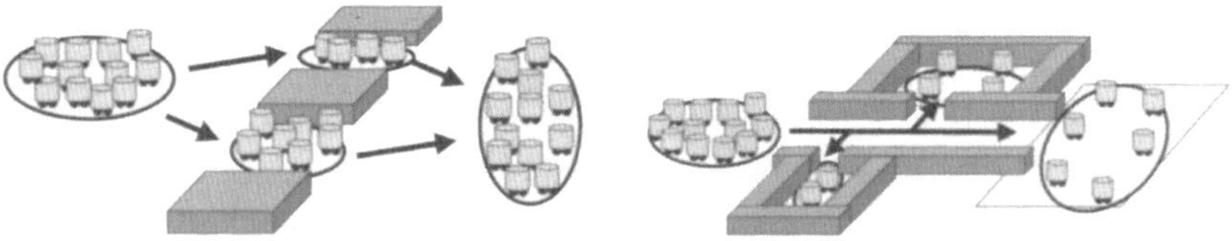

Fig. 4. Scenarios where groups of robots perform split and merge behaviors for avoiding obstacles and exploring environments.

In this paper, we use a deliberative approach where each blimp has a plan for its group. This plan is represented by a directed graph in which each node contains a desired pose and shape for the group (g_{des}, ρ_{des}) and the edges indicate the ordering of the goals. Here, these plans are specified manually, but variations of automated techniques such as grid based approaches or voronoi diagrams could be used.

Merge

The merge process occurs when two groups that are heading to the same goal are close enough to be shepherded by a single blimp. In this case, one of the blimps will escort all the robots while the other blimp will be dismissed and will return to its base. Let's consider two groups γ_1 and γ_2 moving to a common goal and being shepherded by two blimps b_1 and b_2 respectively. When a robot $i \in \gamma_2$ is detected inside the visibility region of b_1, b_1 will create and start to track a *virtual group* $\gamma_{1,2}$ composed by the robots of both groups. When all robots from γ_2 are inside the visibility region of b_1, the two groups are merged into γ_1, γ_2 is removed from the hierarchy, and the blimp b_2 is dismissed. The blimps must coordinate themselves to do the merging process, but, as mentioned, the amount of communication necessary for this is very small. Basically, b_1 inquires robot i to discover who is its shepherd blimp. Then b_1 and b_2 communicate to check if they are going to the same goal, and if this is the case, b_1 creates the virtual group and start tracking it, receiving information from b_2 regarding the pose and shape of its group. To track both groups, b_1 will probably have to fly at a higher altitude in order to increase its visibility area. This will be demonstrated in the simulations of Section 4. When all robots are inside its visibility region, b_1 finishes the merging processes: it broadcasts a message to the robots of γ_2 informing them that they now belong to a new group and have a new shepherd, and also sends a message to b_2 releasing it from its group.

Split

The split processes is initiated by a blimp when its group reaches some pre-determined goals according to its plan. If the out-degree (q) of a certain node (goal) in its plan is greater than 1, the group will split in q subgroups. The shepherd blimp is responsible for deciding how many and which robots should be assigned to each of the new groups, and for requesting help from other blimps to escort the new groups. Let's consider that a group, γ_1, shepherded by blimp b_1 has reached a goal and must be split in two groups γ_2 and γ_3, that will move to different goals. The blimp will compute the number of robots (n_j) that should be assigned to each group γ^j based on the shape of the next desired goals on the graph. Then, the n_j robots that are closer to one of the next goals are assigned to the group that will move towards that goal. After this division, b_1 will create a virtual group $\gamma_{2,3}$ and start escorting both groups. At the same time, b_1 will broadcast a message to the other blimps to see if there is an available blimp that can shepherd one of the new groups. If there is such blimp, say b_2, it communicates with b_1, receives information about the pose and shape of the new group, and starts moving towards it. If there is no other blimp available or if b_2 takes too much time to reach the group, probably b_1 will reach its maximum altitude if the groups are moving apart. In this situation, b_1 will abandon one of the groups and shepherd the other to its goal. The abandoned group will stop and wait for the arrival of b_2 or for the return of b_1, in the case that no other blimps were available.

4 Experiments

The proposed architecture was implemented and tested using MuRoS, a multi-robot simulator that we have developed for cooperative robotics [3]. Implemented for the MS Windows environment, MuRoS has a friendly user interface and can be easily extended with the development of new inherited classes defining new robots, controllers and sensors.

The simulation presented here demonstrates the initial distribution and the merge and split behaviors described in the previous section. Figure 5 shows some snapshots of this simulation where 50 robots are shepherded by two blimps. Again, the small circles represent the robots, the dashed circles represent the blimps and the large solid circles represent the visibility area of each blimp, which varies with its altitude z. The two rectangles are obstacles and the (barely visible) crosses represent the group's current and desired poses and shapes. The robots are initially divided in two groups (Figure 5(a)), that are allocated to two blimps after the application of the EM algorithm. The blimps start to shepherd the groups towards the first goal (b) and, when the groups are sufficiently close, they merge and become escorted by a single blimp, while the other return to its base (c). After moving to the second goal (Figure 5(d)), the group have to split to move to different goals. In this case, as mentioned, the blimp creates a virtual group and tries to shepherd both groups until the arrival of a second blimp (e). The last snapshot (f) shows the two blimps shepherding their groups to their final destinations.

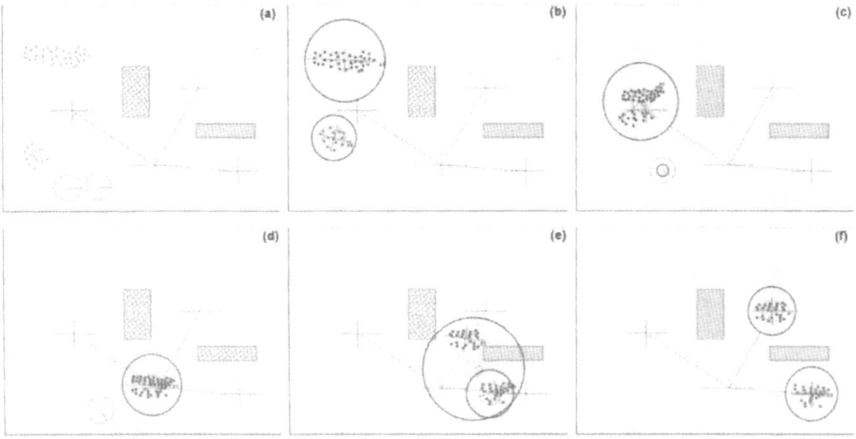

Fig. 5. Simulations where two UAVs coordinate a group of 50 UGVs.

The graph of Figure 6 shows the altitude (z) of both blimps (b_0 and b_1) during the simulation. The labels T_m and T_s mark the times in which the groups merged and split. Basically, the blimps initially fly from the base to their initial groups and start to shepherd them to the first goal. As shown in Figure 5(b), the group on the top, that is shepherd by b_0, is larger and has to compress its shape en route to the first goal while the group on the bottom has to expand. Consequently, b_0 can decrease its altitude while b_1 must fly higher in order to escort its group. This can be observed in the graph of Figure 6, just after time 5. Just before time T_m, b_0 creates a virtual group and increases its altitude to track both groups. After the merging, b_0 lowers its altitude again while shepherding the merged group to the first goal and b_1 returns to its base and lands. The merged group shepherd by b_0 then moves to the second goal where it should split. After the split (T_s), b_0 has to increase its altitude again to shepherd both groups while b_1 takes off and goes after one of the groups to help b_0, as can be seen on Figure 5(e). Finally, b_1 takes over one of the groups, b_0 decreases its altitude, and both blimps stabilize and escort their groups to their final goal.

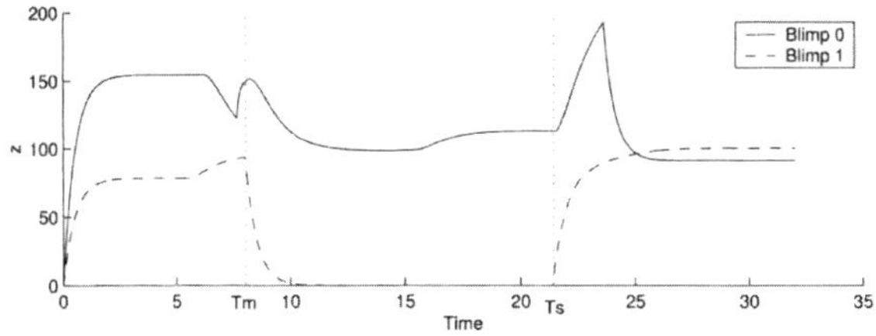

Fig. 6. Blimps' altitude (z) during the execution of the task.

These results, in spite of not providing rigorous measurements of performance, demonstrated that this hierarchical architecture can be used for coordinating UAVs and swarms of UGVs in a scalable manner. It is important to mention that other simulations were performed on different scenarios, using a different number of robots, blimps, and groups, also leading to similar results.

5 Conclusion

In this paper, we presented a hierarchical architecture in which a few number of UAVs is used to command, control and monitor swarms of UGVs. Basically, the individual UGVs are assembled together into groups that are shepherded by medium altitude UAVs (blimps). The initial assignment of blimps to groups is done in a distributed manner using the EM algorithm and the blimps are responsible for controlling the groups' shape, pose and motion. An important feature of this architecture is that the communication requirements grow linearly with the number of vehicles, making it scalable to tens and hundreds of robots. We described behaviors that allow blimps to coordinate the splitting and merging of groups during the task execution and demonstrated these concepts through simulations.

Future work is directed towards improving this architecture to deal with some specific situations. For example, if one group is too large to be shepherded by a single blimp, we would like to have two blimps coordinating themselves in order to escort this group. We are also interested in executing simulations to obtain more detailed performance measurements of this architecture, mainly in regard to communication and control.

References

1. C. Belta, G. A. S. Pereira, and V. Kumar. Abstraction and control for swarms of robots. In *Proceedings of the 11th International Symposium of Robotics Research (ISRR'03)*, 2003.
2. J. A. Bilmes. A gentle tutorial of the EM algorithm and its application to parameter estimation and for gaussian mixture and hidden markov models. Technical Report TR-97-021, Dept. of Computer Science – UC Berkeley, 1998.
3. L. Chaimowicz, M. Campos, and V. Kumar. Simulating loosely and tightly coupled multi-robot cooperation. In *Proceedings of V SBAI – Brazilian Symposium on Intelligent Automation*, November 2001.
4. L. Chaimowicz, B. Grocholsky, J. F. Keller, V. Kumar, and C. J. Taylor. Experiments in multirobot air-ground coordination. In *Proceedings of the 2004 IEEE International Conference on Robotics and Automation*, pages 4053–4058, 2004.
5. L. Chaimowicz and V. Kumar. A framework for the scalable control of swarms of unmanned ground vehicles with unmanned aerial vehicles. In *Proceedings of the 10th International Conference on Robotics and Remote Systems for Hazardous Environments*, 2004.
6. A. P. Dempster, N. M. Laird, and D. B. Rubin. Maximum likelihood from incomplete data via the EM algorithm. *Journal of the Royal Statistical Society, Series B (Methodological)*, 39(1):1–38, 1977.
7. J. Desai, J. Ostrowski, and V. Kumar. Modelling and control of formations of nonholonomic mobile robots. *IEEE Transactions on Robotics and Automation*, 17(6):905–908, 2001.
8. A. Elfes, M. Bergerman, J.R.H. Carvalho, E.C. Paiva, J.J.G. Ramos, and S.S. Bueno. Air-ground robotic ensembles for cooperative applications: concepts and preliminary results. In *Proceedings of the International Conference on Field and Service Robotics*, pages 75–80, 1999.
9. D. G. Ibanez, B. Grocholsky, F. Hager, J. F. Keller, J. W. Kim, V. Kumar, P. Srivastava, and C. J. Taylor. An experimental test-bed for air ground coordination. Submitted to the 2004 American Control Conference, 2004.
10. J. W. Kim, J. F. Keller, and V. Kumar. Design and verification of controllers for airships. In *Proceedings of the 2003 IEEE International Conference on Intelligent Robots and Systems*, pages 54–60, 2003.
11. S. Lacroix, I-K. Jung, A. Mallet, and R. Chatila. Towards cooperative air/ground robotics: issues related to environment modeling. In *10th International Conference on Advanced Robotics*, Budapest (Hungary), 2001.
12. T. Stentz, A. Kelly, H. Herman, P. Rander, and R. Mandelbaum. Integrated air/ground vehicle system for semi-autonomous off-road navigation. In *Proceedings of AUVSI Symposium on Unmanned Systems*, 2002.
13. G. S. Sukhatme, J. Montgomery, and R. T. Vaughan. Experiments with aerial-ground robots. In T. Balch and L.E. Parker, editors, *Robot Teams: From Diversity to Polymorphism*. AK Peters, 2001.
14. S. Sukkarieh, B. Grocholsky, E. Nettleton, and H. F. Durrant-Whyte. Information fusion and control for multiple UAVs. In *NRL Workshop on Multi-Robot Systems*, volume 2. Kluwer, March 2003.

Dispersing robots in an unknown environment

Ryan Morlok and Maria Gini

Department of Computer Science and Engineering, University of Minnesota, 200 Union St. S.E., Minneapolis, MN 55455-0159 {morlok,gini}@cs.umn.edu

Summary. We examine how the choice of the movement algorithm can affect the success of a swarm of simple mobile robots attempting to disperse themselves in an unknown environment. We assume there is no central control, and the robots have limited processing power, simple sensors, and no active communication. We evaluate different movement algorithms based on the percentage of the environment that the group of robots succeeds in observing.

1 Introduction

The problem we address is that of dispersing a group of mobile robots in an unknown environment. We assume the robots do not know how many other robots are operating in the same environment, where those robots are located, and where those robots have been.

The primary motivation for this work comes from the Scout project [9]. The scouts are small, two wheeled robots with extremely limited processing capability. In general, the scouts are deployed by being hauled or launched into the environment by a larger robot. Their job is then to disperse throughout the environment so that it can be effectively monitored. Currently the scouts are teleoperated, but they can also perform some autonomous operations, such as hiding and watching for motion [9], by proxy processing over a radio link.

One of the major issues with these type of robots is communication. Since the robots are small, and therefore do not have a great deal of available electrical power, it can be difficult (and in some cases impossible) to generate a signal strong enough to communicate with all other robots. This problem is worsened by the fact that the scouts are designed to explore hostile environments, which may have physical characteristics that further hamper any sort of radio based communication. Because of this, we will constrain our algorithms not to require any explicit communication, and to use the sensors to communicate implicitly by observing cues from the environment. This type of communication, which is called *stigmergy* in the biology literature, is common in swarm approaches to robotics [1]. We will further assume that the robots have enough local processing power, so all computation is done locally.

2 Related Work

Coverage of terrain during motion is important in many application domains, such as floor cleaning, lawn mowing, harvesting, etc. A recent survey [2] classifies the existing algorithms for terrain coverage.

Wagner et al. [11] formalize the terrain covering problem and propose two algorithms. one called mark and cover (MAC), the second called probabilistic coverage (PC), both for single and multiple robots. They show how several cooperating robots can obtain faster coverage. Algorithms inspired by insect behaviors, such as ants, are becoming popular both for terrain coverage [6], where robots leave trails and cover the terrain repeatedly, and for optimization of paths [8].

The study by Hsiang et al. [4, 5] is the closest to our work. In their work, they examined methods for dispersing robots from fixed locations to cover the entire environment. They assume a continuous stream of robots would be entering the environment through specific, predetermined locations. The goal of the robots would then be to position themselves such that the entire area of the accessible space is covered. While this work has great properties/guarantees, it is not immediately applicable to the problem we are investigating in this paper. The reason is that while each robot only has extremely local knowledge of the environment, through the use of the deterministic movement and infinite supply of robots the information available at the point at which any given robot is located is sufficient to guarantee that the robot will make the correct choice. In our investigation, it would be impractical to assume that there are enough robots to cover the entire map and to guarantee that every robot can remain within sensor range of the other robots. In our experiments, we assume there are at most 50 robots present in the environment. In environments such as the Hospital World (shown later in Figure 2), this would allow possibly two robots per room explored. Clearly there are not enough robots to make Hsiang's algorithm feasible.

3 Motion algorithms

The purpose of this study is to examine how the selection of the movement algorithm for a multi-robot system affects the coverage of robot observation in a variety of environments.

We considered four distinct movement algorithms, all of them reactive in nature. Each movement algorithm controls two types of movements: forward/backward and turning. Turning can occur in place or while the robot is moving. The sensors available to the movement algorithms are 16 sonar range

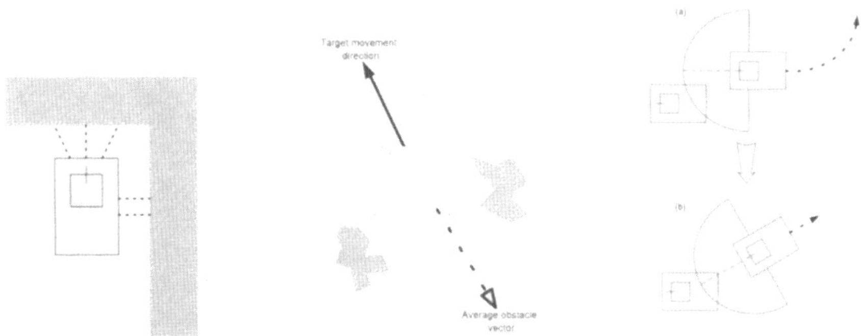

Fig. 1. *Left:* A robot using the FOLLOWWALL algorithm navigates a corner. *Center:* A robot using the SEEKOPEN algorithm calculates the average obstacle vector and moves in the direction opposite from this vector. *Right:* The FIDUCIAL movement algorithm. (a) A robot detects another robot behind it within its sensor range, and begins to adjust its forward motion to turn so the detected robot will be immediately behind. (b) The detector robot has successfully positioned the detected robot behind it; the detector robot will now continue with straight forward movement.

finders, each of which returns the distance of the nearest object detected in the direction in which the range finder is pointing. A fiducial range finder is also used to determine the location of other robots, which is needed for the FIDUCIAL algorithm. Robots have only local knowledge, they are not under global control (no central source knows the state of all of the robots), and do not have any knowledge of the environment other than what they can detect with their sensors.

3.1 Random Walk

The RANDOMWALK movement algorithm is the most basic of the algorithms we examined. A robot using this algorithm can be in one of two states: random forward movement or obstacle avoidance. In random forward movement, the robot moves forward with a small random turn factor between -10° and 10° (the robot's path is curved) which is changed at random intervals, ranging between 10 s and 15 s. The amount of the turn is constrained to ensure the robot does not end up going in small circles. Once the robot detects that it has encountered an obstacle, it enters the obstacle avoidance state. In this state, the robot will stop, turn a random amount (in the range 120° to 240°), and transition back to the standard forward movement state.

3.2 Follow Wall

The idea behind the FOLLOWWALL algorithm comes from the fact that in many indoor environments, if a robot could find an outer wall of the building and follow it, the robot would be led through much of the structure. A robot using the FOLLOWWALL algorithm will search for an obstacle (presumably a wall or large object in the environment) and then proceed to follow that wall indefinitely. In this algorithm, a robot has four states: find wall, align to wall, follow wall, and navigate corner. If the robot believes that it has lost the wall in any of the three non-find-wall states, it will reset back to the initial find wall state and search for a new wall to follow.

The major problem of this movement algorithm is that it assumes every obstacle encountered is a wall, rather than trying to determine if the observed entity is something smaller, such as a robot at close range. Because of this, when many robots using this algorithm are together, they will tend to perceive each other as walls and try to align themselves to each other. This is wasteful, since the alignment procedure will not effectively spread the robots out in the environment.

3.3 Seek Open

The SEEKOPEN movement algorithm causes a robot to move toward open areas in the map. The motivation for the SEEKOPEN algorithm is similar to the fiducial robot avoidance algorithm (discussed next). According to the assumptions of the experiment, all robots start out in the same general area, grouped fairly close together. Because of this, all the robots tend to have objects (generally other robots) close to themselves at the beginning of a run. The goal of the seek open algorithm is to motivate the robots to disperse as quickly as possible.

SEEKOPEN is executed by first calculating the average obstacle vector for all obstacles in sensor range. The average obstacle vector is computed by summing the vectors pointing to all of the objects within sensor range and dividing by the number of vectors summed. The magnitude of the vector must be large for objects close to the robot and small for objects far away. This is accomplished by setting the magnitude of a perceived obstacle vector equal to the maximum range of the sensors minus the perceived distance a given obstacle is from the robot, or by using some other function which decreases with distance, as done when using artificial potential fields for navigation [7]. After the average obstacle vector is computed, the goal of the robot becomes to move in the opposite direction of the average obstacle vector. The robot turns toward the direction of the negative obstacle vector. The rate of turn is determined by the magnitude of the average obstacle vector. This allows the SEEKOPEN algorithm to not run into walls as well as disperse from other robots. This is illustrated in Figure 1.

3.4 Fiducial

The FIDUCIAL movement algorithm was inspired by the idea that the robots would be able to recognize other robots, and therefore move away from them. The original concept involved a simple signal (possibly a weak radio signal) that each robot would emit, so that another robot could detect the signal, and determine an approximate distance to the originating robot based solely on signal intensity. The problem of moving away from other robots would then become a goal of finding areas in which signal intensities are low, which can be done by any hill-climbing algorithm. Unfortunately, there was no straightforward way to implement such as system within Stage, and therefore an alternative method was sought.

The solution to the simulation problem was to use a fiducial device (generally used to find beacons in the Stage simulator) and attach a beacon to every robot. This allows a given robot to know the polar coordinates of other robots within sensor range (sensor range is a semi-circle of fixed radius) with respect to its own position and orientation. The information can be used to steer away from other robots.

With the fiducial information, implementation of an avoidance algorithm is straightforward. Whenever a robot detects another robot within sensor range, the robot adjusts its movement so that it is moving away from the detected robot. When no robots are in sensor range, a robot simply moves according to the random walk algorithm. If at any time a robot encounters a physical obstacle such as a wall, the obstacle avoidance technique takes precedence over whatever movement algorithm the robot is currently executing.

4 Simulation Environment and Data Analysis

To compare the algorithms, we performed a large number of experiments within the Player/Stage [10, 3] simulator. The virtual robot used for experiments is a rectangular, four-wheeled, Pioneer-like robot. Although the motivation for this work comes from the scout project, the scouts are not currently modeled in Stage, and this motivated the change in platform. The robots used in the simulations have 16 sonar range finder sensors, a pan-tilt-zoom camera with blob finding software capabilities (not used for any experiments), a laser fiducial finding device, and a truth device (a device used in the simulator to extract information about the robot's position status to record experimental data). The dimensions of the simulated robots are 33x44 cm.

The experiments were carried out by executing each of the four movement algorithms in five different simulated environments with different numbers of robots (10, 20, 30, 40, and 50) and two different durations (5 min and 10 min). The environments are described in Table 1 and shown in Figure 2. In each experiment the robots started out clustered together near the center of the map, each robot facing a random direction (except for the house world where the robots began in the left most room).

Experiments were carried out within the Player/Stage [10, 3] architecture. The Player Java client library was used to control the robots for all experiments. Each robot's control ran on its own thread, and all robot control code was executed on the same machine as the Stage simulation. This was done

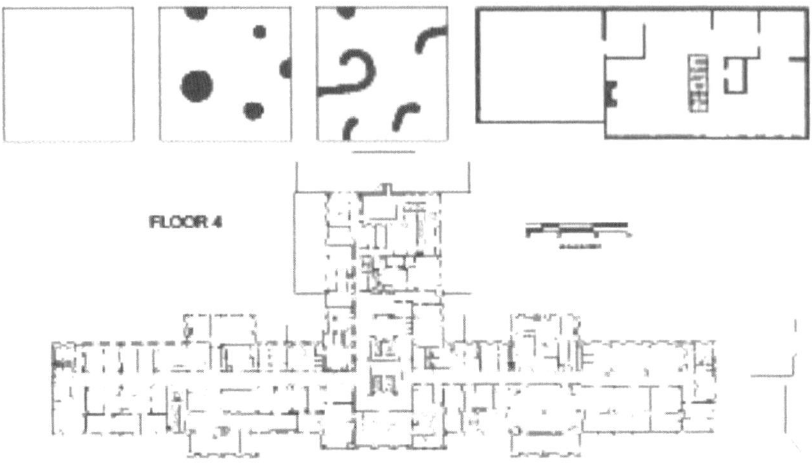

Fig. 2. The simulated worlds used in the experiments. *Top:* Square, Convex, Concave, and House worlds. *Bottom:* Hospital World.

World	Size	Key Features
Square	30x30 m	Simplest world, large open area designed to provide a baseline for other worlds
Convex	30x30 m	Simple world with basic, convex obstacles; no locations where robots can get trapped
Concave	30x30 m	More extended, concave barriers in which robots can get trapped if they are unwilling to backtrack
House	41x16 m	World modeled after a simple house blueprint; robots interact with a simplified map of a real world environment
Hospital	140x54 m	Complex world (packaged with Stage) designed to test robots in complicated world with many corridors and rooms

Table 1. Simulated test environments; see Figure 2 for visual reference.

primarily because of the difficulty of starting all robots simultaneously on multiple machines. Each of the experiments was run four times.

Because the Stage simulator does not provide built in utilities for analyzing the performance of the robots as they observe the environment, snapshots of relevant data were taken from the simulation for later off-line analysis. This was done by attaching a Truth device (a simulated device in Stage) to each robot. The Truth device is a device that can either get or set the world coordinates/orientation of any object to which it is attached. For each robot, an additional thread running on the same JVM queried the robot via the Truth device and recorded its position/orientation once per second. Data for all robots for a given experiment was written out to the same file. The data could then be used later (in combination with the world map) to determine the observation coverage of the environment.

For each experiment, a single, large file containing the position coordinates of all the robots was created. In order to determine observation coverage of the robots, the data was analyzed in combination with the world map used in the experiment. The procedure was implemented in Matlab.

First, all data points are loaded from the file. A binary bitmap of the same size as the world bitmap is created. Each pixel in the binary bitmap represents a discrete location in the world, which can be in one of two states: observed (1) or not observed (0). Initially, all pixels are not observed. For each location that is taken from the robot position file, all of the pixels within a set radius are set to observed. This is repeated for all the locations in the position file.

After all locations in the position file are processed, the observed region is oversized. Some areas are marked as observed when they are in fact unobserved. This is because the observed region includes pixels that are in fact obstacles, as well as those that are outside the accessible region of the world (a closed region from which the robot cannot escape). To account for this, a logical AND is performed on the observed binary bitmap and the interior region of the world. The interior region of the world is found by performing a flood fill, beginning at the start location of one of the robots. This leaves the observed bitmap as the locations that have fallen within the robots' observed radii at some point, that are valid, and accessible points in the world.

Note that this procedure is not a perfect model of what the robots could actually observe, especially in blueprint-like environments. Consider the situation shown in Figure 3. Suppose the elongated rectangle represents a wall separating two rooms (both of which are accessible to the robot). Here the algorithm will mark the area beyond the wall as observed, where clearly it is not. We developed a method to deal with this problem, but we did not use it in the data reported, since it significantly increases the time for data analysis and only leads to a minor improvement in accuracy for a relatively small view radius of the robots. Because of this, it should be noted that data reported for the House and Hospital worlds are slight overestimates of the actual values.

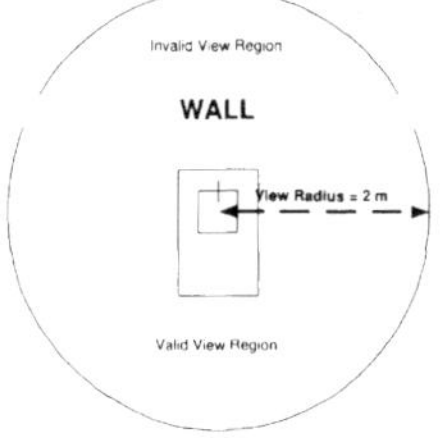

Fig. 3. A problem with the model used to calculate the region observed by a robot.

5 Experimental Results

Table 2 summarizes the results from both the five and ten minute experiment runs for the four algorithms in all five environments with different numbers of robots. The percents values in the table indicate the percentage of the accessible area of the environment that was observed by the robots.

	5 Minutes					*10 Minutes*				
	10	20	30	40	50	10	20	30	40	50
	Square Environment									
Fiducial	61.0%	77.4%	81.6%	85.5%	83.4%	81.2%	95.8%	96.9%	96.9%	97.1%
Follow wall	28.8%	57.4%	52.8%	53.8%	57.3%	56.6%	76.1%	77.0%	68.4%	78.6%
Random walk	50.2%	65.7%	76.4%	84.8%	79.8%	73.3%	90.6%	96.6%	95.8%	97.7%
Seek open	18.7%	36.7%	43.9%	47.3%	44.4%	41.8%	72.7%	70.7%	87.4%	77.0%
	Convex Environment									
Fiducial	52.2%	66.1%	71.6%	75.1%	75.6%	74.7%	82.9%	93.7%	94.4%	92.0%
Follow wall	22.7%	41.8%	36.2%	36.0%	37.5%	28.8%	64.6%	61.1%	59.8%	55.4%
Random walk	44.2%	58.5%	62.4%	64.9%	65.9%	69.8%	82.4%	84.0%	88.2%	93.1%
Seek open	18.9%	30.2%	40.9%	38.4%	33.6%	36.6%	52.8%	59.9%	65.9%	56.9%
	Concave Environment									
Fiducial	46.2%	58.7%	64.5%	67.3%	69.0%	67.8%	85.9%	85.8%	90.7%	88.5%
Follow wall	14.9%	34.6%	37.1%	35.7%	35.5%	35.5%	53.1%	58.3%	52.4%	56.3%
Random walk	33.8%	48.2%	56.2%	64.8%	59.8%	51.0%	73.4%	78.6%	79.0%	86.4%
Seek open	16.2%	29.4%	35.0%	33.5%	40.8%	36.1%	49.1%	53.4%	54.6%	60.9%
	Home Environment									
Fiducial	37.0%	40.0%	43.0%	40.9%	40.7%	39.7%	46.3%	47.2%	44.5%	43.9%
Follow wall	23.9%	22.3%	27.3%	30.9%	35.2%	31.4%	32.6%	39.1%	37.0%	40.7%
Random walk	33.3%	37.1%	40.0%	38.6%	40.4%	39.2%	41.6%	42.9%	44.4%	44.1%
Seek open	23.8%	31.6%	33.6%	33.4%	35.2%	35.5%	37.6%	36.6%	37.1%	36.9%
	Hospital Environment									
Fiducial	5.6%	6.0%	7.0%	7.7%	8.7%	8.4%	11.0%	10.6%	10.3%	13.3%
Follow wall	3.3%	3.6%	3.1%	3.7%	4.3%	4.1%	6.5%	5.5%	7.0%	5.0%
Random walk	4.7%	4.4%	4.1%	6.1%	5.9%	5.0%	6.5%	4.9%	7.5%	8.0%
Seek open	3.4%	3.5%	3.7%	3.9%	4.8%	3.4%	3.5%	3.8%	4.3%	5.1%

Table 2. Results for all experiments

The data shows that the FIDUCIAL algorithm performs the best in every situation. This is not surprising in that this algorithm has access to more data than the other algorithms. It does, however, indicate that knowledge of the locations of the other robots can help to speed up the exploration process.

The results for the FIDUCIAL algorithm can be seen graphically in Figure 4. This illustration shows that the robots had difficulty observing the more enclosed areas of the map, which is to be expected since the obstacle avoidance portion of the movement algorithm tends to favor regions in which it does not run into things. To get into the enclosed-hook region, the robot would have to intersect the obstacle, and then by random chance be redirected towards the hook region. None of the movement algorithms have the ability to be naturally attracted toward this type of region.

Figure 4 also shows the performance of the FIDUCIAL algorithm in the House world. All robots started out in the garage (the large, left most rectangle that is all green (light colored)) clustered together, facing different, random directions. Here we can see that the robots managed to make it into the house,

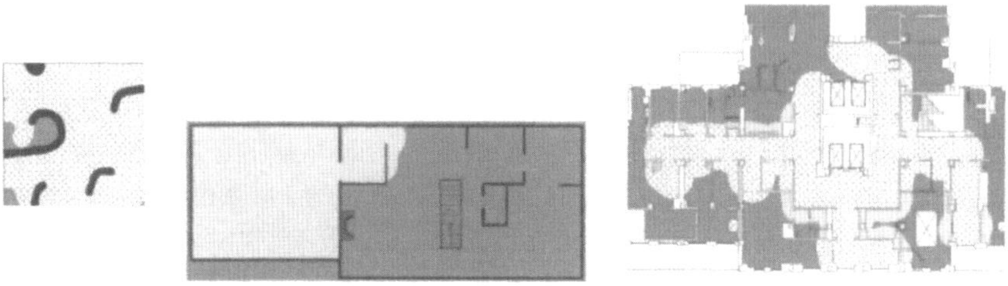

Fig. 4. Results of the FIDUCIAL algorithm in the concave, house, and hospital worlds. Red (dark) indicates an unexplored area, green (light) indicates an observed region. The concave and house results were obtained by running 50 robots for 10 minutes, and the hospital results were obtained by running 50 robots for 1 hour.

but not past the first room. Doors are a natural obstacle for all movement algorithms in this experiment, in that they present a situation where there is obstacle on both sides, with a small opening. This can cause the avoidance logic to move away from the door area when it is encountered, unless the robot hits the door dead on.

What was somewhat surprising was that the RANDOMWALK movement algorithm performed second best among those algorithms tested, and generally close to the performance of the FIDUCIAL movement algorithm. The similarity in performance between them should be expected, since for the majority of the time, both are acting in the same manner. The fiducial algorithm will only act differently than the random walk algorithm early in the simulation, when all of the robots are still clustered together. After the robots have spread out, the movement patterns become identical. The FIDUCIAL algorithm simply speeds this spreading process.

Both FOLLOWWALL and SEEKOPEN have some innate flaws. The flaw in FOLLOWWALL is that it assumes two things. First, it assumes the robot to be in a closed region with no internal, isolated obstacles. Robots using FOLLOWWALL are likely to find an obstacle and orbit it indefinitely. The robot would require some form of self-position estimation to detect when it traverses the same positions over and over again. The second problem with FOLLOWWALL has been mentioned previously. The algorithm has no ability to distinguish between robots and obstacles (namely walls) because of this, in the initial robot cluster, many robots ended up trying to follow each other as walls, and ended up going in circles.

The fundamental problem with the SEEKOPEN algorithm is that it is subject to orbiting around local maxima for "openness." Some robots using this algorithm were observed to be traveling in small circles, remaining in one area of the map. SEEKOPEN also tends to prevent robots from going through narrow passageways such as doors. The hook in the Concave world, as shown, could act as one such narrow passageway.

6 Conclusions and Future Work

We have examined the performance of several movement algorithms at dispersing a group of robots in an unknown environment when starting from a initial cluster. The algorithms have been tested in various virtual worlds varying from a simple open area to a complex real-world building. The results show that even approximate knowledge of the locations of close-by robots helps the robots to spread out. The next step is to move into the real world. The combination of these algorithms and a group of small robots would make for an effective system for automatically exploring unknown worlds.

Acknowledgments

Work supported in part by NSF under grants EIA-0224363 and EIA-0324864.

References

1. E. Bonabeau, M. Dorigo, and G. Theraulaz. *Swarm Intelligence: From Natural to Artificial Systems.* Oxford University Press, 1999.
2. H. Choset. Coverage for robotics - a survey of recent results. *Annals of Mathematics and Artificial Intelligence*, 31:113–126, 2001.
3. B. P. Gerkey, R. T. Vaughan, K. Stöy, A. Howard, G. S. Sukhatme, and M. J. Matarić. Most valuable player: A robot device server for distributed control. In *Proc. IEEE/RSJ Int'l Conf. on Intelligent Robots and Systems*, pages 1226–1231, Oct. 2001.
4. T.-R. Hsiang, E. Arkin, M. A. Bender, S. Fekete, and J. Mitchell. Algorithms for rapidly dispersing robot swarms in unknown environments. In *Proc. 5th Workshop on Algorithmic Foundations of Robotics (WAFR)*, 2002.
5. T.-R. Hsiang, E. Arkin, M. A. Bender, S. Fekete, and J. Mitchell. Online dispersion algorithms for swarms of robots. In *Proc. of the 19th Annual ACM Symposium on Computational Geometry (SoCG)*, pages 382–383, 2003.
6. S. Koenig, B. Szymanski, and Y. Liu. Efficient and inefficient ant coverage methods. *Annals of Mathematics and Artificial Intelligence*, 31:41–76, 2001.
7. J. C. Latombe. *Robot Motion Planning.* Kluwer Academic Publ., Norwell, MA, 1991.
8. D. Payton, M. Daily, R. Estkowski, M. Howard, and C. Lee. Pheromone robotics. *Autonomous Robots*, 11(3):319–324, Nov 2001.
9. P. E. Rybski, S. A. Stoeter, M. Gini, D. F. Hougen, and N. Papanikolopoulos. Performance of a distributed robotic system using shared communications channels. *IEEE Trans. on Robotics and Automation*, 22(5):713–727, Oct. 2002.
10. R. T. Vaughan. Stage: A multiple robot simulator. Technical Report IRIS-00-394, Institute for Robotics and Intelligent Systems, University of Southern California, 2000.
11. I. A. Wagner, M. Lindenbaum, and A. M. Bruckstein. MAC vs PC – determinism and randomness as complementary approaches to robotic exploration of continuous unknown domains. *Int'l Journal of Robotics Research*, 19(1):12–31, 2000.

Embedding heterogeneous levels of decisional autonomy in multi-robot systems*

Jeremi Gancet, Simon Lacroix

LAAS-CNRS, 7 avenue du colonel Roche, 31400 Toulouse
jgancet@laas.fr, simon@laas.fr

Summary. This paper presents an architecture for multi-robot (MR) systems in which robots exhibit heterogeneousness in terms of autonomous decisional capacities. This architecture enables various configurations of decision distribution among the components of the system, according both to the available decision making capabilities of the robots, and to the operational constraints related to the tasks to be performed. This architecture is instanciated in the context of a multi heterogeneous Unmanned Aerial Vehicles (UAV) project, and in this context, details related to a generic executive (supporting the versatility of the architecture) are provided.

1 Introduction

The architecture presented in this paper deals with multi-robot (MR) systems operations, considering a particular, understudied frame: contrary to most of the existing MR architectures, we do not make any definite assumption regarding the autonomous decisional capabilities of the robots composing the system. Different kinds of contexts or applications justify such an approach, like the two following ones :

- when a MR application requires to be able to add new (a priori unkwnon) robots in this MR system, with limited work/adaptation load
- when a MR system's user wants to be able to take control over robots, whatever their decisional capabilities.

We introduce in this paper an architecture concept and associated tools dedicated to the requirements of such *heterogeneous multi-robot systems*, the term "heterogeneous" being essentially related to decisional autonomy capabilities.

Problem statement. MR architectures embrace more concerns than single robot architecture: in particular, designing MR architectures requires to define the decision making scheme and to specify the interaction framework

* This work is partially supported by the Comets IST project (# 34304) of the 5th European Framework Program

among the different robots of the system, which of course influences the design of the individual robot architecture. For instance ALLIANCE [8] provides a behavior-oriented solution, enabling the design of totally distributed, fault tolerant multi-robot systems, whereas Simmons & al. [9] extend the three-layers architecture model [1, 4] within a MR framework, where interactions between robots may occur along the different layers.

These different MR architectures enable coordination and cooperation of several robots, but assume a given, homogeneous level of decisional autonomy for all the system's robots. Although these architectures may allow the integration of physically heterogeneous robots, they can not cope with heterogeneous robots in terms of *decisional capabilities*. Nevertheless, some MR applications clearly require the possibility to involve robots exhibiting quite different decisional abilities. This is typically the case in the COMETS project [7, 3]: COMETS aims at designing a multi-UAV framework enabling cooperative operations, in the context of forest fire monitoring and surveillance applications. In COMETS, some UAVs are directly controlled by an operator, some others are only endowed with *operational autonomy* (their tasks being planned and monitored by a central station), whereas others may have *decisional autonomy* capacities (hence should achieve given missions by themselves). Moreover, depending on the situation, the central station should be able to take control over UAVs endowed with decisional capabilities. COMETS hence requires an architecture integrating both *central decision making* and *distributed decision capabilities*; moreover, this architecture should provide with the possibility to configure dynamically the decisional scheme, depending on the available decisional capabilities of the robots and on the operational context.

Outline. Section 2 details the proposed multi-robot architecture: a taxonomy of robots decisional autonomy is introduced, and used as a foundation to state this architecture. In section 3, we focus on the design and implementation of a generic executive that is suited for various levels of decisional autonomy. Section 4 then provides a general scheme for the higher levels of robots autonomous decisional capabilities, as well as the main directions to implement these features.

2 Robot decisional autonomy architecture

2.1 A taxonomy of decisional autonomy capabilities

Within a multi-robot system, the "decision" encompasses several notions:

- **Supervision and execution**: The *executive* is a passive, reactive management of tasks execution, whereas *supervision* is an active process that manages all the decisional activities of the robot, whatever their extent.
- **Coordination**: It ensures the consistence of the activities within a group of robots. It defines the mechanisms dedicated to avoid or solve possible resource conflicts that may arise during operational activities. This is especially related to trajectories and multi-robot cooperative tasks execution (*e.g.* simultaneous perception of the same target by several robots).

- **Mission refinement, planning and scheduling**: These decisional activities are dedicated to plan building, taking into account on one side models of missions and tasks, and on the other side models of the world: robot's perception and motion abilities, current knowledge related to the environment, etc.
- **Task allocation**: This deals with the way to distribute tasks among the robots. It requires to establish a task assignment protocol in the system, and to define some metrics to assess the relevance of assigning given tasks to such or such robot.

These decisional components can be implemented according to different configurations: they can be gathered within a Central Decisional Node (CDN), or be partially (or even totally) distributed among the robots. We define the "level of autonomy" of a robot as the amount of decisional mechanisms it is endowed with, and consider the following five levels (see table 2.1):

		Supervision and execution	Coordination	Planning	Task allocation
Level 1	Central	X	X	X	X
	Distrib.	-	-	-	-
Level 2	Central	x (supervision)	X	X	X
	Distrib.	x (executive)	-	-	-
Level 3	Central	x (supervision)	x (high level coordination)	X	X
	Distrib.	x (executive)	x (low level coordination)	-	-
Level 4	Central	-	-	-	X
	Distrib.	X	X	X	-
Level 5	Central	-	-	-	-
	Distrib.	X	X	X	X

Table 1. Repartition of the decisional components between the central station and a given robot, regarding the five decisional autonomy levels defined.

- **Level 1**: no autonomy onboard the robot. The robot is only able to directly execute elementary tasks requested by the CDN.
- **Level 2**: executive capabilities (operational autonomy). The robot is able to manage partially ordered sequences of elementary tasks, and to return execution status of the tasks.
- **Level 3**: same as level 2, plus simple interacting capabilities. The robot may manage online simple interactions (synchronizations) directly with other robots endowed with at least the same level of decisional autonomy.
- **Level 4**: deliberative capabilities. High level tasks requests are managed (involving autonomous task planning/scheduling), and the multi-robot tasks coordination is autonomously ensured in a distributed way among robots endowed with at least the same level of autonomy.

- **Level 5**: same as level 4, plus tasks reallocation capabilities. The robot may opportunistically reallocate tasks and accept new tasks from other robots of the system also endowed with this level of autonomy.

This taxonomy is characterized by a large gap between levels 3 and 4: up to the level 3, a CDN is expected to ensure the global consistence of the system's activity: these levels are considered as **"low levels"** of decisional autonomy. Whereas levels 4 and 5 introduce the possibility to delegate coordination and mission refinement activities in a distributed way (**"high levels"** of decisional autonomy), and even, for level 5, to have the tasks allocation configurations dynamically updated.

2.2 Robots decisional architecture

Figure 1 depicts the overall architecture of the system: a CDN (interfaced with users) communicates with robots, exchanging messages which abstraction level depends on each robot's current level of autonomy. Robots are provided with a set of functional components and with a Distributed Decisional Node (DDN) possibly ranging from the simplest decisional autonomy level, up to the highest decisional capabilities.

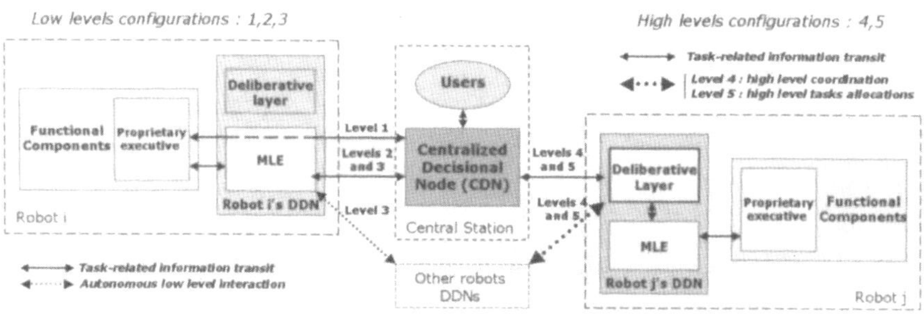

Fig. 1. Left: Low levels (1 to 3) configurations of decisional autonomy. Right: High levels (4 and 5) configurations of decisional autonomy

Regarding low autonomy levels, the DDN is restricted to an executive. This executive being actually common to all levels, we denote it as the *multi-level executive* (MLE). At level 1, the MLE behaves as a transparent connecting point between the CDN and the robot's functional components. However, the full MLE's features are required when considering upper levels: at levels 2 and 3, it manages tasks sequences execution, and at level 3 it enables simple interactions with other robots of the same level.

For higher levels, the MLE relies on the robot's Deliberative Layer (DL) instead of the CDN, tackling higher autonomous decisional capabilities. The DL deals with missions and tasks refinements, as well as coordination activities or task reallocation (when relevant, i.e for the level 5). Stakes and issues related to the DL are further discussed in section 4.

This architecture is currently exploited for the development of the COMETS multi-UAV system, where different, heterogeneous UAVs have to achieve missions related to surveillance and monitoring implying collaborative activities such as cooperative perception.

3 Multi-Level Executive (MLE)

We focus here on the MLE's features, relying on the COMETS project, as the application context: hence robots are UAVs in all the following.

3.1 General task model and assumptions:

Our task model is built around elementary events processing: these events are expected to occur whenever the states of tasks evolves. Events can also correspond to other noticeable activities evolution, such as the reception of a message, or the elapsing of a certain lapse of time. Tasks have a temporal extent: a task starts, then ends after a certain amount of time. The starting event is the only controllable event: all other kinds of events related to a task are contingent, *i.e.* the system can not guarantee that such an event will occur, neither exactly when it may occur. A task can give rise to several, partially ordered contingent events during its execution.

For the low levels of autonomy, the CDN is supposed to be able to elaborate a safe and consistent MR plan, and therefore to provide the UAV with the (already consistent) tasks to be processed, according to a task communication formalism. On the other side, the minimal requirement expected from an UAV is its ability to execute *elementary tasks*, *i.e.* unitary, "simple" tasks that can be handled by robot's functional components. In the COMETS project, the following tasks are expected to be processable by the functional components of a robot integrated in the system: take-off (TO), go-to (GT), take-shot (TS), wait (WT), and land (LD).

Integrating a given UAV in the whole system requires to provide this UAV with a basic interface enabling elementary tasks information transmission (requests, status, exec. results). For that purpose, an *elementary task formalism* has been developed (its specification is not detailed in this paper).

3.2 Executive's mechanisms for the low decisional levels

At the **first level** of decisional autonomy, the MLE only transmits the elementary tasks requested by the CDN to the functional components of the UAV, and then sends back execution status.

Regarding the **second level**, the MLE manages partially ordered sequences of tasks in a consistent way and in a timely and safe manner. Two main mechanisms are involved for this purpose:

- **dynamic tasks insertion:** this enables the possibility to request tasks insertion in the UAV's current task plan, according to an *insertion mode* that characterize the relative order of the newly inserted task versus the current partial order of the tasks already scheduled. Four possible insertion modes are defined:

- **SEQuential (SEQ) mode:** The task has to be provided with a certain number of preconditions (in terms of expected events), which satisfaction can be specified either as *mandatory* or *optional*: in the first case, the satisfiability itself should be permanently satisfied, *i.e.* if the precondition happens not to be satisfiable anymore, then the task is aborted. On the contrary, an *optional* precondition is considered as satisfied (and hence removed from the task's list of preconditions) if it is actually satisfied *or* if it happens that its own satisfiability becomes **unsatisfiable**. In this case, the task is not aborted. Figure 2 illustrates these precondition mechanisms.
- **Very Urgent Task (VUT) mode:** this mode is a way to trigger a priority task, preventing any incompatible task to be executed during this time: the list of incompatible tasks to prevent should be provided as parameters of the task insertion. If an incompatible task is already running, it is interrupted. Otherwise, if an incompatible task is scheduled, then it can be either cancelled or only delayed (its preconditions are updated taking into account the task being inserted in VUT mode).
- **DEPendant (DEP) mode:** it allows to insert a task with as many preconditions as tasks currently scheduled: each precondition is satisfied when the corresponding task triggers its "end of task" event. Moreover, these are *mandatory* preconditions (*i.e.* as defined in the SEQ insertion mode).
- **Non Urgent Task (NUT) mode:** it allows to set as many preconditions as tasks currently scheduled: each precondition is satisfied when the corresponding task triggers its "end of task" event. However, contrary to the DEP mode, these are *optional* preconditions (*i.e.* as defined in the SEQ insertion mode).

The SEQ insertion mode is the most usual mode: it is a natural way to attach preconditions to a new task being inserted in the UAV's current plan. The VUT mode provides with a way to request urgent tasks execution, bypassing the current plan's constraints, but keeping it consistent. Finally, DEP and NUT modes are only shorcuts: a SEQ mode task may as well substitute itself to each of them.

- **dynamic tasks aborting:** this mechanism enables the possibility to request tasks abortions in the current plan. If the task is already running, then the abortion of the task is an interruption. If the task is not yet running, then the abortion is a cancellation (the task is descheduled). The abortion triggers a propagation mechanism, that checks which of the scheduled tasks depend on the aborted task (*i.e.* the tasks having a precondition expecting an event from the aborted task, like a "end-of-execution" event): if the dependence is a *mandatory* precondition, then this task is also aborted, and so on. If the dependence is an *optional* precondition,

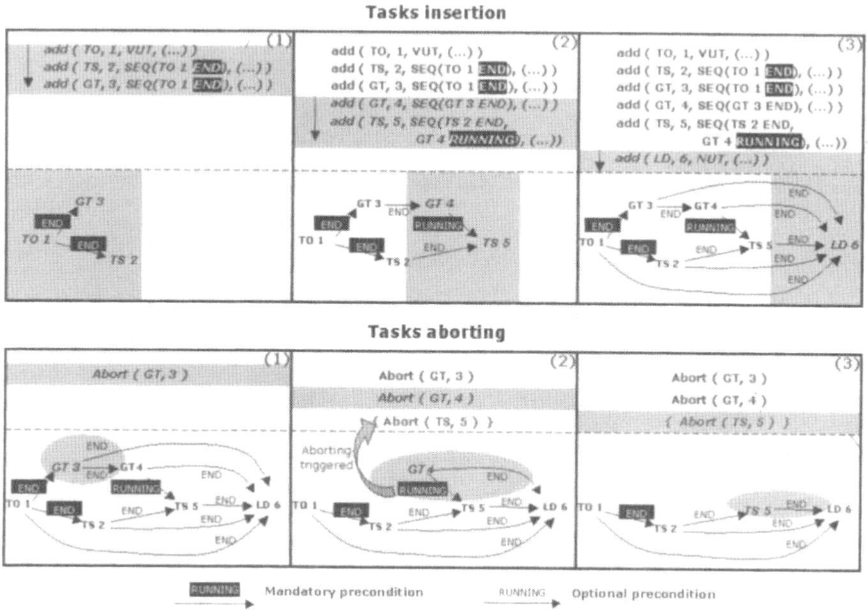

Fig. 2. Top: Examples of tasks insertion and illustration of the corresponding preconditions dependencies. (1): a VUT task and SEQ tasks with single mandatory precondition. (2): SEQ tasks with both mandatory and optional preconditions. (3): NUT task. **Bottom:** Examples of tasks aborting (1) and illustration of abortion propagation toward dependent tasks having mandatory preconditions (2) and (3)

then the dependence is removed as if the precondition was satisfied, and the corresponding task is not aborted.

The **third level** of decisional autonomy deals with an additional mechanism intended to enable autonomous synchronizations between several UAVs' MLEs. A synchronization can be requested to a given MLE as a particular task, that produces events (start, running, end...) in the same way as usual tasks do. It is also possible to insert a synchronization task with particular insertion modes as defined previously. Two "roles" are specified as parameters of a synchronization task: sender ($\mathcal{S}$), and receiver ($\mathcal{R}$): $\mathcal{S}$ and $\mathcal{R}$ are the sets of UAVs considered respectively as senders and receivers of the synchronization message. When a synchronization task is processed, the MLE checks whether its own ID is noticed in the $\mathcal{S}$ or $\mathcal{R}$ sets. Three situations may occur:

- ID $\in \mathcal{S}$ (only): the MLE has to send a synchronization signal to all UAVs which ID belongs to the set $\mathcal{R}$. This signal contains the synchronization task's ID, and also this UAV's ID. From this UAV's point of view, the task is considered achieved.
- ID $\in \mathcal{R}$ (only): the UAV expects to receive synchronization signals from all UAVs which IDs belong to the set $\mathcal{S}$. From the point of view of this

UAV, the synchronization task is considered achieved as soon as all signals are received.

- ID $\in \mathcal{S}$ and ID $\in \mathcal{R}$: the UAV should both send its own synchronization signal then wait for signals from all other UAVs specified in the set $\mathcal{S}$. The synchronization task is considered achieved as soon as all signals are received. If $\mathcal{S}$=$\mathcal{R}$, then the synchronization is a general "rendez-vous" between all UAVs.

Figure 3 illustrates this synchronization mechanisms.

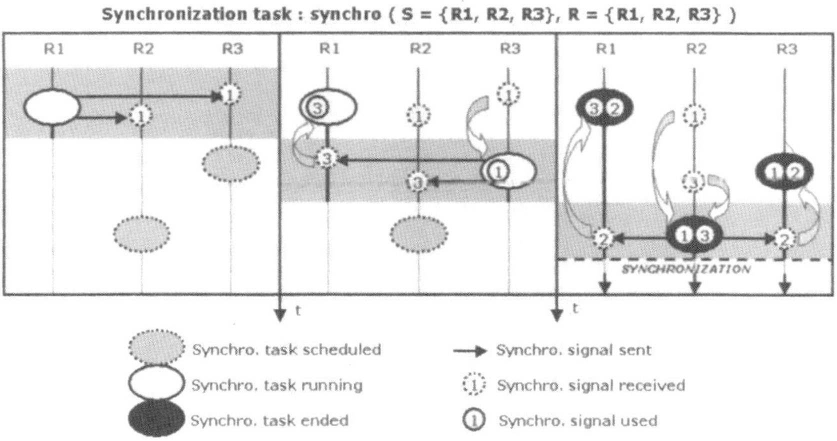

Fig. 3. Illustration of a synchro. task with 3 UAVs, in the case of a "rendez-vous"

4 High decisional autonomy levels

We briefly present here the general framework for the development of the DL, dealing with the higher autonomy levels. Figure 4 exhibits the general structure of the UAVs' DDNs: the main part is the *supervisor*, which is responsible for missions and tasks refinements. It elaborates plans with the support of a *tasks planner/scheduler* and the *specialized refiners*. It also triggers negotiation sessions for coordination purposes and tasks reallocations (when relevant, i.e. at level 5 only), through the *negotiation manager*.

- **Task planning / scheduling**: given a requested mission, the deliberative layer is expected to build executable plans, which execution would be straightforward in an ideal context. Actually, contingencies during missions execution may require to repair dynamically parts of the plans, hence a plan is never considered as definite: it should be reprocessable, online, taking into account latest contingencies. This is partially devolved to the *task planner / scheduler*, manipulating symbolic tasks models. The *specialized refiners* are intended to bridge the gap between abstract, symbolic plans and actual data of the world, known through various models (namely environment model, UAV and communication models): the "*path*" refiner provides the trajectories, expected times and resources consumptions in order

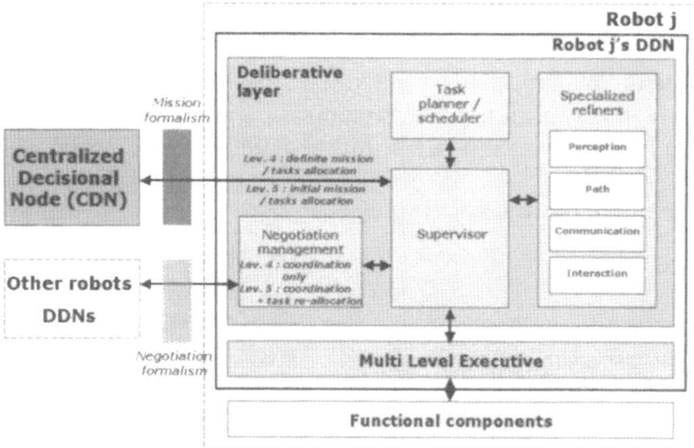

Fig. 4. Deliberative layer detail

to reach given points. The "*perception*" refiner computes the most relevant perceptions in the current context (mapping, tracking, or fire monitoring, for instance). The "*communication*" refiner computes the valid ranges of communications, taking into account UAVs and environment models. Finally the "*interaction*" refiner takes into account interactions models, and restricts/constraints operational areas according to these models. A preliminary version of these refiners has been developped, and on-going work deal with the integration of a HTN-oriented symbolic planner (D. Nau's SHOP2 [6]) with these refiners.

- **Coordination**: two kinds of coordination can be pointed out: spatial coordination, and interactions-related coordination. The spatial coordination with other UAVs is an activity that should continuously work in background, since it is a vital mechanism (prevents conflictual trajectories). It should overcome all other kinds of deliberative mechanisms (we are currently exploring the possibility to adapt the Plan Merging Protocol [2] to the particular purpose of UAVs spatial coordination). Regarding interactions-related tasks coordination, flexible plans produced by the planner/scheduler should be coordinated with the plans of other involved UAVs, for each given interaction: this mechanism should rely on a relevant negociation protocol, ensuring a consistency of the joint plans of the UAVs. Such a protocol definition and development, based on incremental assessment and exchange of critical plan's sections, is currently on-going.
- **Task reallocation**: task reallocation is a distributed mechanism intended to dynamically improve missions/tasks allocation. Task reallocation continuously requires to identify which tasks or parts of plans would be worth to be reassigned to other UAVs: this can be solved using an opportunistic mechanism, pointing out "expensive" tasks (metrics to be defined), or tasks looking consistent with regards to other UAVs plans (criteria to be defined). For the reallocation process itself, we intend to extend the Contract Net Protocol pardigm [10][5].

These directions are guidelines for our next developments related to the distributed decisional capabilities for the COMETS' UAVs: all the mechanisms and decisional features briefly introduced in this section deal with current, on-going investigations and developments. We plan to provide much more details related to these purposes in future publications.

5 Current results and future work

The MLE's low level mechanisms (levels 1 to 3) introduced in section 3 have been tested both in simulation and in actual experimentation, with several UAVs (up to 3 UAVs: 2 helicopters and 1 blimp). During these tests, a MLE was attached to each of the UAVs, and high level decision making was performed within a control center (the CDN) interfaced with human users. Furthermore, some preliminary simulated tests have also already been led for some of the DL components.

The design and development of higher decisional capabilities are on-going, to meet our objectives of applying the whole versatile architecture to an actual multi-UAV system (e.g. the COMETS' UAVs fleet).

References

1. Alami R., Chatila R., Fleury S., Ghallab M., Ingrand F.(1998). An Architecture for autonomy. In: Int. Journal of Robotics Research. Vol.17, p.315-337
2. Alami R., Ingrand F., Qutub S.(1998). A Scheme for Coordinating Multi-Robot Planning Activities and Plans Execution. European Conf. on Artificial Intelligence (ECAI'98)
3. COMETS project's official web site: http://www.comets-uavs.org
4. Gat E.(1991). Integrating planning and reacting in a heterogeneous asynchronous architecture for mobile robots. In SIGART Bulletin 2, 17-74
5. Lemaire T., Alami R., Lacroix S.(2004). A Distributed Tasks Allocation Scheme in Multi-UAV Context, to appear in Int. Conference on Robotics and Automation (ICRA'04)
6. Nau D., Au T.C., Ilghami O., Kuter U., Murdock W., Wu D., Yaman F. (2003). SHOP2: An HTN planning system. Journal of Artificial Intelligence Research 20:379-404
7. Ollero A., Hommel G., Gancet J., Gutierrez L.G., Viegas D.X., Forssen P.E., Gonzalez M.A. (2004). COMETS: A multiple heterogeneous UAV system. To appear in SSRR 04, Bonn, Germany
8. Parker L.(1998). ALLIANCE: An architecture for fault-tolerant multi-robot cooperation. In IEEE Transactions on R.& A. 14(2):220-240
9. Simmons R., Smith T., Dias M.B., Goldberg D., Hershberger D., Stentz A., Zlot R.M. (2002) A Layered Architecture for Coordination of Mobile Robots. In: Multi-Robot Systems: From Swarms to Intelligent Automata, Proc. of the 2002 NRL Workshop on Multi-Robot Systems, Kluwer Academic Publishers
10. Smith R.G.(1980). The contract net protocol: High-level communication and control in a distributed problem solver. IEEE Transactions on Computers
11. Volpe R., Nesnas I., Estlin T., Mutz D., Petras R., Das H.(2001). The claraty architecture for robotic autonomy. In: Proc. of the 2001 IEEE Aerospace Conference, Big Sky, Montana

Part VII

Distributed Problem Solving

Collective Energy Distribution: Maintaining the Energy Balance in Distributed Autonomous Robots using Trophallaxis

Chris MELHUISH[1] and Masao KUBO[2]

1. IAS lab., CEMS Faculty, University of the West of England, Chris.Melhuish@uwe.ac.uk
2. National Defense Academy, and IAS lab., CEMS Faculty, University of the West of England, masaok@nda.ac.jp

Abstract Truly autonomous robots, either as single units or in groups will be required to generate and manage their own energy. This paper explores the idea of robot 'trophallaxis' whereby one robot can donate an amount of its own internal energy reserve to another. Different strategies for energy transfer are considered within the test simulation scenario of a dust cleaning task. Successful strategies are shown to confer benefits including survivability, task performance and offer the potential to inform design parameters of the storage media.

1. Introduction

Truly autonomous systems will be required to generate and manage their own energy budget [1,9]. This is true of single robots as well as a group of distributed autonomous robots. Research has shown how a group of 'food foraging' robots can benefit when employing a strategy which involves them replenishing their energy from the collective repository (nest) created by the individual foragers [6]. In contrast to this form of 'central sharing' it is observed that in the world of social insects, such as ants, the phenomenon of food sharing – trophallaxis - is also employed to distribute the collective energy resource owned by the group members [2]. It is reasonable to ask if such a mechanism might be beneficial to a collective of autonomous robots. Already, some researchers are exploring the prospect of creating robots which will be able to generate their own energy from the environment employing, for example, sugar [10] as well as natural substrates [8,5,3]

These robots have rechargeable energy storage systems and each type of system has its own characteristics. For example, most Lithium ion batteries have high energy density but require a 'long' time to charge. Fuel cells, such as Direct Methanol Fuel Cells, are very 'clean', have 'low' energy density and are now becoming portable. The capacity of such cells as well as super capacitors and polymer battery cells is also continuously improving. Roughly speaking, energy devices are becoming more portable, quicker to charge, and with higher energy densities. The improvements might be able to be employed in a manner which allows robots to transfer energy between them. How then will these characteristics influence the choice of design and behaviour of a robot collective? Such considerations will be important to future designers employing collective robot solutions. It appears that, currently there are few practical examples of multiple robots systems. We think that one of the reasons is that there is no methodology for the maintenance and management of energy for multi robots. A single Trilobite™ carpet cleaner is said to work for an hour with fully charged Ni-H batteries and takes 2 hours to re-charge. How would such a system scale up when cleaning a larger environment? Are there benefits from allowing robots to share energy rather than having no other choice than to return to the charging point?

This short paper reports on some of the initial work conducted in addressing this issue. In this first, relatively simple, study we look at the scenario of floor cleaning by a robot collective in order to understand the effects of parameters such as battery size, charge time and power density. We required an example scenario in which robots could carry out their task while moving without prescribed trajectories which would give plenty of opportunities for collisions. Each collision offered the opportunity for the necessary proximity for energy transfer to occur between a pair of robots. The task and locomotion would incur an energy cost. We therefore employed a scenario in which robots are tasked to clean a floor in a room where dust is continuously falling on the surface. Power is required to move the robots as well as clean the floor. By manipulating some of the elements above, such as the number of robots and the battery capacity size, we explore the relationship among these fundamental factors and parameters. In particular, we introduce a collective energy management system based on energy transfer between robots. We show that this strategy could offer considerable advantages.

The paper is arranged in 4 sections; in section 2, we define the model of simulation and describe a simple robot swarm; scenario 1 (1 charger and n robots without energy transfer). In this section, we demonstrate the limitations of such a simple collective robot system. In section 3, we show that active energy distribution, employing 'robot trophallaxis' confers benefit. Concluding remarks are presented in section 4.

2. Defining the Simulation

We introduce a simple model for a collective system of robots. We suppose that there is a homogeneous group of robots R and an energy charger C on rectangle grid field A whose edge is equal to m. The task of each robot is to remove the dust which is constantly falling on A. The amount of dust on A is represented by D. The system is:

$$A = \langle R, C, D \rangle, \quad D = \{D_{xy} > 0\} \quad x, y = 1, \cdots, m \tag{1}$$

The amount of dust on each grid on A is increased by ΔD at each time step. The dust level is restricted to a maximum level $D_{\max}$. The dust at each grid point is:

$$D_{x,y} \leftarrow Minimum(D_{x,y} + \Delta D, D_{max}) \tag{2}$$

In the following all experiments, the amount of stacking dust per time step is set as $\Delta D = D_{max}/10000$. The initial dust level of each grid is determined from 0 to D_{max} at random.

The total dust condition on A is referred to as the **dust value** V_D, and is defined as:

$$V_D = \sum_x \sum_y (D_{x,y}/D_{max})/m^2, \quad 0 \le V_D \le 1 \tag{3}$$

V_D represents the average of the level of dust on A. V_D=0 means that surface A is dust free whereas with V_D=1 A at its maximum dust level.

Each robot is considers as follows. The i-th robot $r_i \in R$, $|R| = n_R$ is defined as:

$$r_i = < p_i, h_i, v_i, e_i, \eta_i, \theta_i, \omega_i, g_i, \zeta_i, \Delta eg_i > \tag{4}$$

where $p = \{p_x, p_y \in 1, \cdots, m\}$ represents a grid on A. h shows the heading of the robot, where $h \in \{N, S, W, E, NE, NW, SE, SW\}$. In the following section, we abbreviate index i whenever we refer to a single robot. The remaining parameters are explained below.

Firstly, a robot selects action v from a set of actions ω using by function η. Next, according to action v, a robot changes its position or direction. In this simulation, we assume that each robot has 4 candidates for action

$$\omega = \{Forward, Backward, TurnRight, TurnLeft\}. \tag{5}$$

If a robot selects Forward or Backward, it tries to move the corresponding grid among its Moore neighbors (cardinal+diagonal elements [4]). The robot can carry out the action if the destination grid is unoccupied by another robot or charger. When movement is prevented the behaviour 'fails' with no resulting change to the robot's direction and position. The robot's heading h is changed when behaviour v is either TurnLeft or TurnRight in which a turn of 45^0 is executed.

The execution of the chosen behavior demands an energy cost. After the completion of action v, the energy is updated as follows:

$e \leftarrow Maximum(e - \theta(A, v), 0)$ where the function θ generates the energy consumption cost for the execution of the action. (6)

For simplicity, every execution requires the same energy, Δe, whether the execution of the action is successful or not. However, while recharging the battery, no energy consumption is required since the robot would be physically restricted by its charger. The energy cost function is defined as:

$$\theta(A, v) = \begin{cases} \Delta e & for \quad movement \;\&\; dust\_collection \\ 0 & charging \end{cases} \tag{7}$$

In this paper, Δe=1. Each robot has an energy accumulator, which we refer to simply as the **battery**, which can store energy up to a maximum e_{max}. A robot can move until $e = 0$.

Each robot has a suction pump. Each pump has a 'capability' g which describes the ability to collect dust (amount dust collected/time step). The suction pump is always on. The actual amount of dust collected/unit time is between 0 and g dependent on how much dust in present. Therefore, the amount of dust remaining at x,y is:

$$D_{xy} \leftarrow Maximum(D_{xy} - g, 0) \tag{8}$$

Of course, when a robot employs its suction pump, it consumes energy. The pump runs continuously and the associated energy cost Δsp every time step is included in Δe of Eq.7.

It may arise that a robot exhausts itself of energy. Let $R_{ne} = \{r_i \mid e_i \leq 0.0, r_i \in R\}$ be the set of such robots at that time. The number of surviving robots n_s is:

$$n_s = n_R - |R_{ne}|. \tag{9}$$

The rate (and therefore amount) of energy which can be transferred between a charger and a robot, is described by Δeg, a constant which refers to the maximum transfer of electrical current of the battery. ζ denotes the transfer rate when energy is transferred between robots. This issue is discussed later.

Next, we describe the energy charging unit which will simply be referred to as the "charger". The charger is conceptually very similar to a robot but its energy level is constant. Let the j-th charger $c_j \in C$ $|C| = n_C$ where:

$$c_j = < p_j, h_j, v_j, w_j, \eta_j, \omega_j, s_j, \zeta_j > \tag{10}$$

In the following section, we ignore index j whenever we refer to a single charger. As same as robot r, p indicates a grid position on A, h represents direction of charger. Our study has looked at the case of mobile chargers but this case is not discussed in this paper due to limitations of space.

Charger can recharge 'neighboring' robot's battery. We suppose that it is possible to recharge s robots on its Moore neighborhood simultaneously. Function ζ sets the amount of energy per unit time that can be transferred between for each robot. However, for clarity in this paper, we suppose that a charger gives each robot the same amount of energy w in total and s=1. Usually, some time steps are required for recharging. The time for a robot with energy e to be recharged fully is:

$$e = (e_{\max} - e)/\Delta eg \quad (w \geq e_{\max}) \tag{11}$$

During recharging the robot cannot move. Also, charger has no cleaning ability. η is the function which chooses an action from ω. We suppose η is a random function. At each time step, robots choose their action randomly.

Let us now consider a number of different scenarios involving differing numbers of robots, chargers and energy transfer mechanisms. A table of symbols is provided in Table 1.

Table.1 Summary table with main symbols

Symbol	Meaning	Symbol	Meaning
n_R	The number of robots (i.e. $n_R = \|R\|$)	m	The edge size of the quadratic arena A.
e	Current remaining energy	n_s	The number of survivor (i.e. e>0).
$e_{\max}$	The maximum capacity of battery	V_D	Dust Value (V_D=1 means A is fully dusty)
$g/\Delta D$	Cleaning ability (g is suction pump power, ΔD indicates stacking dust per time step)		

2.1 Test Scenarios

Perhaps the most obvious approach when employing a collective robot system for large floor cleaning, could be as follows: (1) the large floor is divided into several smaller regions. (2) a charger and n robots are set on each region. Therefore, we examined two test case, a group of robots not using energy transfer and using energy transfer.

2.2 Scenario 1: static charger n_C=1, worker n_R>1

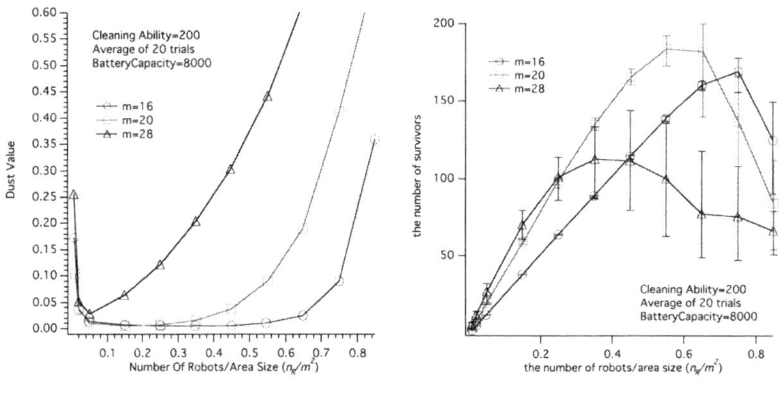

Fig.1 **(a)** the average of dust value of scenario 1. **(b)** the average of the survivor of scenario 1.

Scenario 1 employs a single static charger and multiple robots in a field A. The most important and different thing is that robots with no energy are ***not*** removed. Exhausted robots are considered as obstacles. Fig.1(a) shows the dust value in this scenario. We employ 3 different arenas m={16,20,28} and a maximum of (m^2-1) robots can be allocated in each arena. The density of robots n_R/m^2 is adopted for x axis. Each robot has a low cleaning ability ($g/\Delta D$=200), and sufficient battery capacity size, $e_{\max}$=8000 except for m=28. Each point on the graph is an average of 20 trials. Each trial is executed for 90,000 time steps.

Fig.1(a) shows that even for a small number of robots, the dust value is appreciably reduced. After that, the dust value slightly improves before finally becoming worse as the number of robots increases further. For example, when m=20, for a robot density of 0.25 (the number of robots is 100), dust values are almost zero. Fig.1(b) shows the survivability of the collective. The y axis is the number of survivors n_s. Fig.1(b) demonstrates that number of survivors increases if each robot has sufficient battery capacity but the limiting factor is the number of robots – as the robots increase in number the survivability becomes worse. We also have observed that as collisions increase the 'working' range from the charger is reduced.

To summarize this section we note, firstly, that low dust levels are achieved by the increase in the number of robots. However, an increase in the number of the robots makes it increasingly difficult for a robot to survive because it is forced to spend a 'extra' energy dealing with collisions. This means that each robot in the collective has to employ a larger battery than that used by a robot working on its own. Therefore, the result indicates that this simplest form of collective robot system does not improve upon a single robot system. A better mechanism is required and energy transfer between robots is discussed in the next section.

3. Energy Transfer between Robots (E.T.)

We introduce the notion of energy transfer rules and discuss their effects. We begin by assuming that the simplest rules basically depend on each robot's store of energy. In this paper, we limit the discussion to energy transfer between two robots, ie each robot in a group can only receive or donate energy from one other robot at any one time. The function ξ in section 2, corresponds to these energy transfer rules. Firstly, we suppose that energy transfer begins when two robots collide, and after simple arbitration, one of them requests energy (recipient, r_R) from the other (donor, r_D). We constrain the systems such that the donor cannot reject the request and ***must*** donate an amount of energy defined by the energy transfer rule ξ_S where the recipient is identified by the index R and the donor the index D. Generally speaking, in nature, energy flows from higher place to lower. However, in a robot system we are not constrained and this is reflected in some of the rules. The rule sets used in the trials are outlined below.

Type 1: (relative steady policy) If the donor has larger energy than the recipient, the donor sends a fixed proportional amount energy.

If $e_D > e_R$, then $\begin{array}{l} e_R \leftarrow e_R + e_D \cdot k_1 \\ e_D \leftarrow e_D(1-k_1) \end{array}$ where k_1={0.5,0.25,0.125} ==> ET 1, 2, 3

Type 2: (Ruthless Policy) If the donor has almost no energy and is below a threshold, the recipient gets ***all*** the donor's energy.

If $e_D < k_3 \cdot e_{\max}$ and $e_D < e_R$ then $\begin{array}{l} e_R \leftarrow e_R + e_D \\ e_D \leftarrow 0 \end{array}$ ====> ET 4

Type 3: (Greedy) The recipient gets energy to fill its battery up to $e_{S\max} k_7$.

$\begin{array}{l} e_R \leftarrow e_{R\max} k_7 \\ e_D \leftarrow e_D - (e_{R\max} k_7 - e_R) \end{array}$ where k_7={1.0,0.8,0.6,0.4} ==> ET 5, 6, 7, 8

All of these energy transfer rule have to satisfy the conditions, $0 \le e_D \le e_{D\max}, 0 \le e_R \le e_{R\max}$. Therefore, after transfer, if they do not satisfy this condition, the excess of energy is returned to the suitable robot. Also, we assume that the extra energy consumption for energy transfer Δet =0.

3.1 Scenario 2: static charger n_C=1, worker n_R>1 with E.T.

In this scenario we consider a static charger n_C=1 and multiple workers n_R>1 employing different energy transfer rules (ζ_S). In this paper we will consider the case of an area A with m=20 for convenience of explanation. Although we tested 8 different battery capacity sizes, in this section due to page limitations, we will only show one of them; the collective robots group with $e_{\max}$=500. The number of survivors of the group using the above energy transfer rules (ET) n_s are shown in Fig.2. The x axis shows the number of robots n_R. Also, the dust values are presented in Fig.3. For comparison scenario 1's number of survivors n_s and dust value also are shown by dashed line labeled as ET0. The following observations are highlighted.

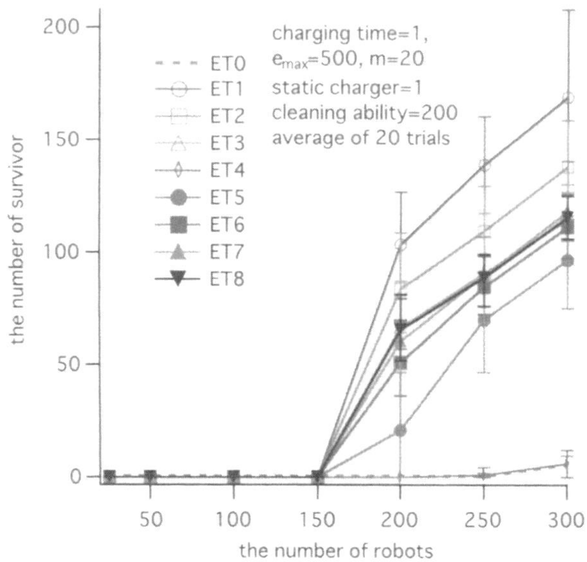

Fig.2 Energy Transfer ($e_{\max}$ =500): Average of the number of survivors

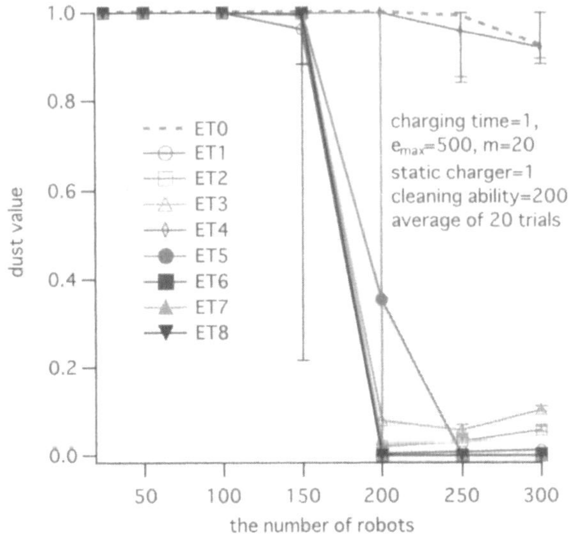

Fig.3 Energy Transfer (e_{max} =500): Average of Dust Value V_D

1 All strategies, except for ET4, produce a dramatic improvement in the number of survivors and dust value when compared to not using an ET strategy. Of course, at the extreme when the number of robots n_R is small and have very low battery capacity (e.g. n_R=50,100 in Fig.3), there is only marginal improvement. However, unlike scenario 1, an increase of the number of robots was accompanied by an increase in the number of survivors n_s (e.g. n_R>100 in Fig.3).

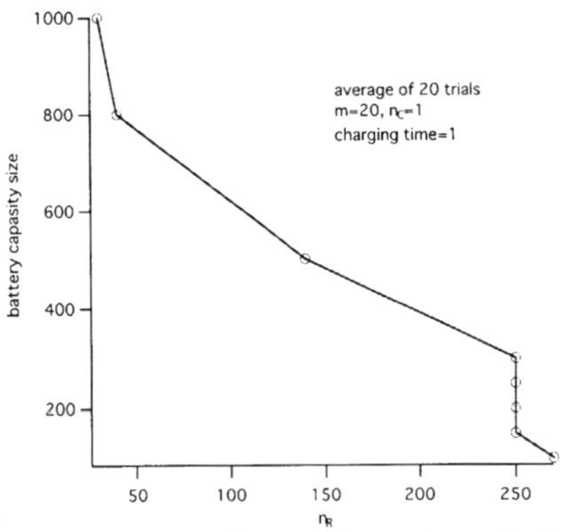

Fig.4 The minimum battery size and effect on Energy Transfer Rules

2 It appears that a minimum limit of robots is required in order for the E.T. strategy to become effective. For example, in this simulation for the case of robots with e_{max}=500, Figures 2 and 3 suggest that about 150 robots should be deployed. Fig.4 shows the relation between the minimum number of robots and battery capacity size when m=20. The x axis indicates the number of robots. The y axis depicts the battery capacity size required for the survival of at least one robot. Fig.4 implies that as one deploys increasingly larger number of robots, the battery size of each robot progressively decreases. We argue that if the density of robots is low, a robot's battery would be empty before the robot attempts to transfer energy. Although the graph falls quickly down from n_R=250, this is not unexpected since an increasing number of robots will increase the number of collisions. Perhaps the most important observation is that (up to a point) the increase in the number of robots reduces the battery size and this size is very much smaller than that of scenario 1.

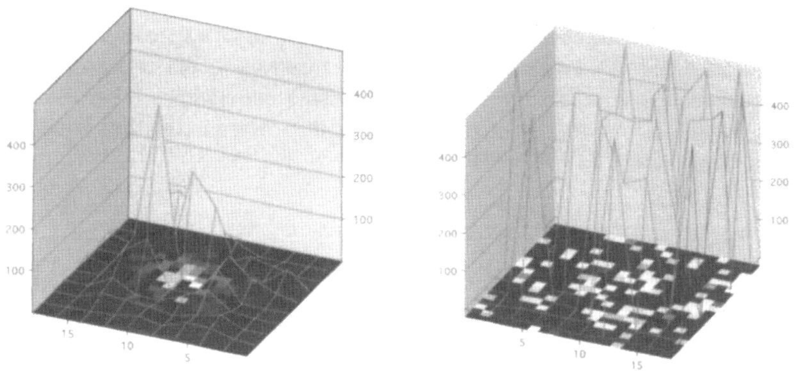

Fig.5 **(a)** Geographical distributions of robots' energy at the end of trial (ET1) **(b)** Geographical distributions of robots' energy at the end of trial (ET5)

3 From Fig.2, one observes that ET4, closely followed by ET0 are markedly worse than any other ET rule. Moreover, the dust value of ET5 is better than almost of others when the density of robots is high although the survivability of ET5 is worst except for ET4 and ET0. These results suggest that different energy transfer rules create different energy distributions patterns in collective robots system. Fig.5(a) shows the 'geographical' distribution of energy using ET1 at the end of a trial. Robots near the charger have appreciable energy while the energy of robots gradually decreases according to the distance from the charger. A white cell on the floor in this figure indicates a higher energy robot. On the other hand, a lower energy robot is shown by black. Fig.5(b) shows the geographical distribution of energy of robots with ET5. It is noticeable that the surface is very variable and robots with high energy are scattered around the field. The key observations from this section as follows: 1) if a sufficient number of robots with an appropriate energy transfer rule is deployed, then these robots do not need large batteries and 'large' suction pumps. 2) Different energy transfer rule can generate particular energy distribution within the robot group and thus the rule set would need to be tuned for the task. 3) Collective task performance was improved when employing an energy sharing strategy as opposed to a 'selfish' system where energy is not shared.

4. Conclusion

This paper represents an early step in the development of energy management strategies for a collective of autonomous mobile robots. Although a simple simulation was employed, the results indicate that employing an energy transfer policy (robot trophallaxis) may confer benefit for a task requiring multiple robots. This benefit may be in the form of task completion and performance, survivability of individual robots and have an impact on the type of energy storage media such as weight, energy density and capacity. Additionally, a comparison scenario 2 with scenario 1 also suggests the following resource allocation implications: For a given set of resources, such as robots and battery cells, an energy transfer strategy (ET) will favour the employment of many robots with fewer battery cells over a single robot (or a small number) using all the cells. This, of course, could have an impact on the design constraints for mobile autonomous robots. We argue that energy management is at the core of practical distributed autonomous robotics and offer some indication why trophallaxis has evolved in social insects. We hope to continue this work further on real robots. *

References

1. Holland O (1998) Towards true autonomy. In 29th International Symposium on Robotics (ISR98), International Federation of Robotics, Birmingham, UK
2. Hölldobler B, Wilson E (1990) The Ants. Springer-Verlag
3. Ieropoulos I, Greenman J, Melhuish C. (2003) Imitating Metabolism: Energy Autonomy in Biologically Inspired Robotics. In Proceedings of the AISB '03, 2nd Int. Symposium on Imitation in Animals and Artifacts, Wales, 191-4
4. Jackson EA (1991) Perspectives of Nonlinear Dynamics 2, Cambridge Univ. Press
5. Kelly I, Holland O, and Melhuish C(2000) Slugbot: A Robotic Predator in the Natural World In Proc. of 5th Int. Symposium on Artificial Life and Robotics.
6. Krieger MJB, Billeter JB (2000) The call of duty: Self-organised task allocation in a population of up to twelve mobile robots, Robotics and Autonomous Systems 30,65-84, Elsevier Science
7. McFarland D, Spier E (1997) Possibly Optimal Decision Making und Self-Sufficiency and Autonomy, Journal of Theoretical Biology, 189 317-331
8. Melhuish C, Greenman J, Bartholomew K, Ieropoulos I, Horsfield I (2002) Towards Robot Energetic Autonomy using microbial Fuel Cells. submitted to the special edition on Biofuel Cells of the *Electrochemica Acta* Journal
9. Pfeifer R(1996) Building Fungus Eaters: Design Principles of Autonomous Agents. In Proceedings of the Fourth International Conference on Simulation of Adaptive Behavior, Cambridge, MA, MIT Press/Bradford Books, 3-12.
10. Wilkinson S (2000) Gastrobots– Benefits and Challenges of Microbial Fuel Cells in Food Powered Robot Applications, Autonomous Robots 9, 99-111

---

* **Acknowledgements:** The authors would like to thank the reviewers for their insightful and helpful comments.

Building Blocks for Multi-robot Construction

Justin Werfel

Massachusetts Institute of Technology, Cambridge, MA, USA jkwerfel@mit.edu

Summary. One notable capability of social insect colonies that has traditionally inspired distributed robot systems is their construction activity. In this paper, I describe a system of simple, identical, autonomous robots able to build two-dimensional structures of arbitrary design by rearranging blocks of building material into desired shapes. Structure design is specified compactly as a high-level geometric program; robots translate this program into physical form via their fixed behavioral programming. Robots are interchangeable both within and between construction projects, and need not be individually reprogrammed between dissimilar projects. Such a construction team could be used as the first stage in a system for remote building of structures, laying out the floor plan that a more sophisticated system could extend upwards.

1 Introduction

A primary inspiration for distributed multi-robot systems is the set of orders of social insects, notably ants, termites, and bees, whose swarms or colonies accomplish many complex high-level tasks through the collective actions of lower-level agents. One of the most characteristic of these tasks is the robust construction of large-scale, complicated structures, despite the insects' own small size and limited complexity. A corresponding research pursuit is the engineering of multi-robot systems that build specifi c desired structures, while retaining advantageous features of the insect systems that inspire them (flexibility, robustness, etc.). The possible uses for structure-building teams of robots are many and far-ranging, from automating the production of low-cost housing to allowing construction and related activities in settings where human presence is dangerous or problematic. This latter class in turn ranges from uses in disaster areas, to the construction of fi rst-stage bases of operations to await the arrival of pioneers in, for example, underwater or extraterrestrial environments.

In this work, I describe the design and simulation of a system of simple, identical, autonomous robots able to build structures in the shape of arbitrary Jordan curves (i.e., non-crossing and closed, in the horizontal plane), by rearranging blocks of building material into desired shapes on a grid. The shape is specifi ed compactly by a high-level geometric program stored in a separate beacon, which serves as the reference point around which all robot activity occurs. Robots receive the program for the structure shape from the beacon at short range during the course of the construction project, and translate it into the appropriate arrangement of blocks via their behavioral programming. Thus the same robots can be used in any construction project without needing to be reprogrammed. The intended method of operation is to scatter a handful of generic robots in the vicinity of suffi cient building material, place a beacon preprogrammed with the desired structure design, and let construction proceed without further intervention. This system is an example of those for which the goal is to robustly generate prespecifi ed global behavior from local interactions among myriad unreliable components [1].

1.1 Previous work

Most previous work on autonomous construction teams has focused on other aspects of the problem. In [20], robots build a linear wall out of blocks held together by Velcro of alternating polarity. Their multi-robot simulations focus on the benefi t of explicit communication, showing that when robots broadcast one bit indicating the polarity of the last block placed, the number of attempts to place blocks of inappropriate polarity is reduced. However, they do not address the issue of specifying more complex structures, nor consider more extensive communication in their building strategies. [10] describes a system of physical robots with force sensors only, that work without explicit cooperation or communication to clear an area of material, by pushing it to the edges of a gradually expanding clearing. [8, 9] describe minimalist approaches to sorting and construction, which have the advantage of simplicity but are typically slow, probabilistic (relying on the correction of frequent errors), and relatively inflexible in the range of structures they can generalize to building. [5] outlines a project whose goal is robots that build 3-D arches and walls at human scale; its robots are intended to work independently rather than collaboratively, and its primary concern is with mechanical engineering considerations, with no reference to the question of controlling high-level building design. Its approach is that of [3,4], whose simulations consider the inverse problem of studying the kinds of structures that result from different simple rules for agent behavior, but do not address the issue of generating prespecifi ed high-level structures.

A related topic is the regulation of formations of agents. Such approaches can be applied directly to construction if building blocks themselves are mobile robots. Some approaches to formation control require continuous global knowledge about all agents, and/or user intervention [2, 6, 15]; others can generate crystalline formations, but do not lend themselves to the design of high-level forms [14]; reconfi guration algorithms for modular robots create two- or three-dimensional forms out of agents which are not arbitrarily mobile, but remain always in contact with one another [18, 19].

In contrast to the preceding, this work focuses on a system of mobile robots with local knowledge and local interagent communication. These arrange passive building materials in the horizontal plane into arbitrary Jordan curves, which can be easily prespecifi ed by the user. Mobility and structural requirements are separated, allowing the design of each class of elements to be specialized, the more sophisticated elements (robots) to be reused for multiple projects, and the passive elements (building materials, which after installation need never move again) to be of minimal manufacturing diffi culty and cost.

2 Component capabilities

Objects in the world of this system are mobile robots, a fi xed beacon, and passive, movable blocks, all of which are initially scattered at random over the workspace.

Robots are assumed to possess the following abilities: move in any direction unless obstructed, and detect if intended movement in a direction is impeded due to some obstacle; pick up, carry, and put down blocks (carrying a block may increase the 'footprint' a robot occupies, which in turn may affect how it must plan trajectories in some cases); recognize blocks and other robots when close to them (in the simulations described here, 'close' was four body-lengths); communicate with other robots within that distance, exchanging information and commands; and detect and evaluate the direction and strength of a signal emitted by the beacon. With the exception of that latter long-distance signal, robots are restricted to local information about their immediate surroundings only.

The beacon can broadcast a long-range, low-bandwidth signal which can be detected by robots from anywhere in the workspace. It cannot obtain long-distance information from robots about their status or the progress of the task; thus the primary utility (and motivation) of the broadcast is as a reference to orient to. The beacon can communicate with robots that are near enough, just as they communicate with each other.

Blocks in this work are taken to be identical, so that robots need not be confronted with the additional problem of determining how to manipulate heterogenous blocks in varying circumstances.

3 Methods

The simulation was written in Swarm, a free objective-C-based system available at `http://www.swarm.org`. Many details of the model were simplifi ed away for this preliminary study. Most immediately, the simulation took place on a two-dimensional cellular grid; thus robots and blocks each occupied exactly one cell, robots were restricted to move in the four cardinal directions, and issues of fi ne position adjustment were sidestepped.

Before deployment of the system, the beacon is programmed with the design for the desired fi nal structure. This program takes the form of a list of corners; each specifi es its distance from the beacon, the angle (positive or negative) to the next corner, and whether the wall between the two is to be curved (perpendicular to the signal gradient everywhere) or straight (Fig. 1A). Such a list completely specifi es the structure's geometry, though not its orientation; if robots or the beacon are equipped with a compass, additionally establishing a desired building orientation is trivial.

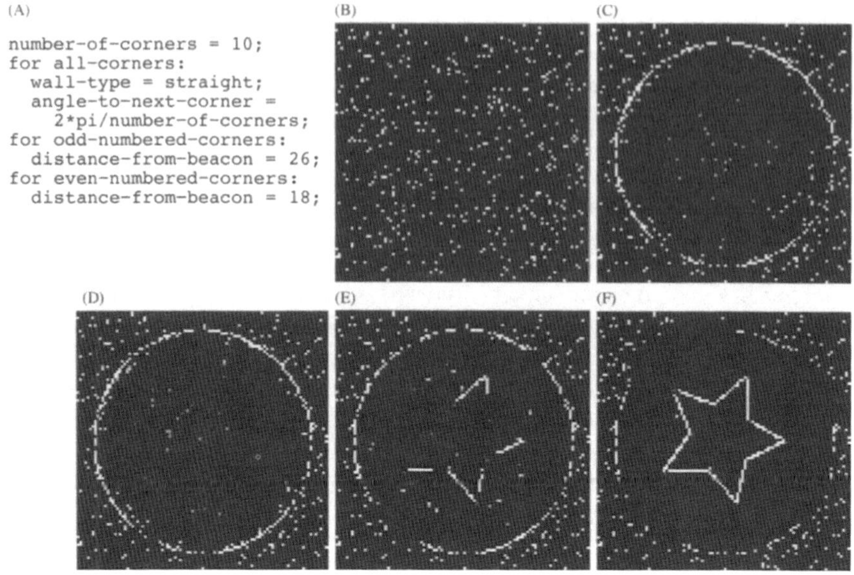

Fig. 1. Example of a structure program and snapshots of several steps in the construction of the associated structure. **A**: pseudocode program for a star-shaped building. **B-F**: stages in the construction of that building. **B**: initial state, with blocks (white) and robots (light gray) scattered randomly, and beacon (not shown) at center. **C**: First the robots clear a space to work. **D**: Some robots take on the role of embodying corners (dark gray) and begin to localize themselves according to the building program. **E**: When corners are placed, the remaining robots begin to build walls between them. **F**: Final structure (only blocks and corners shown).

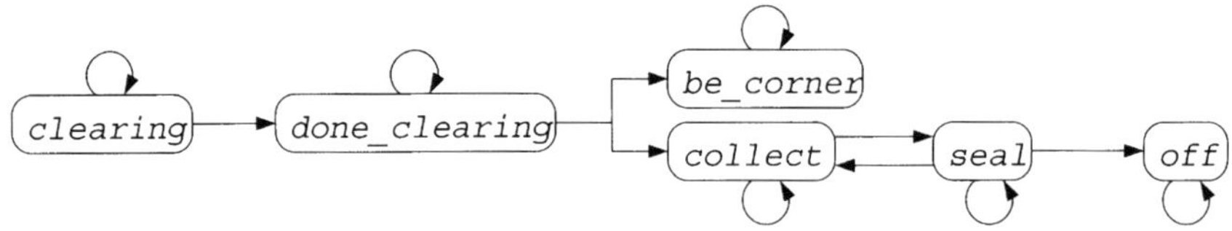

Fig. 2. FSM for behavioral mode. Robots start in the *clearing* state.

3.1 Robot behavior

The algorithm the robots follow can be described as follows (also see Fig. 1). A fi nite state machine (FSM) specifi es each robot's high-level behavioral mode (Fig. 2). All robots start in *clearing* mode: they follow the signal to the beacon (noting its position), then spiral outwards. If a robot encounters a block at any point along the way, it picks it up and carries it directly outward until the signal strength from the beacon falls below some predefi ned threshold; the robot then returns to the beacon and repeats the process. If, while spiraling out, it reaches that signal threshold without encountering any blocks, or if it encounters a robot in any mode other than *clearing*, the robot enters *done_clearing* mode: it spirals back inwards, in the opposite direction to increase the number of *clearing* robots encountered, to bring the entire robot population onward to the same mode and avoid the problem of having some robots working on building the structure while others work just as hard to clear it away. Upon reaching the beacon, the *done_clearing* robot receives a new assignment.

The beacon contains the program for a C-corner structure, as described above. The first C robots that come to it in *done_clearing* mode are assigned to act as successive numbered corners, and enter *be_corner* mode. The first of these moves outward from the beacon to the appropriate radius, using odometry and beacon signal strength to estimate distance, and immobilizes itself there. Each succeeding corner is specified in relation to the previous one; the corresponding robot circles at the radius of its predecessor, until it finds that previous robot fixed in its final location, or encounters another robot that knows that location; it then calculates its own destination location on that basis[1], and goes and immobilizes itself there.

A robot after the first C that reaches the beacon receives the building design, chooses a pair of successive corners at random to build a wall between, and enters *collect* mode: First it must know the locations of its selected corners, which it finds either by seeking them out itself or by being told their locations by robots it encounters which already know. During this stage it circles in the opposite direction to other robots, again to increase the rate of unique encounters. Next, it goes out beyond the outskirts of the cleared area to find a block, takes it to the first of its two corners, and follows the line between the two (straight or curved as appropriate) until it finds a valid unoccupied position to place its block. It does this by calculating the location of the nearest point to itself on the desired wall, i.e., the perpendicular to the line or arc connecting the two corners, based on the known positions of those corners and the type of wall desired. It then moves within sensor range and looks to see if that cell is occupied.[2] If not, it goes on to try to place the block there; otherwise, it moves along the direction of the desired wall, and will check the corresponding new perpendicular location on the next time step.

Robots repeat this process until they reach the second corner without finding a place that needs a block, at which point they enter *seal* mode; they return back along the wall to the first corner, making sure there are no gaps they missed the first time. More elaborate future versions of the system might have robots, for instance,

---

[1]Note that each robot must by necessity maintain its own private coordinate system. In general, each robot's coordinate system may permissibly differ from those of the others by rotation, translation, and scaling; common reference points can be used, whenever two robots exchange information, to calculate the appropriate linear transformations to convert between the two systems for that interchange. See also the discussion on localization in §4.

[2]In the present instantiation, robots do not distinguish between occupation by carried blocks, placed blocks, or other robots; this may lead to temporary bypassing of locations that would have opened up a few time steps later when the blocking robot moved on, but it also helps avoid traffic jams (the robot in the way may in turn be waiting for the first robot to get out of its own way so it can leave the area), and the gap can be filled in during a later pass by any robot; also, distinguishing between carried and placed blocks would require more sophisticated identification capabilities in a hardware implementation of this system.

spray some sealant over the blocks for airtightness during this stage. If the robot finds a gap, it fills it with a block and returns to *collect* mode; if it reaches the first corner still in *seal* mode, it records that wall as completed, reenters *collect* mode, and chooses another pair of corners to work on the wall between, until in the end it has personally verified that all walls are completed. At that point, the robot enters *off* mode: it heads away from the beacon to the outskirts of the workspace, to avoid interfering with any other construction that may still be ongoing, and ceases to be active. In more complicated situations, where other construction tasks on other parts of a more complex building still remain, or the structure is subject to damage and requires constant maintenance, etc., the appropriate behavior would be to continue to *collect* rather than turning *off*.

Additions to this basic algorithm handle special situations. If a robot wants to place a block somewhere but is prevented from doing so for too long, it will give up and move on. An robot unable to move at all for too long will send out a signal to all robots within range, on receipt of which robots will shuffle around at random for several time steps, in the hopes of breaking up a traffic jam if that was causing the problem (as can occur when more than a few robots are at work in the same area).

While robots are capable of locating gaps in and adding blocks to a wall from either side, their behavioral algorithm favors construction from the side away from the beacon, and the supply of free blocks is located on the outskirts of the building area. Consequently, in complicated structures with corners at different radii, early completion of the more outlying walls can interfere with subsequent work on the inner ones. This problem is materially avoided by having robots choose first to work on walls adjoining the structure's smallest-radius corners before they move on to those of larger radius. Other exceptions to the basic algorithm above respond to an environment that may change in significant ways between the time when a robot begins an action and the later time when it completes it; for example, a robot heading to claim a block which another picks up before the first reaches it will return to its previous goal and continue to search.

4 Results and discussion

By changing the geometric program stored in the beacon, the system can quickly and easily be made to produce a wide variety of closed 2-D structures with non-crossing walls. Fig. 3 shows several examples, giving a sample of the system's flexibility and range. The time course of construction with the same building program but different initial conditions was similar across runs with independent random seeds (Fig. 4A).

A distributed system may derive its effectiveness simply from its intrinsic parallelism; or it may take advantage of explicit cooperation between multiple agents. The system described here takes the former approach, for the most part, with agents largely ignoring one another while going about their behaviors. We might then naïvely expect the building task to be completed, for an N-agent system, in $1/N$th the time it would take a single agent. However, this incremental advantage is diminished as the number of agents increases, since any task has a limit to how many

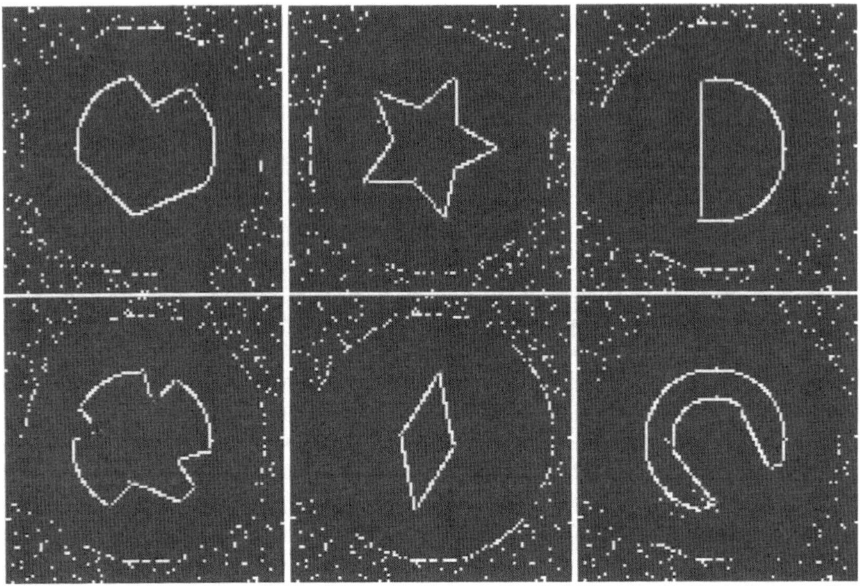

Fig. 3. Different structure programs can direct the system to build a variety of shapes (in this set of examples, the marshmallow shapes of General Mills's Lucky Charms cereal). Only blocks and robots acting as corners (dark gray) are shown. All simulations were conducted on a 100 × 100 lattice with a team of 30 robots.

agents can usefully contribute to its completion at one time, and agents begin to get in one another's way (Fig. 4B, C). Communication between agents, in general, can help reduce such interference [20]. Here, communication was useful for alleviating traffi c jams, in that robots unable to move for too long signaled any nearby to shuffle their positions, often breaking impasses; for coordinating the operating mode, to keep robots from working in direct opposition (e.g., one bringing blocks in, another clearing them away); and for fi nding the locations of corners, which robots obtained more quickly through the team's distributed search than they would have on their own. Modulating the beacon signal, on the basis of information that robots bring to the beacon during the course of construction, could potentially be an alternative way of facilitating the latter two functions.

A further advantage of communication could be used to address an important limitation of this model, the diffi culty in real systems of localization. Odometry alone is unreliable, as sensors are noisy, actuators are imperfect, and an isolated robot's estimate of position becomes increasingly unreliable as errors accumulate. Methods have been developed for individual robots for slowing [13] and bounding [17] this drift. What is more, the multi-robot nature of the system can itself be taken advantage of; robots exchange position and orientation estimates whenever they encounter one another, and using the information provided by the other, each can improve its own estimate to obtain a signifi cant decrease in uncertainty [12]. The ubiquitous signal

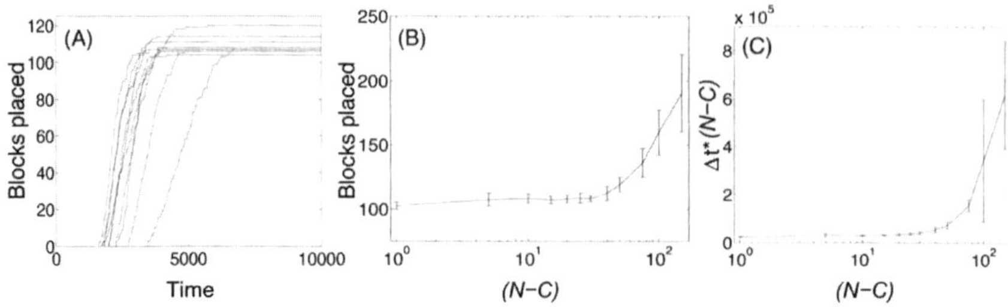

Fig. 4. Aggregated results from many independent runs building the diamond shown in Fig. 3. **A**: For 30 robots, number of blocks placed by robots as part of the structure (not simply dropped at the periphery after being cleared away) as a function of time, for 10 runs with different random seeds. The variation in final numbers is due to occasional remaining gaps and blocks extraneously placed, as are visible in Fig. 3.
B, C: Interference between robots affects number of extraneous blocks placed and proportional time taken to complete the task. With N robots and $C = 4$ corners, $(N - C)$ is the number of robots available to manipulate blocks after the clearing stages. Each data point represents 10 runs. **B**: Total number of blocks placed as part of the structure. **C**: Time between when the first block of the structure was placed and that when 95% of all such blocks had been placed, multiplied by $(N - C)$. For fewer than about 40 robots, interference was small or negligible.

from the beacon will provide another cue that can be used to improve the position estimate; and the beacon itself, and (once in position) the robots that embody corners, represent fi xed landmarks that a robot can use to correct its estimate whenever it comes near them, which it will do frequently in the course of construction.

A clear motivation for the use of distributed systems in general is to improve robustness. While this issue has not yet been studied in these simulations, we can discuss how this system would withstand component failure, and how its response could be improved in those cases where the instantiation described here would do poorly.

- Loss of individual unspecialized robots (i.e., those not acting as corners) would have no signifi cant effect; because they are interchangeable and working independently, loss of one or several would slow construction comparatively little.
- The loss of corner-robots would be more problematic, and without some added system response, construction could halt. To deal with the risk of loss of a corner-robot before it had found its fi nal destination, a suffi cient approach would be to specify that if a robot circles too many times without fi nding the corner it seeks, it takes on the task of embodying that corner for itself (fi rst returning to the center to notify the beacon, so that if it had previously been tasked with embodying another corner, that task can be reassigned). If, on the way to embodying a corner, a robot encounters another one that has already planted itself at the appropriate location as that corner (or if it learns of such a robot from a third party), it reverts to acting as an unspecialized robot. As for corner-robots that fail after positioning themselves, these should not pose a signifi cant threat to the task, since all corner-

robots must do is act as a landmark; if another robot comes looking for that corner and finds a robot which is in the right place but not communicating, the newcomer need only rely on its own sensors rather than the corner-robot's account of its location, and the only cost is a possible increase in positional uncertainty in the vicinity of that corner.

- Loss of the beacon at first seems more fatal still; without its signal as a constant reference, robots will have to rely on their position estimates alone in planning trajectories, and the final construction will be more irregular at best, incomplete at worst. Moreover, if the beacon is lost early enough, corners may go unassigned, the building program may never be communicated to the robots, and robots may have no common basis even for a position estimate. A more robust approach, then, would be to build the potential to act as a beacon into each robot. Rather than having the robots receive the building program in the course of construction, they could receive it before being deployed, when all are close together, via a general broadcast. Then, at the start of the construction process before any beacon has yet existed, each robot can choose to become a beacon at random with low probability per unit time; as soon as one does, the others orient to it and begin the construction process as before. If the beacon later fails, the loss of the long-range signal leads the other robots to put their current tasks aside and head for where it had been; whichever first gets close enough locates the previous beacon, takes its place, and adopts the beacon's role from that point on while the other robots return to work.

In this report, I have described a model system which in simulation allows highly flexible construction of 2-D structures, specified in a simple high-level geometric language, through the distributed actions of many identical, autonomous robots. A straightforward extension of this approach could achieve multiple-room structures; fully three-dimensional ones, a greater challenge, are its ultimate aim. The high-level features of the system described here may be useful to consider in design of hardware implementations of robots intended for autonomous construction projects [5], as well as studies of tasks requiring explicit cooperation between multiple robots [7], heterogenous teams of robots [7, 11], and other related work and its future development.

Acknowledgements

This work was supported by NIH training grant GM07484, with additional support from a Packard Foundation Fellowship (to H.S. Seung). I would like to thank Radhika Nagpal for useful discussions and comments on the manuscript, as well as Gerald Sussman and Hal Abelson.

References

1. Abelson, H., et al. (2001). Amorphous computing. *Communications of the ACM* **43**(5): 74–82.

2. Bahceci, E., Soysal, O. & Sahin, E. (2003). *A Review: Pattern Formation and Adaptation in Multi-Robot Systems* {Technical Report CMU-RI-TR-03-43, Carnegie Mellon Univ.} Pittsburgh, PA, USA.
3. Bonabeau, E., Dorigo, M. & Théraulaz, G. (1999). *Swarm Intelligence: From Natural to Artificial Systems.* Oxford University Press Inc.
4. Bonabeau, E., Theraulaz, G., Deneuborg, J.-L., Franks, N., Rafelsberger, O., Joly, J.-L. & Blanco, S. (1998). A model for the emergence of pillars, walls and royal chambers in termite nests. *Phil. Trans. R. Soc. Lond. B* **353**: 1561–1576.
5. Bowyer, A. (2000). *Automated Construction using Co-operating Biomimetic Robots* {Technical Report, University of Bath Department of Mechanical Engineering}. Bath, UK.
6. Fredslund, J. & Matarić, M. (2001). *A General, Local Algorithm for Robot Formations* {IRIS Technical Report IRIS-01-396}. Los Angeles, CA, USA.
7. Gerkey, B. & Matarić, M. (2002). Pusher-watcher: an approach to fault-tolerant tightly-coupled robot coordination. In *Proc. IEEE Int. Conf. on Robotics and Automation*, Washington, D.C., USA: 464–469.
8. Melhuish, C., Holland, O. & Hoddell, S. (1998). Collective sorting and segregation in robots with minimal sensing. In *5th Conference on Simulation of Adaptive Behaviour*, Zurich, Switzerland.
9. Melhuish, C., Welsby, J. & Edwards, C. (1999). Using templates for defensive wall building with autonomous mobile ant-like robots. In *Proc. Towards Intelligent Autonomous Mobile Robots 99*, Manchester, UK.
10. Parker, C., Zhang, H. & Kube, R. (2003). Blind bulldozing: multiple robot nest construction. In *Proc. IROS 2003*, Las Vegas, USA.
11. Parker, L. (2003). The effect of heterogeneity in teams of 100+ mobile robots. In *Multi-Robot Systems Volume II: From Swarms to Intelligent Automata*, Kluwer: 205–215.
12. Roumeliotis, S. & Bekey, G. (2000). Distributed multi-robot localization. In *Proc. 5th International Symposium on Distributed Autonomous Robotic Systems (DARS 2000)*, Knoxville, TN, USA: 179–188.
13. Roumeliotis, S., Sukhatme, G. & Bekey, G. (1999). Smoother based 3-D attitude estimation for mobile robot localization. In *Proc. 1999 IEEE Int. Conf. in Robotics and Automation*, Detroit, MI, USA: 1979–1986.
14. Spears, W. & Gordon, D. (1999). Using artificial physics to control agents. *IEEE Int. Conf. on Information, Intelligence, and Systems*, Bethesda, MD, USA.
15. Sugihara, K. & Suzuki, I. (1996). Distributed algorithms for formation of geometric patterns with many mobile robots. *J. Robotic Systems* **13**(3): 127–139.
16. Théraulaz, G. & Bonabeau, E. (1995). Modelling the collective building of complex architectures in social insects with lattice swarms. *J. Theor. Biologie* **177**: 381–400.
17. Thrun, S. (1999). Learning maps for indoor mobile robot navigation. *Artificial Intelligence* **1**: 21–71.
18. Vassilvitskii, S., Yim, M. & Suh, J. (2002). A complete, local and parallel reconfiguration algorithm for cube style modular robots. In *Proc. 2002 IEEE Int. Conf. on Robotics and Automation*, Washington, DC, USA: 117–122.
19. Vona, M. & Rus, D. (2001). Crystalline robots: self-reconfiguration with compressible unit modules. *Autonomous Robots* **10**(1): 107–124.
20. Wawerla, J., Sukhatme, G. & Matarić, M. (2002). Collective construction with multiple robots. In *Proc. 2002 IEEE/RSJ Int. Conf. on Intelligent Robots and Systems*, Lausanne, Switzerland.
21. Werger, B. & Matarić, M. (1999). *Exploiting Embodiment in Multi-Robot Teams* {IRIS Technical Report IRIS-99-378}. Los Angeles, CA, USA.

Coordinating Aerial Robots and Sensor Networks for Localization and Navigation

Peter Corke[1], Ron Peterson[2], and Daniela Rus[3]

[1] CSIRO ICTC Centre,
Brisbane, 4069 Australia
peter.corke@csiro.au

[2] Dartmouth Computer Science Department,
Hanover, NH 03755 USA
rapjr@cs.dartmouth.edu

[3] Computer Science and Artificial Intelligence Lab
Cambridge, MA 02139
rus@csail.mit.edu

We consider multi-robot systems that include sensor nodes and aerial or ground robots networked together. We describe two cooperative algorithms that allow robots and sensors to enhance each other's performance. In the first algorithm, an aerial robot assists the localization of the sensors. In the second algorithm, a localized sensor network controls the navigation of an aerial robot. We present physical experiments with a flying robot and a large Mica Mote sensor network.

1 Introduction

We wish to develop distributed networks of sensors and robots that perceive their environment and respond to it, anticipating information needed by the network and by users of the network, repositioning and organizing themselves to best acquire and deliver that information. These networks, thousands of small sensors, equipped with limited memory, sensing, communication, and actuation capabilities will autonomously organize themselves and move to track a source and convey information about its location to a human user, and to the rest of the system.

In this paper we discuss the cooperation between a ground sensor-network and a flying robot. We assume that the flying robot is connected by point-to-point communication with a ground sensor network. The nodes of the sensor network are simple and they support local sensing, communication, and computation. The communication range of all nodes is limited, but the resulting mobile sensor network supports multi-hop messaging. The flying robot makes the sensor network localization easy by providing access to GPS data to all nodes. In turn, the sensor network helps the navigation of the flying robot by providing information outside the robot's immediate sensor range. In our previous work [2] we discuss the details of robot-assisted localization. In this paper we summarize robot-assisted localization, discuss why it is hard, and present a new sensor-assisted robot guidance algorithm. We also describe new experimental results with an integrated testbed for robot-assisted localization and sensor-assisted guidance.

2 Robot-assisted localization

Ground sensor networks are usually deployed on-demand, so that the location of the sensors may not always be presettable by the deployment system. Once on the ground the sensors acquire location information autonomously and form a distributed system. The individual sensor nodes are too simple to include complex sensors such as GPS, or to support complex information processing tasks estimating location from range measurements.

The node localization problem has been previously discussed by others and usually requires estimates of inter-node distance, a difficult problem. Simić and Sastry [8] present a distributed algorithm that localizes a field of nodes in the case where a fraction of nodes are already localized. Bulusu etal. [1] propose a localization method that uses fixed beacons with known position. Galystyan etal. [3] described a constraint-based method whereby an individual node refines its position estimate based on location broadcasts from a moving agent. We wish to address the sensor localization problem in a uniform and localized way, without relying on beacons or pre-localized nodes, while minimizing the number of broadcasts required.

In [2] we introduced the idea of robot-assisted localization, an approach to localization that is orthogonal to this previous work, does not require inter-node communication, and is suitable for sensor networks deployed outdoors. For a large sensor network the location requirement could be limiting since it would be impractical (for reasons of cost and power consumption) for each node to have GPS capability. However, a mobile aerial robot equipped with a GPS system can assist the sensors to localize. The aerial robot sweeps across the area of the sensor network, for example along a random path or a path defining a grid, broadcasting GPS coordinates. The sensors process all broadcasts they hear and estimate their location. If the mobile node beams messages containing its position $p_i = (x_i, y_i)$ any sensors receive the message with signal strength s_i, a simple averaging procedure can estimate a sensor's location as the centroid of the set of GPS locations heard over time. Other methods discussed in [2] include taking just the strongest received signal, a signal strength weighted mean, a median, a set intersection approach as suggested by Galystyan etal. [3]. The latter requires a parameter which is the notional reception range of the radio, assumed to be circular. Note that algorithms **mean**, **wmean** and **median** can be modified so that the estimate is only updated when $s_i > s_{min}$ which artificially reduces the size of the radio communications region.

2.1 Challenges with Distributed Localization

In the robot-assisted localization algorithm, the robot regularly broadcasts its location. When within the reception range of the sensor, these broadcasts provide input to the localization algorithm. The reception range is not symmetrical due to the lobe shape of both the transmitting and receiving radios involved, terrain, etc. Since the asymmetry depends on the relative orientation of both antennas it will vary from encounter to encounter, which highlights two problems.

1. The asymmetry is not known apriori, so the best we can do is to approximate the center of the radio reception range, i.e., assume the sensor is at the center of the radio reception range. Node 7 in Figure 3(a) shows an extreme case of directional reception.
2. With relatively few measurements occurring within the reception range the estimate of centroid will be biased.

The first problem is not solvable given current radios Multiple encounters at different relative antenna orientations might provide some relief, but would increase the time and cost of any post-deployment localization phase.

There are ways to improve the second problem however:

1. Increase the rate at which position broadcasts are sent, giving more samples within the reception range, and improving the estimate of the centroid.
2. Increase the size of the reception range in order to acquire more samples. One way to do this would be to relay messages between close neighbors, perhaps based on a hop-count estimate of distance. A disadvantage of this method is that the asymmetry problem is likely to be exacerbated.
3. Decrease the size of the reception range, perhaps combined with improvement #1, so that those broadcasts that are received originate very close to the sensor.

In early simulation studies we observed that the localization result is strongly dependent on the path of the robot with respect to the deployed nodes. To sidestep this dependence while testing observations (1–3) above our simulation uses a fixed serpentine robot path and 100 sensors deployed randomly with a uniform distribution in a square region 100 × 100m (mean inter-node spacing is 17m). The robot starts at the origin in the lower-left corner, moves 100m to the right, up 20m, 100m to the left, then up another 20m and repeats the cycle. The total time to execute this path is 1 unit, and we investigate the effect of changing the broadcast interval. The radio propagation model assumes that signal strength decreases with distance and becomes zero at the maximum distance parameter which we also vary.

For each set of simulation parameters, such as radio range or position broadcast rate, we compute mean and maximum localization error. We repeat the experiment 100 times, and compute second-order statistics. For each experiment, for each node, we run the 5 localization algorithms previously described. Figure 1 shows some of the results.

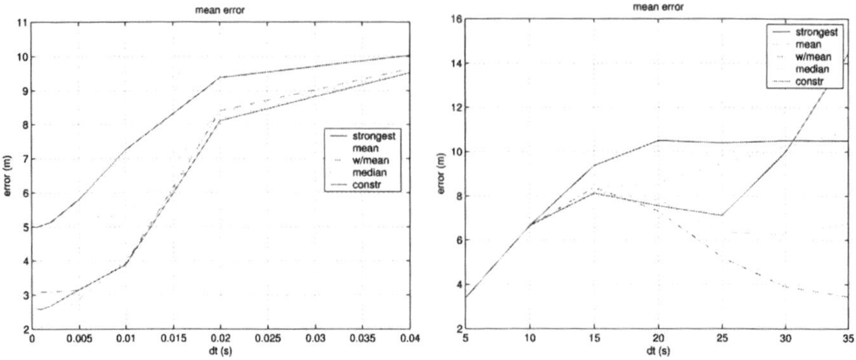

(a) (b)

Fig. 1. Mean localization error from the Monte Carlo study using the five methods of [2]. (a) Effect of varying the broadcast interval (transmit range = 15m). (b) Effect of varying the transmission radius (dt=0.02).

We observe that as the number of broadcasts increases (ie. broadcasts are closer together) the localization error decreases and reaches a plateau at around 5m or better. The method **strongest** performs least well while **constr** performs best. For a given number of broadcasts, 50, along the path we investigate the performance of the methods for varying transmit radius. We see that the method **constr**, previously a strong performer, breaks down when the actual and assumed transmit radii are not equal. The best performer in this test is **wmean**, though **mean** also behaves well.

3 Sensor-assisted navigation

A localized set of sensors can facilitate the aerial robot's navigation by encoding path information which provides the robot with point-by-point navigation directions using networking. The path can be communicated to the robot by the ground sensors. The sensor network can employ mapping algorithms such as those described in [6] to compute adaptive, time-varying paths to events. One application of this approach is in the area of monitoring and surveillance, where the sensor network may detect something that requires further investigation with a more complex sensor, say with the camera on board of the flying robot.

Suppose a path is stored in the sensor field. Sensor-assisted navigation for the flying robot has two phases: firstly getting to where the path starts, and secondly being guided along the path. In some situations the first phase may not be needed (e.g., the path may always be computed to include the known location of the robot or the robot could always be told where the start of the path is.)

One important goal in this first phase is to avoid flooding the entire network with messages in an attempt to discover location. Algorithm 1 summarizes a method for guiding the robot to the path. For example, for the robot to find the path, first one (or

Algorithm 1 The FindPath algorithm to get the robot to the start of the path.

The sensor does this to announce the location of a path start to the robot.

```
if Incoming message is a PathMessage AND this sensor is at the start of a path then
    Broadcast FindRobotMessage with 0 degree heading to MAXRANGE distance
    Broadcast FindRobotMessage with 120 degree heading to MAXRANGE distance
    Broadcast FindRobotMessage with 240 degree heading to MAXRANGE distance
else if Incoming message is a FindPathMessage then
    if This sensor is storing a path start location then
        Broadcast a PathStartMessage
else if Incoming message is a PathStartMessage then
    Compute distance to vector from path start to robot.
    if distance < PathMessage.PathWidth then
        // Forward message towards the robot.
        Rebroadcast PathStartMessage
```

The robot does this to find the start of the path.

```
while forever do
    // Seek the path start
    Broadcast FindPathMessage with 0 degree heading to MAXRANGE distance
    Broadcast FindPathMessage with 120 degree heading to MAXRANGE distance
    Broadcast FindPathMessage with 240 degree heading to MAXRANGE distance
    if A PathStartMessage is received then
        Store location of start of path.
        Head for start of path.
        break
```

all) of the sensors that know they are near the start of the path send out three messages that contain the location of the start of the path. The messages also each contain a heading, set 120 degrees apart[4], a width for the vector they will travel along, and a maximum range beyond which they are not intended to travel. The messages are forwarded out to that range in each of the three directions. The sensors that forward the messages store the location of the start of the path.

The robot at some later time sends out the same sort of messages in three directions. If the robot and path start are in range of each other's messages, the message paths will cross (due to using a 120 degree dispersal angle.) The sensor(s) at the crossing will have a stored location for the start of the path and a location for the robot and can send a directional message (perhaps with a gradually increasing width since the robot may have moved slightly) back to the robot telling it where the start of the path is. In this way only the sensors along specific lines extending to a maximum range carry messages instead of the entire network.

[4] Other patterns of radiation (a star pattern of 72 degrees) might increase the likelihood of intercepts occurring, though they also increase the number of sensors involved.

Algorithm 2 The QueryPath algorithm for robot guidance.

while forever **do**
  // Seek path information from the sensors
  Broadcast a *QueryOnPath* message
  Listen for the first sensor to reply
  if a sensor replies with an *OnPathAck* message **then**
    Send a *QueryPath* message to that sensor
    // The sensor should reply with a list of *PathSegments* it is on
    if that sensor replies with a *QueryAck* message **then**
      Store the *PathSegments* from the *QueryAck* message in order of precedence.
  // Guide the robot
  if Robot has reached current *Waypoint* **then**
    Get next *Waypoint* from list in order of precedence
    Head for next *Waypoint*

After the initialization phase that places the robot on the path, the navigation guidance algorithm summarized as Algorithm 2 is used to control the motion direction of the robot. The robot starts by broadcasting a `QueryOnPath` message which includes the sender's id and location. A sensor on the path that receives this message replies with a `QueryAck` message which includes the path section, some consecutive way points, and an indication of where these way points fit into the path. By gathering lists of segments from multiple sensors the entire path can be assembled incrementally as the robot moves. Paths that cross themselves allow for some fault tolerance in the robot's knowledge of the path, since if the robot loses the path, it may have a future segment of it already stored if it has passed an intersection.

Once the robot has acquired path segments from a sensor, it can then arrange them in order of precedence and follow them in order. Thus the path itself is independent of the sensor's own location and can be specified to any level of precision needed.

4 Experiments

4.1 Experimental Testbed

The experiments were carried out on September 17, 2003 in the Planetary Robotics Building at CMU. We implemented the robot-assisted localization algorithm and the sensor-assisted guidance algorithm on an experimental testbed consisting of a sensor network with 54 Mica Motes [4, 5] and a flying robot. The flying robot consists of 4 computer controlled winches (implemented using Animatics Smart motors) located at the corners of a square with cables going up to pulleys at roof height then down to a common point above the 'flying' platform. The crane is controlled by a server program running on a PC. Commands and status are communicated using the IPC protocol(see www.cs.cmu.edu/afs/cs/project/TCA/www/ipc). The platform comprises a single-board Pentium-based computer running Linux, with an 802.11

Fig. 2. The experimental testbed consisting of 49 Motes on the ground and the flying robot.

link and an on-board serially connected basestation Mote, to communicate with the sensor field. The robot has a workspace almost 10 m square and 4 m high.

We used a 7x7 grid of sensors, laid out with a 1 meter spacing. The center of the grid was (0,0) and the sensors were placed starting 0.5 meters from the center, see Figure 3(a) where the diamonds represent the surveyed positions of the motes.

The Flashlight sensor interface [7] was used to adjust the RF power of the sensors in the grid to an optimal level for communication with the robot as it traveled 1-2 meters above the sensors (this was a trial-and-error adjustment, gradually incrementing the mote power until the robot was getting good communications) The motes ran TinyOS 0.6 with long (120 byte payload) messages.

4.2 Localization results

During localization the flying robot followed a preprogrammed serpentine path, see Figure 3(a). Once per second the flying computer obtained its current coordinate from the control computer using IPC over the 802.11 link, and broadcast this via the basestation mote. Each ground mote used the broadcasts to compute a centroid based location for itself. Figure 3(a) shows the robot path and the location of each of the broadcasts the motes received. It is clear that the motes do not receive messages uniformly from all directions, motes 6 and 7 are good examples of this. We speculate that this is due to the non-spherical antenna patterns for transmitter and receiver motes, as well as masking by the body of the flying platform itself. The motes received between about 2 and 16 broadcasts each as can be seen in Figure 3(b) with a median value of 10. Figure3(c) shows a histogram of the distances over which the broadcast messages were received, a maximum of 3m and a median of 1m.

Each mote computes its location using the centroid of all received broadcasts, but can store up to 200 localization broadcasts for download and analysis. Figure 4(a) shows the true and estimated mote locations. We can see a general bias inward

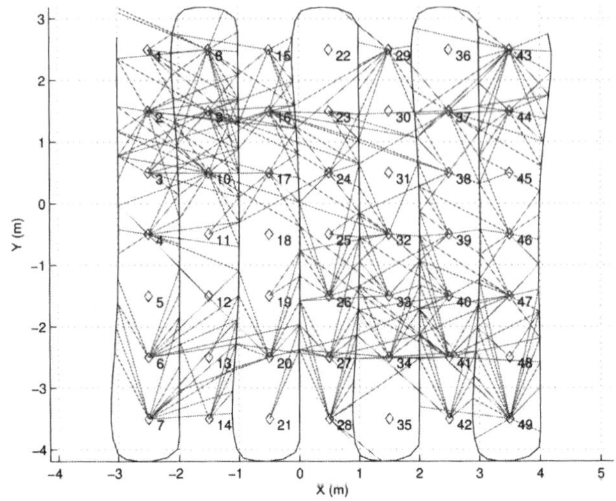

(a)

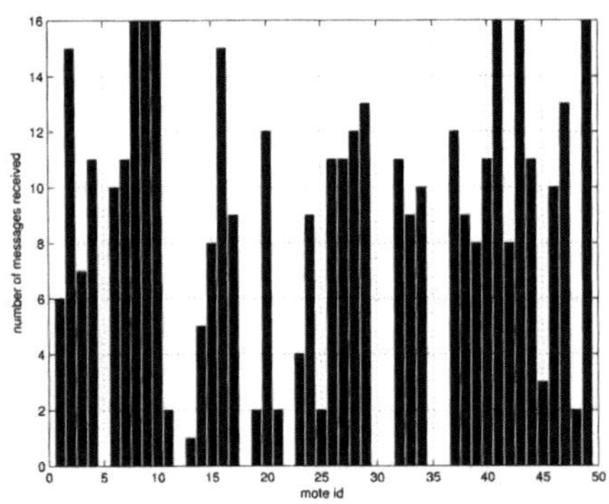

(a)

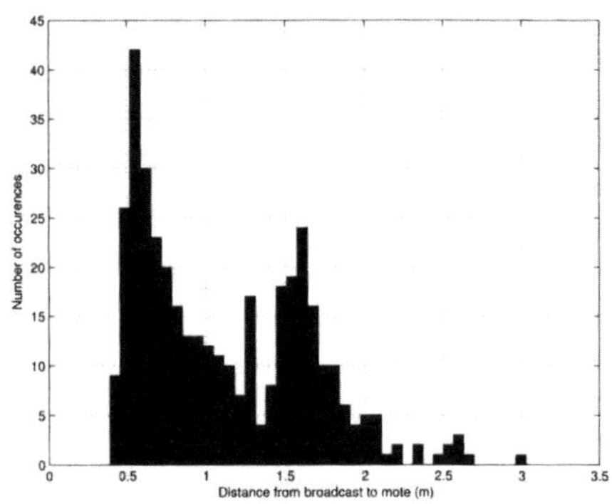

(b)

Fig. 3. Localization results. (a) Mote field showing path of robot and broadcast positions, and all broadcasts received. (b) Number of localization messages received by each node. (c) Histogram of distances from mote to broadcast.

and this would be expected given the the bias in the direction from which broadcasts were received. Figure 4(b) shows a histogram of the error magnitudes and indicates a maximum value of 1.4m and a median of 0.6m which is approximately half the grid spacing. This level of performance matches our previous results obtained with experiments with a real helicopter and differential GPS [2].

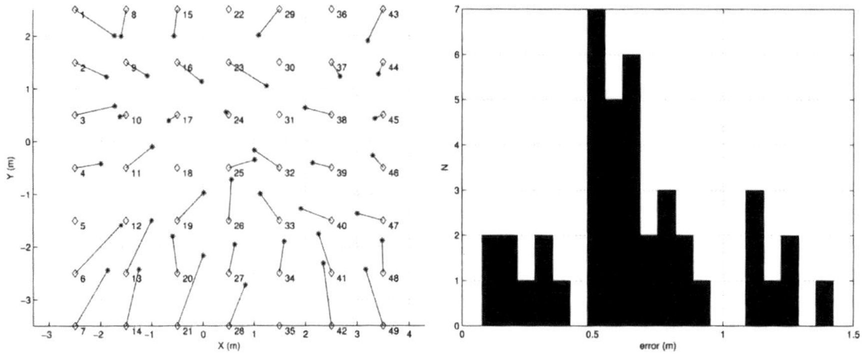

(a) (b)

Fig. 4. Localization performance using centroid method. (a) Actual (◇) and estimated (*) location. (b) Histogram of error vector length.

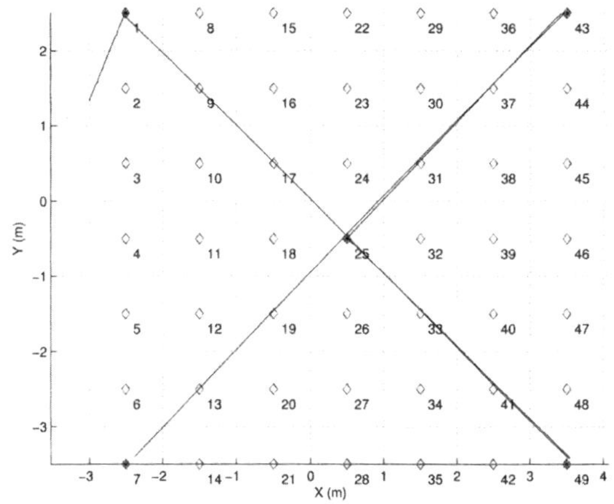

Fig. 5. Path following performance. The actual path followed by the robot is shown in black, and the asterisks indicate waypoints. The path started at node 7.

4.3 Path following results

Once localized, a PATH message was sent from a basestation to establish a path through the mote field. The PATH message propagated using the algorithm described in [2]. Then the robot was turned loose in a path following mode, using the path following algorithm in [2]. It queried for path waypoints and built up a list of waypoints as it followed the path. We experimented with a square path (around the border of the grid) and an X shaped path (corner to center to corner). The robot followed both types of path perfectly. Even though the localization of the motes was not perfect, it was sufficient to support the geographic routing of the PATH message with a 1 meter width. The actual path itself was stored as perfectly precise information in these motes and hence the robot was able to get precise waypoints to follow, resulting in perfect path following (within the tolerances of the system) as shown in Figure 5.

Since there were multiple motes along each segment of the path, there was redundant information in the sensor field in case any of the motes were not working (and as it later turned out about 6-7 of them were not during each test, either due to defunct radios, or due to not hearing any messages for other reasons.)

5 Conclusion

We have described how robots and sensor networks can function synergistically to perform tasks such as localization and guidance. Simulation studies provide insight into the achievable performance of various localization methods, and experimental results are provided. The localization approach does not require inter-sensor communications. New algorithms for path following are presented along with experimental validation.

Acknowledgments

This work is done as part of the First Responder Project at the Institute of Security Technology Studies at Dartmouth College and is a collaborative project between the labs at Dartmouth, MIT, and CSIRO. The authors would like to thank Dr. Sanjiv Singh for facilitating the physical experiments at CMU. Support for this work has also been provided in part through the NSF award 0225446, ONR award N00014-01-1-0675, the Darpa TASK program, MIT project Oxygen, and Intel.

References

1. N. Bulusu, J. Heidemann, and D. Estrin. Adaptive beacom placement. In *Proceedings of the* 21st *Conference on Distributed Computing Systems*, Phoenix, AZ, 2001.
2. P. Corke, R. Peterson, and D. Rus. Networked robots: Flying robot navigation with a sensor net. In *Proc. of the 2003 International Symposium on Robotics Research*, Siena, Italy, 2003.
3. A. Galstyan, B. Krishnamachari, and K. Lerman. Distributed online localization in sensor networks using a moving target. submitted to 2003 acm senssys.
4. J. Hill, R. Szewczyk, A. Woo, S. Hollar, D. Culler, and K. Pister. System architecture directions for network sensors. In *ASPLOS*, 2000.
5. Jason Hill, Philip Bounadonna, and David Culler. Active message communication for tiny network sensors. In *INFOCOM*, 2001.
6. Qun Li, Michael de Rosa, and Daniela Rus. Distributed algorithms for guiding navigation across a sensor net. In *Proceedings of MobiCom 2003*, 2003.
7. Ron Peterson and Daniela Rus. Interacting with a sensor network. In *Proceedings of the 2002 Australian Conference on Robotics and Automation*, Auckland, NZ, November 2002.
8. S. Simic and S. Sastri. Distributed localization in wireless sensor networks, Available at citeseer.nj.nec.com/464015.html, 2002.

Pervasive *Sensor-less* Networks for Cooperative Multi-robot Tasks

Keith J. O'Hara and Tucker R. Balch
{kjohara, tucker}@cc.gatech.edu

The BORG Lab
College of Computing
Georgia Institute of Technology
Atlanta, Georgia 30332–0250

Summary. A number of researchers are investigating the use of embedded sensor networks to facilitate mobile robot activities. Previous studies focus individual tasks (e.g. navigation to a goal) using networks of several to tens of expensive ($\approx$ \$100) nodes placed by the robots themselves or in predetermined geometric grids. In this work we explore the use of tens up to hundreds of simple and cheap ($\approx$ \$10) sensor-less nodes placed arbitrarily to support a complex multi-robot foraging task. Experiments were conducted in a multi-robot simulation system. Quantitative results illustrate the sensitivity of the approach to different network sizes, environmental complexities, and deployment configurations. In particular, we investigate how performance is impacted by the density and precision of network node placement.

1.1 Introduction and Related Work

We are interested in the application of low-cost, pervasively distributed network nodes to support cooperative multi-robot tasks. In this work we consider a heterogeneous system composed of small, embedded, immobile sensor-less communication nodes and larger mobile robots equipped with sensors and manipulators. The embedded network serves as a pervasive communication and computation fabric, while the mobile robots provide sensing and actuation. We refer to the embedded nodes as forming a *sensor-less* network to distinguish the approach from those where the network nodes also have sensors. In our work the embedded nodes provide only modest computation and communication for the team.

As noted above, we depart from the usual approach where the embedded nodes are equipped with sensors. There are a number of arguments in favor of sensor-less embedded networks. First, the cost and power requirements for simpler embedded nodes is lower, thus enabling us to distribute more of them. Second, it is likely that even for a network with sensor nodes, certain activities for which the nodes do not have sensors will need to be conducted. For instance, we may want to utilize a network that has already been deployed (e.g. wildfire monitoring), to support a new application (e.g. search and rescue). Unanticipated applications can be addressed on deployed networks by equipping the mobile robots with appropriate sensors. In this way application-specific hardware is concentrated on the smallest portion of the team – the mobile platforms.

The algorithm for guiding the mobile robots essentially works as a distributed variant of the popular wave-front path planning algorithm, or a breadth-first search from the goal, propagating paths from the goal location. The embedded nodes make up the vertices of the path planning graph, and the network connections between them are the edges of the graph. Mobile robots can then use reactive navigation to traverse the graph by visiting the vertices (i.e. the embedded nodes) to the goal. The sensor-less network creates navigation networks for supporting mobile robots in various tasks such as coverage, recruitment, and path planning. We demonstrate all three of these distributed skills in a multi-robot foraging task. We show that a pervasive network of embedded sensor-less nodes can enable a team of mobile robots to accomplish complex tasks effectively. We analyze the sensitivity of the approach to different team sizes, environmental complexities, and deployment configurations.

Parunak et al developed a technique for coordinating multiple unmanned air vehicles (UAVs) using synthetic pheromones. [1, 2] Inspired by pheromone communication in insects, they create potential fields for guiding the UAVs around threats to goal locations in a distributed manner. We do not assume the embedded nodes are arranged uniformly in any structure. We also rely on a distributed dynamic programming solution to the path planning problem, rather than an approach based on the dynamics of insect pheromones.

Like Parunak, Payton et al present an approach for large scale multi-robot control referred to as "Pheromone Robotics" inspired by biology. [3] They use a system based on virtual pheromones, by which a team of mobile robots use short-range communication to accomplish cooperative sensing and navigation. In Payton's work "virtual pheromones" are communicated over an ad hoc network to neighboring robots. In contrast, in our approach information is not distributed by the mobile robots, but rather by the relatively static, embedded nodes scattered throughout the environment.

Both Batalin et al [4, 5] and Li et al [6] have developed similar approaches using heterogeneous teams composed of mobile nodes and an embedded network. The network of embedded nodes, creates a "Navigation field" [4], which mobile nodes can use to find the their way around. They differ in how they compute this navigation field. Batalin et al use Distributed Value Iteration [4]. In their approach, the embedded nodes use estimated transition probabilities between nodes to compute the best direction to suggest to a mobile robot for moving between a start and goal node. These transition probabilities are established during deployment and both the robots and sensor nodes have synchronized direction sensors (e.g. digital compass). Our approach does not require the nodes to store transition probabilities, instead we rely on the communication network to establish the navigation paths. Also, in our approach the mobile robots only need a local sense of direction in order to move toward the correct embedded node. Neither the robots or the embedded nodes need any shared sense of direction.

Li et al are able to generate an artificial potential field for navigation based on the obstacles and goals sensed by the network. [6] This potential field is guaranteed to deliver the mobile node to the goal location via an danger-free (obstacle-free) path. The field is created by the embedded nodes propagating goal-ness or danger to neighboring nodes. In our approach the embedded nodes do not have sensors, this capability is provided by the mobile nodes, and thus can not sense obstacles directly. We assume the obstacles are sensed indirectly by the resulting communication topology. The later three of these approaches, as well as ours, use distributed dynamic programming [7] to create the navigation field.

Koenig [8] and Wagner [9, 10] also devise some related methods for doing parallel coverage using simple ant robots that communicate indirectly by leaving indicators in the environment. Batalin et al [5] also use communication nodes as "markers" in aiding mobile robots in the exploration problem. The embedded nodes offer a suggested un-explored direction for the mobile robots to follow. Unlike our approach, their embedded nodes do not communicate with each other, but only to mobile robots.

1.2 Problem Statement and Assumptions

Our system is composed of mobile robots with sensors and actuators supported by an embedded immobile network of nodes without environmental sensors. We assume the embedded network nodes have the following capabilities:

- **Limited computation and memory,** on the order of a PIC microprocessor with 2K ROM and 256 bytes of RAM operating at 4 MHz.
- **Short range communication** with adjacent nodes up to 4 meters distant.
- **Communication is blocked by navigation obstacles.**

We assume the robots and embedded nodes communicate using a short-range medium that is occluded by the same objects that occlude navigation (e.g. walls). Line of sight between nodes implies open space for navigation.We have implemented a hardware platform to these specifications, the GNAT (see Fig. 1(a)). The GNAT is a low-power, omni-directional, infrared device costing about 30 dollars to build. The mobile robots in our system are somewhat more capable. We assume they support:

- Communication with embedded nodes;
- Relative bearing estimation to nearby embedded nodes;

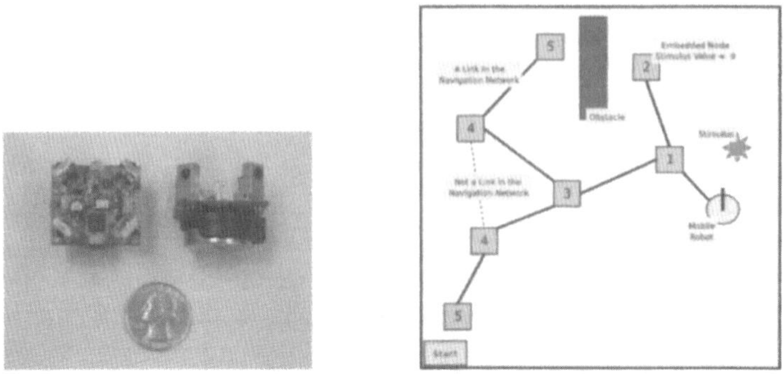

(a) The GNAT (b) Navigation Network

Fig. 1.1. (a) The GNAT is a low-power, omni-directional, infrared device. (b) An illustration of a navigation network.

- Local attractor and obstacle sensing; (e.g. food objects);
- Attractor object grasping.

These robot capabilities are sufficient for cooperative foraging in the presence of an embedded network.

Significantly, there are a few assumptions we do not make. In particular, **we do not assume localization or mapping capabilities** on the part of the robots or the embedded nodes. No mobile robots or embedded nodes are expected to perform localization or mapping. Furthermore **we do not assume the environment is static**. Obstacles to navigation can appear and disappear. We expect the network to automatically adapt to dynamic conditions.

In this work do not address the deployment of the embedded nodes. We assume they have already been placed in the environment, but their positions are unknown and the uniformity of their placement can vary. In fact, one objective of this work is to assess the impact on performance with respect to different network sizes, environment complexities, and deployment configurations.

Given the system of robots and network nodes described above, we would like to solve a multi-robot foraging problem. Foraging is a well-studied, canonical multi-robot task [11, 12]. In this task a robot team is initialized at a "homebase" location, from which they should begin to explore the environment in search of attractor (food) objects. Once a cache of attractor objects is discovered, this information should be disseminated to the other robots, along with a means for them to navigate to the cache. Finally, all of the attractor objects should be collected by the robots and delivered to homebase. We have decomposed the overall problem into the following sub-problems:

- Cooperative coverage: enable a team of mobile robots to completely explore the area covered by embedded nodes. For efficiency, robots should avoid traveling through areas already explored by other robots.
- Recruitment: alert the team to new, critical information. In the context of a foraging task, the discovery of a food cache is an example recruitment situation.
- Path planning: Without requiring localization capabilities, provide an efficient route for each robot, located anywhere in the environment, to a goal location.

1.3 Approach

The sensor-less network creates navigation networks for supporting mobile robots in various tasks such as coverage, recruitment, and path planning. We use navigation networks to accomplish three different steps in the task: 1) directing the robots to visit uncovered areas, 2) directing the robots to a discovered attractor cache, and 3) directing them home. Navigation networks for each of these tasks are present in the sensor-less network simultaneously. A mobile robot can then follow whichever navigation network corresponding to it's current sub-task goal. A navigation network is illustrated in Fig. 1(b). We are able to use navigation networks to complete the multi-robot foraging task in complex environments without mapping or localization.

We follow Payton's virtual pheromone technique [3] and assume the communication paths are similar to the navigation paths, and use this to propagate navigation information. By using a short-range communication medium that is occluded by obstacles to navigation, the communication paths carve out free-space. As also pointed out by Payton [3] and Li [6], this results in a kind of distributed physical path-planning. To create a navigation network for a particular goal we use a distributed dynamic programming approach; specifically, we apply the distributed Bellman-Ford algorithm. The Bellman-Ford equation [13] for finding the shortest path from i to j is:

$$D(i,j) = \min_{k \in neighbors} d(i,k) + D(k,j)$$

Where $D(i,j)$ is the path cost from i to j, and $d(i,k)$ is the distance between i and k. It can be used to find the shortest path to a destination from all nodes. The distributed version of Bellman-Ford was created for network routing protocols [14]. In the distributed network routing version, neighbors share their path costs and the distance between nodes is usually measured in hops. We use distributed Bellman-Ford to effectively create a tree of shortest paths from every node to the goal – this tree is the navigation network. The embedded network can be thought of as "routing" the mobile robots to their destination. However, note that the embedded nodes do not know the global, or local, position of their neighbors, so they are not directing the robot in any direction. Instead, the mobile robot greedily approaches the lowest-valued node currently in its communication range. As it closes in on the node it will come within communication range of that node's parent. The robot continues this until it comes within sensing distance of the goal.

As the mobile robots discover attractors, they broadcast this information to neighboring embedded nodes and navigation trees are created with roots near the attractor. As the information is propagated throughout the network by the embedded nodes, the hop-count, or path-cost, increases. Any mobile robot can then approach the network, access the relevant navigation network, and descend to the root of the tree, eventually reaching the original goal.

Using a static sensor-less network provides several advantages over fully mobile networks and static sensor networks. The first benefit is cost. By having the bulk of our system be composed of cheap communication and computation nodes, we lower the cost and power requirements of the entire team. In addition, the embedded network can be used generally, for instance, when the desired sensor for the application is not known beforehand. Another related advantage is that the bulk of the application-specific hardware is concentrated in the smallest portion of our team – the mobile platforms – allowing the network to be used in a general manner. Another advantage is that the majority of the system is relatively static and connected. We say relatively because the topology can indeed change, but we assume for the most part, the network will be fully connected and fairly static. When using all mobile nodes, as done by Payton [3], we must assure the network stays connected. Because the propagation algorithms use a distributed dynamic programming solution, they can fail when the network becomes disconnected for long periods of time.

The attractor navigation network allows a robot from anywhere in the environment to find a path to an attractor that was sensed by another mobile robot. An illustration of the navigation network for a discovered food-source is shown in Fig. 2(b). It is obvious that if we filled the space with embedded nodes arranged in a grid, and gave the nodes range sensors, we would see a picture very similar to many grid-based planning approaches. The path planning space is approximated by the communication network. But how many nodes are required for this approximation to hold, and how uniformly do they have to be arranged? We address these questions in the experiments below.

The home navigation network is an instance of an attractor navigation network, with the homebase being the attractor. This creates a tree rooted at the homebase, assuring the robots can return home. An illustration is shown in Fig. 2(c).

Next, we consider a mechanism for building a coverage navigation network. It is a straightforward extension of the attractor navigation network, where each unvisited node is an attractor. The mobile nodes are then offered paths to reach the closest unvisited nodes. This approach assures all nodes will be visited. If a node has been visited, it uses the default scheme of propagating one of its neighbor's values. This results in the visited embedded nodes directing the mobile robots into unexplored areas.

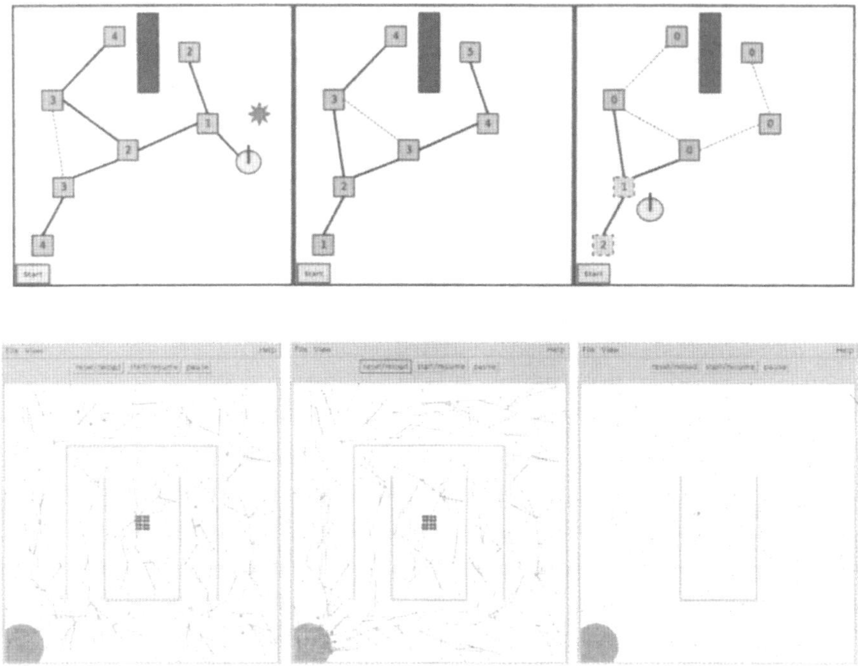

(b) Attractor Navigation Network

(c) Home Navigation Network

(d) Coverage Navigation Network

Fig. 1.2. The components of the foraging task. The illustrations depict the embedded nodes as squares, labeled with their value, the mobile robot as the green circle, and the attractor cache as the red star. The solid lines connecting the nodes make up the navigation network, the dotted lines are connections that aren't included in the navigation network. For the screenshots of the TeamBots simulation environment, the blue circles in the center represent the food source, the blue circle in the bottom left corner is the robots' starting position, and the gray lines are walls. (a) An example navigation network for the attractor cache. (b) The homebase navigation network. (c) The coverage navigation network.

Although the coverage solution generated is not optimal in the sense of shortest circuit to visit all the nodes (i.e. the traveling salesman problem) it does assure all nodes are visited. Depending on the configuration of the embedded nodes, this can assure systematic coverage of the terrain, even though neither the robots or the embedded nodes have any localization capabilities or a map. In addition, both algorithms can be used in dynamic coverage scenarios by changing their state to unvisited after a certain amount of time has passed. The algorithm works well for both single and multi-robot exploration. An illustration of the coverage algorithm is given in Fig. 2(d).

1.4 Experiments

We combined the three navigation networks (attractor, homebase and coverage) to complete a multi-robot foraging task. We implemented this system in the TeamBots multi-robot simulation environment. The control systems were encoded using the Clay behavioral architecture [15]. In all the experiments we used 8 mobile robots with grippers, and 16 attractors in a group to represent the attractor cache to be exploited. All the robots had limited sensing and communication ranges of 4 meters that were occluded by obstacles. We tested the technique in three different $36x36m^2$ environments of increasing complexity. The three environments are shown in Figs. 3(a), 3(b), and 3(c).

Two key factors impacting how the system performs are the number of embedded nodes and how they are placed. As mentioned previously, when we have a large number of embedded nodes deployed uniformly we effectively have a real grid-world and the navigation networks are accomplishing distributed path-planning. Much work has dealt with trying to optimally deploy a sensor network for these tasks. In this research, however, we assume that the network is approximately uniformly distributed, but with random placement error. Placement error in a real system could be due to error in deployment or changes over time. Since our approach does not depend on the embedded nodes being localized, it is robust to changes in placement.

We ran experiments with 81, 121, 169, 225, 289, and 361 embedded nodes. Additionally, we varied the error in placement using the following technique. First, we placed the nodes uniformly across the space, then added error to each node's position by some random amount, the average distance from original position was varied: from 0, .5, and 2, to 10 meters. In the case of 10m average error, placement is essentially uniform random. Each experimental configuration was run 10 times. The graphs show mean performance, with errorbars denoting standard deviations.

We present the results of the complete foraging task. Space limitations preclude presenting the individual analyses of the coverage and delivery subtasks. The results show the average time, in timesteps, to deliver each of the 16 attractors. A delivery time of 7200 timesteps (simulation timeout) were used for undelivered attractors. The results of the first obstacle free environment are shown in Fig. 4(a). The results of the second and third more complex environments are shown in Figs. 4(b) and 4(c). We see that as we increase the error in placement and the complexity of the environment, more nodes are needed to maintain the same level of performance. This is due to the fact that with a small number of nodes and large amount of error, the navigation network is disconnected and isn't able to guide the robots in the foraging task. Instead, they must rely on a random walk to cover the space and purely local reactive navigation.

We presented a technique for using a pervasive network of embedded sensor-less nodes to support multi-robot exploration and navigation. We showed that with enough embedded nodes, the distributed physical path plan-

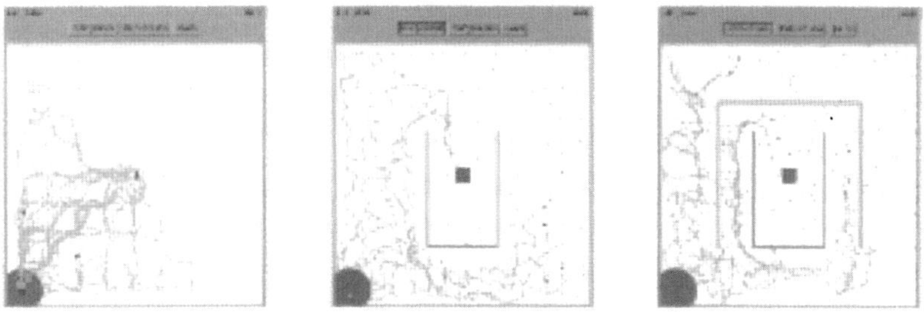

(a) Map 0 (b) Map 1 (c) Map 2

Fig. 1.3. The simulation environments. (a) Map 0 is the first obstacle-free environment. A sample configuration is shown with 81 embedded nodes distributed uniformly throughout the environment without any error in placement. (b) Map 1 is a more complicated environment with a box canyon. A sample configuration is shown with 225 embedded nodes distributed uniformly with .5 m of error in placement. (c) Map 2 the most complicated environment with two box canyons. A sample configuration is shown with 289 embedded nodes distributed uniformly with 10m of error in placement.

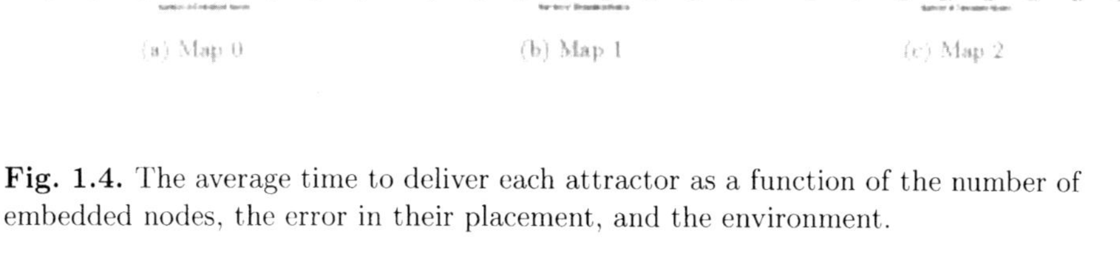

(a) Map 0 (b) Map 1 (c) Map 2

Fig. 1.4. The average time to deliver each attractor as a function of the number of embedded nodes, the error in their placement, and the environment.

ning works even with very random, non-uniform, deployment of the embedded nodes in complex environments. In contrast, without enough nodes to form a connected network, the approach does not work since the network can not guide the robots. This approach also fails when the communication paths and navigation paths differ. One possible solution would be to use real navigation experiences to reinforce the paths in navigation networks. We are currently evaluating the technique on real robots.

References

1. H. V. D. Parunak, S. Brueckner, and J. Sauter, "Synthetic pheromone mechanisms for coordination of unmanned vehicles," in *Proceedings of First International Conference on Autonomous Agents and Multi-Agent Systems*, 2002, pp. 449–450.
2. H. V. D. Parunak, M. Purcell, and R. O'Connell, "Pheromones for autonomous coordination of swarming uavs," in *Proceedings of First AIAA Unmanned Aerospace Vehicles, Systems,Technologies, and Operations Conference*, 2002.
3. D. Payton, M. Daily, R. Estowski, M. Howard, and C. Lee, "Pheromone Robotics," *Autonomous Robots*, vol. 11, pp. 319–324, 2001.
4. M. Batalin and G. Sukhatme, "Sensor network-based multi-robot task allocation," *Proceedings of International Conference on Intelligent Robots and Systems (IROS 2003)*, October 2003.
5. ——, "Coverage, exploration and deployment by a mobile robot and communication network," *Telecommunication Systems Journal, Special Issue on Wireless Sensor Networks*, 2003.
6. Q. Li, M. DeRosa, and D. Rus, "Distributed algorithms for guiding navigation across a sensor network," *The 2nd International Workshop on Information Processing in Sensor Networks*, 2003.
7. D. P. Bertsekas, "Distributed dynamic programming," *IEEE Transactions on Automatic Control*, vol. 27, no. 3, pp. 610–616, 1982.
8. S. Koenig, B. Szymanski, and Y. Liu, "Efficient and inefficient ant coverage methods," *Annals of Mathematics and Artificial Intelligence*, vol. 31, pp. 41–76, 2001.
9. I. A. Wagner, M. Lindenbaum, and A. M. Bruckstein, "Smell as a computational resource - a lesson we can learn from the ant," *Proceedings of the ISTCS'96 - 4'th Israeli Symposium on Theory of Computing and Systems*, June 1996.
10. ——, "Distributed covering by ant-robots using evaporating traces," *IEEE Transactions on Robotics and Automation*, vol. 15, no. 05, pp. 918–933, October 1999.
11. T. Balch and R. C. Arkin, "Communication in Reactive Multiagent Robotic Systems," *Autonomous Robots*, vol. 1, no. 1, pp. 27–52, 1994.
12. D. Goldberg and M. Mataric, *Robot Teams*. A K Peters Ltd., 2002, ch. Design and Evaluation of Robust Behavior-Based Controllers for Distributed Multi-Robot Collection Tasks.
13. R. E. Bellman, *Dynamic Programming*. Princeton, NJ: Princeton University Press, 1957.
14. L. Ford and D. Fulkerson, *Flows in Networks*. Princeton, NJ: Princeton University Press, 1962.
15. T. Balch, "Clay: Integrating motor schemas and reinforcement learning," Georgia Institute of Technology, Tech. Rep. GIT-CC-97-11, 1997.

Part VIII

Group Behavior

Communication Strategies in Multi-robot Search and Retrieval: Experiences with MinDART

Paul E. Rybski, Amy Larson, Harini Veeraraghavan, Monica LaPoint, and Maria Gini

Department of Computer Science and Engineering, University of Minnesota, 200 Union St. S.E., Minneapolis, MN 55455-0159
{rybski,larson,harini,mlapoint,gini}@cs.umn.edu

Summary. To explore the effects of different simple communications strategies on performance of robot teams, we have conducted a set of foraging experiments using real robots (the *Minnesota Distributed Autonomous Robotic Team*). Our experimental results show that more complex communication strategies do not necessarily improve task completion times, but tend to reduce variance in performance.

1 Introduction

Designing a distributed robotic system using simple units is an attractive engineering solution for many reasons [2]. Each robot in the swarm uses simple local rules to decide its actions without needing any command from a central controller or from any other robot. Obvious advantages to this approach are robustness to individual failure, ability to scale with minimal tractability issues, low unit complexity, and decreased costs.

In this study, we are interested in determining what level of improvement in task performance we can expect by adding simple communications capabilities to the robots in the swarm. In order to explore this question, we built a group of simple robots, the *Minnesota Distributed Autonomous Robot Team* (MinDART) shown in Figure 1, to perform a foraging task and we enhanced them with communication capabilities. We conducted a series of experiments with these robots and compared their performance when using different simple communication strategies.

Although many tasks can serve as a testbed, we chose foraging, which is well studied, so that solutions and results can be compared more easily. In our version of the task, robots locate a target in an enclosed arena, pick up the target, and drop it off at a designated home base. The arena contains some obstacles, and the distribution of the targets varies. The performance

Fig. 1. The Minnesota Distributed Autonomous Robotics Team (MinDART). The MinDART robots searched for the infrared emitting targets in a search and retrieval task. Landmarks were used for homing and localization.

criterion we selected is the time to complete the task, i.e. the time to collect the entire set of targets and return them to home.

All our experiments use physical robots, as opposed to simulated robots. As eloquently explained in [6], we believe that a rigorous study of swarm intelligence warrants physical robots, as opposed to simulated robots.

The simplest control strategy for foraging is random walk. We use reactive behaviors to avoid obstacles and random direction changes at random intervals to increase the probability of complete coverage. This strategy is an attractive choice for simple robotic hardware because it is easy to program and requires little sensor bandwidth.

We analyzed random walk versus control strategies that use communication. The communication methods we chose are forms of indirect communication based on cues from the environment (this is called *stigmergy* in the biology literature). We studied two types of communication (reflexive communication and deliberate communication) and studied how the duration of deliberate communication (10, 20, and 30 seconds) affected the time to complete the task. Our experimental results show that for simple robots such as the MinDART, deliberative strategies help in decreasing the variance of the team's performance. However, this decrease in variability does not correspond to a significant decrease in the mean time to solve the task. Instead of spending time wandering randomly, the robots spend time recruiting other robots.

2 MinDART Hardware and Software

Each MinDART robot is constructed out of LEGO Technic blocks. The robot is 29 cm long by 24 cm wide by 37 cm tall and has a dual-treaded skid-steer chassis that allows the robot to turn in place and translate at a speed of

0.1693 m/s. The gripper is an articulated cargo bay that grasps and transports targets. Bumpers are used for obstacle avoidance. The robot's infrared sensors can detect the target IR beacons at a range of approximately 70 cm. A light-bulb beacon serves as binary form of communication among the robots. To detect this beacon, and to identify landmarks for homing, a CMUCam [13] is mounted on top of a servo-controlled turret. The camera processes images at 2-3 frames/second. The robot is controlled by a Handyboard.Power is provided to the camera and Handyboard by two 9.6V NiCad battery packs.

The MinDART's control software consists of a finite state machine. Each state in the controller solves one subtask in the robot's overall task. There are three subtasks comprising the search and retrieval task, which are find a target, grab a target, and return a target to the home base. In the initial state, FINDTARGET, a robot searches for targets, or heads toward an activated light-bulb beacon. Once a target is detected with the robot's infrared sensors, the control system switches to the GRABTARGET state which is responsible for maneuvering the robot such that the target fits into the gripper. If the robot successfully grabs the target, the control system switches to RETURNTARGET, which returns the robot to the drop-off location.

3 Communication Experiments

Figure 2 shows a view of the experimental setup. The area was 7 m x 8 m and contained uniformly distributed obstacles. The targets were distributed in a single non-uniform distribution in the corner of the environment furthest from the drop-off location. All experiments were run with four robots. Robots communicated by turning on their light-bulb beacons. Beacons could be seen at a maximum range of 2.9 m.

The goal of communication is to reduce the target search time by attracting robots to the area of a sensed target. The experiments were designed to test the robot's abilities to lead each other to a single clump of targets. The communication method chosen was an attracting light-bulb beacon, which would direct the other robots toward the targets. Communication varied by intent and duration as follows:

- **No Communication**. This was used as a baseline experiment.
- **Reflexive Communication.** A robot turned on its light-bulb beacon while trying to pick up a target (i.e. while in the GRABTARGET state). Once the robot grabbed the target, the beacon was deactivated. We consider this a statement of action, not a request for help. This strategy would be of no use in an environment with a uniform distribution of targets, but we hypothesized it would help when targets are clumped and harder to find by random walk.
- **Deliberative Communication.** A robot turned on its light-bulb beacon when a target was sensed, but the robot was unable to pick it up. This form

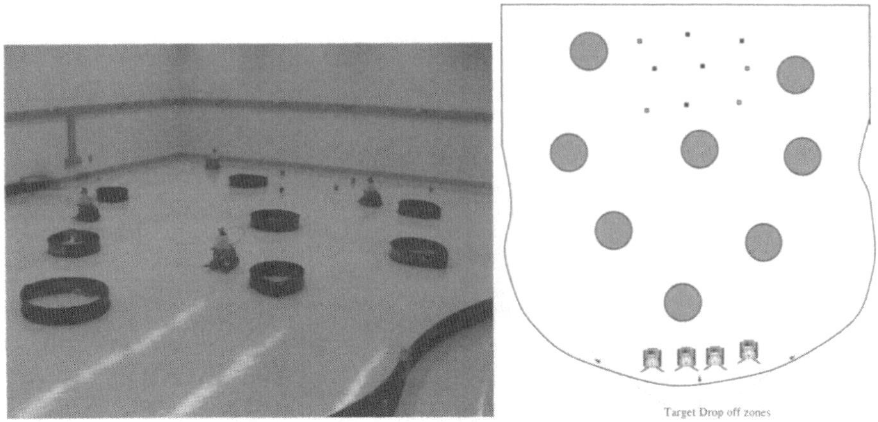

Fig. 2. Diagrams of the 7 m x 8 m experimental environment showing obstacles, initial placement of targets, and initial starting point for the robots. All experiments contained nine targets and eight obstacles. The obstacles were relatively low and did not block a robot's view of the landmarks or of each other. However, they did block a robot's view of the targets. The drop-off area is the same as the robot starting point.

of communication was used if a robot encountered a target while on its way to the home base to drop off one that it had in its gripper. The robot would stay motionless for a fixed amount of time as a deliberate request for assistance. We tested three fixed durations: 10 s, 20 s, and 30 s.

We hypothesized that any form of communication would provide a performance enhancement, due to the reduced time spent in random search, and that some form of communication would provide better results than no communication. Similar findings are reported in the simulation work of [1].

We also predicted that deliberative communication would provide the most benefit and that there would be a peak or plateau in the duration, as seen in the simulation work of Sugawara [19]. In other words, we predicted that there would be an ideal communication duration that would maximize performance, and any duration longer than that would not enhance performance any further. This is because the longer the beacon is left on, the better chance the other robots would see it. However, deliberative communication requires a robot to stay stationary while recruiting others. There is a tradeoff between this delay and the time spent doing random search.

Since one of the goals of the experiments was to measure the effect of the amount of time the light-bulb beacons were on, and since in the reflexive communication experiments the beacons were on for different amounts of time, we recorded the light-on time for each communication occurrence. The average light-on time was approximately 16 s with a standard deviation of 11.6 s, but the distribution is not Gaussian, as seen in Figure 3. Instead, it clusters around

5 and 10 s (the mode of the distribution is 5 s). We will see later how this is correlated with the variance in the experimental results.

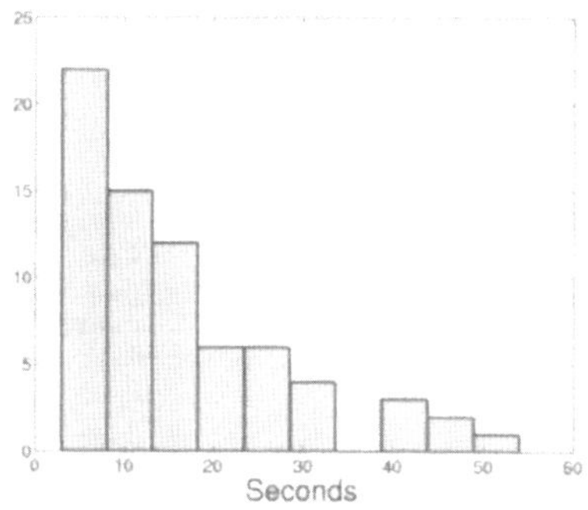

Fig. 3. Histogram of the times the light-bulb beacon was on, observed in the reflexive communication experiments. The majority was in the 5 to 10s range. This can be compared to the deliberative communication experiments, where beacon-on times were fixed.

For each of the experiments, we recorded the time a robot returned a target to the drop-off zone. The results were averaged over five runs. Each experiment was run until all nine targets were retrieved. We compared the times between the dropping off of the first and eighth target, to discount the times in the experiment when communication had little effect. Figure 4 shows the means and standard deviations of these times.

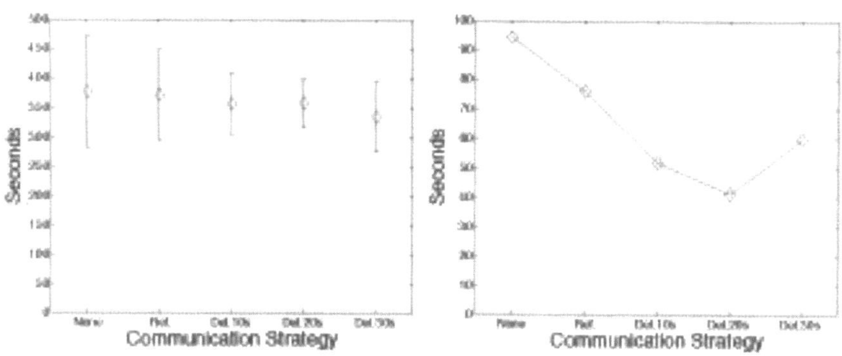

Means with standard deviation ($\sqrt{\text{variance}}$) errorbars

Standard deviations only

Fig. 4. Means and standard deviations of the times to complete the task for each of the communication strategies. The labels on the x axis stand for the different communication experiments. None=none, Ref.=reflexive, Del.10=10 s deliberative, Del.20=20 s deliberative, Del.30=30 s deliberative.

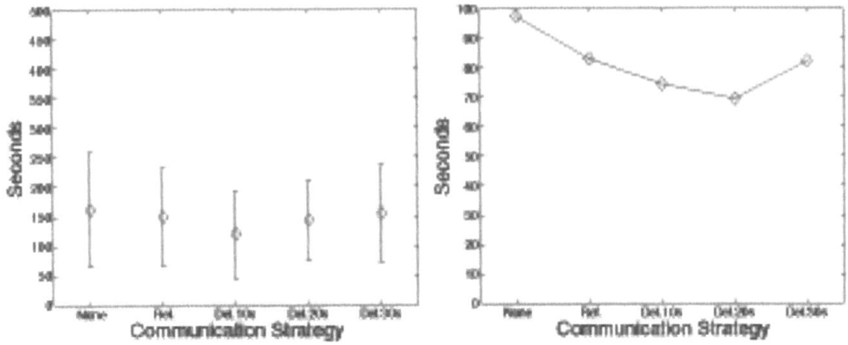

Means with standard deviation ($\sqrt{\text{variance}}$) errorbars

Standard deviations only

Fig. 5. Means and standard deviations of the times the robots took to retrieve a new target after dropping one off (i.e. target search time) for each of the communication strategies. The labels on the x axis stand for the different communication experiments. None=none, Ref.=reflexive, Del.10=10 s deliberative, Del.20=20 s deliberative, Del.30=30 s deliberative.

The left graphs of Figure 4 and of Figure 5 plot the means and standard deviations of task completion and of search times, respectively, for the communication experiments. The graphs reflect a slight performance benefit from the use of all forms of communication, but, surprisingly, nothing statistically significant. However, the variance of both show an obvious trend, that can be seen more clearly from the right-hand graphs of Figures 4 and 5. Although f tests show no statistical significance of the difference between the variances at the 95% confidence interval, the variance of the 20 second communication trials were very close to being significant (one-tailed, two-sampled f test with p=0.0682 and p=0.0511, for time to completion and target search times, respectively).

The beacon-on times recorded in the reflexive communication experiments (shown earlier in Figure 3) suggest that the correlation between the beacon-on times and the variance in completing the task cannot be explained simply by the duration of the beacon-on times.

Robot-to-robot interference and the specifics of how the robots operate are other important factors. The CMUCam turrets can rotate 360 ° in 5 seconds, but it may take several rotations to detect a beacon. The probability of detection decreases with distance and becomes zero at 2.9 m. A robot can rotate 180 ° in 5 seconds and can translate at a maximum of 0.17 m/s.

Using these ranges and approximating the probabilities for the time to find a beacon, to rotate, and to home in, we calculated the mean interference time and the mean travel time (i.e. the average time a homing robot traveled toward a communicating robot once it was oriented) for the deliberative experiments:

	10 seconds	20 seconds	30 seconds
Mean Interference Time	0.9971	5.8964	13.8634
Mean Travel Time	2.8500	11.7006	21.6406

Short communication durations (times less than 5 s) do not give enough time to the robots to see the beacon and change their heading to travel towards it. Longer communication times increase the probability of robot interference (mean interference time doubles from 20 s to 30 s). Travel time also increases with long communication.

4 Discussion

To explain the results, we have analyzed some of the failure points during communication. The probability of successful communication (i.e. the ability to see the beacon) is inversely proportional to distance. Even when two robots are in close proximity, successful communication depends upon their relative headings. If the homing robot is facing away from the communicating robot it may not be able to orient itself before the beacon is turned off. If a robot does successfully home in on a communicating robot, the target may be occluded. If the two robots make contact, the homing robot may turn away from the target as it executes obstacle avoidance. Finally, a common source of noise is inter-robot interferences. This becomes particularly troublesome when robots are drawn to the same area by some attractor, such as a beacon.

We could claim that these points of failure for communication are implementation details that can be addressed with more sophisticated hardware or better engineering, but discounting implementation details raises an important issue. These implementation details are precisely why we think real robots are necessary for this type of analysis. It is too easy to discount or underestimate the effects of even simple implementations on real hardware. For example, in [1] communication was shown to improve performance, but nearly all of the experiments were done in simulation where the effects of specific actions on the performance of the system (such as cooperative carrying or consuming of a resource) can be abstracted away. The details involved in physically implementing a system which can carry heavy objects or can consume liquid from a spill may affect the performance of the team in ways that those results did not illustrate. Considerable engineering effort may be necessary before the robots would be able to effectively achieve their tasks at the rates reported in this work.

As a point of comparison, consider a MinDART robot that executes a collection of behaviors to align itself to a target when in the GrabTarget state. The time that it takes a robot to pick up a target is heavily dependent on the interaction between the robot and its environment. To better quantify this, the times the beacons were turned on in the reflexive communication experiments are the same as the times the robots spent in the GrabTarget state. These times (see Figure 3) were quite variable. This illustrates the complexity that can arise from a simple operation implemented on real robots.

We believe our findings are validated by work done by others in simulation, particularly by Balch and Arkin [1] mentioned above. They concluded that simple communication often provides the best performing robots, but sometimes no communication performs just as well. We believe that once you carry robotics into the real world, some improvements in performance found in simulation get reduced by the noise and errors of implementation.

5 Related Work

Most research with multiple robots has focused on various forms of collaborative work [3]. While collaboration may be essential for some tasks, we are interested in studying tasks that can be done by a single robot, but where using multiple robots can potentially increase performance by decreasing the time to complete the task and/or by increasing the reliability. Sample tasks include placing a sensor network [16], cleaning up trash [12], pushing boxes [9], or detecting odors [5].

Foraging is a widely used testbed for distributed systems, but there are differences in the way the task is defined. In most studies the goal is to collect a fixed number of objects (roughly half) [4], in other cases objects continue to appear probabilistically and the duration of each experiment is fixed [8]. In our experiments the task is completed when all the objects have been collected, which makes the task more difficult since it is harder for the robots to find targets when they are very sparse. In our previous work we studied the effect of the number of robots [14] and of localization on performance [15]. In addition to these experimental studies, predictive models of foraging behaviors [11] and of robot interference during foraging [10] have been proposed.

There have been a handful of studies to evaluate the efficacy of communication strategies applied to the foraging task. Our work on communication strategies has been inspired mostly by the theoretical model proposed by Sugawara [18, 19] and by the simulation work of Balch and Arkin [1]. Sugawara's model accounts for the effects of indirect communication in foraging tasks. He performed simulation studies and some limited experiments with physical robots to support his model [19]. An interesting aspect of the model is that it predicts that the duration of the communication affects performance, and that there is a critical duration at which the performance is maximized, below and beyond which team performance deteriorates. Our communication experiments were designed to test this specific aspect of the model.

The study by Balch and Arkin [1], which evaluates the effects of various communication strategies on three different tasks, including foraging, was mostly conducted in simulation. The study predicts that communication improves performance by reducing the time spent wandering around. Our communication experiments were designed to verify this improvement and to quantify it. It is reasonable to assume that communication will assist in foraging, since it is a strategy that has evolved in nature. It is widely known that bees "dance" to communicate the direction of pollen sources [17] and ants communicate the location of prey with pheromone trails [7]. To our knowledge, biologically-inspired communication strategies for foraging on small scale robots have yet to provide performance improvements as predicted by the above mentioned work.

6 Conclusions and Future Work

We studied the effects of communication on a robotic team doing foraging. We compared this against a baseline of a random-walk search strategy.

We hypothesized that communication would decrease the time the robots spent randomly searching their environment and would improve overall performance, but we did not find a statistically significant improvement compared to the baseline. Instead, what we found was a decrease in the variance of the task completion times. We attribute the decrease in variance to the reduction of random search for targets. With communication capabilities, robots have to randomly wander into the communication range of another robot, but are then drawn directly to targets when attracted by a communicating beacon. Analysis of the average homing distances and interference times supports our conclusion that a 20 s communication duration represents a minimal point of variance for our experimental setup. Durations beyond this increase the probability of robot interference which negatively impacts performance.

For future work, we will explore how robots might dynamically adapt to their environment and tune their communication durations to optimize the team's overall performance. This learning capability would require upgrades in the processing and communication systems of the robots. Such upgrades would facilitate a robot's ability to share more information such as intentions, therefore teams could collaborate at a higher level.

7 Acknowledgments

Work supported in part by the Doctoral Dissertation, the Louise T. Dosdall Fellowships of the University of Minnesota, and by NSF under grants EIA-0224363 and EIA-0324864. We would like to thank Heather Metcalf and Devon Skyllingstad, who were supported by the CRA Distributed Mentor Program, for their help in constructing the electronics for the robot's beacons.

References

1. T. Balch and R. C. Arkin. Communication in reactive multiagent robotic systems. *Autonomous Robots*, 1(1):27–52, 1994.

2. E. Bonabeau, M. Dorigo, and G. Theraulaz. *Swarm Intelligence: From Natural to Artificial Systems*. Oxford University Press, Oxford, England, 1999.
3. G. Dudek, M. Jenkin, and E. Milios. A taxonomy for multi-agent robotics. In T. Balch and L. E. Parker, editors, *Robot Teams: From Diversity to Polymorphism*, chapter 1. A K Peters Ltd, Natick, MA, 2002.
4. D. Goldberg and M. J. Matarić. Design and evaluation of robust behavior-based controllers. In T. Balch and L. E. Parker, editors, *Robot Teams: From Diversity to Polymorphism*. A K Peters Ltd, Natick, MA, Apr. 2002.
5. A. T. Hayes, A. Martinoli, and R. M. Goodman. Distributed odor source localization. *IEEE Sensors*, 2(3):260–271, 2002.
6. O. Holland and C. Melhuish. Stigmergy, self-organisation, and sorting in collective robotics. *Artificial Life*, 5:173–202, 2000.
7. B. Hölldobler and E. Wilson. The multiple recruitment systems of the african weaver ant *oecophylla longinoda* (latreille). *Behavioral Ecology and Sociobiology*, 3:19–60, 1978.
8. C. Jones and M. J. Matarić. Adaptive division of labor in large-scale minimalist multi-robot systems. In *Proc. IEEE/RSJ Int'l Conf. on Intelligent Robots and Systems*, pages 1969–1974, Las Vegas, Nevada, Oct. 2003.
9. R. C. Kube and E. Bonabeau. Cooperative transport by ants and robots. *Robotics and Autonomous Systems*, 30(1/2):85–101, 2000.
10. K. Lerman and A. Galstyan. Mathematical model of foraging in a group of robots: effect of interference. *Autonomous Robots*, 13:127–141, 2002.
11. A. Martinoli, A. Ijspeert, and F. Mondada. Understanding collective aggregation mechanisms: From probabilistic modelling to experiments with real robots. *Robotics and Autonomous Systems*, 29:51–63, 1999.
12. L. E. Parker. On the design of behavior-based multi-robot teams. *Journal of Advanced Robotics*, 10(6):547–578, 1996.
13. A. Rowe, C. Rosenberg, and I. Nourbakhsh. A low cost embedded color vision system. In *Proc. IEEE/RSJ Int'l Conf. on Intelligent Robots and Systems*, Lausanne, Switzerland, 2002.
14. P. E. Rybski, A. Larson, M. Lindahl, and M. Gini. Performance evaluation of multiple robots in a search and retrieval task. In *Workshop on Artificial Intelligence and Manufacturing*, pages 153–160. AAAI Press, Aug. 1998.
15. P. E. Rybski, A. Larson, A. Schoolcraft, S. Osentoski, and M. Gini. Evaluation of control strategies for multi-robot search and retrieval. In *Proc. Int'l Conf. on Intelligent Autonomous Systems*, pages 281–288, Marina Del Rey, CA, March 2002.
16. P. E. Rybski, S. A. Stoeter, M. Gini, D. F. Hougen, and N. Papanikolopoulos. Performance of a distributed robotic system using shared communications channels. *IEEE Trans. on Robotics and Automation*, 22(5):713–727, Oct. 2002.
17. T. Seeley. The honeybee colony as a superorganism. *American Scientist*, 77:546–553, 1989.
18. K. Sugawara and M. Sano. Cooperative acceleration of task performance: Foraging behavior of interacting multi-robots system. *Physica*, D100:343–354, 1997.
19. K. Sugawara and T. Watanabe. Swarming robots – foraging behavior of simple multi-robot systems. In *Proc. IEEE/RSJ Int'l Conf. on Intelligent Robots and Systems*, Lausanne, Switzerland, 2002.

Value-Based Communication Preservation for Mobile Robots

Matthew Powers and Tucker Balch

Borg Lab
College of Computing
Georgia Institute of Technology
Atlanta, Georgia 30332–0250
mpowers@cc.gatech.edu tucker@cc.gatech.edu

Summary. Value-Based Communication Preservation (VBCP) is a behavior-based, computationally efficient approach to maintaining line-of-sight RF communication between members of robot teams in the context of other tasks. The goal of VBCP is, at each time step, to reactively choose a direction in which to move that provides the best communication quality of service with the rest of the team. VBCP uses information about other robots, real-time quality of service measurements and an a priori map of the environment to approximate an optimal direction in an efficient manner. Here, VBCP maintains communication between members of a robotic team while traversing an urban environment in formation. Quantitative and qualitative results are demonstrated in simulation and physical robot teams.

1 Introduction

This work addresses the task of maintaining line-of-sight RF communication between the members of a team of robots in the context of other tasks. Exact requirements vary from mission to mission, but systems dealing with issues of coordinated group behavior often must maintain communication between team members [3]. In the context of a multi-robot surveillance mission, it might be required that all members of the team share information, throughout the mission in a dynamic and noisy environment. The members must react to their teammates' actions in order to maintain a signal in an urban or otherwise RF-unfriendly environment. Simultaneously, each robot must also go about its surveillance mission. This work uses motor schemas, a behavior-based architecture, to preserve line-of-sight communication between members of a team of robots.

Value-Based Communication Preservation (VBCP) is a navigation behavior that takes into account shared locations of its teammates, measured communications signal quality and map-based predictions of communications signal quality to calculate movement vectors. These movement vectors serve to direct the robot along paths that tend to conserve communication between teammates. This behavior can be used in conjunction with other behaviors, in the context of a motor schema based architecture, to create complex, behavior-based robotic team behavior.

While the VBCP behavior takes into account the current position of all the robot's teammates, it does not use multi-agent planning to calculate movement vectors. It does use map information to estimate communication quality one step, or a short distance, away. The computational complexity of calculating the communication quality at a point one step away is kept to a minimum by assuming that all the robot's teammates remain in the same position one step into the future. By estimating the communication quality at several positions a short distance away in several directions, an estimation of a communications quality gradient can be calculated for the current position. This work is based on the hypothesis that following this gradient will tend to preserve communication quality within the team

Current applications of VBCP are based around military surveillance or terrain coverage missions. This work is funded under the DARPA MARS Vision 2020 program. Simulation tests were run on models of the Military Operations in Urban Terrain (MOUT) facility at Marine Corps Base Quantico, in Quantico, Virginia. Hardware tests were run at the U.S. Army's McKenna MOUT site at Ft. Benning, Georgia. These tests showed promising results for the utility of this behavior in military applications. While the emphasis of this research is of a military nature, it is expected that this behavior be found to be of use in a wide variety of multi-robot applications.

2 Related Work

This work is based on the motor schema approach to reactive robotics presented by Arkin [1]. Arkin presents an architecture for choosing mobile robotic actions. In this architecture, behaviors are defined for every sub-goal of the mission. These vectors are then combined to create a movement vector that the robot then acts on.

In [2], Balch and Arkin present a motor schema approach to multi-robot formation maintenance. Balch and Arkin defined several formations, including line, column, wedge, and diamond. Building on their work on formations, Balch and Arkin presented a study on the effect of formations on line-of-sight communication in a cluttered environment [3]. Balch and Arkin concluded that column formations allow teams to maintain communication more easily than line formations.

In 2003, Redi and Bers presented a hardware platform for mobile ad-hoc networks [4]. This platform, besides routing data between nodes, provides a user at a particular node with real-time quality of service metrics, including measures of signal strength to neighboring nodes, the number of hops required to reach nodes in the network and the identities of other nodes in the network. The algorithm explained below relies on this type of information.

Finally, in 2003, Stroupe and Balch presented Value-Based Observation for Robot Teams (VBORT) [5]. In this work, a team of robots use a one-step lookahead heuristic to maximize certainty of a group of targets' locations. VBCP should be viewed as an evolution of VBORT, using a similar approach applied to the communication realm.

3 Approach

VBCP is a behavior-based approach to communication preservation. At every time step, each robot in the team chooses its next action according to its current state and several predicted states, based on current observations and rules based on an a priori map of the environment. In order to choose an optimal next action with respect to communication preservation, it would be necessary to evaluate all possible next steps in combination with all possible next steps for each teammate. In order to remain computationally tractable for large robot teams, the VBCP behavior approximates the optimal next action.

VBCP reduces computational complexity on two fronts. First, rather than predicting every possible next step, VBCP predicts just a small set of possible next steps. Each next step is equally spaced around a radius representing the distance to be traveled in the next time-step. An approximated best next step is calculated by summing vectors in the direction of each candidate next step, respectively scaled with respect to their predicted communication quality. This is equivalent to approximating the predicted gradient of communication quality at the robot's current position. It is possible to make this approximation if quality of communication is assumed to be relatively smooth. The line-of-sight pathloss model used in this work is discontinuous when obstructions block the signal between two robots. However, in this case the model of communication quality can simply be assumed to decline steeply between regions of continuous communication quality.

VBCP further reduces complexity by approximating the future positions of a robot's teammates. When predicting the quality of communication at each next step, all teammates are assumed to remain still in the next step. Thus, computation of the behavior reduces from exponential to linear complexity, with respect to the number of robots in the team. Stroupe and Balch point out that, given no information about the teammates' intentions, their current positions represent an average predicted next position [5].

Communication quality between multiple teammates is evaluated according to a value function that can be crafted to reward behaviors defined in the mission specification. For this work, it was assumed that each robot must maintain connectivity with at least two teammates. Additionally, it was considered advantageous to maintain two signals with similar quality over two signals with disparate quality. A continuous function provides smoother behavior with respect to changes in communication quality, and relatively large "dead zone", in the range of communication quality considered to be adequate, allows robots to move relatively independently when communication quality is good enough.

The following function satisfies all the above constraints:

$$v = \frac{1}{(1 + e^{-C_1(\frac{r_1+r_2}{100} - C_2)})(1 + e^{C_3(\frac{r_1-r_2}{100} - C_4)})} \tag{1}$$

where r_1 and r_2 are the strongest and second strongest predicted or measured signals with teammates. (In the case of two-robot teams, r_2 is set equal to r_1.) C_1, C_2, C_3 and C_4 are positive constants. Increasing C_1 decreases the dead zone with respect to the added strength of r_1 and r_2. Increasing C_3 decreases the dead zone with respect to the difference of r_1 and r_2. C_3 and C_4 affect the steepness of the function. Figure 1 is a plot of v where r_1 and r_2 range between 0 and 100 and $C_1 = 9$ and $C_2 = 4$. There are undoubtedly many ways to evaluate communication quality according to the above specifications. The above function falls under no claims of being the best. It does, however, provide reasonable behavior in the context for which it was designed.

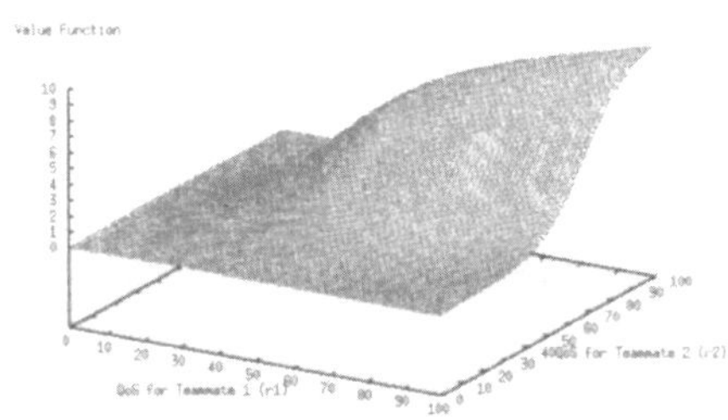

Fig. 1. Equation 1 plotted as r_1 and r_2 vary between 0 and 100.

The full algorithm VBCP uses to calculate a movement vector at each time-step follows:

1. The current signal strength with each teammate is measured. The current overall communication quality is calculated using Equation 1 , where r_1 and r_2 are respectively the two strongest measured signals.
2. The predicted signal strength at possible next steps evenly distributed around a radius representing the distance to be traveled in the next time-step. The overall communication quality is calculated at each next step using Equation 1, where r_1 and r_2 are the two strongest respective predicted signals at each next step. (Figure 2a)

3. A unit vector a_i in the direction of each next step is created and scaled by the overall communication quality at the respective next steps. (Figure 2b)
4. A unit vector b is created in the direction of the sum of the a_i's and scaled by the current overall communication quality. (Figure 2c)
5. Vector b is returned as the best next movement with respect to communication preservation.

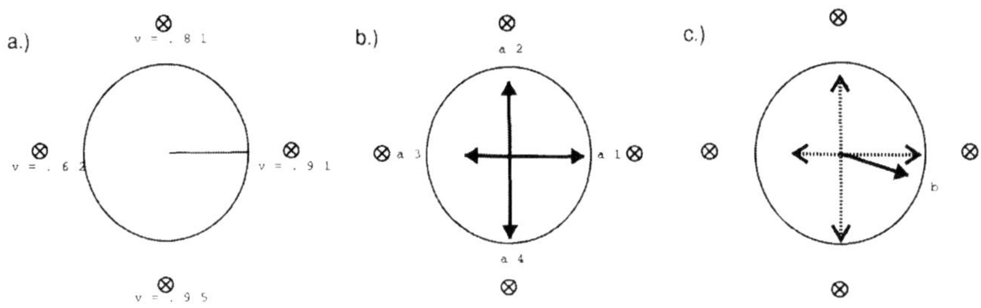

Fig. 2. A graphical representation of the VBCP algorithm. This figure represents steps 2-4 as explained in Section 3.

Measured and predicted values of communication quality are never compared in the computation of a movement vector. Predicted communication quality is used only in computing the direction of the movement vector. Measured communication quality is used only in computing the magnitude of the movement vector. Therefore, the scales of the predicted communication quality and measured communication quality need not match. This makes the task of modeling communication quality easier, as only general trends must be accurately modeled.

4 Experimental Approach

A series of experiments was run to evaluate the effect VBCP has on communication preservation in the context of an overall mission. Simulation experiments were run using the MissionLab [6] behavior specification software. Quantitative statistics were measured and compiled from these simulations. The simulated environment was modeled after the MOUT facility at Marine Corps Base Quantico in Quantico, Virginia. (Figure 3) This environment was chosen because it met the description of the target environment in the mission specification. Communication between robots was modeled using a line-of-sight pathloss model.

These experiments were compared using five metrics:

- Time to Complete Mission - measures the number of simulation cycles required to complete the mission. Real time was not used as simulation cycles take longer to compute as more robots are added to the team.

- Percent of Time as One Network - measures the percent of time every robot in the team has a network route to every other robot. Multi-hop routes are considered equal to one-hop routes in this metric.
- Percent of Time as Fully Connected Network - measures the percent of time every robot in the team has a one-hop network route to every other robot in the team.
- Percent of Time at Least One Robot Alone - measures the percent of time at least one robot from the team has no network routes to any other robot.

The one-network, full-network and lone-robot metrics prove to be trivial for the one-robot case. They are calculated, however, as the one-robot case provides a baseline for all the metrics.

The motor schema used consisted of 4 behaviors, summed proportional to their respective gains. The behaviors and respective gains used follow:

- **preserve-communication** – gain = 1.0
- **move-to-goal** - gain = .4
- **avoid-static-obstacles** - gain = .6
- **maintain-formation** - gain = .4

The move-to-goal, maintain-formation and preserve-communication gains were chosen so that the move-to-goal and maintain-formation behaviors would never force a robot into a position that would compromise its communication quality of service. The avoid-static-obstacles gain was chosen to provide an adequate margin of safety from any obstacles that might pose a danger to the robots, given the gains of other behaviors in use.

Proof-of-concept hardware experiments were run on two laboratory robots in both relatively controlled and uncontrolled environments. The robots used are iRobot ATRV-jr robots. They are equipped with differential GPS receiver, digital compass and gyroscopic accelerometer for localization. An industrial-grade laser scanner provides perception for obstacle detection. As the networking hardware as presented in [4] was not yet available at the time of testing, an IEEE 802.11b bridge on each robot provided the infrastructure for a simple mobile ad-hoc network. Because the target hardware was not available, real-time network quality of service measurements had to be replaced by calculations from the line-of-sight pathloss model. It was at least possible to measure a binary connected/not connected metric to make a coarse judgment about the quality of the network between the robots.

The first set of experiments took place on the intramural athletic fields on the Georgia Tech campus. These large, flat artificial turf fields provide a relatively clean environment, similar to that of the simulations in the preceding section. Temporary obstacles were constructed on field to create a very simple urban environment. These obstacles were visible to the robots' laser scanners and were modeled in the a priori map as communication-obstructing obstacles.

The second set of experiments took place at the McKenna MOUT site at Fort Benning, GA. This urban testing ground provided an experimental environment very close to the target environment. Large cinder block buildings made up the set of modeled communication obstacles. The cinder block walls were a closer fit to the line-of-sight pathloss model than the temporary obstacles on the athletic fields. In addition to modeled communication obstacles, the environment was full of unmapped physical obstacles, such as fire hydrants, cars and trees. While these obstacles pose no serious threat to communication quality, they demonstrate the advantages of a reactive approach.

5 Results

Simulation experiments were run comparing a variety of robot team configurations. Each experimental run consisted of running a team of one, two, three or four robots in a terrain-coverage formation either with or without VBCP across the experimental environment. Three formation configurations were tested: line formation, column formation and no formation. Each unique robot team configuration ran twenty missions from unique starting points in each cardinal direction. (i.e., twenty starting points were evenly distributed on the west side of the environment, moving toward a goal on the east side; likewise on the north, east and south sides.)

Figure 4 shows the effect of number of robots on the communication quality of service in line and column formations, both with and without the VBCP behavior, according to the above metrics. Without VBCP, communication quality of service according to all metrics declines as the number of robots is increased. However, with VBCP, the mean percent of time in one network is greater than 98% for all cases. The percent of time in fully-connected network declines as robots are added to the team, but does not decline as rapidly as without VBCP.

Fig. 3. Experiments being run in the simulated MOUT site, the Georgia Tech intramural athletic fields and the McKenna MOUT site.

In the athletic field environment, teams of two robots were started at one side of the obstacle field and tasked with moving in formation to a point on the other side of the field while maintaining communication quality of service.

Although the temporary obstacles did not necessarily break the robots' communication link according to the line-of-sight pathloss model, they were considered adequate for a qualitative demonstration of a robotic behavior. Because the obstacles were modeled in the a priori map as opaque to the robots' communication signal, and the line-of-sight pathloss model was being used, line-of-sight and a maximum separation should be maintained by the robots at all times. (i.e., because the obstacles were modeled as serious communication barriers, the robots should behave as if they are.)

As the robots moved through the obstacle field, they could, qualitatively, be seen to be maintaining line-of-sight throughout the mission. Qualitative demonstrations of this behavior include robots passing an obstacle on the same side, so as not to allow and obstacle to come between them, and swinging wide around corners, so as not to allow the corner of an obstacle come between the robots. Figure 3 shows two robots swinging wide around a corner to maintain line-of-sight communication.

The experiments at the McKenna MOUT site resembled the athletic field experiments, in an environment that more closely resembles the target environment. As in the experiments on the athletic fields, teams of two robots were tasked with moving in formation across a subset of the environment while maintaining communication quality of service. Again, the robots should maintain line-of-sight and a maximum separation to maintain communication quality. Qualitative results were easier to demonstrate in these experiments over the athletic field experiments since a loss of line-of-sight was likely to cause a real communication failure.

As the robots moved through the urban environment, they could again qualitatively be seen to be preserving line-of-sight communication. Additionally, the robots maintained actual communication throughout the mission. Figures 3 shows two robots finding their way around a building without breaking communication. If the same mission is run without VBCP, the robots will move around opposite sides of the building.

6 Discussion

Overall, VBCP improves communication quality of service within a team of robots, especially as the number of robots in the team is increased. The performance of team configurations without VBCP declines in one-network and full-network metrics as the number of robots in the team is increased. This indicates that the problem of communication preservation gets harder as the number of robots increases. VBCP markedly improves performance in these metrics. The full-network metric declines slightly for the four-robot case with VBCP. However, the value function used in this work rewarded states where robots had double-connectivity. In the four-robot case, all robots can achieve double-connectivity without a fully-connected network. The time required to

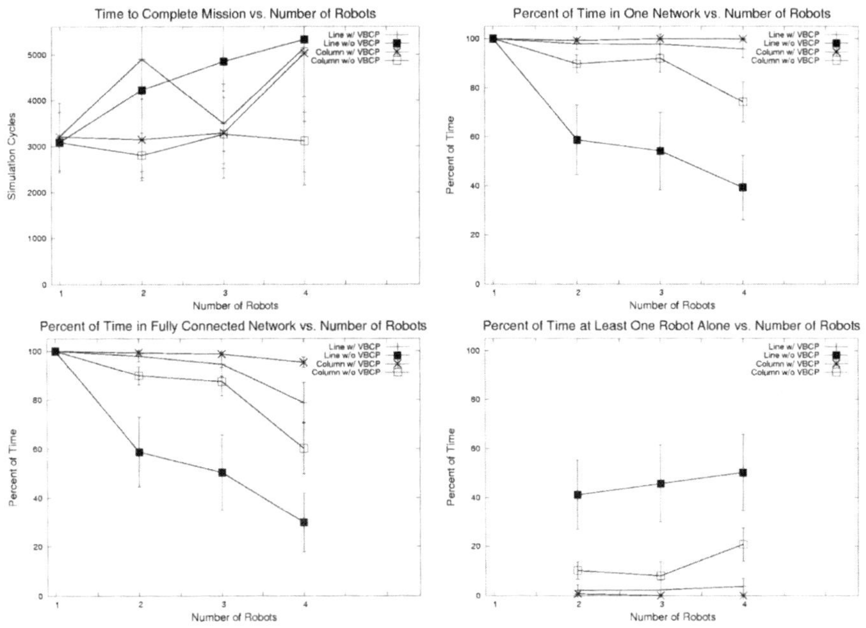

Fig. 4. The effect of the number of robots on the time required to complete a mission, the percent of time the team spends as one network, the percent of time the team is in a fully connected network and the percent of time at least one robot is disconnected from all other robots, both with and without VBCP. Errorbars indicate 95% confidence intervals.

complete a mission increased slightly when VBCP was used. This is both expected and acceptable. Teams using VBCP are likely to take the same path as teams not using VBCP, until this path takes them into a situation that causes a loss of line-of-sight communication. At this point, teams using VBCP will find a different, often longer, path to the goal. However, since the goal of this research is to maintain communication quality of service, an increase in running time is considered acceptable, providing the problem remains tractable.

The percent of missions completed declined slightly when VBCP was used. Of missions that were not completed, with or without VBCP, most were not completed because one or more robots became stuck in local minima. The use of VBCP made robot teams more likely to become stuck in local minima. This is not of major concern, since strategies for keeping reactive systems out of local minima exist [7, 8]. None of these strategies were used in the above experiments, as they do not take into account communication-sensitive strategies. At least one of these strategies will have to be adapted to communication-sensitive missions before this work is deployed into the target environment.

In simulation, the cause of failure of VBCP seems to be incorrect evaluation of possible next steps. This stems from the inherent loss of information in estimating that a robot's teammates will remain still during the next time step. In practice, the effect of this misestimation can be mitigated by increasing the distance the robots look ahead at each step, the trade-off being a loss of responsiveness to smaller fluctuations in communication quality.

In practice, an inherent weakness of VBCP is its reliance on map-accuracy. While a conservative signal strength model can make up for some of a map's shortcomings, at some level, the behavior is only as good as the map provided. Work is currently underway to relieve some of this reliance on an a priori map by learning and/or updating the map as the mission is run.

Acknowledgment

This work is funded under DARPA/DOI contract #NBCH1020012 as part of the MARS Vision 2020 program that is a joint effort of Georgia Tech, UPenn, USC and BBN.

References

1. R. Arkin "Motor Schema-Based Mobile Robot Navigation", *International Journal of Robotics Research*, 8(4), 1989.
2. T. Balch and R. Arkin "Motor Schema-Based Formation Control for Multiagent Robot Teams", *Proceedings of First International Conference on Multi-Agent Systems*, 10-16, 1995.
3. T. Balch and R. Arkin "Communication in Reactive Multiagent Robotic Systems", *Autonomous Robots* 1(1):27-52, 1994.
4. J. Redi and J. Bers "Exploiting the Interactions Between Robotic Autonomy and Networks", *Proceedings of 2003 International Workshop on Multi-Robot Systems*, pp. 279-289, March, 2003.
5. A. Stroupe and T. Balch "Value-Based Observation with Robot Teams (VBORT) for Dynamic Targets", *Proceedings of International Conference on Intelligent Robots and Systems 2003*, September, 2003.
6. Georgia Tech Mobile Robot Laboratory, *User Manual for MissionLab*, Version 5.0, http://www.cc.gatech.edu/ai/robot-lab/research/MissionLab/, January 2002.
7. L. Steels "Exploiting Analogical Representations", *Designing Autonomous Agents*, Macs, P. Ed., MIT Press, 1991.
8. T. Balch and R. Arkin, "Avoiding the Past: a Simple but Effective Strategy for Reactive Navigation", *Proceedings of 1993 IEEE International Conference on Robotics and Automation*, pp. 678-685, May, 1993.

Dynamical Reconfiguration of Cooperation Structure by Interaction Network

Kosuke Sekiyama and Yukihisa Okade

Department of Human and Artificial Intelligence Systems, University of Fukui
3-9-1 Bunkyo, Fukui 910-8507, Japan.
{sekiyama, okade}@dis.his.fukui-u.ac.jp

1 Introduction

Cooperative or competitive relationship in the multi-agent systems is often depicted by a graph, which features a collective behavior. The collective behavior has been analyzed under the specific form of interaction, such as a team play in the soccer agent [1, 2, 3], and a formation in the multi-robots [4, 5]. However, a relational structure among agents is generally assumed beforehand. While, some works deal with a relationship in the group organization by an evolutionary graph network [6, 7, 8], but a functional interpretation of the graph is not discussed in explicit way. In order to represent the functional meaning of the graph and provide an evolutionary mechanism to enhance cooperation structure, we have proposed *Interaction Network* [9]. Hence, the model was rather abstract and a graph node was static. In this paper, we extend the model to a version of mobile nodes, which can exhibit dynamic adaptation of cooperation form in multi-agent behavior. In this paper, we model *Interaction Network* to deal with a team play in the collective game, and propose the decision-making mechanism to enhance efficient organized behavior. We also evaluate the tradeoff and balance between the team play and the individual play.

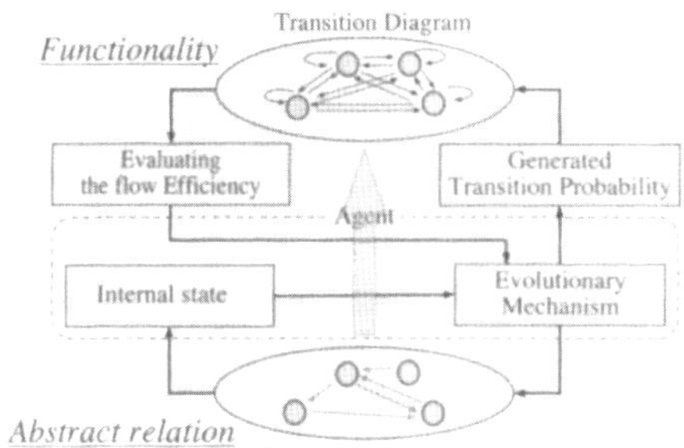

Fig. 1. Concept of *Interaction Network* and Transition Diagram

2 Interaction Network

Interaction Network is defined as a relational structure in cooperation and competition in collective agent systems, which is represented by a bidirectional directed simple graph with N nodes. Each node indicates an autonomous agent, and a directed link corresponds to the interaction between the agents. Figure 1 depicts the conceptual diagram of proposed model, where Transition Diagram represents flow in the system by the relational structure. Features of *Interaction Network* is summarized as follows:

- The internal state of the node is determined by interactions from the neighborhood nodes and the graph structure of local *Interaction Network.*
- There is an action attribute named activation and inhibition in interaction between nodes. Activation increases the internal state of the connected agent; on the other hand, inhibition decreases the internal state.
- The variable number of interactions (out-degree of the node about the graph theory), which are allowed in each agent, depends on the level of internal state of the agent.
- Each agent can update the graph structure of local *Interaction Network* by changing the interaction agents, where the number of interactions is also variable according to the situation.

We define some notations for formulation. Let $U_i, (i = 1, \cdots, N)$ be the numbered agent i, then the position of U_i is defined by vector $\boldsymbol{p}_i(t)$. A set of agent in the R-neighborhood to agent U_i is given by,

$$S_i(t) = \left\{ U_j \middle| \|\boldsymbol{p}_i(t) - \boldsymbol{p}_j(t)\| < R, i \neq j \right\}. \tag{1}$$

Since any relation of interactions is locally recognized from the viewpoint of U_i, we introduce the local description of *Interaction Network.* Assume that there exits $N_i(t)[= |S_i(t)|]$ agents in the neighborhood of U_i, then we define $U_{ij}, (j = 1, \cdots, N_i(t))$ as the jth nearest agent from U_i, hence U_i recognizes itself as U_{i0}. Then, the interaction between U_{ij} and U_{ik} is defined as follows,

$$a_{i,jk}(t) = \begin{cases} +1 & \text{activation,} \\ -1 & \text{inhibition,} \\ 0 & \text{no interaction.} \end{cases} \tag{2}$$

Note that we argue as $N_i = N_i(t)$ in the following sentences. The graph structure of locally described *Interaction Network* of U_i is represented by adjacency matrix $\boldsymbol{A}_i(t)$,

$$\boldsymbol{A}_i(t) = \begin{pmatrix} 0 & a_{i,01}(t) & \cdots & a_{i,0N_i}(t) \\ a_{i,10}(t) & 0 & \cdots & a_{i,1N_i}(t) \\ \vdots & \vdots & \ddots & \vdots \\ a_{i,N_i0}(t) & a_{i,N_i1}(t) & \cdots & 0 \end{pmatrix}. \tag{3}$$

In this model, we do not consider activation and inhibition to itself, so the diagonal elements $a_{i,jj}(t), (j = 0, \cdots, N_i)$ of the matrix $\boldsymbol{A}_i$ are always 0. Moreover, U_i can change the target of interaction agent as well as action

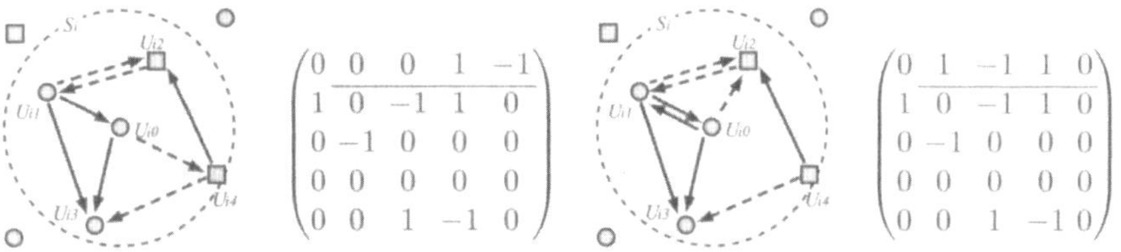

(a) *Interaction Network* before updating. (b) *Interaction Network* after updating.

Fig. 2. Correspondence of the updating local *Interaction Network* to adjacency matrix. The full line arrow and the dashed line indicate activation and inhibition respectively.

attribute. The change of interactions corresponds to the change of 1st row of a adjacency matrix. An example is depicted in Fig.2. The internal state $E_i(t)$ of agent U_i is decided by receiving activation and inhibition from the neighborhood agents in $S_i(t)$ as follows;

$$E_i(t) = \Phi\left(\sum_{j=0}^{N_i} a_{i,j0}(t) + E_{\text{base}}\right), \quad \Phi(\xi) = \begin{cases} E_{\text{max}} & \text{if } \xi > E_{\text{max}}, \\ E_{\text{min}} & \text{if } \xi < E_{\text{min}}, \\ \xi & \text{else.} \end{cases} \tag{4}$$

where E_{base} is default value of the internal state, E_{max} and E_{min} are maximal and minimal internal state respectively. $E_i(t)$ determines the upper bound of the number of interactions to the neighborhood agents. Hence, the actual number of interactions is referred as the interaction degree of freedom (interaction DOF) $b_i(t)$ which takes an integer value set $\{0, 1, \cdots, E_i(t)\}$ according to the surrounding situation. If the agent receives more activating interactions, it can have larger interaction DOF, while if the agent receives more inhibiting interactions, it can be zero. Using these variable $E_i(t)$ and $b_i(t)$, the index $x_i(t) \geq 0$ defines the remaining level of autonomy which means the resource of selfish action as follows

$$x_i(t) = E_i(t) - b_i(t). \tag{5}$$

Therefore, larger value of $x_i(t)$ implies higher the ability of autonomous behavior for itself, but it will reduce interactions to the neighborhood agents. Under the constrain of $E_i(t)$, the balance of $b_i(t)$ and $x_i(t)$ affects the effectiveness of the graph network.

3 Simulation model

3.1 Design of collective game by Interaction Network

Formation play is an easy understanding model to exemplify the functional connectivity. In this paper, we model the collective game, which passes a ball toward the goal using *Interaction Network*. Suppose that there are one ball and N agents on the game field, and agent $U_i, (i = 1, \cdots, N)$ belong to one of the two groups $G^{(1)}$ and $G^{(2)}$. Each agent moves and passes the ball for the goal area, and prevent the opponent group. Finally, the total score of 20

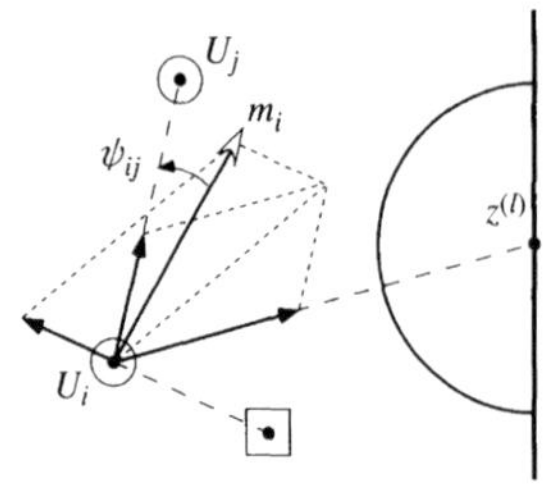

Fig. 3. Relationship between the direction of pass $m_i(t)$ and the angle ψ_{ij}

games is competed. As shown in Fig.1, we define Transition Diagram which is generated by *Interaction Network* as a destination of ball transition among the agents. If agent U_i has the ball, the destination of ball is decided stochastically to the neighborhood of U_i. Note that we argue as $\boldsymbol{p}_i = \boldsymbol{p}_i(t)$ in the following sentences. Let $\boldsymbol{z}^{(l)}$ be a position vector of the center area in the goal of group $G^{(l)}, (l = 1, 2)$. When $U_i \in G^{(l)}$, the intentional direction of pass $\boldsymbol{m}_i(t)$ for U_i is decided as follows,

$$\boldsymbol{m}_i(t) = \frac{\boldsymbol{z}^{(l)} - \boldsymbol{p}_i}{\|\boldsymbol{z}^{(l)} - \boldsymbol{p}_i\|^2} + M_1 \sum_{U_{ij} \in G^{(l)}} \frac{\boldsymbol{p}_{ij} - \boldsymbol{p}_i}{\|\boldsymbol{p}_{ij} - \boldsymbol{p}_i\|^2} - M_2 \sum_{U_{ik} \notin G^{(l)}} \frac{\boldsymbol{p}_{ik} - \boldsymbol{p}_i}{\|\boldsymbol{p}_{ik} - \boldsymbol{p}_i\|^2}, \quad (6)$$

where $M_1, M_2 > 0$ are constant. So the intentional pass direction is decided by the linear combination of three vectors as shown in fig.3. Then, the angle $\psi_{ij}(t)$ of agent $U_j \in S_i(t)$ from the direction of pass $\boldsymbol{m}_i(t)$ is calculated as follows,

$$\psi_{ij}(t) = \arccos\left(\frac{(\boldsymbol{m}_i(t), \boldsymbol{p}_j - \boldsymbol{p}_i)}{\|\boldsymbol{m}_i(t)\| \|\boldsymbol{p}_j - \boldsymbol{p}_i\|}\right). \quad (7)$$

The transition probability w_{ij} of the ball from U_i to U_j is generated by

$$w_{ij}(t) = \frac{\frac{1}{\|\boldsymbol{p}_j - \boldsymbol{p}_i\|} \exp\left[\left(-\sigma\psi_{ij}(t)^2\right) + \left(x_j(t) - x_i(t)\right)\right]}{\sum_{U_k \in S_i(t)} \frac{1}{\|\boldsymbol{p}_k - \boldsymbol{p}_i\|} \exp\left[\left(-\sigma\psi_{ik}(t)^2\right) + \left(x_k(t) - x_i(t)\right)\right]}. \quad (8)$$

Where, the probability reflects the influence of directivity of the pass direction and uncertainty due to the distance between the agents, and $\sigma > 0$ is constant. Also, the transition probability to $U_k \notin S_i(t)$ is $w_{ik}(t) = 0$. Thus, the cooperation structure in the model becomes the variable node probability network, generated by the graph structure of *Interaction Network* and the distribution of agents. Therefore, each agent tries to reconfigure the cooperation structure *Interaction Network* to improve the dominant rate of the ball. To achieve this, each agent behaves according to the following procedure:

(1) Agents are assigned to the Game Field.
(2) $U_i, (i = 1, \cdots, N)$ decides the interaction DOF b_i and remaining level of autonomy x_i, based on *Interaction Network*.
(3) The agent moves on the field, and update the graph structure of *Interaction Network* in the neighborhood by $b_i(t)$.
(4) Transition probability is generated by $x_i(t)$.

(5) Destination of the ball is decided stochastically by transition probability.

Where (1) is an initialization of a collective game. Then, repeat from (2) to (5), until a score enters.

3.2 Decision-making mechanism of agent

Evaluation on positioning

The positioning of agent depends on dominance of the ball. Hence, let the position of the ball be $\boldsymbol{r}(t)$ and the candidate of motion of U_i be $\tilde{\boldsymbol{p}}_i$. Then, we define the evaluation function for positioning of agent $f_i(\tilde{\boldsymbol{p}}_i)$ as follows,

$$f_i(\tilde{\boldsymbol{p}}_i) = \begin{cases} \exp\big(-\alpha\|\boldsymbol{z}^{(l)} - \tilde{\boldsymbol{p}}_i\|^2\big) - V_i\left(\tilde{\boldsymbol{p}}_i\right), & \text{for } U_i \text{ has a ball,} \\ \exp\big(-\beta\|\boldsymbol{r}(t) - \tilde{\boldsymbol{p}}_i\|^2\big) - V_i\left(\tilde{\boldsymbol{p}}_i\right), & \text{for } U_i \text{ doesn't has a ball.} \end{cases} \tag{9a}$$

$$V_i(\boldsymbol{p}) = \sum_{j=1}^{N_i} \exp\big(-\gamma\|\boldsymbol{p}_{ij} - \boldsymbol{p}\|^2\big). \tag{9b}$$

V_i in Eq.(9a) is the evaluation for collision avoidance to U_{ij}.

Evaluation on reconfiguration of the graph structure

As seen in fig.2, reconfiguration of *Interaction Network* in the neighborhood of agent U_i is made by changing only the 1st row of the adjacency matrix $\boldsymbol{A}_i$ in Eq.(3). A candidate of the graph structure $\tilde{\boldsymbol{a}}_i$ and modified adjacency matrix $\widetilde{\boldsymbol{A}}_i$ are given as

$$\tilde{\boldsymbol{a}}_i = (\tilde{a}_{i,01}, \cdots \tilde{a}_{i,0N_i}), \qquad \widetilde{\boldsymbol{A}}_i(\tilde{\boldsymbol{a}}_i) = \begin{pmatrix} 0 & \tilde{a}_{i,01} & \cdots & \tilde{a}_{i,0N_i} \\ a_{i,10}(t) & 0 & \cdots & a_{i,1N_i}(t) \\ \vdots & \vdots & \ddots & \vdots \\ a_{i,N_i0}(t) & a_{i,N_i1}(t) & \cdots & 0 \end{pmatrix}. \tag{10}$$

A candidate of the updated graph structure is evaluated such that it can improve dominant rate of the ball in the near future. In order to obtain such a *predicted condition* for each of candidate of graph in a short calculation, we make use of the stationary solution of stochastic process produced by the local Transition Diagram. Hence, the agent U_i forms a local Transition Diagram according to the decision law of the transition probability of Eq.(8). However, U_i cannot recognize the remaining level of autonomy and the directions of path for $U_{ij}, (j = 1, \cdots, N_i)$ directly, because these are information of internal state of each agent. So, U_i decides to estimate $x_{ij}(t)$ for $U_{ij}, (j = 1, \cdots, N_i)$ from Eq.(4), Eq.(5), and Eq.(10) as follows,

$$x_{ij}(t) = \Phi\left(\sum_{k=1}^{N_i} a_{i,kj}(t) + \tilde{a}_{i,0j} + E_{\text{base}}\right) - \sum_{k=0}^{N_i} |a_{i,jk}(t)|. \tag{11}$$

The first term of Eq.((11)) is estimated from each agent's internal state based on Eq.(4), and the second term is interaction DOF from the adjacency matrix $\widetilde{\boldsymbol{A}}_i$. Also, the direction of pass $\boldsymbol{m}_{ij}(t)$ for U_{ij} is estimated by Eq.(6), (7), and the angle $\psi_{i,ij}(t)$ of agent U_{ij} from the direction of pass $\boldsymbol{m}_{ij}(t)$ is calculated by Eq.(7). Therefore, the transition probability $w_{i,jk}(t), (j,k = 0,\cdots,N_i)$ of a ball from U_{ij} to U_{ik} is generated by

$$w_{i,jk}(t) = \frac{\frac{1}{\|\boldsymbol{p}_{ik}-\boldsymbol{p}_{ij}\|} \exp\left[\left(-\sigma\psi_{i,jk}(t)^2\right) + (x_{ik}(t) - x_{ij}(t)) \right]}{\sum_{n=0}^{N_i} \frac{1}{\|\boldsymbol{p}_{in}-\boldsymbol{p}_{ij}\|} \exp\left[\left(-\sigma\psi_{i,jn}(t)^2\right) + (x_{in}(t) - x_{ij}(t)) \right]}. \tag{12}$$

The local Transition Diagram of U_i is represented as the transition probability matrix $\boldsymbol{W}_i(\tilde{\boldsymbol{a}}_i)$ as follows,

$$\boldsymbol{W}_i(\tilde{\boldsymbol{a}}_i) = \begin{pmatrix} w_{i,00}(t) & w_{i,01}(t) & \cdots & w_{i,0N_i}(t) \\ w_{i,10}(t) & w_{i,11}(t) & \cdots & w_{i,1N_i}(t) \\ \vdots & \vdots & \ddots & \vdots \\ w_{i,N_i0}(t) & w_{i,N_i1}(t) & \cdots & w_{i,N_iN_i}(t) \end{pmatrix}. \tag{13}$$

Since $w_{i,jk}(t) > 0$ for $\forall j,k$ in $\boldsymbol{W}_i$, the transition process is modeled as the regular Markov chain, so a unique stationary distributions exists for the transition process of W_i and is indicated as follows

$$\boldsymbol{\pi}_i(\tilde{\boldsymbol{a}}_i) = \boldsymbol{\pi}_i(\tilde{\boldsymbol{a}}_i)\boldsymbol{W}(\tilde{\boldsymbol{a}}_i), \text{ for } \boldsymbol{\pi}_i(\tilde{\boldsymbol{a}}_i) = (\pi_{i0},\cdots,\pi_{iN_i}), \sum_{j=0}^{N_i} \pi_{ij} = 1. \tag{14}$$

$\pi_{ij}, (j = 0,\cdots,N_i)$ is interpreted as the expected value of the acquisition probability for U_{ij}. From a location of agents and evaluation of the efficiency of pass, the evaluation function $h_i(\tilde{\boldsymbol{a}}_i)$ for $U_i \in G^{(l)}(U_i \notin G^{(L)})$ is defined by the followings according to the case of offense mode and defense mode.

$$h_i(\tilde{\boldsymbol{a}}_i) = \begin{cases} \displaystyle\sum_{U_{ij}\in G^{(l)}} \pi_{ij}\exp(-\alpha\|\boldsymbol{z}^{(l)} - \boldsymbol{p}_{ij}\|^2), & \text{for offense mode,} \\ \displaystyle\sum_{U_{ij}\in G^{(L)}} \frac{1-\pi_{ij}}{N_i - 1}\exp(-\alpha\|\boldsymbol{z}^{(L)} - \boldsymbol{p}_{ij}\|^2), & \text{for defense mode.} \end{cases} \tag{15}$$

4 Search algorithms of candidate set

Agent U_i searches the best solution from candidates of positioning $\tilde{\boldsymbol{p}}_i$ and graph structure $\tilde{\boldsymbol{a}}_i$ using the following genetic algorithms (GA) for every time step. Firstly, ten individuals to candidate of $\tilde{\boldsymbol{p}}_i$ and $\tilde{\boldsymbol{a}}_i$ are created respectively. Secondly, the adaptation value is calculated for the candidate set of individual. According to adaptation value, the selected candidate sets are reproduced in the next generation by the genetic operation. In this genetic operation's parameter, crossover probability and mutation probability are 0.5 and 0.1 respectively. The best solution is selected from candidate sets when this search is repeated to the tenth generation.

4.1 Adaptation function for GA search

Eq.(9a) and Eq.(15) evaluate positioning and pass-play, respectively. The adaptation function is defined according to Eq.(9a) and Eq.(15) as follows,

$$y_i(\tilde{\boldsymbol{a}}_i, \tilde{\boldsymbol{p}}_i) = \varphi_i h_i(\tilde{\boldsymbol{a}}_i) + (1 - \varphi_i) f_i(\tilde{\boldsymbol{p}}_i). \tag{16}$$

Where, $\varphi_i \in [0,1]$ defines tendency of the strategy between team-play and individual play as strategy parameter.

$\rho(1)$	$\cdots$	$\rho(n)$	$\theta(1)$	$\cdots$	$\theta(n)$

(a) potitioning

Agent Number :	$s(1)$	$s(2)$	$\cdots$	$s(N_i)$
Connection :	$q_{s(1)}$	$q_{s(2)}$	$\cdots$	$q_{s(N_i)}$

(b) graph structure

Fig. 4. The genotype of positioning and graph structure

Algorithm Decoding of individual.
$sum \leftarrow 0$
for $k = 1$ to N_i **do**
  if $sum < E_i$ **then**
    $sum \leftarrow sum + q_{s(k)}$
    if $U_i, U_{is(k)} \in G^{(l)}$ **then**
      $\tilde{a}_{i,0s(k)} \leftarrow +1 \times q_{s(k)}$
    else
      $\tilde{a}_{i,0s(k)} \leftarrow -1 \times q_{s(k)}$
    end if
  else
    $\tilde{a}_{i,0s(k)} \leftarrow 0$
  end if
end for
$b_i \leftarrow sum$
end

Fig. 5. Decoding Algorithm of individual

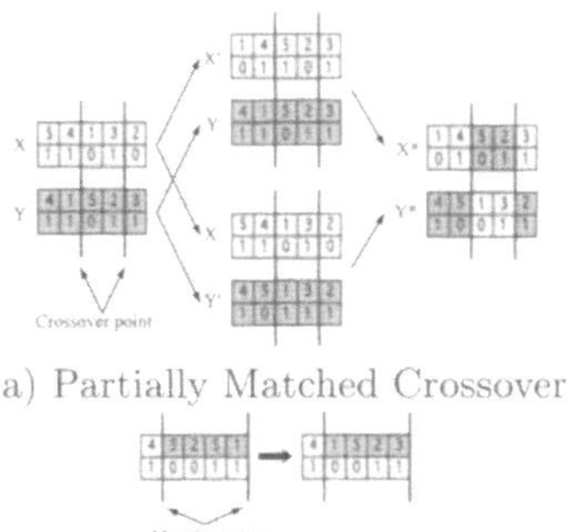

(a) Partially Matched Crossover

(b) Inverse

Fig. 6. GA Operator of individual

4.2 Individual expression and genetic operation of positioning

Fig.4-(a) shows a genotype of the candidate of positioning, where $\boldsymbol{\rho}$ is a binary code of the moving distance v_ρ, and $\boldsymbol{\theta}$ is a binary code of the moving direction v_θ. The conversion rule of decimal number from a binary code $\boldsymbol{\xi}$ repressed as $\mathcal{D}(\boldsymbol{\xi})$, the decoding of a genotype is defined as follows

$$v_\rho = R_v \times \frac{\mathcal{D}(\boldsymbol{\rho})}{2^n - 1}, \qquad v_\theta = 2\pi \times \frac{\mathcal{D}(\boldsymbol{\theta})}{2^n - 1}. \tag{17}$$

where, R_v indicates the maximum moving distance. The candidacy of motion $\tilde{\boldsymbol{p}}_i$ is calculated according to moving distance v_ρ and direction v_θ, as follows

$$\tilde{\boldsymbol{p}}_i = \boldsymbol{p}_i + [v_\rho \cos v_\theta,\ v_\rho \sin v_\theta]. \tag{18}$$

Moreover, the uniform crossover and mutation of bit reversal are used as genetic operation of the reproducing individuals.

4.3 Individual expression and genetic operation to the graph structure

For calculating a best graph structure by GA, the candidate of the graph structure is coded as shown in Fig.4-(b). In this genotype, the agent number, which is possible to interact with U_i is assigned to $s(k), (k = 1, \cdots, N_i)$ by the random sequence, and the lower row element $q_{s(k)}$ is defined by,

$$q_{s(k)} = \begin{cases} 1 & \text{action to } U_{is(k)}, \\ 0 & \text{no action to } U_{is(k)}. \end{cases} \tag{19}$$

As the constrain for candidate of graph structure $\tilde{\boldsymbol{a}}_i$, interaction degree of freedom of U_i is restricted according to the number of agents in the neighborhood

Table 1. Parameter set for simulation

N	number of agent	8
$E_{\max}$	maximal internal state	5
$E_{\min}$	minimal internal state	0
E_{base}	default internal state	3
R	radius of neighborhood circle	20
R_v	maximal distance	5

α	evolution coefficient to goal	0.277×10^{-3}
β	evolution coefficient to ball	1.09×10^{-3}
γ	reflection coefficient to V_i	4.43×10^{-3}
σ	directivity parameter to w_{ij}	0.632

and internal state E_i. Hence, the genotype(s, q) is decoded by the algorithm of Fig.5. In this genetic operator of crossover method, partial matched crossing method is adopted and new individual to candidate of the graph structure X^* and Y^* are reproduced from the individual X and Y (Fig.6-(a)). Also, mutation method adopts the Inversion (Fig.6-(b)).

5 Simulation Results

5.1 Simulation Condition

In this simulation, suppose $G^{(1)}$ and $G^{(2)}$ indicate the offense side and the defense side respectively, and one game is finished when $G^{(1)}$ scores or $G^{(2)}$ clears the ball from the game field. The result is evaluated based on the total score of $G^{(1)}$ in 20 games. Then, we utilize the average result of 20 trial set in the later discussion. $\varphi^{(l)}$ defined in (16) is the parameter for decision-making of agent in $G^{(l)}, (l = 1, 2)$, which implies tendency of the strategy, that is, $\varphi^{(l)} = 1$ corresponds to team-play oriented strategy, and $\varphi^{(l)} = 0$ corresponds to the individual play strategy. By changing the set of parameter values of $(\varphi^{(1)}, \varphi^{(2)})$, we compare the difference of performances in the collective game and evaluate the cooperation structure. Parameters utilized in the simulation are listed in Table1.

5.2 Example of collective behavior in two different cases

We exemplify a typical behavior in the offense group $G^{(1)}$. Snapshots in Fig.7 illustrate collective behaviors in two different cases, $\varphi^{(1)} = 0.1$ and 0.9. Where a sphere and a cube node represents the agent in $G^{(1)}$ and $G^{(2)}$ respectively. In addition, a solid arrow means an activation and a dashed arrow means an inhibition. Also, table in each figure shows the agent number U_i, the remaining level of autonomy x_i and the transition probability w_{ij}.

Individual play oriented Collective behavior

Figure7-(a) shows the case of $(\varphi^{(1)}, \varphi^{(2)}) = (0.1, 0.5)$, where agent U_2 holds the ball and the position of U_1 is between U_2 and the goal area. U_1 receives two inhibitions from U_4 and U_7. Therefore, the internal state E_1 is 1 according to Eq.(4). Also, because U_1 does not give any interaction to the other agents, interaction DOF b_i is 0 and $x_1 = E_1 - b_1 = 1$ by Eq.(5). Moreover, $x_2 = 3$, since U_2 receives one activation and one inhibition. The transition probability of the ball from U_2 to U_1 w_{21} is 0.025, which is smaller than 0.934 of w_{22},

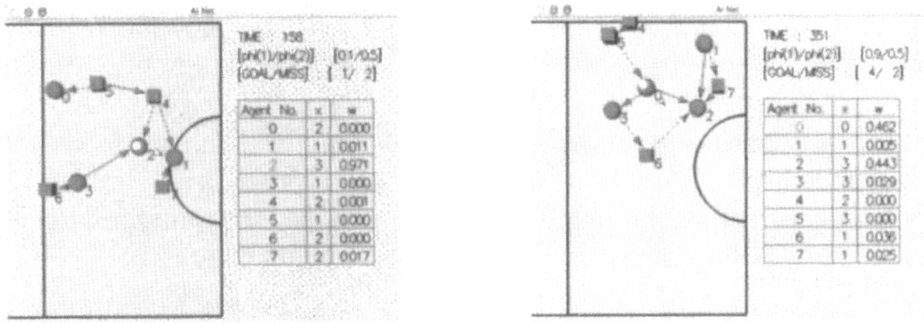

(a) Oriental individual play ($\varphi^{(1)} = 0.1$). (b) Oriental team play ($\varphi^{(1)} = 0.9$).

Fig. 7. Snapshots in collective game for the different strategy parameter φ.

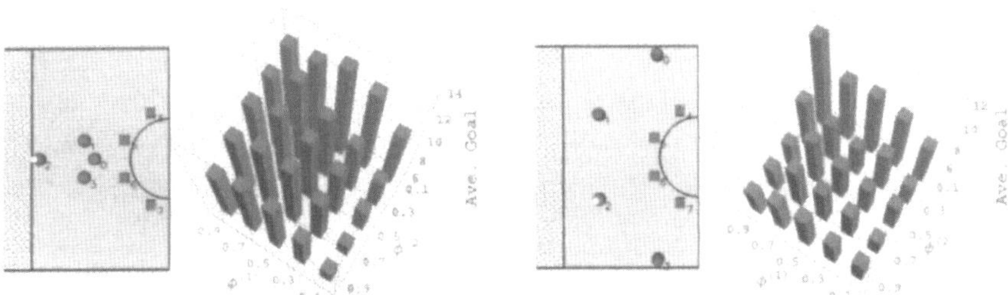

(a) For situation 1. (b) For situation 2.

Fig. 8. Result of collective game for two situations.

therefore U_2 continues to hold the ball and moves to the goal area. It can be seen that this individual play oriented strategy does not produce the efficient flow of ball toward the goal, since the effective cooperation structure is not formed.

Team play oriented Collective behavior

Figure7-(b) shows the case of $(\varphi^{(1)}, \varphi^{(2)}) = (0.9, 0.5)$, where U_2 is located in the near of the goal area. Since U_2 receives two activations and two inhibitions,it does not give any interaction to the neighborhood agents. Therefore, $x_2 = 3$ according to Eq.(4) and Eq.(5). The transition probability w_{02} is 0.443, which is larger than the transition probability of the other defense agents. It can be seen that the effective relational structure is formed such that it produces the efficient flow of the ball toward the goal area.

5.3 Evaluation of Cooperation Structure by *Interaction Network*

As discussed in the previous section, the tendency of strategy defined by φ gives an significant influence over the performance of the collective behavior. In this section, dependency of the strategy parameter φ is investigated for two specific situations. Figure.8 depicts two dimensional histogram for the different initial formations, where the value of each histogram indicates the average goal score of group $G^{(1)}$ according to the different values of $(\varphi^{(1)}, \varphi^{(2)})$. From the result of Fig.8-(a), in case that the strategy parameter $\varphi^{(1)}$ takes a large value, the goal scores of $G^{(1)}$ tends to be high, however this does not mean that the fully team play strategy is the optimal. According to the opponent strategy, there is an equilibrium solution in terms of the minimax strategy in zero-sum game, which is given by $(\varphi^{(1)}, \varphi^{(2)}) = (0.7, 0.9)$. On the other hand, Fig.8-(b) shows that an equilibrium solution is not found as in the case of Fig.8-(a). Depending on the game phase, the suitable strategy parameter must be determined according to an opponent's attitude.

6 Conclusions

In this paper, a collective game is modeled using *Interaction Network*, and the decision-making mechanism is proposed based on evaluation about the surrounding relation and self-behavior. From simulation results, we confirmed that effective cooperation structure is formed, when strategy parameter φ is large. Moreover, when each agent selects a suitable φ according to an opponent's attitude and further more effective cooperation structure is formed. However, in the present work, agents cannot change φ dynamically depending on the game situation. Extension to dynamical adaptation of team play strategy is a future work.

References

1. M. Tambe, J. Adibi, Y. Alonaizon, A. Erdem, G. Kaminka, S. Marsella, and I. Muslea. Building agent teams using an explicit teamwork model and learning. *Artifical Intelligence*, Vol. 110, No. 2, pp. 215–239, 1999.
2. Hitoshi Matsubara, Itsuki Noda, and Kazuo Hiraki. Learning of cooperative actions in multiagent systems: A case study of pass play in soccer. In *Working Notes for the AAAI Symposium on Adaptation, Co-evolution and Learning in Multiagent Systems*, pp. 63–67, 1996.
3. Peter Stone and Manuela M. Veloso. Task decomposition and dynamic role assignment for real-time strategic teamwork. In *Proc. of Agent Theories, Architectures, and Languages*, pp. 293–308, 1998.
4. J. Fredslund and M. Mataric. Robot formations using only local sensing and control. In *Proc. of IEEE International Symposium on Computational Intelligence in Robotics and Automation, 2001*, Vol. 2001, pp. 308–313, 2001.
5. T. Balch and R. Arkin. Behavior-based formation control for multi-robot teams. *IEEE Transactions on Robotics and Automation*, Vol. 14, pp. 926–939, December 1998.
6. W. Aiello, Fan Chung, and L. Lu. Random evolution in massive graphs. In *IEEE Symposium on Foundations of Computer Science*, pp. 510–519, 2001.
7. S. N. Dorogovstsev and J. F. F. Mendes. Evolution of networks. *Advances in Physics*, Vol. 51, pp. 1079–1187, 2002.
8. A.S.Mikhailov and V.Calenbuhr. *From Cells to Societies*, chapter 9, pp. 233–277. Springer, July 2001.
9. K. Sekiyama and Y. Okade. Variable interaction network based on activation and inhibition. In *IEEE International Symposium on Computational Intelligence in Robotics and Automation, 2003*, pp. 1551–1556, 2003.

Collecting Behavior of Interacting Robots with Virtual Pheromone

Toshiya Kazama[1], Ken Sugawara[2] and Toshinori Watanabe[1]

[1] Graduate School of Information Systems, University of Electro-Communications Chofu, Tokyo 182-8585, JAPAN {kazama,watanabe}@sd.is.uec.ac.jp
[2] Department of Liberal Arts, Tohoku Gakuin University Izumi, Sendai, Miyagi 981-3193, JAPAN sugawara@cs.tohoku-gakuin.ac.jp

In multi-robot system, communication is indispensable for effective cooperative working. In this paper, we focus on two types of communication method, and investigate a foraging behavior of multi-robot system. Firstly, we discuss the foraging by physical communication. Next, we propose a virtual chemical system and discuss the foraging by chemical communication. Lastly, we show these behaviors can be treated by a common framework using a state transition diagram.

Keywords collective behavior, foraging, interaction, V-DEAR

1 Introduction

Many kinds of fishes and birds live in groups. Social insects such as ants and bees establish well-ordered societies even in the absence of particular individual intelligence[1]. Multi-cellular organisms depend on the cooperation of cells[2]. In these cases, each element does simple tasks by responding to local conditions without any central control, but the whole system exhibits complex and adaptive functions. By what mechanism does cooperation by many simple elements create qualitatively new behaviors? Many researchers have studied such clusters. Not only scientists but also engineers are interested in such interactive biological systems[3]. Recently, research has also been very active in the area of multi-robot systems, and many researchers are currently studying the behavior of these types of systems[4][5]. One of the most important aspect in a multi-robot system is the ability of several robots to work cooperatively. Working together, they complete tasks that a single robot cannot. For effective cooperative working, communication is indispensable, and we can consider two types of communication method: physical and chemical (Fig.1). In general, it is popular and useful for multi-robot system to communicate directly by physical method such as light signal, sound, radio wave and so on. But chemical method is also utilized in nature. Actually the chemical method is very common in the insect world, and it has some interesting characteristics that physical method does not have.

In this paper, we choose a simple foraging task[6][7] [8][9][10] and investigate the experimental performance of physical communication and "virtual" chemical communication. At first, we discuss the foraging with physical communication. Secondly, the foraging with "virtual" chemical communication is discussed. Lastly, we show these behaviors can be treated by a common framework using a state transition diagram.

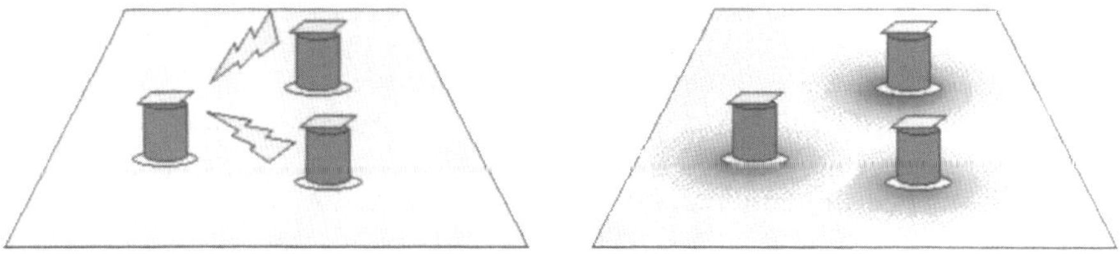

Fig. 1. Communication method. Physical communication (left) and chemical communication (right).

2 Foraging Behavior by Physical Communication

In this section, we discuss foraging behavior by physical communication. Each robot has following five movements: "searching," "attracted," "broadcasting," "homing," and "avoidance" (Fig. 2). In foraging, when a robot in searching state finds a food, it stays there and radiate light to broadcast its location for a constant period. After that, it moves to home and then search again. Here broadcasting period is called "interaction duration."

Fig. 2. Basic behaviors of robot introduced in this experiment.

We conducted experiments in real robots as shown in Fig. 3, which size is 60 mm width × 85 mm length × 92 mm height, and which has a 65-mm-wide forward square-brackets gripper to push objects. Objects were circular metal pucks, 30 mm in diameter and 30 mm high.

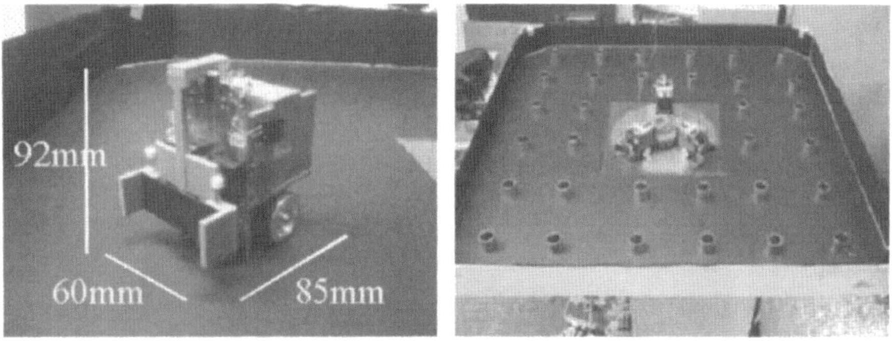

Fig. 3. Robot with a gripper for object gathering.

There is a sensitive touch sensor in the middle of the gripper, and it detects objects to be gathered. The gripper is also connected to a pair of micro-switches, and it can detect the collision to other robots and boundary walls.

In order to indicate its location to other robots, each robot has Infrared(IR) LEDs on the top, and to receive other robots' broadcasts, it has a pair of IR sensors. Home radiates isotropic light signal and they home in using a pair of light sensors at the bottom.

The field for this experiment was an 90×90 cm black surface with wall at the boundary and an yellow "home" circle and IR-LED array at the center. We used 32 metal pucks for this experiment. In this experiment, various puck distributions can be considered, and here we chose homogeneous and localized distributions (Fig. 4).

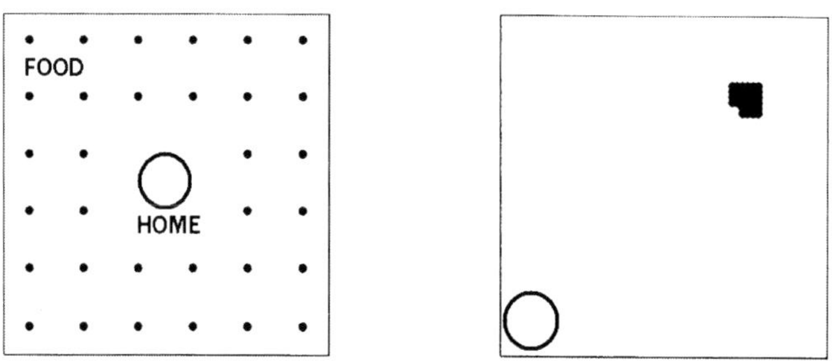

Fig. 4. Schematic of experimental field. Homogeneous (left) and Localized (right).

The efficiency of the group work was measured by the number of collected pucks for constant period. Fig. 5 shows the result of this experiment. This is an average of three trials, in which each trial was run for 10 minutes. In homogeneous field, the robots without communication gather food effectively. On the other hand, in localized field, the performance of the robots with interaction is much better than no interaction.

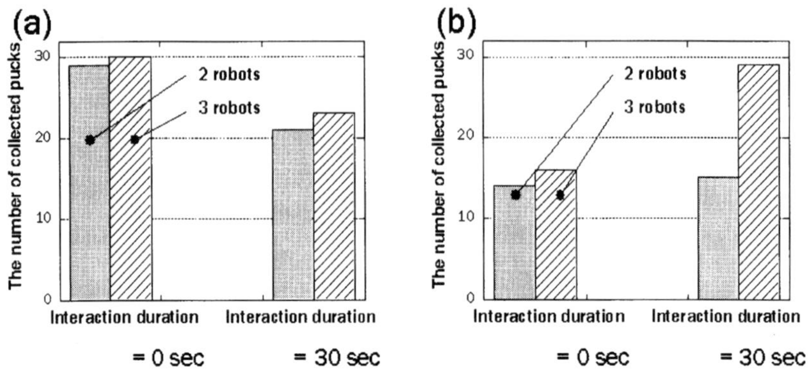

Fig. 5. The number of collected pucks. (a) In homogeneous field. (b) In localized field.

3 Foraging Behavior by Chemical Communication

As described above, direct physical communication method is popular in multi-robot system, but in insect world, indirect chemical communication method is also very common. Chemical signals, which are generally called "pheromone", are utilized for various purposes such as alarm, aggregation, sex attractant, recruitment, defense, trail-making, and so on. It is attractive topic to apply this kind of communication method to real robot system, but there are few researches that treat chemical communication for physical robot system[11]. To treat chemical materials is not easy comparing with the physical material, and it is also not easy to get proper chemical sensors. Moreover, chemical materials, especially gas, are invisible and it is quite difficult to observe how they spread and affect the robots' behaviors.

Here we propose "Virtual Pheromone System" for real robot experiment. In this system, virtual pheromones are drawn by computer graphics. As virtual pheromones are CG, we can avoid the problems described above. In addition, we can arbitrary control the property of the virtual chemical materials. Next, we will explain the virtual pheromone system in detail.

3.1 Virtual Dynamic Environment for Autonomous Robots (V-DEAR)

The schematic and the photo of "Virtual Dynamic Environment for Autonomous Robots (V-DEAR)" are shown in Fig. 6. This is composed of LCD projector to draw the chemical field by CG and CCD camera to track the position of the robots in the field. The robots moving in the field have sensors on the top to sense the color and the brightness of the field, and determine their behavior autonomously based on the light information.

Fig. 6. Virtual Dynamic Environment for Autonomous Robots. Schematic (left) and photo (right).

Combining the position information of the robots from CCD camera and the drawn CG by projector, we can express the dynamic interaction between the environment and the robots.

3.2 Experiment

Basic behavior of the robots here is almost same as shown in previous section. Different point is the robots that find a food move to home leaving "pheromone trail" (Fig. 7).

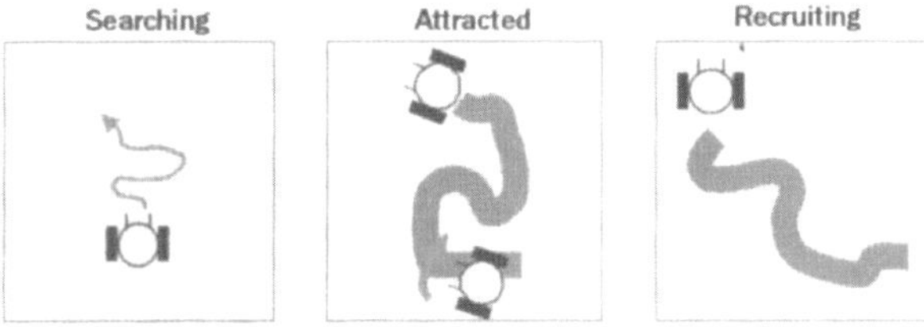

Fig. 7. Basic robot behavior.

The shape of the robot is shown in Fig. 8, which size is 75 mm diameter × 140 mm height. It has 8-directional touch sensors to detect collision, a pair of infrared sensors to return to home, three color-sensors to detect field condition, bottom sensors to detect "Home", and 1 LED to indicate its state. Driving system is same as described in previous section.

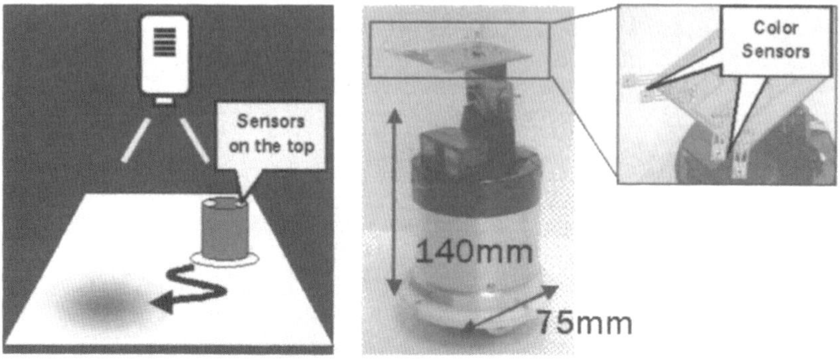

Fig. 8. Robots detect the field condition by sensors on the top.

The field for this experiment was an 90×90 cm black surface with wall at the boundary and a yellow "home" circle with IR-LED array at the center. Dynamics of pheromone concentration was described as $\dot{\rho} = \delta + D\nabla^2\rho - k\rho$, where ρ is the concentration of the pheromone, δ is an injection concentration, D is a diffusion coefficient, and k is a rate of evaporation. In this experiment, we also chose homogeneous and localized distributions (Fig.9).

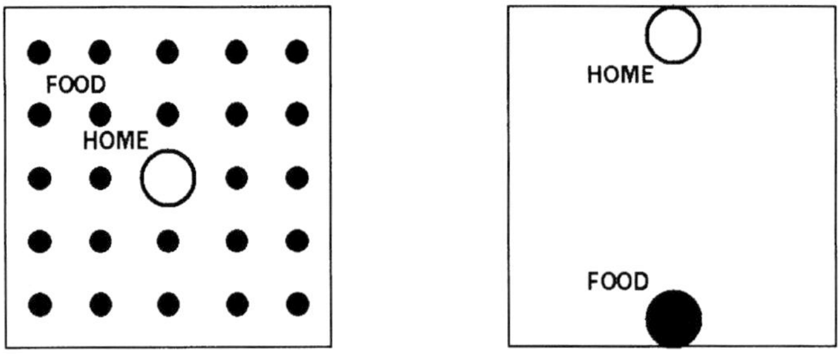

Fig. 9. Schematic of Experimental Field. Homogeneous (left) and Localized (right).

Fig. 10 and 11 are snapshots of experiments in homogeneously distributed field and locally distributed field, respectively. Fig. 12 shows the relation between the rate of evaporation and the number of collected pucks. From these results, we can say that less pheromone is left in the field, more foods are collected in homogeneous field, and that the performance becomes better when a pheromone trail is formed continuously in localized field.

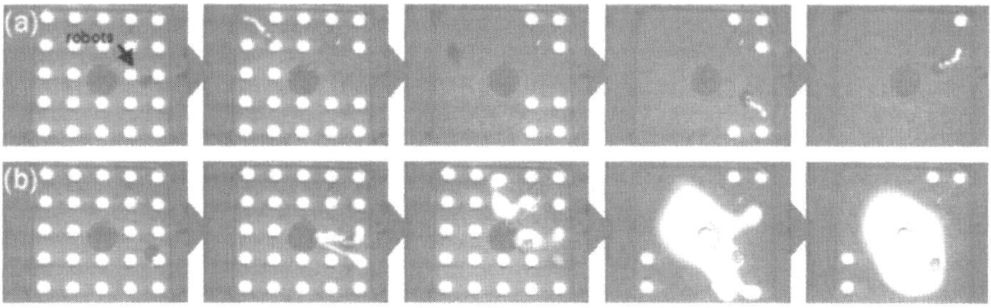

Fig. 10. Snapshot of experiment in homogeneously distributed field. (a) $k = 0.01$ (b) $k = 0.002$.

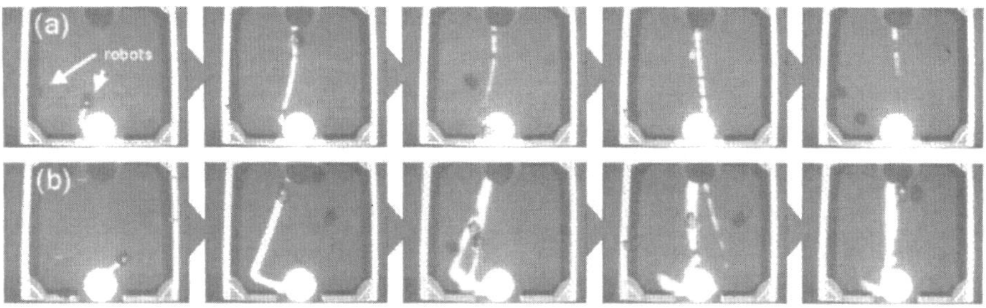

Fig. 11. Snapshot of Experiment in locally distributed field. (a) $k = 0.01$ (b) $k = 0.002$.

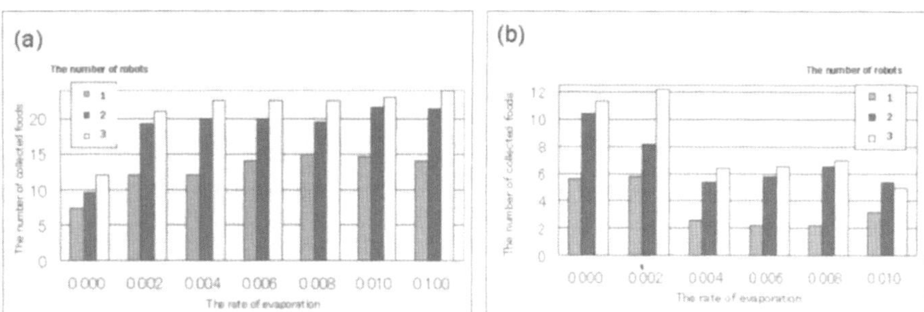

Fig. 12. Relation between the rate of evaporation and the number of collected pucks. (a) Homogeneous field. (b) Localized field.

4 Mathematical Analysis of Foraging Behavior

4.1 Computer Simulation

In previous section, we showed experimental results of foraging behavior of simple interacting robot system with physical and chemical communication. From the experimental results above, we understand their behavior qualitatively, but it is quite difficult to evaluate the efficiency of the group quantitatively because large scale experiment is tough to execute in real system. We simulated the robots' behavior by computer simulation. Details of simulation method and condition have been adequately discussed in previous paper[12]. Here we show the results in Fig. 14(a) and Fig. 15(a).

4.2 Mathematical Analysis by State Transition Diagram

Foraging behaviors of simple interacting robots can be analized by state transition diagram. When a robot is in "Searching"(S), its next state is "Attracted"(A) by finding other broadcasting robots, or is "Broadcasting"(B) finding a puck by chance. The robot in state (B) changes its state to "Homing"(H) after a constant period, and searches again when it arrives at home. A robot in state (A) may find a puck by attraction, but it may take much time because of "Congestion"(C), which depends on the number of robots in the congestion and the condition of puck distribution.

The state transition diagram of the behavior above can be expressed as Fig. 13 and derives a set of equations (1).

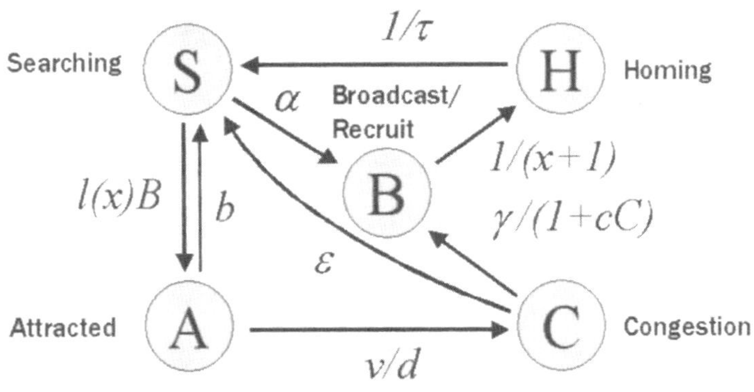

Fig. 13. State transition diagram of foraging behavior.

$$
\begin{aligned}
\dot{S} &= -\alpha S + \tfrac{1}{\tau}H - l(x)\cdot B\cdot S + \delta A + \epsilon C \\
\dot{B} &= -\tfrac{1}{x+1}B + \alpha S + \gamma\cdot\tfrac{1}{1+\varepsilon C}C \\
\dot{H} &= \tfrac{1}{x+1}B - \tfrac{1}{\tau}H \\
\dot{A} &= -\tfrac{v}{d}A + l(x)\cdot B\cdot S - \delta A \\
\dot{C} &= \tfrac{v}{d}A - \tfrac{\gamma}{1+\varepsilon C}C - \epsilon C,
\end{aligned}
\qquad (1)
$$

In the equations, α is the probability of finding a puck independently, τ is the time returning home, x is the interaction duration, d is the average distance between interacting robots, v is the velocity of the robot, γ is the probability of finding a puck by following other robots, δ is the probability of losing the direction to the signal source, ε is the probability of losing the direction to the signal source by congestion, $l(x)$ is the probability of turning to the broadcasted signal source, and ε is a constant.

Fig. 14 is the result of computer simulation and static feature of the equation in case of foraging with physical communication, and Fig. 15 is the case of foraging with chemical communication. As you see, the solution of equation describes the simulation results well.

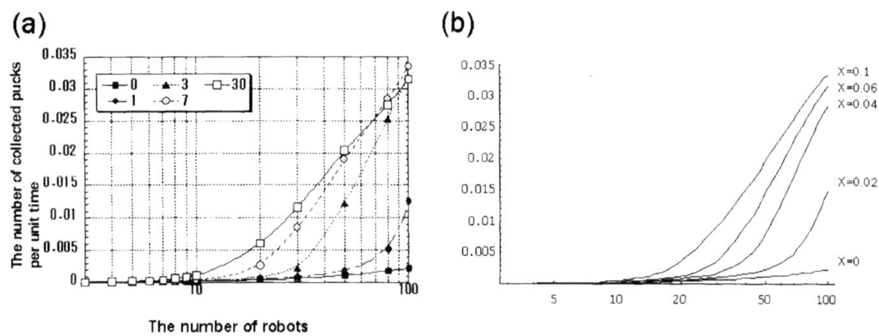

Fig. 14. The result of robot simulation (a) and static feature of the equation (b). Here a=0.0002, b=0.01, d=200, v=10, g=0.02, c=100.

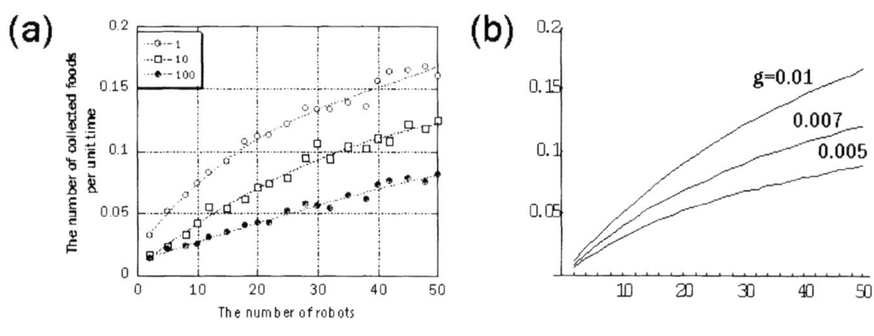

Fig. 15. The result of robot simulation (a) and static feature of the equation (b). Here x=10, a=0.0002, b=0.01, d=50, v=10, c=100.

5 Conclusion

In this paper, we treat two types of communication method, physical and chemical, and investigate a foraging behavior by multi-robot system. Firstly, the foraging by physical communication was discussed. Next, we proposed a virtual dynamic environment for autonomous robots (V-DEAR) to simulate chemical communication system such as pheromone, and discussed the foraging behavior of multi-robot system using this device. Lastly, we showed these behaviors can be treated by a common framework using a state transition diagram.

**

This work was partially supported by PRESTO, JST and a Japanese Grand-in-Aid for Encouragement of Young Scientists from the Ministry of Education, Science and Culture (No.15760291).

References

1. D. M. Gordon and M. Schwengel, *Ants at Work: How an Insect Society Is Organized*, Simon & Schuster, (1999).
2. Alberts, Felix A. Khin-Maung-Gyi, *Molecular Biology of the Cell*, Taylor & Francis, (2001).
3. E.Bonabeau, M.Dorigo and G.Theraulaz, *Swarm Intelligence - From Natural to Artificial Systems*, Oxford University Press, (1999).
4. Y.U.Cao, A.S.Fukunaga and A.B.Kahng, "Cooperative Mobile Robotics:Antecedents and Directions," *Autonomous Robots*, 4, (1997) pp.7-27.
5. L.E.Parker, "Current Research in Multi-Robot Systems", Journal of Artificial Life and Robotics, vol. 7 (to appear).
6. L. Steels, "Cooperation between distributed agents through self-organization," *Proc. of the First European Workshop on Modeling Autonomous Agents in a Multi-Agent World*, (1990) pp. 175-195.
7. A. Drogoul and J. Ferber, "From Tom Thumb to the Dockers: Some Experiments with Foraging Robots," *Proc. Simulation of Adaptive Behavior*, (1993) pp.451-459.
8. T. Balch and R.C. Arkin, "Communication in Reactive Multiagent Robotic Systems," *Autonomous Robots*, 1, (1994), pp. 27-52.
9. R. Beckers, O.E. Holland and J.L. Deneubourg, "From Local Actions To Global Tasks: Stigmergy and Collective Robotics," *Artificial Life IV*, MIT Press, (1994) pp. 181-189.
10. M.J.B. Krieger, J.B. Billeter, "The call of duty: Self-organized task allocation in a population of up to twelve mobile robots", *Robotics and Autonomous Systems*, Vol.30 (2000) pp.65-84.
11. A.T. Hayes, A. Martinoli, and R.M. Goodman, "Distributed Odor Source Localization", *IEEE Sensors*, Vol. 2, No. 3 (2002) pp.260-271.
12. K. Sugawara, M. Sano and T. Watanabe, A Study on a Foraging Behavior of Interacting Simple Robots, Journal of Advanced Computational Intelligence and Intelligent Informatics Vol.7 No.2 (2003) pp.108-114.

Part IX

Swarm Intelligence

Distributed Autonomous Micro Robots: From Small Clusters to a Real Swarm

H. Woern, J. Seyfried

Institute for Process control and Robotics (IPR), Kaiserstr. 12,
Universität Karlsruhe (TH), 76128 Karlsruhe, Germany
woern@ira.uka.de, seyfried@ira.uka.de

Abstract

In classical micro robotics, highly integrated and specialised robots have been developed in the past years, which are able to perform micromanipulations controlled by a central high-level control system. On the other hand, technology is still far away from the first "artificial ant" which would integrate all capabilities of these simple, yet highly efficient swarm building insects.

This has been the motivation of other research fields focusing on studying such swarm behaviour and transferring it to simulation or physical robot agents. Realisations of small robot groups of 10 to 20 robots are capable to mimic some aspects of such social insects, however, the employed robots are usually huge compared to their natural counterparts, and very limited in terms of perception, manipulation and co-operation capabilities.

This paper describes work currently being carried out within two EC-funded projects which aim to take a leap forward in robotics research, in distributed and adaptive systems as well as in self-organizing biological swarm systems. The first project, **MiCRoN**, will establish a small cluster of (up to five) micro robots equipped with on-board electronics, sensors and wireless power supply, while the second project, **I-Swarm**, aims at technological advances to facilitate the mass-production of micro robots, which can then be employed as a "real" swarm consisting of up to 1,000 robot clients. These clients will all be equipped with limited, pre-rational on-board intelligence. The swarm will consist of a huge number of heterogeneous robots, differing in the type of sensors, manipulators and computational power. Such a robot swarm is expected to perform a variety of applications, including micro assembly, biological, medical or cleaning tasks.

Introduction

The Institute for Process control and Robotics (IPR) at the University of Karlsruhe coordinates two European projects on micro robotics. The objective of the first project, **MiCRoN**, is the development of a multi-micro robot manipulating system prototype to handle μm-sized objects as well as smaller nano-scale objects. The system is based on a cluster (5 to 10) of small (cm^3) mobile autonomous robots. These wireless agents, each equipped with onboard electronics, cooperate within a desktop environment to execute a range of tasks associated with assembly and processing from the nano- to the micro-range. Bridging these three orders of magnitude is accomplished by the robots' actuation system capable of motion resolutions down to 2 nm and integrated scanning probe microscopes. The system comprises several subsystems such as a global positioning system to provide accurate position information (resolution of about 1 μm) [1] of each micro robot, advanced manipulating tools and a wireless power supply unit. It also includes user interfaces as well as systems for transporting μm-sized objects into and out of the working range of the robots.

Based on the intermediate promising results of this first project, a second project, **I-Swarm**, has now been launched to achieve the following strategic objectives: Realization of a "real" micro robot swarm, *i.e.*

- collective task execution
- by several hundred to a thousand autonomous micro robots manufactured by mass-production techniques
- by the collective intelligence of these robots in terms of
  - co-operation
  - collective perception
  - using integrated sensors and tools for micro manipulation.

This paper describes the work carried out so far, and the strategies envisioned to achieve the ambitious goals in the next phase to realize a first "real" robotic swarm exhibiting swarm intelligence, swarm behavior and collective perception.

Micro robots

At the IPR, micro robotics research is being pursued since 1995 [2]. A multitude of tethered, piezo-driven micro robots has been developed for various application fields [3], [4], ranging from handling of biological cells, micro assembly tasks to tele-micro manipulation in the vacuum chamber of a scanning electron microscope [5]. Figure 1 shows some of the prototypes. All of these prototypes are connected to the control system by wires. These wires supply the robots' actuators with the driving voltages, the driving signals are generated by the off-board control system. Our micro robots have a strong emphasis on object handling as compared to other projects [6], [7].

Fig. 1. Tethered micro robot prototypes developed at the IPR

Mobile micro robots have several disadvantages compared to well-established stationary micro manipulation devices in terms of speed, integration issues due to their small size and robustness. These disadvantages can only be compensated for if they are used in a cooperative manner: When several mobile micro robots perform transportation, assembly or manipulation tasks *in parallel*, the system as a whole will perform better than a stationary system, given a sufficient number of robots. This is due to the parallelization, specialized robots which make gripper exchange unnecessary, and redundancy if a single robot fails.

Additionally, micro effects prevent handling tasks to be performed as we would expect them to be performed: for part sizes below 1 mm, the surface adhesion forces (proportional to the object's surface) become dominant over the gravitational force (proportional to the object's volume). Therefore, concepts like the "helping hand" have to be employed which base on the use of at least two robots [8].

Sensor system

In order to be able to automate micro handling tasks at all, the use of a sensor system is mandatory. Only with sensor feedback, one will be able to detect handling uncertainties and failures online. For this, force, tactile and vision sensors have been researched, based on optical microscopy [9] or scanning electron microscopy [4]. In the integration of the robot system with a microscopic system (either an optical or a scanning electronic), small mobile micro robots offer obvious advantages over stationary systems. Conventional microscopes will not be outperformed in terms of resolution, magnification and image quality by miniaturized versions in the near future; however, integrated sensors for mobile micro robots are a logical in a context of swarm perception[1], but demanding task. Figure 2 shows an overview over the planned system within the framework of the **MiCRoN** project.

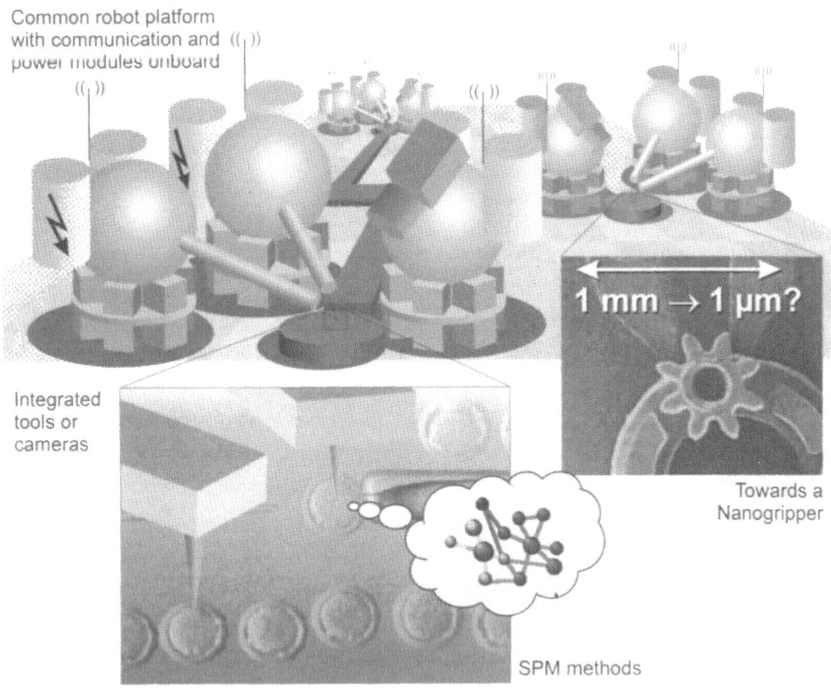

Fig. 2. MiCRoN project overview

Micro robot clusters

To be able to employ several cooperating micro robots in a cluster, the umbilical cables have to be eliminated, which is a demanding task which

[1] Where a single "seeing spot" like a microscope objective would be inappropriate

has also been addressed by other groups [7]. To reach autonomy, processing facilities have to be integrated into the robots, information transmission and power supply have to be performed without cables. To reach this goal, several research groups within the consortium are working together in the **MiCRoN** project to establish wireless power transmission by induction (performed by the Fraunhofer Institute for Biomedical Engineering (IBMT), St. Ingbert, Germany). They are also setting up an infrared communications module which will be the interface to the control and planning system currently being developed by us and the National Technical University of Athens (NTUA), Control Systems Lab, Mechanical Eng. Dept. in Athens, Greece. Other project partners are involved in actuator design [10], robot design and scene understanding. Figure 3 shows a first robot prototype, manufactured by the EPFL Lausanne [11], which is still relying on wires for power transmission, but has already an infrared (IRDA standard) link. The on-board electronics of the final robots will be integrated to make a robot of 1 cm^3 possible.

Fig. 3. First **MiCRoN** prototype (left)

Micro robot swarms

Based on the advances in micro robot autonomy, within the framework of the **I-Swarm** project, a real micro robot swarm will be realized. Here, a huge number of micro robots will act as a swarm, perform collective perception and self organization. Evolution of complex behaviors from simple, but real agents will be pursued. To reach this, an interdisciplinary approach is taken (one consortium member is active in the field of zoology) to broaden robotics research. Technological advances with high potential beyond the project are expected in the following areas:

- miniaturization of highly accurate micro robots
- new actuators
- new micro building techniques
- commercialization potential: first available mass-produced micro robot will revolutionize robotics research.

Today, the world-wide research focuses on highly sophisticated micro robots while the **I-Swarm** approach aims to find the break-even point between robot simplicity and emerging complex behavior by cooperation.

Controlling a swarm of robots

Two architectures for collective perception will be investigated. The first is a mesh or *n*-dimensional hypercube configuration, Fig. 4, left, consisting of a swarm of micro robots with equal perceptive capabilities. This is a true swarm-like configuration where the ability of each individual is equal to its neighbors'. By combining the resources gathered from its neighbors, the individual will have a larger perspective of the situation. The "depth" of perception necessary will depend on the amount of information that is required to complete a task. For example, in the case of terrain exploration, a depth level of one is probably sufficient, where each micro robot unit only needs information from its immediate neighbors. When performing highly co-operative tasks, a larger depth value may be required.

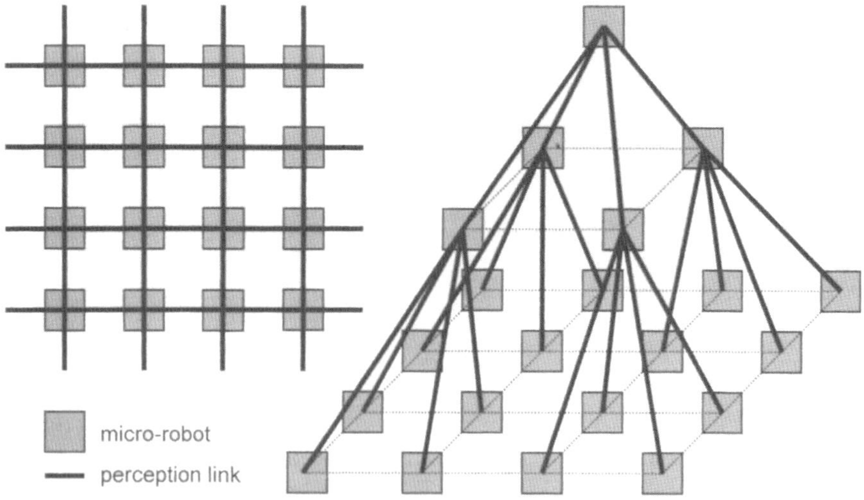

Fig. 4. Control topologies

The second configuration is a pyramidal topology, Fig. 4, right. This moves away from a purely homogeneous configuration to one that is hierarchical. In this configuration, perceptive cues are gathered from micro robots at the lowest level of the pyramid. This information is fed to micro robots higher up the chain, which in turn is again fed to other micro robots even higher up. Information passing is dictatorial, *i.e.* micro robots in a higher plane have a higher level of perception and so are given the right to command micro robots lower down the chain. The world view is held by a single micro robot (or system) at the highest level, which in the context of this configuration will be the most interesting part for investigation. To this end, investigations will be performed on how information passed by individuals lower down the chain may affect the world view at the top.

Availability of the robot hardware is crucial for such a project. Here, the advances generated in the **MiCRoN** project are a perfect starting position: Advanced, precise micro robot hardware capable to perform micro-nano handling tasks (SEM, AFM, etc.) is available and can be employed as a higher-level "queen"-like robot. Furthermore, risk reduction is performed by different approaches in parallel by the evaluation of various locomotion solutions, *e.g.* insect-like piezo or polymer actuators.

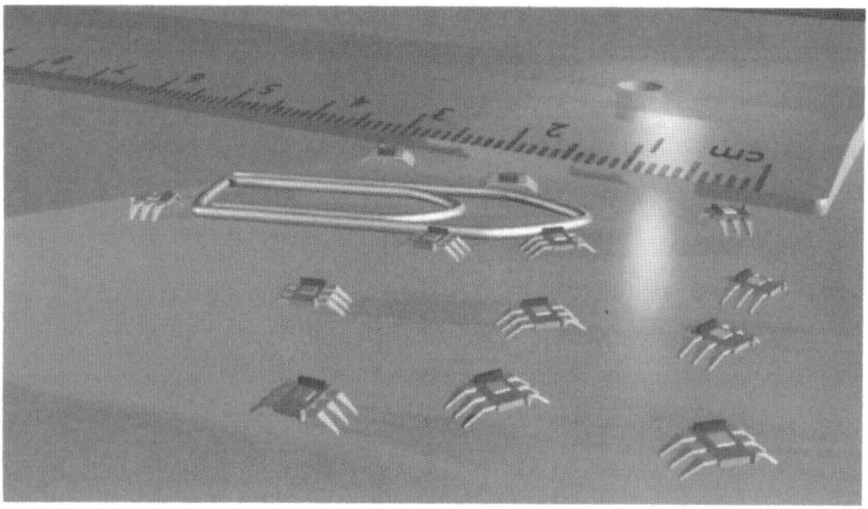

Fig. 5. Environment exploration by some swarm robots

Possible scenarios for such a micro robot swarm comprise:

- environment exploration, *cf.* Fig. 5
- grouping and moving: "queen" robot orders some robots to a specified position

- ordering robots to form geometrical patterns on the working base (*e.g.* a queue)
- surface cleaning
- object collection: swarm is ordered to collect all objects of a specified type and put them to a specific place
- object passing: transportation by passing an object from one robot to the next
- simple assembly tasks performed by co-operating robots

All these scenarios have been researched by the AI community since quite a while (research of swarm robotics, however, is still at the beginning), but implementations in the real world (and not only in simulation) were limited to groups of some tenths of real robots. The establishment of an interdisciplinary research group by European funding has now made it possible to boost swarm research ranging from modeling over software to robot hardware.

Communication in a swarm

Communication technologies which will be evaluated for their fitness to swarm robotics are the following: IR (host-robot, directional), RF (robot-robot, limited distance), transponders, mechanical contact (electrodes, antennae: mechanical forces to transmit information), inductive or capacitive coupling of neighboring (very close) robots. The expected bandwidth will be below 1 kbit/sec (which is achievable by RF), based on the functionality of each robot, and, depending on the control structure, hierarchical (queen/worker robots with different communication capabilities).

Collective perception

Sensing technologies for such a robot swarm will follow one principle: small and simple. Compatibility with the employed micro technology is imperative for the conceivable sensors: low-res image sensors (only a few pixels), silicon arrays and other light sensitive devices (discrete/array). The sensor technologies will range from vision, tactile to integrated position sensors which will give the robots (at least some of the swarm) ego-positioning capabilities (*i.e.* the robot is able to localize itself, real-time tracking of 1,000 robots by a central unit is clearly out of the question). The use of simple, tactile sensors will make scenarios like the one depicted in Figure 6 possible: a group of robots explores the boundaries of an object, and the swarm collects the information to create a representation of the environment the swarm is interacting with.

Fig. 6. Collective perception by tactile sensors

Conclusion

This paper presented work currently going on in two projects funded by the European Commission. **MiCRoN** is now in its final project year and has produced first micro robot prototypes which can be employed in small groups (of up to five robots) working cooperatively. Subsystems tackling problems with wireless power transfer and communications form the stepping stones for the new **I-Swarm** project which aims at the realization of a micro robotic swarm of up to 1,000 robots.

Acknowledgements

This research work has been performed at the Institute for Process Control and Robotics, Computer Science Department, Universität Karlsruhe (TH). The research work is being supported by the European Union (FET-open project **MiCRoN**, IST-2001-33567 and **I-Swarm**, Beyond Robotics project 507006).

References

[1] Ramon Estana, Heinz Woern: "Moire-based positioning system for micro-robots" Proc. SPIE Vol. 5144, p. 431-442, Optical Measurement Systems for Industrial Inspection III; Wolfgang Osten, Malgorzata Kujawinska, Katherine Creath; Eds.

[2] Magnussen, B.; Fatikow, S.; Rembold, U.: "Actuation in Microsystems: Problem Field Overview and Practical Example of the Piezoelectric Robot for Handling of Microobjects"; *Proc. ETFA*, Paris, 1995

[3] Seyfried, J.; Fatikow, S.; Fahlbusch, S.; Buerkle, A.; Schmoeckel, F.: "Manipulating in the Micro World: Mobile Micro Robots and their Applications", *Int. Symposium on Robotics (ISR)*, Montreal, Canada, May, 2000

[4] Woern, H.; Seyfried, J.; Fahlbusch, St.; Buerkle, A.; Schmoeckel, F.: "Flexible Microrobots for Micro Assembly Tasks", *International Symposium on Micromechatronics and Human Science (HMS 2000)*, Nagoya, Japan, October 22-25, 2000

[5] Schmoeckel, F.; Woern, H.: "Remotely controllable mobile microrobots acting as nano positioners and intteligent tweezers in scanning electron microscopes"; *Proc. 2001 IEEE Int. Conf. Robotics & Automation*, ICRA 2001, Seoul, Korea, May 21-26, 2001, pp. 3909-3913

[6] A. Almansa-Martín: "Micro and nanorobotics - present and future". In Proc. of the 35th International Symposium on Robotics, Paris, March 23-26 2004

[7] Martel, Sylvain et al.: "Three-Legged Wireless Miniature Robots for Mass-Scale Operations at the Sub-Atomic Scale"; Proc. 2001 IEEE Int. Conf. Robotics & Automation, ICRA 2001, Seoul, Korea, May 21-26, 2001, pp. 3423-3428

[8] Wörn, Heinz; Schmoeckel, Ferdinand; Buerkle, Axel; Samitier, Josep; Puig-Vidal, Manel; Johansson, Stefan; Simu, Urban; Meyer, Jörg-Uwe; Biehl, Margit: "From decimeter- to centimeter-sized mobile microrobots – the development of the MINIMAN system"; *SPIE's Int. Symp. on Intelligent Systems & Advanced Manufacturing, Conference on Microrobotics and Microassembly*, Boston, MA, USA, October 28 - November 2, 2001

[9] Buerkle, Axel; Schmoeckel, Ferdinand; Kiefer, Matthias; Amavasai, Bala P.; Caparrelli, Fabio; Selvan, Arul N.; Travis, Jon R.: "Vision-based closed-loop control of mobile microrobots for micro handling tasks"; *SPIE's Int. Symp. on Intelligent Systems & Advanced Manufacturing, Conference on Microrobotics and Microassembly*, Boston, MA, USA, October 28 - November 2, 2001, pp. pp. 187-198

[10] *T. Lilliehorn, U. Simu, and S. Johansson*, Multilayer telescopic piezoactuator fabricated by a prototyping process based on milling, Sensors and Actuators A, August 2002

[11] A. Bergander, W. Driesen, T. Varidel, J-M. Breguet: "Development of Miniature Manipulators for Applications in Biology and Nanotechnologies", Proc. Workshop "Microrobotics for Biomanipulation", pp. 11-35, IROS 2003, October 27-31, 2003, Las Vegas, USA

Modeling and Optimization of a Swarm-Intelligent Inspection System

Nikolaus Correll and Alcherio Martinoli

Swarm-Intelligent Systems Group, Nonlinear Systems Laboratory, EPFL
CH-1015 Lausanne, Switzerland
nikolaus.correll|alcherio.martinoli@epfl.ch

Abstract We present a simple, behavior-based, distributed control algorithm to inspect a regular structure with a swarm of autonomous, miniature robots, using only on-board, local sensors. To estimate intrinsic advantages and limitations of the proposed control solution, we capture its characteristics at a higher abstraction level using non-spatial probabilistic microscopic and macroscopic models. Both models achieve consistent prediction on the chosen swarm metric and deliver a series of interesting qualitative and quantitative insights on further, counterintuitive, improvement of the distributed control algorithm. Modeling results were validated by experiments with one to twenty robots using a realistic simulator in the framework of a case study concerned with the inspection of a jet turbine.

1 Introduction

In order to minimize failure of jet turbine engines, the engines have to be inspected at regular intervals. This is usually performed visually using borescopes, a process which is time consuming and cost intensive [5]. One possible solution to speed up and automatize the inspection process is to rely on a swarm of autonomous, miniature robots which could be sent into the turbine without disassembling it. While this idea is intellectually appealing and could pave the way for other similar applications in coverage/inspection of engineered or natural, regular structures, it involves a series of technical challenges which dramatically limit possible designs of robotic sensors. For instance, the shielded, complex, and narrow structure of a turbine imposes not only strong miniaturization constraints on the design, but also prevents the use of any traditional global positioning and communication system. Furthermore, a limited on-board energy budget might prevent computation of a sophisticated deliberative planning strategy and dramatically narrows sensor and communication range of our robots [3].

In this paper, we perform a series of simplifications to the turbine inspection

scenario and present one of the simplest algorithms for such a task. The robots have local, on-board sensors and a simple behavior-based controller that allows them to avoid collisions, follow a blade contour (emulating inspection for blade flaws), and move from blade to blade by exploiting the regularity of the turbine pattern and specific features of the blades (e.g., tips).
Similar to previous case studies concerned with distributed manipulation of objects (see for instance [1, 6]), we make use of non-spatial probabilistic microscopic and macroscopic models in order to understand general properties of the inspection (or coverage) problem, and to estimate and optimize performance and reliability of our approach.

2 The Case Study: Turbine Inspection

In this paper, we are not concerned with the reliable detection of flaws and progress reporting but rather with individual and group motion in the turbine scenario. For the sake of simplicity, we therefore assume that completely circumnavigating a blade is a good emulation of the scanning-for-flaws maneuver.

2.1 Simplification and Simulation of the Turbine Scenario

Figure 1, *left* shows the simulated scenario for this case study. We simplify the real 3D environment by unrolling the axis-symmetric geometry of the turbine into a flat representation with the blades as vertical extrusions. The resulting rectangular arena ($246 \times 186cm^2$) is delimited by walls (emulating the boundaries of the compressor section) on the short edges and a "wrap-around" zone (emulating the continuity of the turbine cylinder) on the long edge.

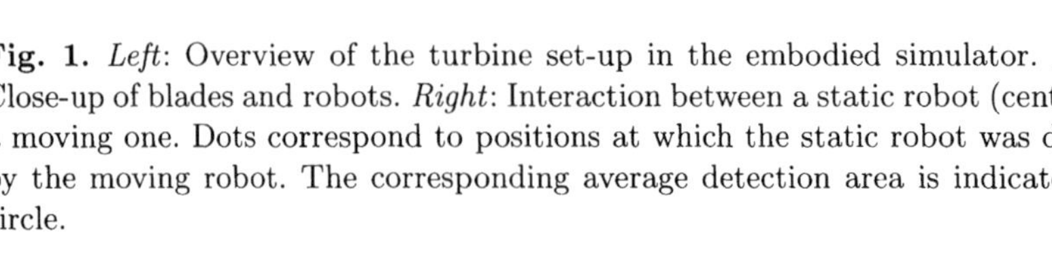

Fig. 1. *Left*: Overview of the turbine set-up in the embodied simulator. *Middle*: Close-up of blades and robots. *Right*: Interaction between a static robot (center) and a moving one. Dots correspond to positions at which the static robot was detected by the moving robot. The corresponding average detection area is indicated by a circle.

We implemented this simplified turbine environment using Webots 4.0 [7], a realistic, multi-robot, embodied, sensor-based simulator. In Webots, simulated sensors and actuators are characterized by precise, user-definable non-linearities and noise—in our simulations, all sensors and actuators on-board are characterized by ±10 percent white noise. As shown in previous publications, this simulator can provide realistic results (e.g. [1, 6]) when compared with real robot experiments. It is worth noting that to allow comparison with real robotic experiments [3], we limit the simulated experimental setup to 16 blades in four stages, and the maximum number of robots to 20.

2.2 The Behavior-Based Robot Controller

The behavior of a single robot is determined by a schema-based controller [2] that tightly links the platform's actions to sensor perception while using as little representational knowledge of the world as possible. For a schema-based controller, behavioral responses are represented by vectors generated from local potential fields, and behavioral coordination is achieved by vector addition. Sequencing of behaviors is achieved by a dynamic action-selection mechanism based on two internal timers which are set and reset by the schemas.

The overall behavior of a robot can be summarized as follows (see Figure 2,

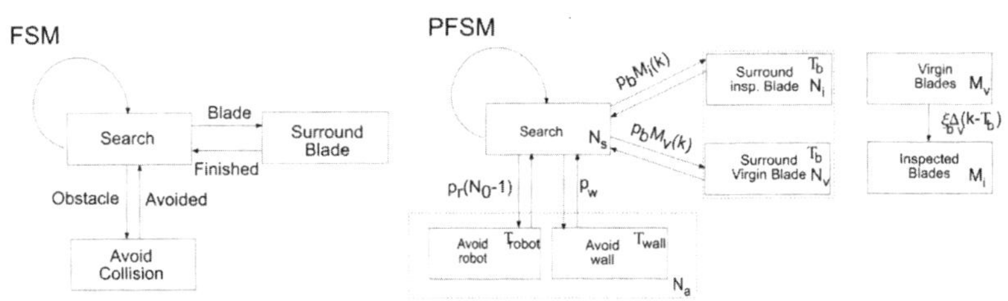

Fig. 2. *Left*: The high-level behavioral flowchart of the robot controller as a deterministic Finite State Machine (FSM). *Right*: The corresponding Probabilistic FSM used in the models, which captures details of interest of the schema-based controller.

left). The robot searches for blades throughout the compressor section. The robot avoids obstacles (teammates and walls) when it is in search mode, and tries to remain on its trajectory while scanning a blade. Teammates (cylindrical shapes) can be reliably differentiated from walls and blades (flat surface) just using on-board distance sensors. We assume that another on-board, local sensor can allow the robot to differentiate between walls (limits of the compressor sections) and blades. As soon as the robot detects a blade, it starts to follow the contour emulating a scanning-for-flaws maneuver and sets an internal timer (T_{max}) which it will later check in order to assess whether or not a pre-established number of tours of the blade has been carried out. In order to bias the robot's trajectory without using any sophisticated navigational mechanisms, the robot can only leave a blade at one of its two tips, which are recognized by a specific sensorial pattern generated by the robot's on-board distance sensors.

3 Microscopic and Macroscopic Models

The central idea of the probabilistic modeling methodology is to describe the experiment as a series of stochastic events with probabilities computed from the interactions' geometrical properties and systematic experiments. Consistent with previous publications [1, 6], we can use the controller's FSM depicted in Figure 2 as blueprint to devise the Probabilistic FSM (PFSM or Markov chain) representing an individual agent at the microscopic level or the whole swarm at the macroscopic level. At the microscopic level, a specific state represents the actual mode a specific individual is in, while a state at the macroscopic level defines the average number of individuals in the same mode. The state granularity can be chosen to capture details of the robot's controller and environment which influence the swarm performance metric; in our case, the time needed to complete the inspection of all the blades. The overall PFSM for the system is represented graphically in Figure 2, *right* using two coupled PFSMs, one representing the robot(s) and one representing the shared turbine environment.
We present the results of the microscopic and the macroscopic models for the following reasons. First, although the microscopic-to-macroscopic mapping is currently linear and therefore no major discrepancies between the predictions of the two types of models can arise, quantization in the number of individuals or blade tours might generate some numerical differences (see Section 4). Second, the time-to-completion metric cannot be captured easily at the macroscopic level due to numerical effects in the integration of the difference equations. Therefore, a precise criterion based on statistics generated by the microscopic model has to be considered in order to obtain good correspondence between the predictions of the two models.

3.1 Modeling Assumptions

As is more extensively detailed in [1, 6], the modeling methodology relies on three main assumptions. First, coverage of the arena by the group of robots is uniform and robots' trajectories and objects' positions in the arena do not play a role in the metric of interest. Second, a robot's future state depends only on its present state and how much time it has spent in that state (semi-Markov property). Third, agents change their state autonomously but synchronously to a common clock whose time step has been chosen to capture, with sufficient precision, all time delays considered in the system as well as changes in the metric of interest. Notice that, although time in the models is discretized, since state changes at the level of individual agents are probabilistic, these models adequately approximate the overall behavior of an asynchronous swarm.

3.2 Characterization of Models' Parameters

All our models are characterized by two categories of parameters: state-to-state transition probabilities and behavioral delays. In contrast with previous publications [1, 6], we do not assume any coupling between these two categories of parameters, and we propose a new way of computing and calibrating them based on the concept of *encountering rates*, as suggested in [4]. Consistent with previous publications, we compute the transition probabilities from one state to another based on simple geometrical considerations about the interaction. However, here we introduce a clear separation between *geometric detection probabilities* and *encountering probabilities*. We call geometric detection probability the probability that a robot is within the detection area of a certain object. The detection area of an object is determined by its physical size, the sensory configuration, and processing used by the robot to reliably detect it (see Figure 1, *right*). After defining the contours of the detection area A_i for an given object i, we calculate its geometric detection probability g_i by dividing A_i by the whole arena area A_a. We can then calculate the corresponding encountering probability, i.e. the probability of encountering the object i per time step, using the corresponding encountering rate r_i (in s^{-1}). The conversion factor from geometric detection probabilities to encountering rates is given by the average robot speed v_r ($6.5\frac{cm}{s}$), its detection width w_r and the detection area A_s of the smallest object in the arena (in our case a robot). The detection width is defined as twice the maximum detection distance of the smallest object in the arena, measured from center of the robot to the center of the object, here given by $2R_s = 15.2cm$ with $A_s = R_s^2\pi$. Equation 1 shows how to compute the encountering probability for the object i given the geometric detection probability g_i:

$$p_i = r_i T = \frac{v_r w_r}{A_s} g_i T, \tag{1}$$

where T is the time step characterizing our time-discrete models. In this paper, we discretize the different average durations of interactions so that changes in our chosen metric are described with sufficient precision using $T = 1s$. Numerical values used for the model parameters can be verified using systematic experiments with real robots [6] or realistic simulations (see [1] and Section 4.1).

3.3 Mathematical Description of the Macroscopic Model

From Figure 2, *right* we can derive a set of difference equations (DEs) to capture the dynamics of the whole system at the macroscopic level. We formulate one DE per considered state (either in the robotic or in the environmental Markov chain) and exploit conservation of the number of robots and

the number of blades to replace two of the DEs respectively.
Given M_0 blades and N_0 robots, the number of robots covering virgin blades N_v, and inspected blades N_i; the number of robots in obstacle avoidance N_a, and the number of robots in search mode N_s are given by equation 2-5 (compare also Figure 2); the number of virgin blades M_v and the number of inspected blades M_i are calculated by equations 6-7:

$$\begin{aligned} N_s(k+1) &= N_s(k) - \Delta_v(k) - \Delta_i(k) - \Delta_r(k) - \Delta_w(k) + \Delta_v(k-T_b) \\ &\quad + \Delta_i(k-T_b) + \Delta_r(k-T_r) + \Delta_w(k-T_w) \end{aligned} \tag{2}$$

$$N_a(k+1) = N_a(k) + \Delta_r(k) + \Delta_w(k) - \Delta_r(k-T_r) - \Delta_w(k-T_w) \tag{3}$$

$$N_v(k+1) = N_v(k) + \Delta_v(k) - \Delta_v(k-T_b) \tag{4}$$

$$N_i(k+1) = N_0 - N_s(k+1) - N_a(k+1) - N_v(k+1) \tag{5}$$

$$M_v(k+1) = M_v(k) - \xi_b \Delta_v(k-T_b) \tag{6}$$

$$M_i(k+1) = M_0 - M_v(k+1) \tag{7}$$

where k represents the current time step (and absolute time kT); $k = 0 \ldots n$, where n is the total number of iterations (and therefore nT the end of the experiment). p_b, p_r, and p_w represent the encountering probabilities of blades, robots, and walls, respectively. T_b, T_r, and T_w define the average times needed for circumnavigating a blade, avoiding a teammate, and avoiding a wall. ξ_b represents instead the percentage of blade coverage when a scanning robot circumnavigates a blade.
The Δ-functions define the coupling between state variables of the model and can be calculated as follows:

$$\Delta_v(k) = p_b(M_v(k) - N_v(k))N_s(k) \tag{8}$$

$$\Delta_i(k) = p_b(M_i(k) + N_v(k))N_s(k) \tag{9}$$

$$\Delta_r(k) = p_r(N_0 - 1)N_s(k) \tag{10}$$

$$\Delta_w(k) = p_w N_s(k) \tag{11}$$

The initial conditions are $N_s(0) = N_0$ and $N_a(0) = N_v(0) = N_i(0) = 0$ for the robotic system (all robots in search mode) while those of the environmental system are $M_v(0) = M_0$ and $M_i(0) = 0$ (all blades virgin). As common use for time-delayed DE, we assume $\Delta_x(k) = N_x(k) = M_x(k) = 0$ for $k < 0$.
For instance, we can interpret the first DE (equation 2) as follows. The average number of robots in the searching state is decreased by those that start to cover a virgin blade or an inspected blade and those that start avoiding either a teammate or a wall; it is increased by all robots resuming searching after either an inspection or an obstacle avoidance maneuver, each of them characterized by a specific duration. The other state equations can be interpreted in a similar way.
The probability p_l of accidently leaving a blade at one of its tips before T_{max} has expired and the control design parameter T_{max} both influence the mean time needed for inspecting a blade T_b, as well as the percentage of inspection per blade ξ_b. For T_{fb}, the average time required to fully circumnavigate a blade once and only once, we can calculate T_b and ξ_b as follows:

$$T_b = -\frac{1}{4}T_{fb} + \sum_{i=0}^{I} \frac{1}{2}T_{fb}(1-p_l)^i, \qquad \xi_b = \frac{T_b}{T_{fb}} \tag{12}$$

with $I \geq 2\frac{T_{max}}{T_{fb}} - \frac{1}{2}$ and $I \in \mathbb{N}$, reflecting the maximal "allowed" tip encounters that are defined by T_{max}. Here, we assume that the robot covers, on average, 25% of the blade it is attached to and has a probability of $(1 - p_l)$ of covering another 50% before encountering the next tip. This process might continue for multiple half tours (from tip to tip) and is only bounded by I. Finally, our metric for evaluating the performance of the swarm is the time to complete the inspection, nT. To compute nT, $M_v(n) = 0$ (all blades are inspected) is an easy condition to apply in the embodied simulator and in the microscopic model. However, in the macroscopic model, this represents a limit condition as $\lim_{k\to\infty} M_v(k) = 0$. Therefore, we solve the DEs numerically for $M_t(n) = \mu$ where μ is the expected values from the microscopic model.

4 Results and Discussion

In contrast with previous experiments in the distributed manipulation class where obstacles were always axis-symmetric and either movable [1] or much smaller [6] than the immovable blades considered in this case study, the metric to evaluate swarm performance in the inspection task appears to be more heavily influenced by the distribution of the robots, while still being essentially non-spatial. For the above reasons, we first validate whether or not the assumptions of our modeling methodology are still valid in this case study and we compare measured with computed model parameters.

4.1 Characterization of Models' parameters

We carried out two series of experiments in order to validate our method of computing geometric detection probabilities and encountering rates, respectively. In the first series of experiments, we measure the ratio of time that a robot spends within and outside a specific area in a fully enclosed obstacle-free arena. This area corresponds to the detection area of the objects later considered in the embodied simulator, blades and robots. We consider different shapes, sizes, and positions in a rectangular arena of the same size as our turbine scenario (see Table 1).

We observe a good match between measured and computed geometric probability of an object with fixed size but varying shape and position. We also notice that the standard deviation in the measurement accuracy is approximately proportional to the object size.

In a second series of experiments, we measured the actual encountering rates using the embodied simulator and compared these with encountering rates computed by Equation 1. Biases due to the interaction duration of the robot (collision avoidance) with specific objects are considered separately in the

Table 1. Comparison of measured and computed geometric detection probabilities for different shapes (square, rectangular, circular) and two different sizes (robot and blade). Measured values represent mean and standard deviation for three different locations during 100h of simulated time. All values are in percent.

Size	Square	Rectangular	Circular	All Shapes	Geo. Prob.
Blade	1.69 ± 0.17	1.72 ± 0.17	1.32 ± 0.62	1.57 ± 0.21	1.52
Robot	0.31 ± 0.04	0.3 ± 0.03	0.32 ± 0.02	0.31 ± 0.03	0.31

model (compare equation 2-5) and have therefore been eliminated from this measurement. We considered four different scenarios: two empty arenas with different boundary conditions (fully enclosed by walls; walls on two sides while "wrap-around" on the other sides); fully enclosed arena with a blade in its center; fully enclosed arena with a robot in its center. The results of this second series of experiments are reported in Table 2.

Table 2. Measured and computed encountering rates for some objects, and their mean interaction time with standard deviation. Each experiment lasted 25h of simulated time.

Object	Measured enc. rate	Computed enc. rate	Interaction time
Half-Wall	$0.0125s^{-1}$	$r_w = 0.0148s^{-1}$	$T_w = 5.38 \pm 1.4s$
Full-Wall	$0.0423s^{-1}$	$r_w = 0.042^{-1}$	$T_w = 5.47 \pm 1.52s$
Blade	$0.0059s^{-1}$	$r_b = 0.0085s^{-1}$	$T_{fb} = 40.29 \pm 8s$
Robot	$0.0021s^{-1}$	$r_r = 0.0017s^{-1}$	$T_r = 3.9 \pm 0.43s$

While in the first series of experiments we observe that the geometric probability is shape and position invariant (see Table 1), the accuracy of prediction of encountering rates is changing for different kind of objects (see Table 2). A possible reason for this observation is that the embodiment of an object causes a partitioning of the robot's distribution over time in the arena and therefore, as a function of object shape and size, its trajectories are affected. When multiple robots are present in the arena (results not shown here), the influence of individual trajectories is weakened, the robot distribution becomes more uniform, and the discrepancies between computed and measured encountering rates tend to vanish.

4.2 Controller Optimization using the Microscopic Model

Before running a whole series of experiments with our embodied simulator, we were interested in evaluating the influence of our control design parameter T_{max} on the swarm metric as a function of p_l, the probability of accidentally leaving a blade at its tips. We observe (Figure 3), that the (intuitive) choice of

$T_{max} = 40s$ has minimized our metric for $p_l = 0$. However, the model suggests that our controller can be further optimized when $p_l \approx 0.2$ for $T_{max} = 40s$ or $p_l \approx 0.4$ with $T_{max} \to \infty$ respectively. Although this results are counterintuitive, one can imagine that leaving blades earlier may be a tradeoff between increased exploration and the risk of prematurely leaving a virgin blade and thus might improve task performance.

4.3 Swarm Performance Metrics

We estimated time to completion for 16 blades for $p_l = 0$ and $T_{max} = 40s$ by performing 10000 runs using the microscopic model and solving numerically the macroscopic model. We considered team sizes from 1–20 robots and used the *computed* rates from Table 2 as model parameters. To validate model predictions, we performed 100 runs each for team sizes of 1,2,5,10,16, and 20 robots in the embodied simulator. In order to come closer to our assumption of spatial uniformity (3.1), robots were initially distributed randomly in the turbine. Figure 3, *right* depicts the resultant completion time as a function of swarm size.

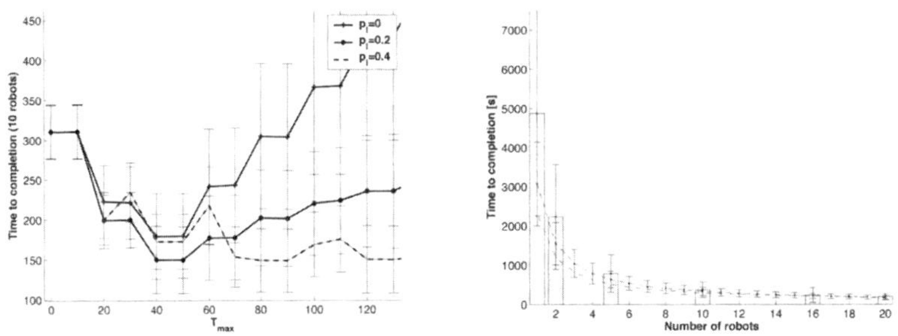

Fig. 3. Results obtained using the microscopic model and the embodied simulator are represented by their mean and corresponding standard deviation. *Left*: Time to completion as a function of T_{max} and different values of p_l. The results have been obtained with the microscopic model. *Right*: Modeling predictions (microscopic and macroscopic) compared with results gathered using the embodied simulator for the time to completion (16 blades) vs. team size (1 to 20 robots).

We observe that for increased swarm size, model predictions for experiments using embodied simulation improve. We believe that this is because in a structured environment an individual robot's trajectory does not satisfy our assumption of spatial uniformity in the distribution of robots. Similar to results of Section 4.1, increasing the team size weakens the effect of individual trajectories and increases the quality of prediction. Note also that, the microscopic model achieves slightly better quantitative results than the macroscopic one when the swarm size is small.

5 Conclusion and Future Work

We have proposed a swarm-intelligent, distributed algorithm for collective inspection of an engineered regular structure. Although the algorithm is fairly simple, its robustness in the presence of noise is remarkable and its computational requirements are extremely low. Furthermore, we demonstrate that our modeling methodology, which has already proven a valuable tool in distributed manipulation experiments, also yields valid predictions in this distributed sensing task. Finally, we explain how models can help to adjust individual (control) parameters to optimize swarm performance. In the future, we would like to use local communication to achieve a more explicit collaboration among the robots and merge the results achieved here with those obtained with real robots [3]. Furthermore, we believe that more detailed information about the geometric structure of the environment should be incorporated in the models in order to design a more effective swarm-intelligent inspection system. For instance, it should be possible to quantify and exploit the regularity of the environment to systematically shift the swarm through the turbine.

Acknowledgments

This work has been initiated at California Institute of Technology, primarily supported by the NASA Glenn Center and in part by the Caltech Center for Neuromorphic Systems Engineering under the US NSF Cooperative Agreement ERC-9402726. Both authors are currently sponsored by a Swiss NSF grant (contract Nr. PP002-68647/1).

References

1. W. Agassounon, A. Martinoli, and K. Easton. Macroscopic modeling of aggregation experiments using embodied agents in teams of constant and time-varying sizes. *Autonomous Robots*, 17(2-3):163–191, 2004.
2. R. Arkin. *Behavior-Based Robotics.* The MIT press, Cambridge, Massachusetts, 2000.
3. N. Correll and A. Martinoli. Collective inspection of regular structures using a swarm of miniature robots. In *Proceedings of the 9th International Symposium of Experimental Robotics (ISER).* Singapure. Springer Tracts for Autonomous Robotics (STAR). To Appear, 2004.
4. K. Lerman and A. Galstyan. Mathematical model of foraging in a group of robots: Effect of interference. *Autonomous Robots*, 2(13):127–141, 2002.
5. K. Martin and C.V. Stewart. Real time tracking of borescope tip pose. *Image and Vision Computing*, 10(18):795–804, July 2000.
6. A. Martinoli, K. Easton, and W. Agassounon. Modeling swarm robotic systems: A case study in collaborative distributed manipulation. *Int. Journal of Robotics Research*, 23(4):415–436, 2004.
7. O. Michel. Webots: Professional mobile robot simulation. *Journal of Advanced Robotic Systems*, 1(1):39–42, 2004.

Scalable Control of Distributed Robotic Macrosensors

Brian Shucker and John K. Bennett

Department of Computer Science
University of Colorado at Boulder
{shucker, jkb}@cs.colorado.edu

Abstract

This paper describes a control mechanism by which large numbers of inexpensive robots can be deployed as a distributed remote sensing instrument, and in which the desired large-scale properties of the sensing instrument emerge from the simple pair-wise interactions of its component robots. Such sensing instruments are called *distributed robotic macrosensors*. Robots in the macrosensor interact with their immediate neighbors using a virtual spring mesh abstraction, which is governed by a simple physics model. By carefully defining the nature of the spring mesh and the associated physics model, it is possible to create a number of desirable global behaviors without any global control or configuration. Properties of the resulting macrosensor include arbitrary scalability, the ability to function in complex environments, sophisticated target tracking ability, and natural fault tolerance. We describe the control mechanisms that yield these results, and the simulation results that demonstrate their efficacy.

1. Introduction

Robots are currently used in a broad range of remote sensing applications. The use of robotic platforms to gather information provides several advantages: robots can be used in hazardous situations without risk of injury to humans; robots are capable of reaching areas that are infeasible to reach with a human observer; and robots can be inexpensive to operate. In many domains, e.g., space exploration and military reconnaissance, robots are an effective tool for discovering and tracking targets of interest.

Advances in integration, actuator design and power management have resulted in the availability of mass-produced, inexpensive robotic components. It is now possible to build small, autonomous robots in large numbers at relatively low cost. The resulting potential to deploy robotic sensors on a large-scale creates the opportunity to explore a new type of remote sensing. Hundreds, or thousands, of robots can potentially be deployed to cover and explore an area, record data, and track targets of interest. However, while it is now relatively simple to deploy large numbers of robotic sensors, the coordination and control of the activities of these robots presents a number of challenging problems, including scalability, autonomy, coverage, flexibility of deployment, fault-tolerance, and security.

We have developed a control algorithm for distributed robotic macrosensors (DRMs) to address these concerns. DRMs consist of large numbers of robots whose individual "instinctive" behavior directs them toward a common goal. The properties of the macrosensor emerge as a result of simple local interactions between individual component robots. By defining these interactions carefully, we create a scalable macrosensor with sophisticated overall behavior. The resulting properties of these macrosensors include the following:

1. Arbitrary scalability - A single macrosensor may contain any number of robots, from a single robot to tens of thousands of robots.

2. Automatic operation - deployment and operation is fully automatic; that is, there is no global control (or even global knowledge from the standpoint of the control algorithm).
3. Starting state independence - The macrosensor can deploy from an arbitrary starting state. For example, robots may or may not start together in a cluster.
4. Exploration - Prior knowledge of the target environment is not required. The macrosensor explores unknown areas automatically, mapping the environment as it extends its deployment.
5. Coverage - The macrosensor deploys in such a way as to efficiently cover the area of interest, regardless of the number of robots deployed.
6. Target tracking and mapping - The macrosensor tracks moving targets within the area of interest. Targets may be discrete (such as a vehicle or person), or diffuse (such as a chemical cloud or fire).
7. Fault tolerance - The macrosensor's overall capabilities degrade gracefully with the loss of an arbitrary number of robots.
8. Extendibility - The macrosensor's capabilities increase gracefully with the addition of new robots of the same or different type.
9. Security - Compromise of individual robots, even in significant numbers, results in little gain for an adversary.

By comparison, centralized control mechanisms do not scale adequately, and existing distributed control mechanisms have serious practical limitations, especially with respect to complex environments and unknown initial distributions. Many existing control methods do not allow new robots to be added easily to a system that is already deployed. To our knowledge, robotic tracking of diffuse targets (as opposed to tracking of discrete objects) has not been addressed previously. Security has been addressed in the related context of sensor networks[6], but existing sensor network security protocols are still in the research stage, and do not address all of the relevant issues pertaining to mobile robots. Our work is intended to overcome these limitations.

We envision a variety of applications for distributed robotic macrosensors. Search and rescue operations can benefit from a mobile, rapidly deployable sensor capable of covering a wide area. Such sensors could discover persons in need of assistance, or map sources of danger such as wildfires. Macrosensors can also be deployed in order to quickly secure an area, or to detect and track intruders or other threats. Large numbers of tiny robots could be dropped from an aircraft or vessel in order to collect information about an area that is difficult to reach. For example, such deployments could be used to monitor a border. These macrosensors could be deployed rapidly into a new area because the robots' inherent mobility removes the need for precise manual placement.

Macrosensors also have significant scientific potential. For example, a macrosensor could perform continuous, high-resolution mapping of atmospheric or oceanic phenomena. Unlike a static sensor network, the nodes in a robotic macrosensor network need not be manually placed, so the macrosensor may be deployed rapidly in order to monitor an emergent event. Macrosensors can also be deployed in areas that are largely unreachable by humans, such as the surface of Mars.

2. Virtual Spring Mesh Based Control

Our approach to deployment and coordination of DRMs employs a virtual spring mesh as the underlying control mechanism. Virtual spring meshes are an extension of virtual physics-based control. Virtual physics-based robot control is inspired by natural phenomena and has been investigated primarily in the context of swarm robotics [10,11,14,25]. The general idea is that each robot is treated as a particle in a simulated physical system, complete with virtual forces and rules of motion. While the forces only exist in simulation, the robots act in the real world as if the forces were real. The object is to define virtual forces and rules of motion in such a way that the local interactions between robots result in desirable global behavior.

Such systems exhibit several desirable characteristics. Each robot chooses its actions based only upon local information and a simple set of rules, but tends to act in a manner that contributes to a common goal. Thus, there is global cooperation without any global control, which makes the system both fault-tolerant and scalable. In our case, the macrosensor itself emerges as a result of the large-scale interactions of individual, nearly stateless components.

Previous attempts at virtual-physics based systems have generally focused on potential fields of some kind, using force fields analogous to gravity or the electromagnetic force. In contrast, the proposed virtual spring mesh only makes use of explicit connections between robots. More precisely, if robots are represented as vertices in a graph and force is transmitted through edges, the spring mesh is not a fully connected graph (even locally). Instead, virtual springs are created to transmit force only between selected adjacent pairs of robots. As we will describe further, besides providing efficiency gains, this approach allows for sophisticated control over the behavior of the group of robots that comprise the macrosensor. Since the spring mesh is created through a series of only local interactions between robots, robots are not required to have any explicit knowledge of the environment, or of other robots, beyond their immediate vicinity. All of the desired large-scale properties of the macrosensor emerge from these local interactions.

We chose the spring metaphor in part because the spring virtual force increases with error, which results in desirable control properties, and because it is beneficial for each connection to have natural length. If a pair of robots is too close together, the spring between them acts to push them apart; if they are too distant, it acts to pull them closer. This corrective force is directly proportional to error, which is defined as the difference between the spring's current length and its natural length. As with real springs, each virtual spring in the mesh has a natural length and a spring constant (that represents the "stiffness" of the spring). These parameters may be different for each robot.

2.1 Spring Formation

Since the proposed approach requires the explicit creation (and destruction) of virtual springs to transmit virtual forces between pairs of robots, the success of the macrosensor relies heavily upon the algorithm used for determining which pairs of robots to connect. We have developed and, using simulation, have evaluated several candidate algorithms for this purpose. An approach that we have found to be both effective and computationally efficient is based upon simple geometric relationships between robots. We describe this algorithm below.

A robot R will create a spring connection with its neighbor S if for every other neighbor T, the interior angle RTS is acute. This relation is simple to compute, is symmetric, and ensures a (locally) connected and planar graph. If all springs have the same natural length and stiffness, the mesh will have a total energy of zero when the robots form a hexagonal lattice, with the distance between each pair of robots equal to the springs' natural length. This is the only such formation where all angles are acute and all lengths are equal, so it is the only zero-energy configuration. Formal proofs of all these properties do exist, but are omitted here for the sake of brevity.

A hexagonal lattice is in fact the shape that is observed when running the algorithm in simulation; results are typically within 2% of optimum, as measured by the mean spring length of the final state. A hexagonal lattice is a practical arrangement with which to cover an area, because it provides uniform spacing between the robots. Figures 1 and 2 depict simulation output in which a random initial distribution of robots is drawn into a hexagonal lattice.

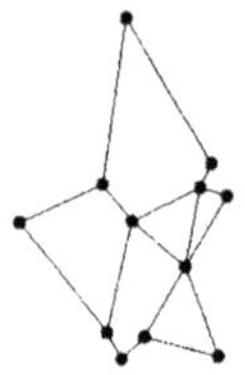

Fig. 1. Example of spring formation in a small, randomly distributed group of robots

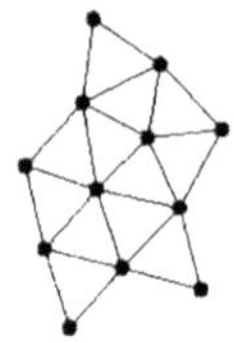

Fig. 2. Final configuration of the same group of robots

We also employ a damping force that acts against the motion of each robot. The damping force serves to improve convergence time, i.e., the time it takes for the robot macrosensor to reach a stable state (at least temporarily). For a sufficiently large area, the basic spring and damping forces are sufficient to ensure convergent behavior--that is, the total amount of energy will decrease over time, and the system will reach a static state within a short period. More precisely, the total energy will decay exponentially, with the decay constant determined by the magnitude of the damping force (which is an adjustable parameter). Figure 3 depicts simulation output of an experiment in which 50 robots start in a tight cluster, and then deploy with a desired spring length of 50. Our simulator enforces a top speed for each robot, so the observed kinetic energy is being additionally reduced early in the simulation run, when the spring model would otherwise result in much higher energies. Also, the simulation includes a velocity "dead-band" such that any speed less than a certain limit is set to zero. This causes the robots to reach zero speed exactly, rather than just asymptotically approaching zero. A potential side effect is that the final configuration is slightly sub-optimal (from the standpoint of coverage), since very small adjustments are ignored. The final state in this example was a stable hexagonal lattice.

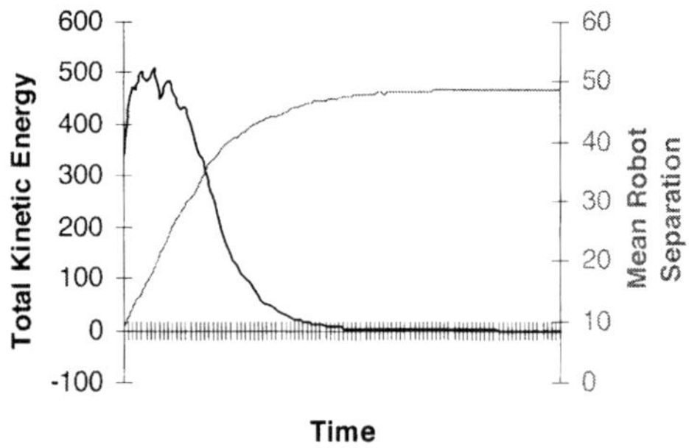

Fig. 3. Kinetic energy and mean robot separation as a function of time

It is possible to create formations other than a hexagonal lattice, either by adjusting spring parameters, altering the spring creation algorithm, or dividing the robots into different types of virtual particles, with rules of interaction that depend on the types of the participants. Various formations may have advantages for different applications; for example, square formations may be useful inside a building where the obstacles tend to have rectangular shapes.

2.2 Exploration

Real-world environments can be complex, containing walls and other impassable barriers. Exploring and mapping such environments has been the focus of a considerable amount of research [13,14,16,18,24,26]. Our approach to robot deployment builds upon this prior work, but differs from this work in that we seek to explore complex environments without the use of global information.

Environments that include concave spaces challenge most existing control algorithms. A narrow hallway may conceal a large unexplored room on the other side. In this case the locally optimal solution of simply spreading the robots out is not desirable, since this approach would not move many robots through the hallway.

Rather than building an explicit map and directing robots through a high-level protocol, we incorporate an additional local force, called the "exploration force." The exploration force draws robots towards areas that are visible but not occupied by another robot. As a robot is drawn towards such open areas, it "pulls" other robots with it through its spring connections. This causes the mesh to expand into unexplored areas in a manner similar to a fluid flow. This approach effectively creates a high potential in areas that are well covered, and a low potential in areas that are unexplored. As the robots flow into a new area, the potential of this area increases until the flow stops. Robots will continue exploring until the potential is equal in all areas (that is, there is no visible area that is not covered), which will occur when the environment is covered uniformly.

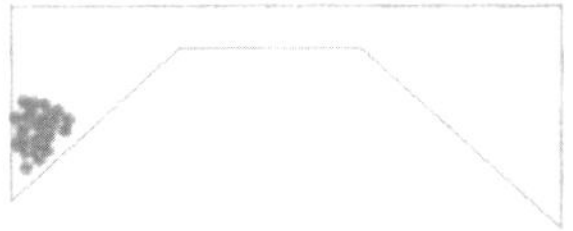

Fig. 4. Robots in a cluster in a concave environment: initial configuration

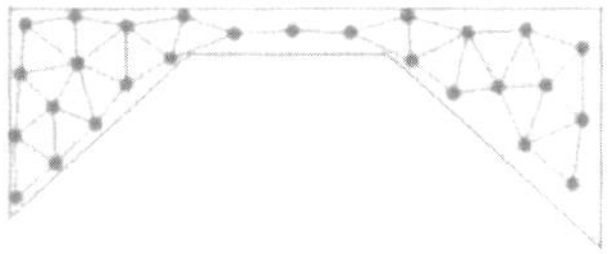

Fig. 5. Final configuration

The exploration force may take on several forms. Our current implementation simply adds a force in the direction of the longest apparent unobstructed path. Other choices include forces based on information-gain criteria, as well as adjustments to spring constants in order to draw "follower" robots. The exact nature of the exploration force influences global properties of the macrosensor, including the speed of deployment and the response to a situation in which there are insufficient robots to cover the entire area of interest. Characterizing various choices for the exploration force is a promising area for future research. Figures 4 and 5 depict simulation results of the simple exploration force in action, demonstrating that this simple concept handles concave areas.

2.3 Target Tracking

While the exact definition of a "target" is application-dependent, it is possible to classify targets into two basic categories: point targets and diffuse targets. Point tracking involves the detection and monitoring of a discrete target or targets; that is, targets with positions (which may change) that can be represented as points. In applications such as security, search and rescue and military reconnaissance, the people, vehicles or other objects are considered to be point targets. Point targets can be easily addressed within the spring mesh framework. We simply add an attractive force that draws robots toward these targets.

Since intercepting the targets is likely to be appropriate in many applications, our current implementation of the point target force is designed to match the robot's velocity with a vector that will intercept the target.

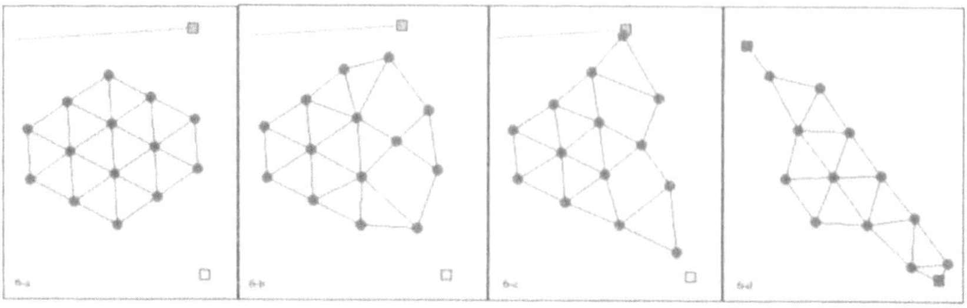

Fig. 6. Target tracking sequence.

Figure 6 shows simulation results of a small group of robots intercepting two point targets. The upper target is moving from right to left along the indicated path, at a speed that is 60% of the top speed of the robots. As seen in this example, targets are intercepted aggressively, and the spring mesh deforms in order to keep all of the robots connected with the desired spacing.

Diffuse target tracking is somewhat more complex, as these targets do not have a well-defined location. Heat plumes from a wildfire, or radioactive or chemical clouds are examples of diffuse targets. Here, in addition to identifying the location of the target, we also want to know both the extent of the target substance and its density gradient. We also may wish to increase the density of sensor coverage in the vicinity of a diffuse target. For example, when tracking a chemical plume, we may want to precisely map the plume extent in areas of significant concentration, while sacrificing detailed information about areas of lower concentration or about areas outside the plume. The spring mesh model is well suited to adaptive robot density, since each robot can control its own spring parameters. Robots that detect the desired diffuse target simply shorten their springs, thus drawing in their neighbors to areas of higher concentration. Figure 7 depicts simulation results showing this process in action.

With a diffuse target it may be desirable to penetrate the target (as in the chemical plume example), or only to tightly surround the target and contain it (as with a fire). Either of these behaviors can be achieved by making the target attractive or repulsive, respectively. Attractive targets will draw a dense group of robots into the target area. Repulsive targets will cause a tight group of robots (with shortened springs) to stay just outside of the target area, since they attract each other but are repelled from the target itself.

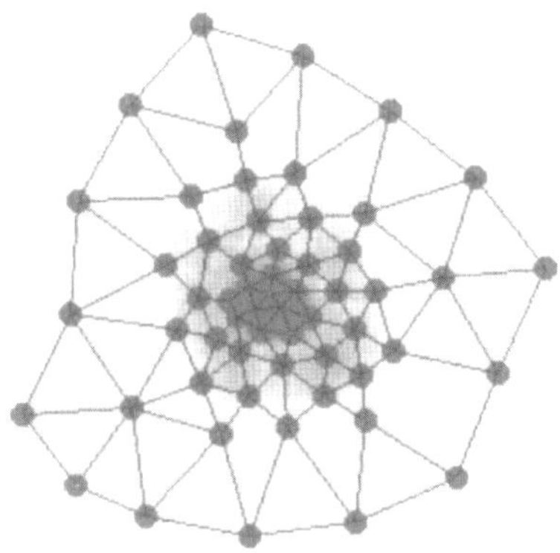

Fig. 7. Robots in the target area shorten their springs to increase density in the area.

3. Discussion

The virtual spring mesh approach is inherently scalable. The complexity of the algorithm running on each robot increases with the number of locally visible robots and targets (that is, the number of neighboring robots and targets that are currently in sight), but significantly, *complexity is independent of the total number of robots*. In fact, robots are not explicitly aware of the existence of any other robot that is not locally visible. For this reason, there is no particular limit on the number of robots that may be members of a single macrosensor. Additionally, the total area covered by the macrosensor increases linearly with the number of member robots, which satisfies our goal of extensibility.

Communication overhead of the virtual spring mesh is low. Since the spring-formation algorithm and force computations are symmetric, it is not necessary for robots to communicate with each other in order to form springs and execute the virtual physics model. Communication is only required for diffuse target mapping (since spring length adjustments must be announced), and possibly for fault detection. Even in these cases, it is only necessary to send messages to immediate neighbors; multi-hop routing is not required.

A significant advantage of the virtual spring mesh approach is that fault tolerance and attack resistance are inherent properties of the macrosensor. Since there is no hierarchy or centralized control, there are no single points of failure or obvious points of attack. Additionally, individual robots are nearly stateless, so recovery from robot failures is simple and rapid. Figure 8 shows a simulated example of a cluster of robots that experiences multiple simultaneous failures. As shown, the remaining robots automatically re-form a smaller spring mesh without the failed units. This recovery happens quickly and without any error handling beyond the detection of failed robots. The major cost associated with fault tolerance is the periodic communication required to assess the health of neighboring robots.

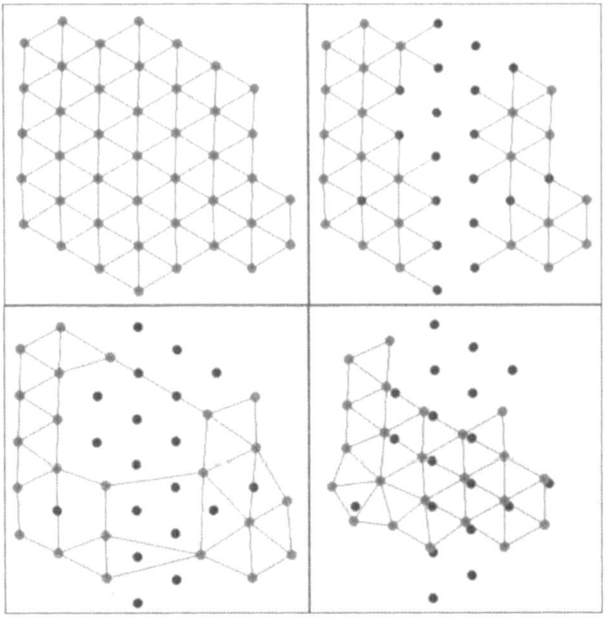

Fig. 8. Massive failure situation. "Dead" robots are represented as dark points. The top right figure is immediately after the robots failed; the lower left is one simulation cycle later. Final configuration is shown on the lower right.

4. Related Work

Explicitly coordinated exploration and mapping was examined in the "Cover Me!" [13] project, which defines a coverage metric and then uses an incremental greedy algorithm to deploy robots into locally optimal locations. This scheme is not designed to scale to large numbers of robots, as it makes use of global information and only deploys one robot at a time. Similar work by Simmons et al. [24] computes desired deployment locations by attempting to minimize overlap in information gain. Explicit loosely-coupled robot coordination for arbitrary goals (not just exploration) is implemented in the ALLIANCE system [20,21], which uses behavior-based task selection. RETSINA [9] operates a team of robots through a shared plan, which is communicated and refined over time. DINTA [3] takes a hybrid approach and uses a static sensor network to assign tasks to a set of mobile robots. This approach has many advantages, but requires a pre-deployed infrastructure.

Target tracking has been addressed within some of these systems. Targets are tracked in ALLIANCE [22] through a combination of local virtual forces and high-level behavior-based selection. Jung and Sukhatme [15] also use a multi-layered approach; their system computes a local solution for tracking groups of targets within the same field of view, operating within a framework that distributes robots into regions according to target density. A different approach has been proposed by Gage [8], who suggests randomized search strategies that use inexpensive systems and make detection very probable.

Fully distributed control based upon simple local behaviors has been used in several contexts. Brooks [4] has investigated behavior-based control extensively. Balch and Hybinette [1] suggest the use of "attachment sites" that mimic the geometry of crystals; this is used to create formations with large numbers of robots. A variety of projects have made use of "swarm robotics" e.g., [23,2] to carry out simple tasks such as light tracking. Gage [7] investigated the use of robot swarms to provide blanket, barrier, or sweep coverage of an area. Several researches have used models based on the interactions of ants within a colony [16,23].

Distributed control based on virtual physics (also called "artificial physics" or "physicomimetics") has also been investigated, although not in the manner that we propose. Howard, Mataric and Sukhatme [14] model robots as like electric charges in order to cause uniform deployment into an unknown enclosed area. Spears and Gordon [10,11,25] use a more sophisticated model analogous to the gravitational force, but make the force repulsive at close range. Both of these models use fully connected graphs, although the latter model cuts off interactions beyond a maximum range.

5. Conclusion

We have developed a scalable approach toward control of distributed robotic macrosensors comprised of large numbers of potentially heterogeneous robotic sensor platforms. Such macrosensors can be tasked with automatically exploring complex and unknown environments, and can track targets of interest within that environment. In addition to their inherent scalability, these macrosensors are extensible and fault tolerant. Our analysis and simulation results indicate that the virtual spring mesh is an effective mechanism for controlling such macrosensors, and that macrosensors built using this approach exhibit the desired properties.

We have demonstrated through simulation that, in principle, a virtual spring mesh can be used to explore and track targets in an unknown environment that includes nontrivial obstacles. We are currently constructing distributed robotic macrosensors that will be used to test the viability of virtual spring mesh based control in practice.

6. References

1. Balch, T. and Hybinette, M. "Behavior-based coordination of large-scale robot formations." *Proceedings of the Fourth International Conference on Multiagent Systems (ICMAS '00)*. July 2000. pp. 363 - 364.
2. Baldassarre G. Nolfi S. Parisi D. "Evolving Mobile Robots Able to Display Collective Behaviors." In *Proceedings of the International Workshop on Self-Organisation and Evolution of Social Behaviour*, pages 11-22, Monte Verità, Ascona, Switzerland, September 8-13, 2002.
3. Batalin, M. and Sukhatme, G.S. "Sensor Network-based Multi-Robot Task Allocation." *2003 IEEE/RSJ International Conference on Intelligent Robots and Systems (IROS 2003)*, October, 2003, Las Vegas, Nevada.
4. Brooks, R.A., "Integrated Systems Based on Behaviors", *SIGART Bulletin* (2:4), August 1991, pp. 46–50.
5. Delin, K "The Sensor Web: A Macro Instrument for Coordinated Sensing." *Sensors*. Vol 2, 2002. pp. 270-285.
6. Deng, J. Han, R. Mishra, S. "A Performance Evaluation of Intrusion-Tolerant Routing in Wireless Sensor Networks," *IEEE 2nd International Workshop on Information Processing in Sensor Networks (IPSN '03)*, 2003, Palo Alto, California.

7. Gage, D. 'Command Control for Many -Robot Systems." *Proceedings of AUVS-92*. Huntsville, AL. June 1992.
8. Gage, D. 'Randomized Search Strategies with Imperfect Sensors." *Proceedings of SPIE Mobile Robots VIII*. Boston. September 1993. vol 2058, pp 270-279.
9. Giampapa, J. and Sycara, K. 'Team -Oriented Agent Coordination in the RETSINA Multi-Agent System. Tech report CMU-RI-TR-02-34. Robotics Institute, Carnegie Mellon University. December 2002.
10. Gordon, D. Spears, W. Sokolsky, O. Lee, I. 'Distributed Spatial Control, Global Monitoring and Steering of Mobile Physical Agents." *IEEE International Conference on Information, Intelligence, and Systems*. November 1999.
11. Gordon-Spears, D. and Spears, W. "Analysis of a Phase Transition in a Physics -Based Multiagent System." In *Proceedings of the FAABS '02 Workshop*. 2002.
12. Hong, X. Gerla, M. Kwon, T. Estabrook, P. Pei, G. Bagrodia, R. 'The Mars Sensor Network: Efficient, Energy Aware Communications" *Proceedings of IEEE MILCOM*. McLean, VA. 2001.
13. Howard, A. and Mataric, M. J. 'Cover Me! A Self-Deployment Algorithm for Mobile Sensor Networks." *Proceedings of the IEEE International Conference on Robotics and Automation (ICRA2002)*, 2002.
14. Howard, A., Mataric, M.J., and Sukhatme, G.S. 'Network Deployment using Potential Fields: A Distributed, Scalable Solution to the Area Coverage Problem." *The 6th International Symposium on Distributed Autonomous Robotic Systems*, pp. 299-308.
15. Jung, B. and Sukhatme, G.S. "Tracking Targets using Multiple Robots: The Effect of Environment Occlusion", *Autonomous Robots*, **13**(3). 2002. pp. 191-205.
16. Koenig, S. and Liu, Y. 'Terrain Coverage with Ant Robots: A Simulation Study." In *Proceedings of the International Conference on Autonomous Agents,* pages 600-607, 2001.
17. Konolige, K. "A gradient method for realtime robot control." *In Proceedings of IEEE/RSJ International Conference on Intelligent Robots and Systems*. 2000.
18. López-Sánchez, M. Esteva, F. López de Màntaras, R. Sierra, C. Amat, J. 'Map Generation by Cooperative Low-Cost Robots in Structured Unknown Environments." *Autonomous Robots*, 5. pp. 53-61.
19. Nissanka, B. Chakraborty, A. Balakrishnan, H. 'The Cricket Location -Support System." *6th ACM International Conference on Mobile Computing and Networking (MOBICOM)*. Boston, 2000.
20. Parker, L. "ALLIANCE: An Architecture for Fault Tolerant, Cooperative Control of Heterogeneous Mobile Robots", *Proceedings of the 1994 IEEE/RSJ/GI International Conference on Intelligent Robots and Systems (IROS ' 94)*September 1994. pp 776-783.
21. Parker, L. 'On the Design of Behavior -Based Multi-Robot Teams." *Advanced Robotics*. 10 (6) 1996. pp 547-578.
22. Parker, L. and Emmons, B. "Cooperative Multi-Robot Observation of Multiple Moving Targets." *Proceedings of 1997 International Conference on Robotics and Automation*, Volume 3, pp. 2082-2089.
23. Sahin E. Franks N.R. 'Measurement of Space: From Ants to Robots." In *Proceedings of WGW 2002: EPSRC/BBSRC International Workshop Biologically-Inspired Robotics: The Legacy of W. Grey Walter*, pages 241-247, Bristol, UK, August 14-16, 2002.
24. Simmons, R. Apfelbaum, D. Burgard, W. Fox, D. Moors, M. Thrun, S. Younes, H. 'Coordination for Multi-Robot Exploration and Mapping." *Seventeenth National Conference on Artificial Intelligence (AAAI)*. 2000.
25. Spears, W., and Gordon, D. 'Using Artificial Physics to Control Agents." In *IEEE International Conference on Information, Intelligence, and Systems*, 1999.
26. Thrun, S. Burgard, W. Fox, D. "A Real-Time Algorithm for Mobile Robot Mapping With Applications to Multi-Robot and 3D Mapping." *IEEE International Conference on Robotics and Automation*. San Francisco. April 2000.

Self-Organised Task Allocation in a Group of Robots

Thomas Halva Labella[1], Marco Dorigo[1], and Jean-Louis Deneubourg[2]

[1] IRIDIA, Université Libre de Bruxelles, Brussels, Belgium
{hlabella,mdorigo}@ulb.ac.be
[2] CENOLI, Université Libre de Bruxelles, Brussels, Belgium
jldeneubourg@ulb.ac.be

Summary. Robot foraging, a frequently used test application for collective robotics, consists in a group of robots retrieving a set of opportunely defined objects to a target location. A commonly observed experimental result is that the retrieving efficiency of the group of robots, measured for example as the number of units retrieved by a robot in a given time interval, tends to decrease with increasing group sizes. In this paper we describe a biology inspired method for tuning the number of foraging robots in order to improve the group efficiency. As a result of our experiments, in which robots use only locally available information and do not communicate with each other, we observe self-organised task allocation. This task allocation is effective in exploiting mechanical differences among the robots inducing specialisation in the robots activities.

1 Introduction

The interest for collective robotics of scientists from disciplines as different as biology and engineering has recently been increasing. For instance, biologists have started to use robots for testing their theories about social animals, while engineers see in collective robotics a means for finding solutions to problems that cannot be solved efficiently by a single robot.

In this paper we consider a typical problem in collective robotics: foraging. Robot foraging consists in the cooperative activity of a group of robots whose goal is to retrieve to a target location a set of opportunely defined objects. A well known problem in robot foraging is the reduction in the performance of the group of robots, measured for example as the number of units retrieved by a robot in a given time interval, when the group size becomes bigger because of interferences among robots (Goldberg and Matarić, 1997; Balch, 1999).

A possible solution to this decreasing efficiency problem is to adopt some task allocation[3] mechanism that allows to automatically determine the optimal size of a group of robots that cooperate in a foraging application.

[3] In the collective robotics literature, the term "task" is given two different meanings, depending on whether the common goal involves one or more tasks: in the first

We propose a method inspired by biology to tune the number of foragers. This method, that exploits positive and negative feedbacks as typically done by self-organised systems (Camazine et al., 2001), does not use any form of direct or symbolic communication and does not require human intervention.

Our work is part of the SWARM-BOTS project,[4] whose aim is to develop a new robotic system, a *swarm-bot*, composed of several independent and small modules, called *s-bots*. Each module is autonomous and capable of connecting to other modules to self-assemble into a *swarm-bot*. The control program of each *s-bot* exploits techniques derived from swarm intelligence studies (Bonabeau et al., 1999) and collaboration among the *s-bots* is achieved by means of stigmergic communication (Grassé, 1959; Dorigo et al., 2000).

This paper is organised as follows. Section 2 introduces the problem of foraging, its issues, and illustrates the biological inspiration of our work. Section 3 describes the hardware and the software used in our experiments. Section 4 shows and analyses the results. Section 5 summarises related work, and finally, Section 6 draws some conclusions.

2 Problem Description and Issues

Foraging is considered a canonical test domain for collective robotics (Cao et al., 1997). The terminology we use in this paper is borrowed from biology: for instance we use the term "prey retrieval" as a synonymous for "retrieving an object".

The environment in which a prey retrieval experiment is performed includes: a group of robots, also called a "colony"; objects spread in the environment, called "prey", that may have different sizes or weights, may be fixed or moving, and may appear and disappear randomly; and a special area called "nest". The robots goal is to collect the prey and bring them to the nest.

A colony of robots can solve the problem in shorter time than a single robot, but the efficiency degrades when there are too many robots because of negative interferences. One way to avoid this is to choose how many robots should be engaged in prey retrieval in such a way that the efficiency of the group is maximised. In other words, to use task allocation.

Biologists have developed numerous models to explain how task allocation can be achieved without using direct communication. In this paper, we refer to Deneubourg et al.'s (1987) model, in which the individuals adapt and learn during their life-time. In Deneubourg et al.'s model, each ant is an agent that leaves the nest with probability P_l. If its foraging trip is successful, that is, the ant retrieves a prey, it increases its P_l by a constant Δ. If it is not successful,

case, which is also ours, a task allocation method is meant to find the optimal number of robots to perform the given task (as in Krieger and Billeter, 2000; Agassounon and Martinoli, 2002); in the second case, task allocation is in charge of assigning one robot to each task (as in Gerkey and Matarić, 2003).

[4]For more information on the project see `www.swarm-bots.org`

it decreases it by the same constant. Unfortunately, the authors performed tests only with numerical simulations.

The aim of this paper is to assess the feasibility of using similar mechanism to control a group of robots and to test whether this form of adaptation, which was only shown to be theoretically correct, works in the real world. Therefore, we decided to use real robots and not simulation for the first experiments. Time and technical constraints obliged us also to use only one colony size, leaving as future work the study of further aspects, such as the dependencies on group size and on the prey density.

In a previous work (Labella et al., 2004), we showed that a variant of this model, described in Section 3, can improve the efficiency of the colony by reducing the number of robots looking for prey. Here, we show that the improvement is achieved by means of group-level task allocation which increases P_l in some robots in the colony and decreases it in the remaining ones (Section 4.1). Additionally, in Section 4.2 we show that our allocation mechanism tends to consistently select the same robots to be foragers, which means that the allocation mechanism exploits differences among the robots that make some of them more suited for prey retrieval. The differences we refer to are not intentionally implemented in the robots, but come from the fact that two artifacts can never be perfectly identical.[5] The mechanism that we propose is not based on direct communication among robots, is completely decentralised and does not require human intervention. It can therefore be considered as *self-organised*.

3 Hardware and Software

The *s-bots* were still in the prototyping phase at the time of the experiments. Therefore, we decided to run the experiments using robots built with Lego Mindstorms™. The resulting robot, that we name *MindS-bot*, is presented in Fig. 1(a). *MindS-bots* use tracks to move. They have two arms, placed symmetrically with respect to the centre of the robot, that they use to grip the prey. Two light sensors are on the top of the *MindS-bot*: one on the front, which is used to sense prey, and one on the back, which is used to search for and go to the nest. Two bumpers, which are also placed on the front and on the back side, are used to avoid obstacles.

Figure 1(b) depicts the control program of the *MindS-bots*. Different states deal with the different phases of prey retrieval, as follows:

Search: the *MindS-bot* looks for a prey and avoids collisions with other *MindS-bots*. If a prey is found, the *MindS-bot* grasps it. If it has spent too much time searching for a prey without finding any, it gives up.

Retrieve: the *MindS-bot* looks for the nest and pulls the prey toward it.

[5]For instance, we observed during the experiments that the motors of some robots were faster and that some robots could grasp stronger than others, although the motors were the same models and the robots were built in the same way.

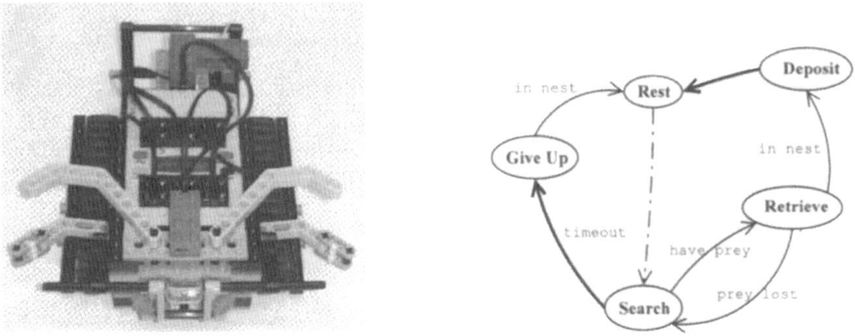

(a) Front view of a *MindS-bot* (b) Schema of the control system

Fig. 1. Hardware and software of a *MindS-bot.* In (b), states represent different phases of the retrieval. The labels on each edge represent the conditions that let the transitions to other states occur. The bold edges show when the probability to leave the nest is updated. The edge between **Rest** and **Search** is dash-dotted to indicate that the transition occurs probabilistically with probability P_l.

Algorithm 1 Variable Delta Rule (VDR). P_l is the probability to leave the nest, *succ* and *fail* are the number of consecutive successes and failures.

initialisation: succ $\leftarrow$ 0; fail $\leftarrow$ 0; $P_l \leftarrow P_{\text{init}}$

if success **then**
  succ $\leftarrow$ succ + 1; fail $\leftarrow$ 0
  $P_l \leftarrow min\{P_{\max}, P_l + \text{succ} \cdot \Delta\}$
end if

if failure **then**
  succ $\leftarrow$ 0; fail $\leftarrow$ fail + 1
  $P_l \leftarrow max\{P_{\min}, P_l - \text{fail} \cdot \Delta\}$
end if

Deposit: the *MindS-bot* leaves the prey in the nest and positions itself for the next foraging trip.
Give Up: the *MindS-bot* looks for the nest and returns to it.
Rest: the *MindS-bot* rests in the nest before restarting the search.

Transitions between states occur on the basis of events that are either external (e.g., finding a prey or entering the nest) or internal to the robot (e.g., a timeout). The labels on the edges in the graph of Fig. 1(b) show the conditions that must be true for the transitions to occur.

The *MindS-bots* change from **Rest** to **Search** with probability P_l, whose value is updated during the transitions from **Search** to **Give Up** (henceforth called *failure*) and from **Deposit** to **Rest** (henceforth called *success*). The update is done as shown in Algorithm 1, named Variable Delta Rule (VDR). The algorithm increments or decrements P_l by a constant Δ multiplied by the number of consecutive successes or failures (not present in the model of Deneubourg et al.). It then bounds P_l in the range $[P_{\min}, P_{\max}]$.

Fig. 2. Snapshot of an experiment. Four *MindS-bots* are looking for three prey. The nest is indicated by a light in the centre. One *MindS-bot* is resting in the nest, two are exploring the environment, and the fourth is retrieving a prey to the nest.

4 Experiments and Results

We performed ten experiments using a circular arena (Figure 2) with a diameter of 2.40 m. Each experiment lasted 2400 s. A light bulb was placed over the centre of the nest area. Walls and floor were white painted to be more reflective, prey were black cylinders.

The timeout was set to 228 s.[6] P_{max} was set to 0.05, which corresponds to a mean idleness time in the nest of 20 s. P_{min} was 0.0015 (mean idleness time: 666.6 s) and Δ was 0.005. At the beginning of each experiment P_l was set to 0.033 (mean idleness time: 30 s). These values were chosen on the basis of a trial-and-error methodology. Prey appeared randomly in the arena with probability 0.006 per second. Their position was selected randomly to be between 0.5 m and 1.1 m from the centre of the arena.[7] The colony size was of four *MindS-bots*, selected out of a pool of six, and some of them were substituted after each experiment.

The next section analyses the task allocation that occurs in the colony, while Sec. 4.2 shows how task allocation takes into account mechanical differences among *MindS-bots*.

4.1 Task allocation

At any given instant t after the beginning of the experiment, the value of P_l in a *MindS-bot* is a random variable. Whether task allocation occurs or not can be observed in the distribution of P_l: if task allocation occurs, then at the end of the experiments some of the *MindS-bots* will have high P_l while the

[6] This value is the estimate of the median time needed by one *MindS-bot* to find one prey when it is alone in the arena.

[7] A computer placed next to the arena was used to warn the experimenter, by means of a random number generator, when and where a new prey should appear.

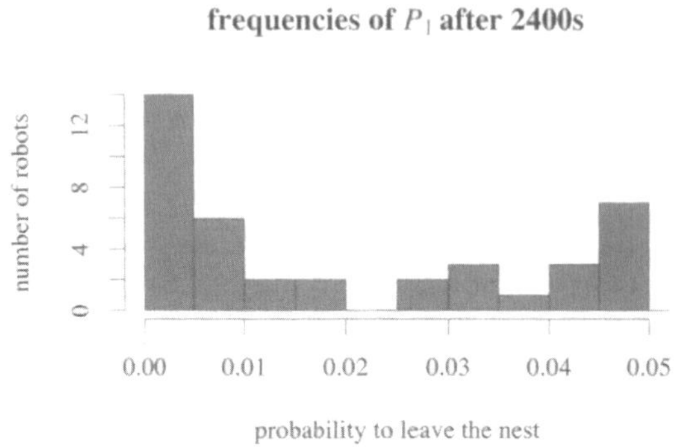

(a)

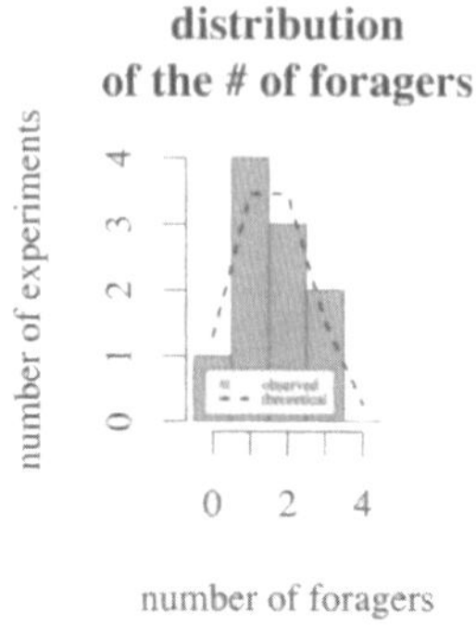

(b)

Fig. 3. (a) Frequency of P_l observed 2400 s after the beginning of experiments. The two peaks demonstrate the occurrence of task allocation. We classify the *MindS-bots* in two groups using 0.025 as a threshold: 40% of the observations are above it. (b) Distribution of the number of foragers ($P_l > 0.025$) observed in each experiment compared with the theoretical binomial distribution with $p = 0.4$.

others will have a low P_l, and the distribution of P_l will present two peaks; otherwise it will have only one peak.

We recorded the value of P_l for each *MindS-bot* during the experiments and estimated the distribution. Figure 3(a) shows the result after 2400 s and its two-peak shape confirms that task allocation has occurred. We classify therefore the *MindS-bots* in two groups: those with P_l higher than 0.025 are called *foragers*, while those with P_l lower than 0.025 are called *loafers*.

One might object that the peak on the right of the distribution in Figure 3(a) could be the result of a few experiments in which all the *MindS-bots* happen to be foragers. To see that this is not the case, it is enough to look at the number of foragers which were present in each experiment, and how this number is distributed. From the data in Figure 3(a), we know that 40% of the population are foragers. Therefore, we expect that the number of foragers in each experiment follows a binomial distribution with $p = 0.4$. Figure 3(b) shows that the profiles of the theoretical and the observed distributions are very similar and suggests that further experiments will confirm the matching.

It is interesting to note that both positive and negative feedbacks are present in the colony. Positive feedback is given by the fact that the higher the P_l of a *MindS-bot*, the shorter the time the *MindS-bot* remains in the nest and, therefore, the shorter the time until it finds a new prey and increases its P_l again. Negative feedback is given by competition among *MindS-bot*: every prey taken by one *MindS-bot* decreases the probability that the others can successfully retrieve. These two forms of feedback are likely to contribute to the occurrence of task allocation.

The evolution of the distribution of P_l over time (Figure 4) shows that the group of foragers starts forming later than the group of loafers (the former

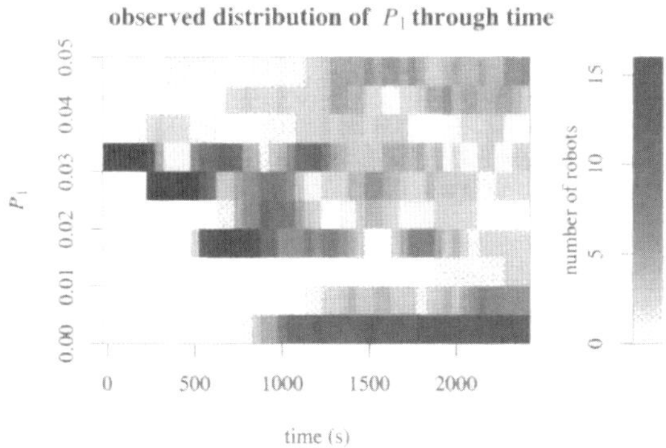

Fig. 4. Dynamics of the observed frequency of P_l. The darkness of a cell in position (t, p) is proportional to the number of *MindS-bots* with $p = P_l$ after t seconds from the beginning of the experiment. The relationship is given by the bar on the right. At $t = 0$ all the *MindS-bots* have $P_l = 0.033$ (see the black stripe on the left). After 1000 s the number of *MindS-bots* with low P_l (the loafers) drastically increases (see the dark stripe on the bottom). Similarly, after 1500 s, the number of robots with high P_l (the foragers) increases, although slowly and reaching a lower value than the loafers (top-right part of the plot).

at 1500 s, the latter at 1000 s). At the beginning, some *MindS-bots* become loafers because they are not successful, while the others alternate successes with failures. The fewer *MindS-bots* are foraging, the fewer competitors are present and the higher is the probability that the foraging *MindS-bots* will remain foragers.

4.2 Exploitation of mechanical differences

Given the stochastic nature of the experiments, we can model the fact that a given *MindS-bot* i is a forager at the end of an experiment as a random event. As in our experiments we use groups of 4 robots selected out of a pool of $N = 6$ robots, the probability of this random event may depend on the specific group G_k, $k \in \{1, \ldots, \binom{6}{4}\}$, to which i belongs in a given experiment: we denote this probability by $P_f(i|k)$.

There are two possibilities, depending on whether the following condition is true or not:

$$\exists\, i, k, j : \quad P_f(i|k) \neq P_f(i|j), \quad k \neq j \,. \tag{1}$$

If (1) is true, then the allocation mechanism exploits mechanical differences, which is what we want to prove. On the contrary, if (1) is false, then $P_f(i|k) = P_f(i)$ (that is, the probability of i being a forager is not a function of the group G_k to which it belongs) and we have that either the following condition is true:

$$\exists\, i, j : \quad i \neq j, P_f(i) \neq P_f(j) \,, \tag{2a}$$

in which case, once again, the allocation mechanism exploits mechanical differences, or the following equation is true:

Table 1. For each robot, identified by an unique name, the total number of experiments in which it was used and the number of times it was a forager are reported. Data refers to ten experiment, four *MindS-bots* per experiment.

ID	Tot. Exp.	$\#_{foragers}$
*MindS-bot*1	6	5
*MindS-bot*2	3	2
*MindS-bot*3	9	1

ID	Tot. Exp.	$\#_{foragers}$
*MindS-bot*4	9	4
*MindS-bot*5	3	0
*MindS-bot*6	10	4

$$P_{\mathrm{f}}(i) = P_{\mathrm{f}}(j) \quad \forall\, i, j \in \{1, \ldots, N\}\,, \tag{2b}$$

in which case the allocation mechanism does not exploit mechanical differences (note that (2a) and (2b) are mutually exclusive). If we assume that (1) is false, we can show that also (2b) is false considering the data in Table 1, which reports the number of times each *MindS-bot* was observed to be a forager at the end of the experiments. In fact, a statistical analysis of this data[8] shows that (2a) is true with confidence 95%.

We are therefore in a situation in which either (1) or (2a) is true, which means that the allocation mechanism exploits mechanical differences of the *MindS-bots*. However, there is not enough data to determine which of the two conditions is verified.

5 Related Work

Other approaches to the issues described in this paper can be found in the literature. We list here a few ones.

Gerkey and Matarić (2003) review and compare some of the main task allocation methods used in the literature, where task allocation is intended as the problem of assigning tasks to one robot. They analyse the methods, that need inter-robot communication and are based either on a solution to the optimal assignment problem or on a market/auction schema, in terms of the complexity of the computation required and of the costs of communication.

The threshold-model (Bonabeau et al., 1996) is widely used in bio-inspired robotics. For instance, Agassounon and Martinoli (2002) use it for a puck-clustering problem as a means to find the optimal number of robots. Krieger and Billeter (2000) use it in a retrieval task and analyse how the performance of the group changes when increasing the group size or when communication is used.

Other works in retrieval tend to focus on the reduction of the interferences by using communication and co-ordination (Balch and Arkin, 1994), by coding territoriality in the control systems of the robots (Schneider-Fontán and Matarić, 1996), or by using heterogeneous groups (Balch, 1999).

[8] χ^2 test with the null hypothesis that (2b) is true and (2a) as alternative hypothesis. This test can be used only if the data sets are independent, which is granted by assuming that (1) is false.

6 Conclusions

We showed that a simple adaptation during the life time of an individual can lead to self-organised task allocation in the colony. Individuals that are mechanically better for retrieval are more likely to be selected. Future work will try to understand better these phenomena, especially in those cases in which the colony has to deal with changing environments.

Our work is also relevant for biologists. Usually, division of labour in animal colonies is explained by looking at the dimorphism of individuals, at class segregation, or also at genetic differences. However, some biologists claim that adaptation, or learning, plays an important role too, but their arguments are usually only theoretical. Our work can therefore be used to give more concreteness to their theories by using real objects in a real environment.

Acknowledgements

This work was supported by the "SWARM-BOTS Project", funded by the Future and Emerging Technologies programme (IST-FET) of the European Commission, under grant IST-2000-31010. The information provided is the sole responsibility of the authors and does not reflect the Community's opinion. The Community is not responsible for any use that might be made of data appearing in this publication. Marco Dorigo acknowledges support from the Belgian FNRS, of which he is a Senior Research Associate, through the grant "Virtual Swarm-bots", contract no. 9.4515.03, and from the "ANTS" project, an "Action de Recherche Concertée" funded by the Scientific Research Directorate of the French Community of Belgium. Jean-Louis Deneubourg acknowledges support from the Belgian FNRS, of which he is a Research Associate.

References

W. Agassounon and A. Martinoli. Efficiency and robustness of threshold-based distributed allocation algorithms in multi-agent systems. In C. Castelfranchi and W.L. Johnson, editors, *Proceedings of the First International Joint Conference on Autonomous Agents and Multi-Agent Systems (AAMAS-02)*, pages 1090–1097. ACM Press, New York, NY, USA, 2002.

T. Balch. The impact of diversity on performance in multi-robot foraging. In O. Etzioni, J.P. Müller, and J.M. Bradshaw, editors, *Proceedings of the Third International Conference on Autonomous Agents (Agents'99)*, pages 92–99. ACM Press, New York, NY, USA, 1999.

T. Balch and R.C. Arkin. Communication in reactive multiagent robotic systems. *Autonomous Robots*, 1(1):27–52, 1994.

E. Bonabeau, M. Dorigo, and G. Theraulaz. *Swarm Intelligence: From Natural to Artificial Systems.* Oxford University Press, New York, USA, 1999.

E. Bonabeau, G. Theraulaz, and J.-L. Deneubourg. Quantitative study of the fixed threshold model for the regulation of division of labor in insect societies. *Proceedings of the Royal Society of London, Series B-Biological Sciences*, 263:1565–1569, 1996.

S. Camazine, J.-L. Deneubourg, N.R. Franks, J. Sneyd, G. Theraulaz, and E. Bonabeau. *Self-Organisation in Biological Systems*. Princeton University Press, Princeton, NJ, USA, 2001.

Y.U. Cao, A.S. Fukunaga, and A.B. Kahng. Cooperative mobile robotics: Antecedents and directions. *Autonomous Robots*, 4(1):7–27, 1997.

J.-L. Deneubourg, S. Goss, J.M. Pasteels, D. Fresneau, and J.-P. Lachaud. Self-organization mechanisms in ant societies (II): Learning in foraging and division of labor. In J.M. Pasteels and J.-L. Deneubourg, editors, *From Individual to Collective Behavior in Social Insects*, volume 54 of *Experientia Supplementum*, pages 177–196. Birkhäuser Verlag, Basel, Switzerland, 1987.

M. Dorigo, E. Bonabeau, and G. Theraulaz. Ant algorithms and stigmergy. *Future Generation Computer Systems*, 16(8):851–871, 2000.

B.P. Gerkey and M.J. Matarić. A framework for studying multi-robot task allocation. In A.C. Schultz, L.E. Parker, and F.E. Schneider, editors, *Multi-Robot Systems*, pages 15–26. Kluwer Academic Publishers, Dordrecht, The Netherlands, 2003.

D. Goldberg and M.J. Matarić. Interference as a tool for designing and evaluating multi-robot controllers. In *Proceedings of the 14th National Conference on Artificial Intelligence (AAAI-97)*, pages 637–642. MIT Press, Cambridge, MA, USA, 1997.

P. P. Grassé. La reconstruction du nid et les coordinations inter-individuelles chez *Bellicositermes natalensis* et *Cubitermes*. La théorie de la stigmergie: essai d'interpretation des termites constructeurs. *Insectes Sociaux*, 6:41–83, 1959.

M.J.B. Krieger and J.-B. Billeter. The call of duty: Self-organised task allocation in a population of up to twelve mobile robots. *Robotics and Autonomous Systems*, 30(1-2):65–84, 2000.

T.H. Labella, M. Dorigo, and J.-L. Deneubourg. Efficiency and task allocation in prey retrieval. In A.J. Ijspeert, D. Mange, M. Murata, and S. Nishio, editors, *Proceedings of the First International Workshop on Biologically Inspired Approaches to Advanced Information Technology (Bio-ADIT2004)*, Lecture Notes in Computer Science, pages 32–47. Springer Verlag, Heidelberg, Germany, 2004.

M. Schneider-Fontán and M.J. Matarić. A study of territoriality: The role of critical mass in adaptive task division. In P. Maes, M.J. Matarić, J.-A. Meyer, J. Pollack, and S.W. Wilson, editors, *From Animals to Animats 4, Fourth International Conference on Simulation of Adaptive Behavior (SAB-96)*, pages 553–561. MIT Press/Bradford Books, Cambridge, MA, USA, 1996.

Distributed Algorithms for Dispersion in Indoor Environments Using a Swarm of Autonomous Mobile Robots

James McLurkin[1,2] and Jennifer Smith[1]

[1] iRobot Corporation,
Burlington, MA 01803
{jamesm, jsmith}@irobot.com
[2] Massachusetts Institute of Technology
Computer Science and Artificial Intelligence Laboratory
Cambridge, MA 02139
jamesm@csail.mit.edu

We describe a set of distributed algorithms used to disperse a large group of autonomous mobile robots efficiently throughout an indoor environment. Only local inter-robot communication and processing is used. Ad-hoc communications network topologies formed by gradient floods spread messages and guide robot motion. Special attention has been given to doors, hallways, and other constrictions. The network maintains a route to chargers to allow self-charging.

1 Introduction

Almost every application of swarms of robots requires them to disperse throughout their environment. Exploration, surveillance, and security applications all require coverage of large areas. In this work, we present algorithms for dispersing a large swarm of robots into an enclosed space. In order for a dispersion algorithm to work on physical robots, it must take into account engineering concerns – maintaining

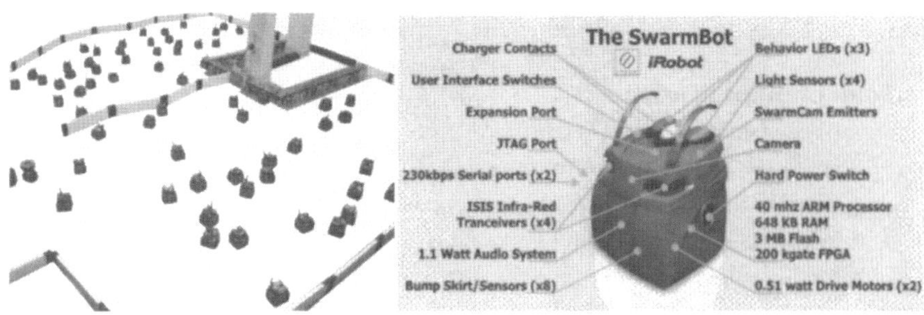

Fig. 1. The iRobot Swarm is comprised of over 100 SwarmBots™, 16 charging stations and navigational beacons. Each SwarmBot is roughly a 5" cube and has a suite of sensors, communications hardware, and human interface devices. Hands-free operation is important, thus the Swarm supports remote downloading and autonomous self-charging.

network connectivity, allowing for robot and communications failures, and providing an infrastructure for the robots to maintain their battery charge.

1.1 Swarm Hardware

The iRobot Swarm is shown in Fig. 1. Each SwarmBot™[1] contains an ARM Thumb CPU, a FPGA, a unique ID (UID) chip, a radio modem, serial ports, eight bump sensors, four light sensors, and a camera. Three extra-large LEDs and a 1.1 watt audio system provide user feedback.

The "Robot Ecology"™

Hands-free operation is critical, as even simple tasks, such as turning the robots on, become time consuming. The "Robot Ecology" infrastructure provides resources for centralized control, autonomous charging, and long-range navigation.

The ISIS™ Infrared Communication System

The primary sensor on the SwarmBot is an infrared inter-robot communication, location, and obstacle avoidance system called ISIS. Each robot has four ISIS transceivers, one in each corner. Nearby robots can communicate and determine the bearing, orientation, and range of their neighbors. The location system has a resolution of 1 cm and 2º at 30 cm range and a maximum range of 250 cm. A smaller range, $\mathbf{r_{safe}}$, is the maximum distance that provides reliable positioning. Reflected packets are used to determine the location of nearby obstacles and walls.

The Neighbor Cycle

Robots periodically transmit their externally visible state at the end of each neighbor cycle. This information includes their UID, what tasks they are performing, and any gradient messages they are relaying. We use the Aloha[1] protocol at the link layer. The period of the neighbor cycle, $\mathbf{t_n}$, is the same for all robots. This implies that each robot will receive only one communication from each of its neighbors during this period. These messages are collected and processed in a batch operation. This transforms the asynchronous distributed system into a synchronous distributed system, which greatly simplifies algorithm design. The period $\mathbf{t_n}$ is 250 ms, which allows for smooth robot motion control based on neighbor positions.

Behavior System

Swarm software is written as behaviors that run concurrently[2]. Each behavior returns a variable that contains actuator commands, i.e. motor velocities and light patterns.

Gradient Communication

A gradient-based multi-hop messaging protocol provides long-range communication using ISIS messages relayed from robot to robot. Gradients are used in many routing protocols to find optimal routes for messages through a network [3]. We use them to spread information and to guide robots through the network.

A source robot creates a gradient message that is relayed throughout the network in a breadth-first fashion, constructing a tree rooted at the source as it

[1] iRobot, ISIS, SwarmBot, Robot Ecology all copyright iRobot

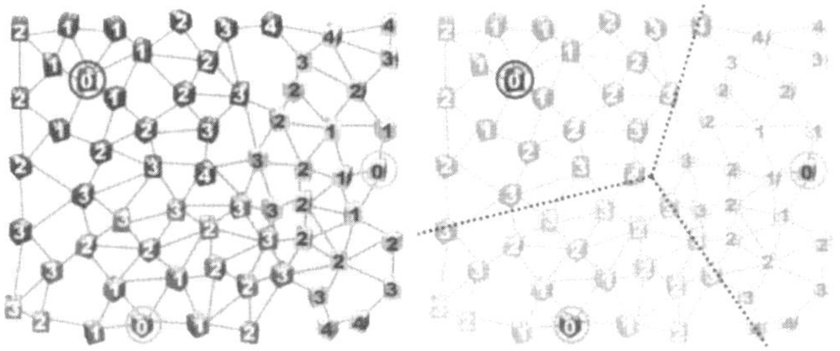

Fig. 2. Robots communicate with their neighbors over the gray lines in the left hand figure. Gradient-based routing protocols are used for long-range communication. Multiple gradient sources of the same type will tessellate the Swarm into Voronoi polygons, shown with the solid black lines in the figure on the right.

propagates. In each neighbor period, each robot processes the messages it has received and relays the one with the lowest hop count[2]. This eliminates cycles, and rebuilds the gradient tree each neighbor cycle to allow it to dynamically respond to changing network topologies as robots move.

With each robot transmitting periodically, but asynchronously, the expected time for any robot to receive a message during the neighbor period is

$$t_p = \frac{t_n}{2}.$$

The expected propagation time for a gradient message to disperse is

$$t_g = \mathrm{diam}(G) \cdot t_p$$

where G is the network graph, $\mathrm{diam}(G)$ is its diameter, and t_n is the neighbor cycle period. In the Swarm, the ideal value for t_p is 125 ms, but communications errors lengthen it to 172 ms. The diameter is dependent on the topography of the environment, but with 100 robots the practical limit is about 40 hops, resulting in a maximum propagation time of around 7 seconds. Multiple sources of the same gradient type will tessellate the Swarm as shown in Fig. 2.

Each robot retransmits messages every neighbor cycle. The total number of messages sent in an execution is given by:

$$n_m = \left(c_{\min\,msgs} + n_g\right)\frac{t}{t_n}n$$

where t is the total running time of the algorithm, n is the total number of robots, n_g is the number of types of gradient messages, and $c_{\min\,msgs}$ is the minimum number of messages sent by a single robot, currently 4. Minimizing the number of messages per cycle per robot is an important design goal.

[2] If there are multiple packets with the same hop count, then the UID of the source and finally the UID of the sender is used as a tiebreaker. This deterministic tie-breaking procedure helps robots select the same neighbors to consider over multiple neighbor cycles, which reduces dithering between multiple equivalent neighbors.

Fig. 3. **Left:** A dispersion into a small test space used to characterize the performance of different dispersion algorithms. This environment is approximately 6 m x 6 m, with several walls and "rooms". **Right:** A dispersion into a large room, note the person in the upper-left corner. It took about 20 minutes for them to achieve this dispersion, but they had to travel through a narrow hallway to get to this space, slowing their progress.

2 Directed Dispersion

The goal of the Directed Dispersion algorithm is to spread robots throughout an enclosed space quickly and uniformly, while keeping each robot connected to the network. The optimal running time occurs when each robot moves from a central start location to its final position along the shortest possible path at maximum velocity. Letting $\mathbf{E}$ be the closed polygon representing the environment, the minimum time for a dispersion is given by:

$$\mathbf{t}_{\mathbf{min}} = \frac{\mathbf{diam}(\mathbf{E})}{\mathbf{v}_{\mathbf{max}}}$$

where $\mathbf{diam}(\mathbf{E})$ is the maximum of shortest paths between any two points in the environment $\mathbf{E}$ and $\mathbf{v}_{\mathbf{max}}$ is the maximum velocity of the robots. This minimum time is used to normalize the results of different algorithms.

The dispersion is accomplished by using two algorithms that alternate running on the swarm: `disperseUniformly` and `frontierGuidedDispersion`. The `disperseUniformly` algorithm spreads robots evenly, using boundary conditions to limit the dispersion. The `frontierGuidedDispersion` algorithm directs robots towards unexplored areas, and is designed to perform well both in open environments and in environments with constrictions.

2.1 Uniform Dispersion - The `disperseUniformly` Algorithm

The `disperseUniformly` algorithm disperses robots uniformly throughout their environment. A thorough treatment of this technique is presented in [4]. Physical walls and a maximum dispersion distance between any two robots of $\mathbf{r}_{\mathbf{safe}}$ are used as boundary conditions to help prevent the Swarm from spreading too thin and fracturing into multiple disconnected components.

The algorithm works by moving each robot, $\mathbf{robot_i}$, away from the vector sum of the positions $\mathbf{p} = \{\mathbf{p}_1, \ldots, \mathbf{p}_c\}$ of their $\mathbf{c}$ closest neighbors $\mathbf{nbr} = \{\mathbf{neighbor}_1, \ldots, \mathbf{neighbor}_c\}$. The magnitude of the velocity vector that is given to the motor controller is:

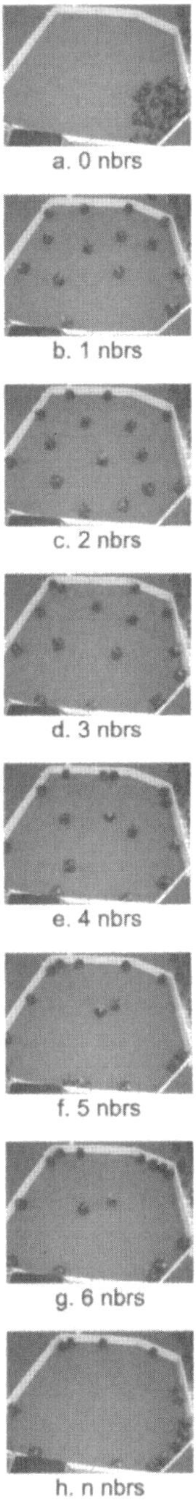

$$\mathbf{v} = \begin{cases} -\dfrac{v_{max}}{c \cdot r_{safe}} \sum_{i=1}^{c} \mathbf{p}_i & |\mathbf{p}_i| \leq r_{safe} \\ 0 & |\mathbf{p}_i| > r_{safe} \end{cases}$$

where v_{max} is the maximum allowable velocity output by this behavior. This vector directs the active robot away from its c nearest neighbors. The drive velocities are:

$$v_{rot} = v \cdot \cos(\mathbf{nbr_i.bearing}), \quad v_{trans} = v \cdot \sin(\mathbf{nbr_i.bearing})$$

where, $\mathbf{nbr_i.bearing}$, $\mathbf{nbr_i.range}$ is the bearing and range to $\mathbf{nbr_i}$.

This is a relaxation algorithm; imagine replacing graph $\mathbf{G}$ with its Delaunay triangulation $\mathbf{G}'$, and then placing compressed springs between connected robots. This will tend to expand the Swarm to fill the available space, but once the space is occupied, robots will position themselves to minimize the energy in the springs. Total group energy is minimized by minimizing local contributions, which happens when all the inter-robot distances are roughly equal. Fig. 4c and Fig. 3 show the robots uniformly dispersed in variously sized spaces.

The neighbors in $\mathbf{G}'$ are also Voronoi neighbors, the neighbors of the adjoining Voronoi cells of $\mathbf{robot_i}$. However, the robots are able to communicate across Voronoi cells, so the graph $\mathbf{G}$ usually has many edges that are not in the triangulation. This means that ISIS neighbors of a robot are not always Voronoi neighbors. Determining the set of Voronoi neighbors $\mathbf{nbr}$ from the set of ISIS neighbors, $\mathbf{nbr_{ISIS}}$, in real-time, is computation-intensive,[5] so an approximation is used. The closest ISIS neighbor will always be in the set $\mathbf{nbr}$. However, avoiding a single robot results in hectic movement as sensor errors can cause the position of this neighbor to change radically. Adding successively further neighbors to the set $\mathbf{nbr}$ cancels some of the position errors, but can also include non-Voronoi neighbors and cause the dispersion errors shown in Fig. 4a-h. When all elements of $\mathbf{nbr_{ISIS}}$ are added to $\mathbf{nbr}$, the robots are forced against the walls because the forces from distant neighbors from the other side of the circle are unbalanced.

In practice, using the two closest neighbors worked the best.

Fig. 4. The `disperseUniformly` algorithm is designed to spread the robots evenly. Instead of computing the closest neighbors (the neighbors of adjoining Voronoi polygons) to determine which robots to avoid, it avoids the n closest neighbors, sorted by range. Figs. a-h show the results of avoiding an increasing number of neighbors, with h showing the limit. Avoiding the two closest neighbors worked best in practice.

There are some cases in which second-closest neighbor is not a Voronoi neighbor, caused when the farther neighbor is "shadowed" by the closer neighbor. This case causes the robot to move in the same direction it would if only avoiding one neighbor, which does not cause errors, but does increase jitter. This "shadowing" effect is usually short lived, as the robot will typically encounter another neighbor or obstacle quickly.

2.2 Exploring New Areas - frontierGuidedDispersion

The goal of frontierGuidedDispersion is to guide robots towards areas they have yet to explore. Practical considerations require that the Swarm cannot fracture into disconnected components, as there must always be a route back to the chargers. The algorithm must self-stabilize to equalize voids and concentrations as robots enter and leave the network to charge. We also desire a termination condition to know when the Swarm is fully dispersed.

The frontierGuidedDispersion algorithm uses robots that are on the frontiers of explored space to guide the Swarm into unoccupied areas, similar to [6], but with support for multiple frontiers. The efficiency goal can be achieved if all the frontier robots move along their optimal path, leading the rest of the Swarm into their final positions.

Frontier Determination

Robots identify themselves as occupying one of three positions in the network: wall, frontier, or interior. "Wall" robots are those that detect an obstacle with the ISIS system. "Frontier" robots are those that have no neighbors and no walls within a large angle on any side; i.e., they are on the edge of an open space. The remainder are "interior" robots, as illustrated in Fig. 6. However, tight hallways require robots to become frontiers even when they detect walls. Fig. 6 shows how including the wall in the calculation of unoccupied space can correct this problem.

```
frontierDetermination() returns integer
edgeNbrSet ⇐ set-of-all-neighbors
if ISISWallSignalStrength > VIRTUALNEIGHBORWALLTHRESHOLD
   edgeNbrSet ⇐ edgeNbrSet ∪ createVirtualNbr(ISISRadar.bearing)
endif

edgeNbrSet ⇐ sortNbrsByBearing(edgeNbrSet )
maxAngle ⇐ edgeNbrSet[1] + (360 - edgeNbrSet[length(edgeNbrs)] )
for i ⇐ 2 to length(edgeNbrSet) − 1
   a ⇐ edgeNbrSet[i] - edgeNbrSet[i - 1]
   if a > maxAngle
      maxAngle ⇐ a
   endif
endfor

if maxAngle > EDGEANGLE
   return FRONTIERROBOT
else if radar.range < WALLRANGE
   return WALLROBOT
else
   return INTERIORROBOT
endif
```

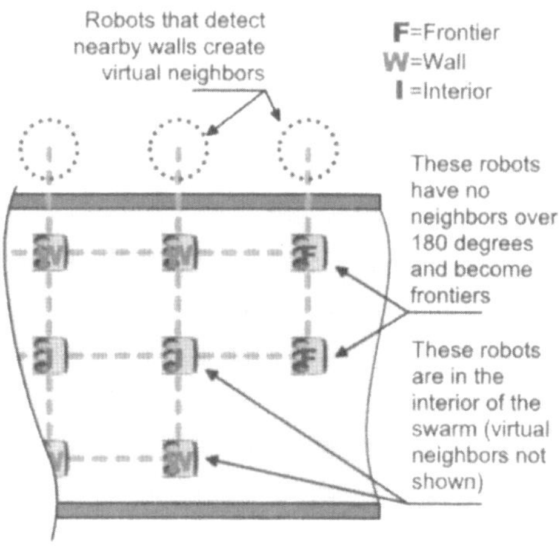

Fig. 5: Robots select their job as either frontier, wall, or interior robot based on the positions of their neighbors and nearby walls.

Lines 1-4 create a virtual neighbor if the global system variable ISISRadarSignal is greater than the VIRTUALNEIGHBORWALLTHRESHOLD. Lines 5-12 find the largest angle between any two adjacent robots. It does so by sorting the robots by bearing, then computing the difference in angle between adjacent elements. Lines 13-19 return the appropriate constant indicating the robot's position.

Swarm Motion - `disperseFromLeaves`

Once the robots know their positions in the network, the frontier robots source a gradient message. The trees created by these gradients ("frontier trees") guide the Swarm towards the frontier robots. It is possible to let the frontier robots "pull" the rest of the Swarm behind them by having the swarm cluster onto the frontier gradient source. However, any algorithm that is based on pulling robots over multiple hops can cause newly discovered frontiers to pull robots away from previously explored areas. This causes a frontier to re-appear at the old location and pull the Swarm back, creating oscillations, or fracturing the swarm and disconnecting robots from the chargers.

Instead, robots move away from children in the frontier tree. In order to build a reliable network, robots only move if they are in contact with at least two children in the frontier tree. This increases the min-cut of the network to two while the robots are dispersing, and helps deal with voids created by corners or robots heading home to charge.

```
disperseFromLeaves(beh)
1. childNbrSet ⇐ all-children-on-frontier-gradient-tree-closer-than-RSAFE
2. siblingNbrSet ⇐ all-siblings-on-frontier-gradient-tree-closer-than-RSAFE
3. if size(childNbrSet ) > 2
4.    avoidManyRobots(beh, (childNbrSet ∪ siblingNbrSet), d)
5. endif
```

Lines 1-2 create sets of children and sibling neighbors on the frontier tree that are closer than $\mathbf{r_{safe}}$. This limits the maximum dispersion to $\mathbf{r_{safe}}$. Line 3 requires a min-cut of 2 between this robot and its children. The `avoidManyRobots` behavior in line 4 takes the vector sum of the positions of the input set, and moves the robot in the opposite direction, i.e away from both sets of neighbors.

Leaves of the frontier tree remain stationary, which leaves robots in place to provide a route to the chargers and to mark previously explored areas. Essentially, the leaves become "anchors" and then limit the dispersion of robots away from them to a distance of $\mathbf{r}_{\mathbf{safe}}$.[3] As robots move away from the leaves, they move closer to their upstream robots, causing a chain reaction that eventually moves all the robots towards the frontiers.

Multiple frontiers often form as the Swarm explores the environment. Their gradients tessellate the Swarm based on hop count as shown in Fig. 2. This is useful because progress of distant frontiers will be slowed as interior robots disperse towards frontiers with smaller hop counts, allowing these closer frontiers to catch up. This tends to make the Swarm explore the building in a breadth-first fashion.

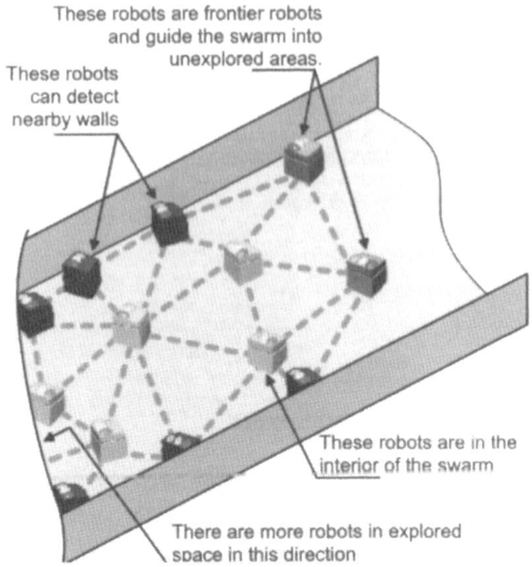

Fig. 6: Frontier robots guide the Swarm into unexplored areas by propagating a gradient that forms a tree rooted at the frontier robot. All robots then move away from their children in this tree. Leaves on the tree do not move, allowing previously dispersed robots to remain stationary.

2.3 Directed Dispersion

```
directedDispersion(beh)
1. if frontierDetermination() = FrontierRobot
2.    gradientSource(FrontierGradient)
3. endif
4. if FrontierGradient.isActive = True
5.    disperseFromSource(beh)
6. else
7.    disperseUniformly(beh)
8. endif
```

Lines 1-3 source a FrontierGradient if this robot is on a frontier. Line 4 checks to see if there are any frontier gradients in the network, including the one from this robot. If so, line 5 runs the disperseFromLeaves behavior. Otherwise, disperseUniformly runs and equalizes inter-robot spacing. The "pressure" from disperseUniformly tends to push robots into open spaces and tight constrictions, which can cause new frontiers to form. This activates the disperseFromLeaves behavior on the rest of the swarm, which causes a directed dispersion towards the frontiers. The disperseFromLeaves behavior stays active until all frontiers encounter walls or move to the interior of the swarm. Termination of the combined

[3] Another way to think about this is to imagine that any robot that is not maximally dispersed from its children will head towards the frontier, causing its parent to move towards the frontier, etc. This results in a "wave" of motion that the frontier "surfs" forward .

algorithm is defined when the frontier behavior stays inactive for a specified amount of time. Unfortunately, complex environments, sensor noise, and robots leaving to charge can make it difficult to quantify this time. We used ten seconds for the experimental results.

3 Experimental Results

Experiments were conducted in February 2004, at iRobot in Burlington, Massachusetts. Fifty-six robots were used with a reduced ISIS communications power setting to explore the small environment shown on the left side of Fig. 3. There were three goals placed at varying distances from the start location. The Swarm was released and times required to reach the three goals and full dispersion were recorded. Five algorithms were compared.

`idealGasMotion`: Robots move in straight lines but turn when they collide with each other or with a wall. The network often breaks into disconnected components. Inter-robot interference is a problem, with robots colliding often. There is no termination condition, and dispersion is rarely uniform.

`disperseFromSource`: A robot near the base station sources a "disperse" gradient. Robots move a distance $\mathbf{r}_{\mathbf{disperse}}$ away from parents in the "disperse" tree. Network connectivity is maintained during the dispersion process if $\mathbf{r}_{\mathbf{disperse}} \leq \mathbf{r}_{\mathbf{safe}}$. Uniform, complete coverage only occurs if the environment area is known in advance and $\mathbf{r}_{\mathbf{disperse}}$ is set accordingly; otherwise robots either bunch up at boundaries or do not completely fill the area. However, the dispersion is very efficient, quickly reaching all goals and full dispersion.

`avoidClosestNeighbor`: Robots move away from their closest neighbor at constant velocity if $\mathbf{r} < \mathbf{r}_{\mathbf{disperse}}$. Network connectivity can be maintained if $\mathbf{r}_{\mathbf{disperse}} \leq \mathbf{r}_{\mathbf{safe}}$. There is no termination condition. This is very similar to `disperseUniformly`, and the results are also similar. Dispersion is uniform, but robots oscillate back and forth between closest neighbors.

`disperseUniformly`: As described above in section 2.1. This algorithm runs more slowly than `avoidClosestNeighbor`, but the motion is smoother. It has very uniform dispersion and maintains network connectivity. Robots remain stationary after dispersion.

`directedDispersion`: As described in section 2.3. The robots rarely head in the

Table 1. Dispersion Efficiency vs. Location

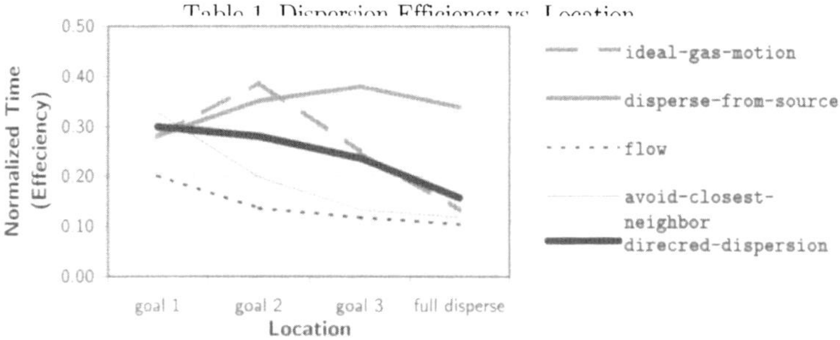

wrong direction, and effectively push frontiers to the boundaries. The algorithm terminates with uniform coverage and robots remain stationary after dispersion.

Additional tests were conducted at a government-run experiment in a empty military schoolhouse in January 2004. A swarm of 108 robots dispersed into 3000 ft^2 of indoor space in about 25 minutes, located an object of interest, and led a human to it. Multiple room configurations were tested. The robots ran almost continuously for six hours, demonstrating the value of a number of features of the iRobot Swarm system: single-command activation, single-command return to base, fully integrated automatic recharging behavior, and the ability to "bulk-reprogram" the robots in the field.

4 Conclusion

Directed dispersion allows robots to explore large, complex, indoor environments. The robots use the information in the graph in which they are embedded to modify this same structure. Path planning and directed motion algorithms become easier to develop when the primary input is the positions of other nearby robots. Practical dispersion algorithms can be designed to meet efficiency, robustness, scalability, and correctness constraints.

Acknowledgments

This work was supported by DARPA IPTO under contracts SPAWAR N66001-99-C-8513 and SMDC DASG60-02-C-0028.

References

1 N. Abramson. "The Aloha System - Another Alternative for Computer Communications". In Proc. Fall Joint Cornput. Conf., AFIPS Conf., page 37, 1970.

2 R. Brooks. "A robust layered control system for a mobile robot". In IEEE Journal of Robotics and Automation, RA-2, pp.14-23, 1986.

3 C. Intanagonwiwat, R. Govindan and D. Estrin. Directed diffusion: A scalable and robust communication paradigm for sensor networks. In Proc. Sixth Annual International Conference on Mobile Computing and Networks, 2000.

4 J. Cortes, S. Martinez, T. Karatas, and F. Bullo. Coverage control for mobile sensing networks. In Proceedings of the IEEE International Conference on Robotics and Automation, pages 1327--1332, Arlington, VA, May 2002.

5 S. Arya and A. Vigneron. "Approximating a Voronoi Cell". HKUST Theoretical Computer Science Center Research Report HKUST-TCSC-2003-10, Hong Kong University of Science and Techology, available at www.comp.nus.edu.sg/~antoine/avn.pdf, 2003.

6 D. Payton, M. Daily, R. Estowski, M. Howard, and C. Lee. "Pheromone Robotics". In Autonomous Robots, vol. 11, pp.319-324, 2001.

Part X

Motion Coordination

Optimal Design Methodology for an AGV Transportation System by Using the Queuing Network Theory

Satoshi Hoshino[1], Jun Ota[1], Akiko Shinozaki[2], and Hideki Hashimoto[2]

[1] Dept. of Precision Engineering, School of Engineering The University of Tokyo Bunkyo-ku, Tokyo 113-8656, JAPAN {hosino, ota}@prince.pe.u-tokyo.ac.jp
[2] Mitsubishi Heavy Industries, LTD. Sagamihara-shi, Kanagawa 229-1193, Japan

Abstract. In this paper, we propose an optimal design methodology for an AGV transportation system by using the queuing network theory. In this study, we deal with an actual transportation system as a combinatorial optimization problem. Therefore, some kind of paths and working multi-agents, such as Automated Guided Vehicles (AGVs), Automated Transfer Cranes (ATCs), and container cranes, are included in this system as design objects. We describe how to derive these design parameters (i.e., design solutions) by the performance evaluation of an AGV transportation system.

1 Introduction

In an automated port transportation system, timeliness is always an important constraint. Therefore, this system offers improvements in efficiency. In this paper, we propose an optimal design methodology for a transportation system with AGVs. This design problem can be considered as combinatorial optimization problem. Therefore, it is necessary to consider the following design elements: (1) an optimal number of working agents to satisfy the requirement, and (2) an optimal number of paths between agents.

Conventional researches relating to the design of transportation systems are generally divided into two categories: (1) A method based on the analysis of the local aspects of the transportation system [1][2], and (2) A method based on solving problems with simulation [3][4]. Abe *et al.* [1][2] proposed a design method using the open queuing network for optimal system design. In this method, a belt conveyor was used at a coaling wharf. However, a transportation system using a transport agent, such as a belt conveyor, which may move constantly, is inadequate for a multi-agent transportation system that includes AGVs. On the other hand, Chiba *et al.* [3] proposed an integrated design methodology in AGV transportation systems. In this study, they designed an optimal number of AGVs and transport routes based on a simulation. Liu *et al.* [4] developed a microscopic simulation model for the purpose of designing, analyzing, and evaluating some different Automated Container Terminals (ATCs).However, these two design methodologies are called simulation-based optimization; therefore, they sometimes need a significant amount of computational time to achieve an optimal design.

For this study, an AGV transportation system shown in Fig.1 was designed. An objective of this study is to establish a specific methodology with the following effects:

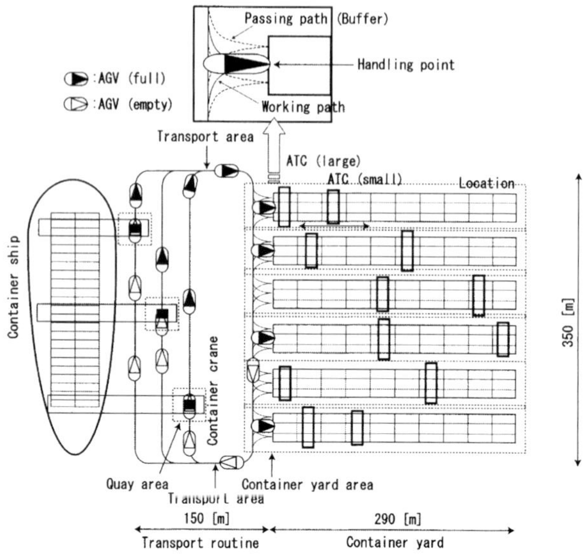

Fig. 1. Vertical layout of AGV transportation system

1. It is possible to model and design the transportation system as a multi-agent system, globally and optimally.
2. The computational time deriving the design solutions is less than that required in the simulation-based optimization method.
3. It is possible to evaluate and analyze the system proposed here.

To achieve the effects outlined above, we applied a closed queuing network, which is used to analyze and design large-scale computer systems for transportation systems. However, because the transportation time changes when the number of AGVs in the system changes, the following problem arises when only the queuing network theory is used.

- It is impossible to design the system by using only the queuing network theory.

For this challenging point, we aim to integrate the queuing network theory into the simulation-based method and iterate a design process; then, we will propose a methodology which can exactly simulate the transportation delay and thus estimate the efficiency of the proposed methodology.

2 Transportation System in a Port Container Terminal

2.1 AGV transportation system

In this study, the AGV transportation system is divided into three areas, namely, the quay, transport, and container yard areas (Fig.1). In this system, AGVs continue to circulate until they successfully complete all tasks by the following procedure:

step1 The Container cranes working in the quay area load a container on the AGVs from the container ship.

step2 A location is assigned as the destination in the container yard area at the time when the AGV leaves the quay area.

step3 The AGV arrives at the assigned location. If it encounters another AGV in its working path, this AGV goes onto the passing path; then, this passing path becomes a working path.
step4 If the ATC is already at the handling point, the AGV will transfer the container to ATC. Otherwise, if the ATC is already engaged, the AGV will wait at this point.
step5 The ATC goes to storage point in the assigned location to store the container.
step6 The AGV that has handled a container goes back to the quay area. (back to step1)

In this system, two different types of ATCs are working at the same location. Therefore, they can cross each other with working.

2.2 Combinatorial optimization problem

Design objects

The parameters of the design object in this study are described as the following:

- The number of AGVs
- The number of ATCs
- The number of passing paths (buffers)

All containers should successfully be transported from the container ship to the container yard area within a limited amount of time. In these constraints, the minimum number of agents that satisfy the requirement is used as a performance function.

Assumption of the transportation system

In this study, each location becomes the destination of a container by a certain probability for the sake of simplicity. As the assignments are made, any location without an engaged ATC becomes the priority destination. Additionally, the general working time of each container crane depends on the position allocated to each container in the container ship. However, we provide a fixed working time for the sake of simplicity. Moreover, three fixed container cranes are used due to the fixed scale of a berth.

3 Queuing Network Theory

3.1 Cyclic queuing network

In the closed queuing network, the number of agents is constant since agents can neither arrive nor leave the system, but, rather, circulate repeatedly through the various nodes at all times [5]. Thus, a closed queuing network, which includes N queues in tandem, i.e., a series of N queues arranged cyclically in such a way that agents proceed sequentially through the cycle, returning to the first node after being serviced at node N, is called a cyclic queuing network [6]. Fig.2 indicates the state of transition in the system from step n (Fig.2a) to step $n+1$ (Fig.2b).

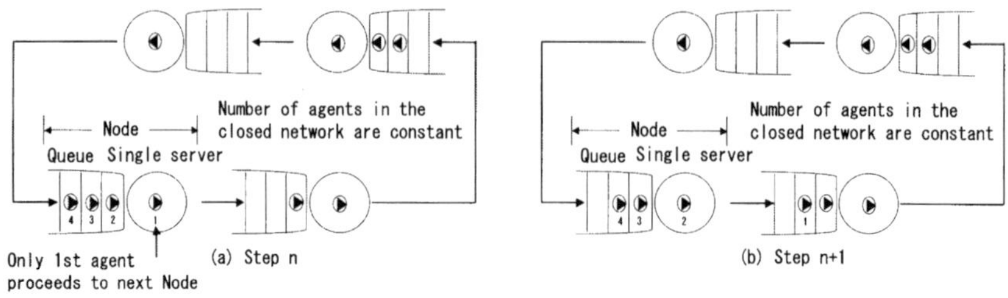

Fig. 2. A state transition diagram in the cyclic queuing network composed of a single server. Only the first agent proceeds to the next node; then, the next queuing agent (the second agent in that queue) goes into the single server, and the third and fourth agents proceed forward in the queue.

3.2 Performance evaluation method

Ottjes *et al.* and Duinkerken *et al.* used some performance indicators when they designed the ACTs using a specific transport simulator [7][8]. Similarly in our study, working AGVs in the system are defined as network agents. The number of nodes, the service time at each node, the number of servers in the nodes, the traffic parameter, and the number of relative arrivals of AGVs at the node are input parameters. After that, (a) traffic intensity(Eq.1), (b) throughput(Eq.2), and (c) the average number of AGVs at the node(Eq.3, 4) are obtained. The following are the performance evaluation criteria: (a) is used to locate bottlenecks in the system, (b) is used to determine whether or not the system satisfies the requirement, and (c) is used to design the number of buffers.

$$\alpha_{j1}(K) = \rho_{j1}\frac{G(K-1)}{G(K)} \tag{1}$$

$$\tau_{j1}(K) = h_{j1}\frac{G(K-1)}{G(K)} \tag{2}$$

$$\phi_{j1}(K) = h_{j1}\frac{G(K-1)}{G(K)} \tag{3}$$

$$\phi_{j1}(K) = \frac{1}{G(K)}\sum_{0\le x_j\le K} x_j q_j(x_j) G_{[j]}(K-x_j) \tag{4}$$

$$G_{[j]}(K) = \sum_{x_1+\cdots+x_{j-1}+x_{j+1}+\cdots+x_N=K}\ \prod_{i=1,i\neq j}^{N} q_i(x_i) \tag{5}$$

where,
K: The Number of AGVs
ρ_{j1}: The traffic parameter
h_{j1}: The number of relative arrivals of AGVs
$G(K)$: Normalization constant
$G_{[j]}(K)$: Normalization constant of j-complement in the network
N: The number of nodes
x_j: The number of AGVs around the node$j$

$q_j(x_j)$: Convolution parameter

where the ρ_{j1} is given by {the number of relative arrivals of AGVs at a certain node j} × {the service time at a certain node j} and the h_j is the number of relative arrivals of AGVs at node j. In this study, the number of relative arrivals of AGVs is the same for each node because the design object is modeled by a single cyclic queuing network (Fig.2). Therefore, the number of tasks is equal to the number of relative arrivals of AGVs. These parameters can be obtained from the system specifications. The function $G(K)$ is defined so that all the $P(x_1, x_2, ..., x_N)$ add up to one. The j-complement network is equal to the normalization constant given by removing the jth node in the closed queuing network, that is, $G_{[j]}(K)$ is obtained by procedure of the $G(K)$ described below [5]:

Define the $G(K)$ and $q(K)$ arrays; then, execute the following procedure after initializing these arrays:

$$G(x) \leftarrow \begin{cases} 1, \; x = 0 \\ 0, \; x \neq 0 \end{cases}$$

for (j = 1; j <= N; j++) {
    for (x = 0; x <= K; x++) {

$$q(x) \leftarrow \begin{cases} \dfrac{h_j^x}{x!}, & \|x\| \leq S_j \\ \dfrac{\|x\|}{S_j! S_j^{\|x\|-S_j}} \dfrac{h_j^x}{x!}, & Sj < \|x\| \end{cases}$$

    }
    for (k = K; k >= 0; k− −) {

$$G(k) \leftarrow \sum_{0 \leq y \leq k} q(y) G(k-y)$$

    }
}

where, S_j is the number of servers working at each node.

4 System Design

4.1 Modeling of the transportation system

Fig.1 is modeled as shown in Fig.3 based on the cyclic queuing network. The three areas in Fig.1 are assigned from nodes 1 to 4, and the number of cranes and ATCs in the quay area and container yard area are the number of servers at nodes 1 and 3. AGVs circulate through those nodes in the network until their transportation tasks are completed.

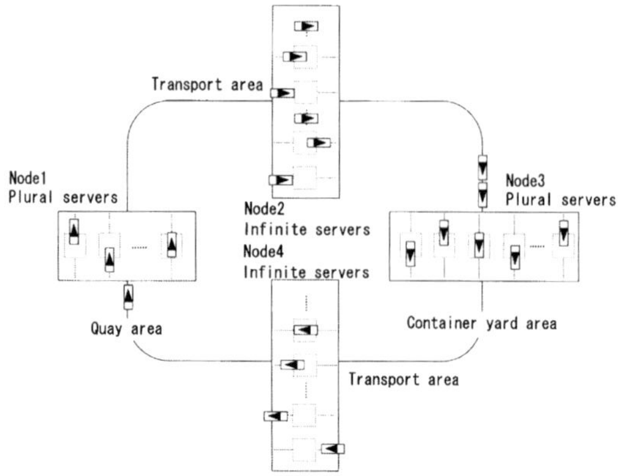

Fig. 3. Modeling the transportation system

4.2 Transport specifications of AGV and ATC

Table1 indicates the specifications for the AGV and ATC. Fig.4 is the transportation route for the AGVs. It is assumed that the AGVs go through the central path in the quay area for the sake of simplicity. Thus, in cases in which there is no traffic congestion, the time at node2 (A to B) and node4 (B to A) is calculated: 165 [s] and 122 [s], respectively. On the other hand, the loading/unloading and handling time of the container crane and ATCs is fixed because of the above assumption; therefore, in this study, the time at nodes 1 and 3 are given as 60 [s] and 30 [s], respectively. However, if the ATC is not at the handling point when the AGV arrives, the AGV will need to wait at this point.

Table 1. Specification of the AGV and ATC

	AGV(full)	AGV(empty)	ATC(full)	ATC(empty)
Max. velocity [m/s]	5.56	6.94	2.25	2.0
Rotational velocity [m/s]	1.39	1.39		
Acceleration [m/s^2]	0.15	0.15	0.1	0.1
Deceleration [m/s^2]	0.63	0.63	0.4	0.4

4.3 Design algorithm

Fig.5 indicates the design algorithm that we propose. In this design algorithm, a transportation simulator is used to (1) simulate a designed system, and (2) calculate the transportation delay by the AGV friction.

The service time at each node and the number of the container cranes and ATCs are inputted as initial parameters. After that, the throughput is obtained by the eq.(2). The throughput is evaluated based on certain requirements. If the throughput satisfies the requirements, the minimum number of AGVs is derived as the optimal number of AGVs based on the performance function. However, if it does not satisfy the requirement, the number of ATCs is increased by two, and the design process is then iterated. In this study, the number of AGVs is designed not to exceed 30 in order to avoid adding the AGVs recklessly.

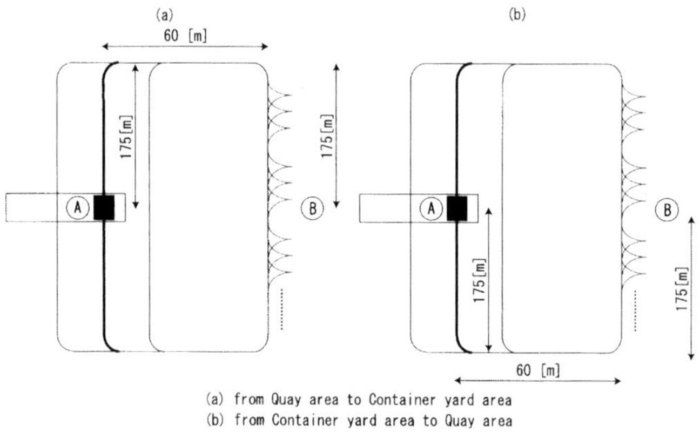

Fig. 4. Modeling of transport routing

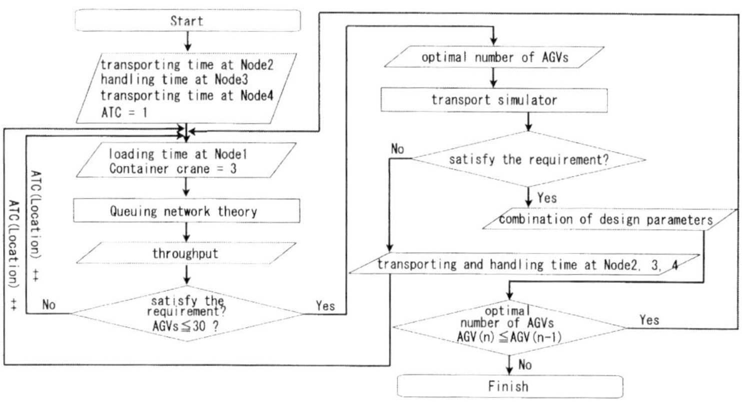

Fig. 5. Design algorithm by using the queuing network theory

The transportation simulator then operates based on the derived number of AGVs to evaluate whether or not this theoretical result also satisfies the requirement. If the simulation result also satisfies the requirement, the combination of optimal design parameters is obtained, as well as a design process in which the number of ATCs is changed and the process is iterated. Otherwise, the time at nodes 2, 3, and 4 are calculated by a simulator, and then the calculated time can be used as input parameters.

This design process will be iterated until the derived number of AGVs in step n is not lower than the derived number of AGVs in step $n-1$.

4.4 Requirement setting

In this study, one of the constraints is the time of berthing at a port container terminal; this time is equal to the time required to complete the transport. This is, {Transporting Requirement} $\leq$ {System Throughput}. In this design process, the number of transportation tasks is 600, and the required throughput is 120, i.e., the system has to successfully complete all tasks within 5 hours.

4.5 The combination of design solutions

Table2 indicates the number of AGVs at each node in the case of the design solutions are obtained. The average number of AGVs at node3 is less than the number of locations (6 and 7). Therefore, the number of buffers is designed as "0" (See Table3).

Table 2. Average number of AGVs at each node

Case	Node1	Node2	Node3	Node4
a	6	4	3 < Location:6	4
b	5	4	3 < Location:7	4

Table3 indicates each combination of optimal design solutions and time cost at each node (See the case a and b). The increase in the time is noticeable as the number of AGVs increases. Here, there is a trade-off between the throughput that depends on the number of AGVs and the time needed by the node. Therefore, there are cases in which increasing the number of AGVs worsens the transport efficiency.

Table 3. The combination of design parameters and time cost at each node

Case	ATC	AGV	Buffer	Transporting time Node2, 3, 4 [s]
a	12	17	0	178, 45, 144
b	14	16	0	170, 36, 140

4.6 Consideration

Consideration of design solutions by the throughput

Fig.6 indicates the throughput obtained by this design result. It can be confirmed that more than the optimal number of AGVs derived by this proposed design algorithm could satisfy the requirement. From the simulation result, the throughput is confirmed to be (a) 120.1 and (b) 126.6. From this result, we consider that the proposed design methodology is available.

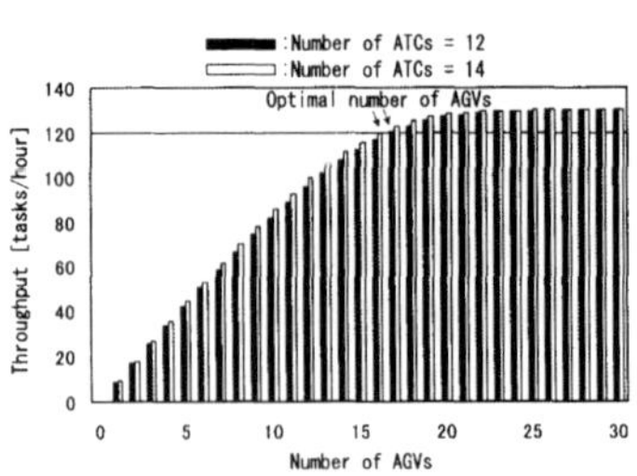

Fig. 6. System throughput

Consideration of computational cost

If we solve this combinatorial optimization problem by using the all search algorithm of the simulation-based optimization method, we must consider all the combinations of design solutions: 30 AGVs × 10 ATCs × 2 buffers equal to 600 times of simulation cost is needed. Correspondingly, in our proposed method, the total simulation cost required until the solutions are derived is just 15 times. From this result, this proposal design methodology is able to derive the solutions with a few simulations.

5 System Performance Evaluation

5.1 Traffic intensity

The traffic intensity at nodes 1 and 3 are evaluated in each design solutions. As shown in Table4, in the system which is designed by the derived solution, it can be located that bottleneck is in the quay side.

Table 4. Traffic Intensity at Nodes 1 and 3

Case	Node1 [%]	Node3 [%]
a	92.4	44.5
b	91.9	33.7

Fig.7 indicates the traffic intensity curve at nodes 1 and 3 while the number of AGVs in the system increases. It has been confirmed that the traffic intensity of node 1 approximates 100 [%] faster than that of node 3. This shows that more container cranes are needed to obtain much more throughput.

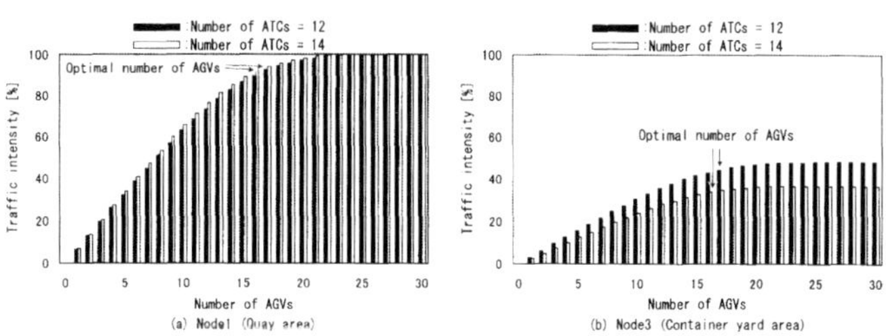

Fig. 7. Performance evaluation by the traffic intensity curve

5.2 Average number of AGVs at each node

Fig.8 indicates the average number of AGVs at each node while the number of AGVs increases. Thus, Fig.8 indicates the rough behavior of AGVs. From Fig.8(b) and (d), we can derive the number of transporting AGVs at nodes 2 and 4. From Fig.8(a) and (c), we can derive the number of loading/unloading and handling AGVs and of queuing AGVs at nodes 1 and 3. This shows that adding more AGVs to obtain more throughput leads the system to breakdown due to the existence of bottleneck.

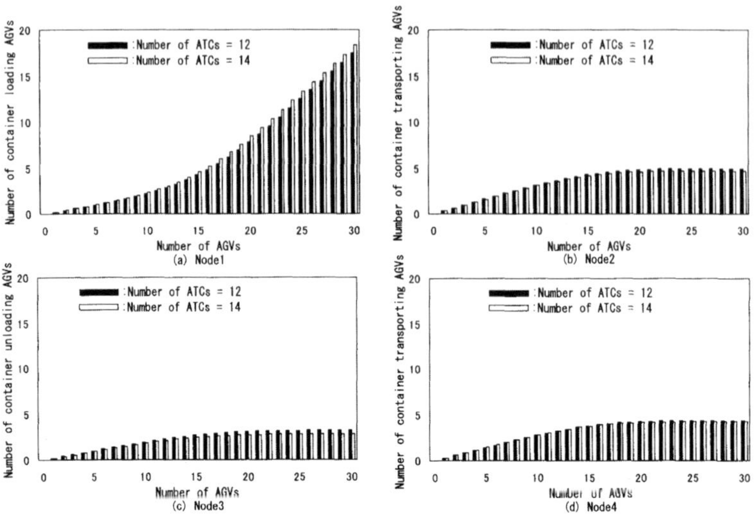

Fig. 8. The number of AGVs at each node

6 Conclusion and Future Work

In this paper, the design methodology and performance evaluation of an actual AGV transportation system are described. For this purpose, we integrated the queuing network theory into the simulation-based optimization method. The methodology was then proposed, which can simulate the transport delay exactly and evaluate the efficiency of the proposed methodology.

As future work, it would be necessary to design actual multi-agent transportation systems by modeling agents other than AGVs on the basis of a certain probabilistic distribution of the rough behavior of such agents. In addition, this methodology will be applied at actual container terminals so that the efficiency of the system can be verified.

References

1. Abe M. *et al.* : The Optimum Design for Materials Handling-Carrying System in Coaling Wharf (1st Rep), Proc. of Int. Conf. on Materials-Handling Equipment and Logistics, pp. 133-143, 1991.
2. Abe M. *et al.* : The Optimum Design for Materials Handling-Carrying System in Coaling Wharf (2nd Rep), Proc. of Int. Conf. on Materials-Handling Equipment and Logistics, pp. 144-157, 1991.
3. Chiba R. *et al.* : Integrated Design with Classification of Transporter Routing for AGV Systems, Proc. 2002 IEEE/RSJ Int. Conf. Intell. Robots and Systems, pp. 1820-1825, 2002.
4. Liu C.-I. *et al.* : Design, Simulation, and Evaluation of Automated Container Terminls, IEEE Tran. on Intelligent Transportation Systems, Vol. 3, No. 1, pp. 12-26, 2002.
5. Buzen J.P. : Computational algorithms for closed queueing networks with exponential servers, Comm. ACM, 16, 9, pp. 527-531, 1973.
6. Gordon W.J. *et al.* : Closed queuing systems with exponential servers, Oper. Res. 15, 2, pp. 254-265, 1967.
7. Ottjes J.A. *et al.* : Simulation of a New Port-Ship Interface Concept for Inter Modal Transport, Proc. of the 11th European Simulation Symposium, 1999.
8. Duinkerken M.B. *et al.* : A Simulation Model for Automated Container Terminals, Proc. of the Business and Industry Simulation Symposium, pp. 134-149, 2000.

Control of Vehicle Cooperative Behavior in Non-Signalized Intersection

Yusuke Ikemoto[1], Yasuhisa Hasegawa[2], Toshio Fukuda[1], and Kazuhiko Matsuda[3]

[1] Dept. Micro System Engineering, Nagoya University, Japan
ikemoto@robo.mein.nagoya-u.ac.jp
[2] Dept. of Mechanical and Systems Engineering, Gifu University, Japan
[3] Fuji Heavy Industries Ltd., Japan

Abstract This paper proposes a control method of vehicle cooperative behavior in an intersection and junction without infrastructures such as a signal system, a road-vehicle communication system, and so on. The vehicle cooperative behavior enables vehicles to pass through an intersection one by one or to interflow at junction one by one without vehicle collision like a weaving and zipping manner. These behaviors are achieved by generating a special-temporal pattern of a reaction-diffusion system, where a mutual communication between vehicles is realized only by IVC device. The mutual communication between vehicles can be expressed in the diffusion system of a certain morphogen so that only simple broadcasting communication could be enough for vehicle communication. Van der Pol model is used as one of the reaction-diffusion system. Finally, the proposed algorithm for vehicle cooperative behavior is experimentally verified using actual autonomous mobile robots.

1 Introduction

Growth in car traffic in an urban area has recently brought a lot of social problems such as a heavy traffic jam, an environmental pollution, and a traffic accident. To cope with such problems, many research projects on an intelligent transportation system (ITS) have been started and many works are reported from a viewpoint of both administrative and technological subjects. An automated transportation is one of the desired system in order to improve the traffic efficiency and to reduce traffic accidents. Most researches of an automated vehicle have an approach to develop an intelligent system that automatically drives a car instead of human or that assists an driver. These are indispensable techniques for the automated transportation, but not enough for smooth traffic flow control. Some system should arrange several or more automated vehicles in a limited area or even provide traffic information of the area that is not sensible for an vehicle since each vehicle controls its own speed, steering only based on local information such as distance from the former car, road conditions and so on. For example we empirically know that a driver can drive more smoothly by observing traffic flow not only proximate area but also distant area. An additional infrastructure for the automated vehicle is therefore required in order to provide information about a limited area not global area such a VICS, or to control traffic flow like a signal at an intersection or junction.

Some researches deal with vehicles behavior control by using an infrastructure [1][2][3]. However, the sites that requires such an infrastructure is not limited to an intersection or a junction. The place where vehicles cross each other or interflow into one line exists everywhere; parking lot, exit of garage and so on. It is economically impossible to set an equipment where it is necessary. The automated system for vehicle behavior control should be independent on the infrastructure and then should be installed into the vehicle.

The purpose of our study is to realize the smooth traffic flow without the additional infrastructure, embedding communication device into the automated vehicle. There are some researches about controlling vehicles behavior without an infrastructure but with an internal-vehicle communication device(IVC). In their researches, the control method is designed under central control architecture, where a leader of vehicles is supposed to exist or a base station for communication is supposed to set. Besides the system is not robust because vehicles are controlled by only one agent that might happen to break down or break away from a network of IVC. Therefore, a global order such a rhythm should be generated through local communication among vehicles in the limited area and they behave following the global order. This distributed autonomous approach could have advantages in viewpoint of install cost and system robustness.

On the other hand, researches achieving merging in an intersection have already reported the complete distributed method [4][5]. The platoon traveling has been achieved in these researches. However, it is assumed that each vehicle could identify the neighboring vehicles and communicate with them. It seemed to be difficult for traveling vehicles to communicate specified one. A desired communication type is a broadcast in a real world.

In this paper, we propose a self-control method of vehicle cooperative behavior at an intersection or a junction through a local communication with a close-by vehicle without infrastructures. In this method some vehicles in an limited small area generates a spatial-temporal pattern by themselves through a local communication with a close-by vehicle, and then each vehicle behaves itself according to the spatial-temporal pattern so that each vehicle could avoid collision and decrease speed control operations before an intersection or junction. We confirm the utility of the proposed method using several autonomous vehicles.

2 Weaving and Zipping Behavior

We consider cross motion at an intersection and join motion at a junction as examples of cooperative behavior. We named a cross motion "Weaving", and join motion "Zipping" shown in Fig. 1. Zipping is that vehicles that come from two roads alternately enters one road like a zipper, which can be seen in a junction of a highway and at exit of parking lot. Weaving is that vehicles that come from different roads alternately passes a cross-shaped intersection one by one, which can be seen in general intersection. The traffic signal currently works for traffic control only major intersections and junctions but in most cases, the infrastructure is not installed due to cost. A driver drives based on own judgment or drivers communicate each other by eye contact there. In the same way, an automated control system should not depend on some infrastructure device, and it should work only communications with neighboring vehicles.

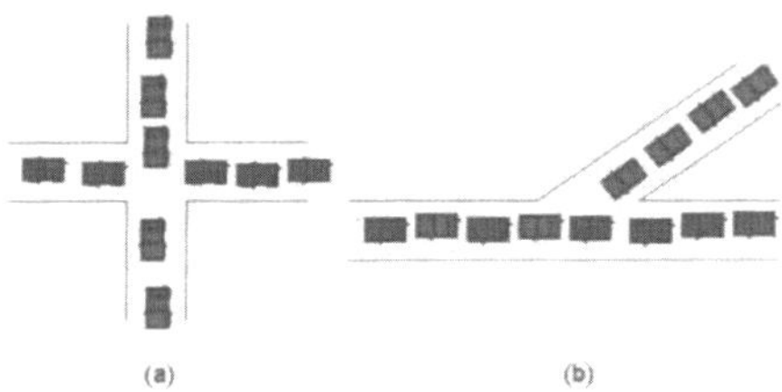

Fig. 1. Weaving and Zipping

In this paper, we consider that a junction for zipping consists of two incurrent roads and one excurrent road and that an intersection for weaving consists of one incurrent roads and one excurrent road. Those roads are supposed to be one-way for simplification.

3 Internal Vehicle Communication System(IVC)

3.1 IVC technology

IVC is an epoch-making system, which enable vehicles to communicate with neighboring vehicles directly. The system could provide safety and convenience to drivers since drivers could get various information about other vehicles such as position, velocity, direction and so on, and vehicles could cooperatively drive using these information. There are some examples as cooperative traveling; stop-and-go control, platoon traveling, collision avoidance, and emergency vehicle support.

The technology of IVC is under developing and there are two types of communication system for IVC. One is milli-wave type and the other is radio wave type. The former type requires a receiver but could exchange advanced information, however, there are some technical problems for implementation on hardware. The latter type can be realized with easy system, but information is limited. In this study, we suppose that the latter type IVC can be available. It means that our proposed algorithm can generate the spatial-temporal pattern based on simple communication way. Actually, we use a wireless LAN in our experiment.

3.2 Problem of IVC

IVC technology has problems of real-time control. One of the severest problems is to build a communication network under a dynamical environment. The allocation of network server and network border is difficult when vehicles in a network approaches to other network or they frequently breakaway from the network. The communication system used IVC should be used in limited conditions as follows;

- The dimensions of the communication information over a IVC network should be low as much as possible.
- The communications system should be free from any network servers.
- The communication type is simple broadcasting.
- The global order formation does not depend on a connection of communication

Under these conditions, we propose a method for cooperative traveling based on a reaction-diffusion system using IVC.

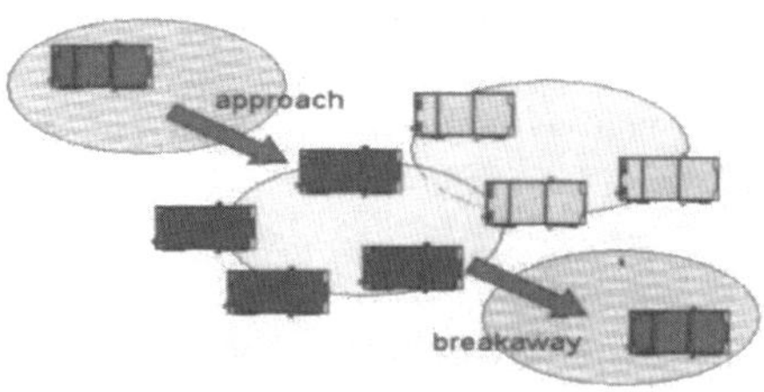

Fig. 2. IVC network

4 Generation of Cooperative Vehicle Behavior

4.1 Van der Pol model

A reaction-diffusion system is used in order to spatial-temporal pattern through simple communication. The reaction-diffusion system has dynamics of reaction and diffusion between morphgens. The diffusion of morphgens corresponds to an interaction between vehicles, which is broadcasting own value in a network group. It is possible to make a global spatial-temporal pattern by setting some factors and equations properly. The synchronization between vehicles can be controlled by digital device level [6]. In this study, van der Pol model[7], that is one of the reaction-diffusion systems, is used to make a temporal pattern. van der Pol model is formulated by equations (1)and (2), as follows;

$$\frac{\partial u}{\partial t} = \mu\{v - (u-p)(u-q)\} + D \cdot \frac{\partial^2 u}{\partial x^2} \tag{1}$$

$$\frac{\partial v}{\partial t} = (\frac{1}{\mu})v, \tag{2}$$

where u and v describe morphogens, respectively. D describes a diffusion coefficient. Parameters p, q and μ are coefficients that determines the dynamics. Using van der Pol model, the dynamics can be easily designed. Figure 3 shows a temporal pattern at one set of parameters. This periodical pattern converges into a state with a certain frequency. It does not depend on initial values of u, v, and network structure between vehicles.

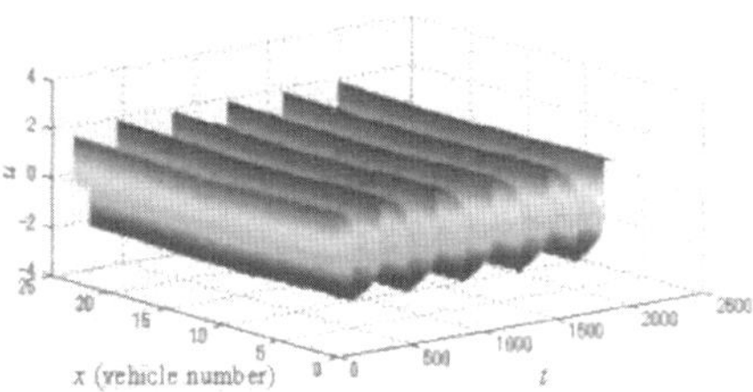

Fig. 3. Temporal pattern by van der Pol model

4.2 Self-organized periodical pattern

In order to generate spatial-temporal periodical patter, a reaction-diffusion system is applied into van del Pol model. The dimensions of van der Pol model is extended into two m and n, and we consider that two parameters interacts each other. The partial differential terms of eqs (1) and (2) are therefore changed to a reaction-diffusion term. The dynamics of two-dimensional system from equations (1) and (2) are expressed by

$$\frac{\partial u_m}{\partial t} = \mu\{v_m - (u_m-p)(u_m-q)\} + D(u_n - u_m) \tag{3}$$

$$\frac{\partial v_m}{\partial t} = (\frac{1}{\mu})v_m \tag{4}$$

$$\frac{\partial u_n}{\partial t} = \mu \{v_n - (u_n - p)(u_n - q)\} + D(u_m - u_n) \tag{5}$$

$$\frac{\partial v_n}{\partial t} = (\frac{1}{\mu})v_n \tag{6}$$

The spatial-temporal periodical patterns of u_m and u_n can be observed and u_m and u_n has the same frequency with a half-phase difference shown in Fig. 4.These periodical patterns can be used for traffic control in an intersection and a junction. The values m and n correspond road number and the values u_m and u_n correspond vehicle control. For example the time when the u_m becomes the maximum could be assigned the moment when one vehicle on the road m can enter an intersection or a junction. The vehicles on the road m and n can alternately enter an intersection because u_m and u_n periodically change with a half-phase difference. IVC is used for exchange the values u_m and u_n between vehicles on the road m and on the road n.

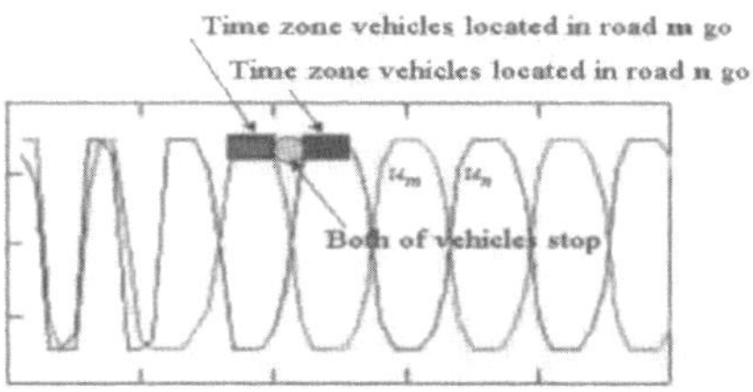

Fig. 4. Self-organized periodical pattern

4.3 Condition for Collision Avoidance

We consider the conditions that each vehicle does not collide each other. A vehicle has its finite size and takes some time to pass an intersection. There are three kinds of time intervals in one cycle when an intersection has two incurrent roads. One is a time t_m when a vehicle on road m passes an intersection. Another is a time t_n when a vehicle on road n pass an intersection. The other is t_s when two vehicles on both road m and n stop in front of an intersection. The total time tp of one cycle is

$$t_p = t_m + t_n + 2t_s \tag{7}$$

The vehicle has to finish passing an intersection during time interval t_m or t_n. Therefore, the following condition is needed to avoid collision;

$$\int_0^{t_m or t_n} v(t)dt > L + l \tag{8}$$

where L and l are width of a road and length of a vehicle, respectively. $v(t)$ is a velocity of a vehicle. Assuming that a velocity of vehicle is constant speed, eq.(8) is transformed to equation (9);

$$t_p > \frac{L + l}{v_{const.}} \tag{9}$$

Actually the parameters of equation (3)-(6) should be designed in order to meet the eq.(9). When these conditions are meeting, each vehicle can pass a intersection without collision. In the cases of zipping and weaving, each value of t_m is different because the length of L is different. The value of t_m becomes bigger since L on zipping is longer than one on weaving.

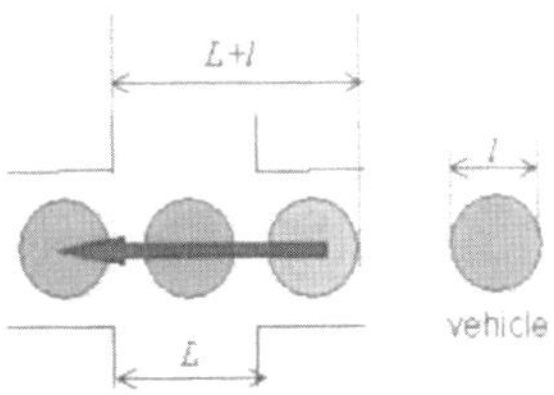

Fig. 5. Condition of collision avoidance

5 Experimental Setting

5.1 Intersection model

The intersection dimensions is determined so that the ratio of actual vehicle dimensions to experimental autonomous vehicle dimensions should be almost same as the ratio of a dimensions of typical intersection shown in Fig.6 to the experimental intersection shown in . Fig.7. A vehicle can communicate with each of other vehicles in communication area. A vehicle can recognize the existence on communication area when infrared sensor detects some landmarks in a side of a road. Actually, vehicles can get information about their location, direction, intersection from the car navigation system with GPS.

5.2 Vehicle configurations

In this experiment, a mobile robot is used instead of a vehicle. Assuming that Actual velocity of vehicles is 20[km/h] when vehicles pass a intersection, a velocity of a mobile robot is calculated 5[cm/s] by the rate of the intersection model. Though the mobile robot is circular form, the ratio of vehicle's size is calculated based on diameter, 6.5[cm]. A velocity is simplified as constant speed without acceleration and braking. Communication is carried out using

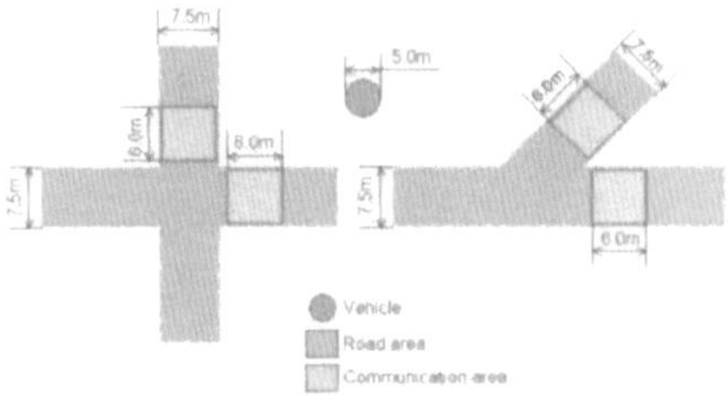

Fig. 6. Typical intersection dimensions

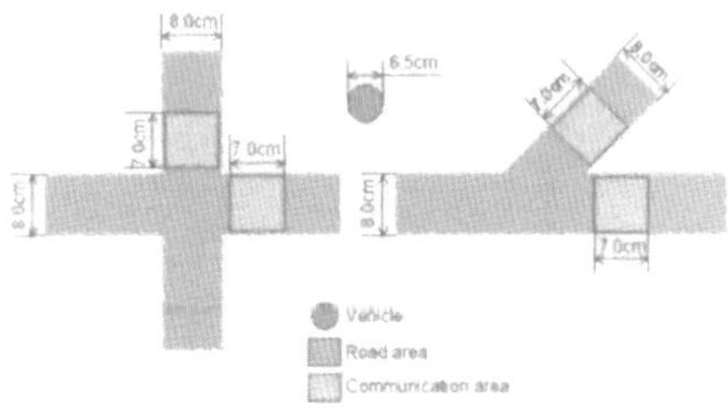

Fig. 7. Experimental intersection dimensions

wireless LAN (FUTABA FRH-SD07T). A time required for vehicles to send and receive one packet is 0.25[s].

Figure 8 shows an autonomous vehicle for the experiment. This vehicle is 8 centimeter high and 6.5 centimeter in diameter. A 16-bit microcomputer (H8/3048), some infrared sensors for line trace, a wireless LAN device to communicate with each robot are equipped. Therefore, it can move along the line on the ground autonomously, sensing the line by infrared sensors. Additional infrared sensors to detect landmarks of communication area and inside of an intersection are attached. Eight vehicles are used in the both zipping and weaving behavior.

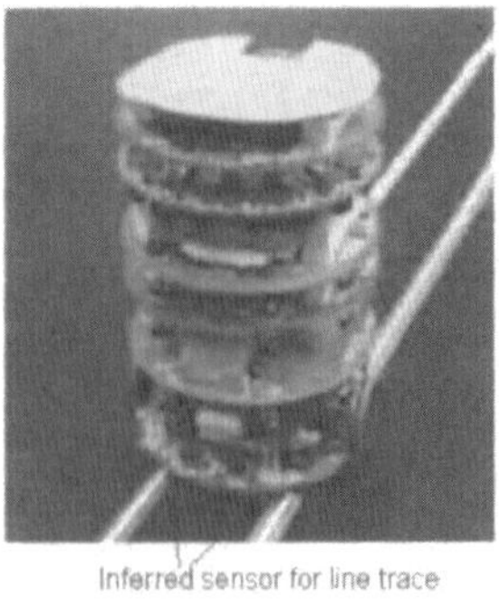

Fig. 8. Autonomous vehicle for experiment

Table 1. Parameter setting

Behavior	$L[cm]$	μ	p	q	D
Zipping	14.5	2.0	1.1	1.1	1.0
Weaving	17.7	3.0	1.1	1.1	1.0

6 Experiment

The parameter shown in table 1 are determined by numerical simulations for satisfying the condition of eq.(9). The experimental results of zipping and weaving are showed in figure 10(a). In both experiments, each vehicle alternately interflows at the junction and passes the intersection.

Additionally, zipping and weaving without vehicle's pause in front of an intersection is achieved by vehicle's keeping longer distance. The distance was longer than intersectional length because for a car from another road has to finish passing intersection before the next car enters the intersection shown in figure 9. The experimental result of zipping and weaving without the pause is showed in figure 10(a).

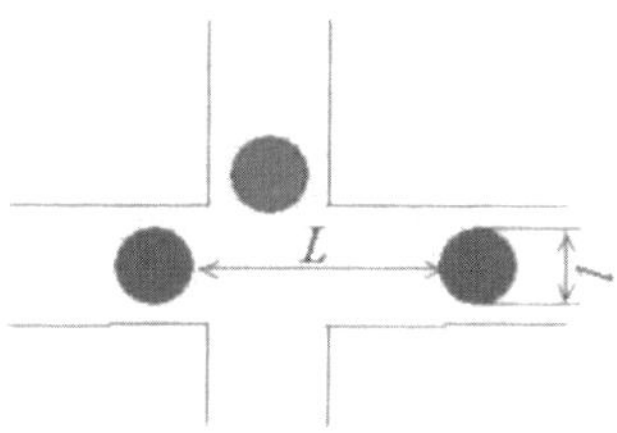

Fig. 9. Condition for weaving without pause

7 Conclusion

In this paper, we proposed the algorithm of a vehicle cooperative behavior control in an intersection and a junction without infrastructures. The algorithm with van der Pol model and the reaction-diffusion system generates a spatial-temporal pattern by exchanging internal values with neighboring vehicles. We applied the algorithm to self-control of traffic flow at an intersection and a junction as a example of cooperative behaviors. A capability of the algorithm is verified by the experiment with automated vehicles in the simplified condition.

We have to investigate our proposed algorithm in more precise environments including vehicle dynamics for its speed control.

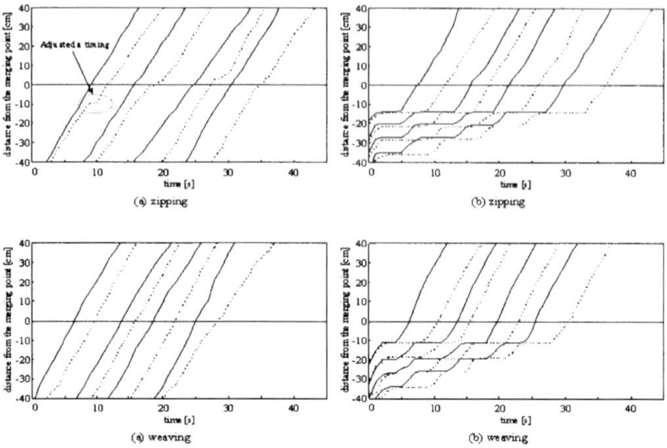

Fig. 10. Experimental results

References

1. Markos Papageorgiou (Ed.), "Merging Control", Cocise Encyclopedia of traffic and Transportation Systems, pages 257-263, 1991
2. Kosuke Sekiyama, Jun Nakanishi, Isao Takagawa, Toshimitsu Higashi and Toshio Fukuda, " Self-Organizing Control of Urban Traffic Signal Network ", IEEE International Conference on Systems, Man and Cybernetics, pages 2481-2486, 2001
3. Masao Sugi, Hideo Yuasa and Tamio Arai, "Autonomous Distributed Control of Traffic Signal Network", Transaction of the Society of Instrument and Control Engineers, 39, 1, pages 51-58, 2003, in Japan
4. Sasayuki Tsugawa, Shin Kato, Takeshi Matsui, Hiroshi Naganawa and Haruki Fuji, " An Architecture for Cooperative Driving of Automated Vehicles ", IEEE International Conference on Intelligent Transportation Systems, Dearborn, USA pages 422-427, 2000
5. Atsuya Uno, Takeshi Sakaguchi and Sadayuki Tsugawa, " A Merging Control Algorithm based on Inter-Vehicle Communication ", IEEE International Conference on Intelligent Transportation Syatems, Tokyo, Japan, pages 783-787, 1999
6. Kimberly Sharman Evans, Cem Unsal, and John S. Bay, " A Reactive Coordination Scheme for a Many-Robot System ", IEEE Transaction on Systems, Man and Cybernetics-Part B : Cybernetics, Vol. 27, No.4, August 1997
7. Steven H. Strogatz, "Nonlinear Dynamics and Chaos". Westiview Press, ISBN 0-7382-0453-6, 1994

High-Level Modelling of Cooperative Mobile Robot Systems[1]

R. Sánchez-Herrera, N. Villanueva-Paredes, E. López-Mellado

CINVESTAV-IPN Unidad Guadalajara, Prol. López Mateos Sur 590, 45090 Guadalajara, Jal. Mexico. elopez@gdl.cinvestav.mx

Abstract— This paper concerns to the specification of interactive mobile robot systems. A Petri net based approach is held: The definition of nLNS, a multi-level net formalism handling nets and/or symbols as tokens is proposed. nLNS supports a modular and hierarchical modelling methodology that is presented through a case study regarding a mobile robot community evolving into a structured environment.

Keywords: Cooperative mobile robots; Modular and hierarchical modelling; Interaction protocols; Petri nets with dynamic tokens.

1. Introduction

The development of coordination systems for large and complex discrete production plants requires, from the earliest stages of design, concise and unambiguous specifications of the activities to be automated and the management of resources. These requirements are fulfilled using formal modelling methods.

Among the existing specification techniques for discrete event systems, Petri nets (PN) have been widely adopted as modelling formalism in the automation community, because of its graphical nature and simple mathematical support allowing clear descriptions of systems exhibiting concurrency [2]. A widely used PN extension named coloured PN (CPN) [6], which handles token identities (colours) represented by symbols in the marking, allows compact models of complex systems.

Later, other extensions were proposed to deal with object oriented software specification: concepts on high level PN and object oriented programming were merged leading to Cooperative Objects [11] and OOPN [8]. These methods bring near the models to software implementation in despite of the loss of clearness of the description.

[1] This work has been partially supported by CONACYT (project 41968-Y)

Recently, R. Valk [12] considers PN as tokens proposing a two-level PN "code-clean" formalism independent to programming languages. Similar definitions have been proposed in [5], [9], and [7].

In this paper we also adopt the approach "nets into nets"; in previous works, the Valk's definition was extended to a less restrictive three-level PN system (NS-3) used to specify mobile software agents [1], batch processes [13], and mobile robot systems [10]. In this paper a definition of multi-level net system (nLNS) is proposed; also it is shown how nLNS copes with the modelling of collaborative tasks of a mobile robot community evolving into a structured environment.

2. A Multi-Level Net System

A model expressed in n-LNS consists mainly of an arbitrary number of nets organized in *n* levels according to a hierarchy. A net may handle as tokens, nets of deeper levels and symbols; the nets of level *n* permit only symbols as tokens similarly to CPN. Interactions among nets are declared through labels associated to transitions.

2.1 n-Level Net System

- ***PN structure.***

Definition. A *PN structure* is a bipartite digraph denoted by a triple $G = (P, T, F)$ where P and T are finite non empty set of vertices called places and transitions respectively, $P \cap T = \varnothing$, and $F \subseteq P \times T \cup T \times P$ is a flow relation of the net. Pictorially, places are represented as circles and transitions as bars or rectangles.

- ***Type nets***

Definition. A type-net of level i is a tuple $typenet_i = (G, TOKEN_i, LABEL_i, VAR_i, \tau, \lambda, \pi)$ for $1 \leq i \leq n$, where:

- G is a PN structure
- $TOKEN_i$ is a finite non empty set of type-nets and symbols permitted into the places of a net level i: $TOKEN_i \subseteq \{typenet_{j,k} \mid i<j\leq n,\ 1\leq k\leq r\} \cup SYMBOL_i$

  n is the number of levels of a multi level net system,

  r is the number of different type-nets allowed into places of a net of level i.
- $SYMBOL_i$ is a finite set of symbols allowed into the places of a net of level i.
- $LABEL_i$ is a finite set of labels defined for a net level i; $LABEL_i \subseteq LABELS$.
- $VAR_i = \{x, y, ...\} \subseteq VARS$ is a finite set of variables defined to a net of level i, where: $Type$: $VAR_i \rightarrow (TOKEN_i - SYMBOL_i)$ assigns types to variables.

- $\tau : P_i \rightarrow 2^{TOKEN\,i} - \varnothing$ is an assignment function of types to places.
- $\lambda : T_i \rightarrow 2^{LAB\,i} - \varnothing$ is an assignment function of labels to transitions, where:
  if i = 1 then $LAB_i = \{\varepsilon\} \times (LABEL_i \cup \{\varepsilon\}) \times \{\varepsilon\}$
  if $2 \leq i \leq n-1$ then $LAB_i = (LABEL_i \cup \{\varepsilon\}) \times (LABEL_i \cup \{\varepsilon\}) \times (LABEL_i \cup \{\varepsilon\})$
  if i = n then $LAB_i = (LABEL_i \cup \{\varepsilon\}) \times \{\varepsilon\} \times (LABEL_i \cup \{\varepsilon\})$
- $\pi : F_i \times LABEL_i \rightarrow M_{VARi \cup SYMBOLi}$ is a weighting function that assigns to every arc, a multi-set of variables and symbols, with respect to transition labels. If $label \notin \lambda(t)$, $\pi((p, t), label) = \pi((t, p), label) = \varnothing$. Moreover if i=n then $VAR_i = \varnothing$, so that $\pi : F_n \times LABEL_n \rightarrow M_{SYMBOLn}$.

A type-net $typenet_i$ is a PN structure with additional information that declares and handles data defined in $TOKEN_i$, according to the *pre* and *post* conditions established by π and λ.

- ***Nets of level i***

Definition. A net of level i is a type-net $typenet_i$ with a marking μ_i : $NET_i = (typenet_i, \mu_i)$; $1 \leq i \leq n$, where $\mu_i : P_i \rightarrow M_{NETSTOKENi \,\cup\, SYMBOLi}$ is a marking function for the type-net of level i. $NETS_{TOKENi} \subseteq \{NET_{i+1}, NET_{i+2}, \ldots, NET_n\}$

- ***Net system***

An n-LNS model, called *net system*, is the set of all the defined nets.

Definition. A n-level net system is a n-tuple NS= $(NET_1, NET_2, \ldots NET_n)$ where NET_1 is the highest level net, *and* $NET_i = \{NET_{i,1}, NET_{i,2}, \ldots, NET_{i,r}\}$ is a set of r nets of level i.

Figure 1 sketches pieces of the components of a 4-LNS. The level 1 is represented by the net NET_1, the level 2 by the nets $NET_{2,1}$ and $NET_{2,2}$, the nets $NET_{3,1}$, $NET_{3,2}$, $NET_{3,3}$ and $NET_{3,4}$ compose the level 3, and the nets $NET_{4,1}$, $NET_{4,2}$, $NET_{4,3}$ form the level 4.

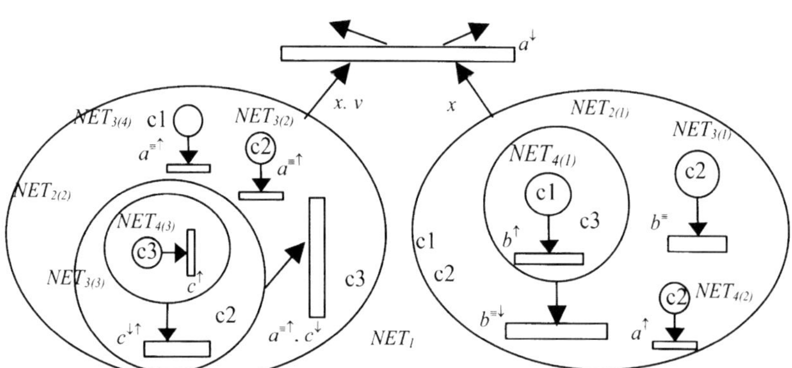

Fig.1. Piece of a 4-LNS

2.2 Net system evolution

The components of a model may interact among them through synchronization of transitions. The synchronization mechanism is included in the enabling and firing rules of the transitions; it establishes that two or more transitions labelled with the same symbol must be synchronized.

In order to define the enabling conditions and firing of transitions we introduce first the notion of variable binding.

Definition: A binding b on a variable set $VARS$ = {x, y, ...} is a function b: $VARS \rightarrow NETS_{TOKENi}$; for a $v \in VARS$, b(v) is a lower level net whose type is $Type(v)$. b_t maps every variable defined on the weight of the input arc to the transition t, with respect to a label. $m_{<b>}$ denotes a multiset of nets resulting of instancing a multiset of variables m with b.

• ***Enabling rule***

Definition. A transition t of a net of level i NET_i is *enabled* with respect to a label $lab \in \lambda(t)$ if:

- There exists a binding b_t: $VAR_t \rightarrow NETS_{TOKENi}$, where VAR_t is the set of variables appearing in all $\pi((p, t), lab)$,
- it must fulfill that $\forall p \in \bullet t$, $\pi((p, t), lab)_{<bt>} \subseteq \mu_i(p)$.
  (The binding $<b_t>$ is not necessary when the level net is n).
- The conditions of one of the following cases are fulfilled:
  Case 1. If $lab = (\varepsilon, \varepsilon, \varepsilon)$. The firing of t is autonomously performed.
  Case 2. If $lab \neq (\varepsilon, \varepsilon, \varepsilon)$ one must consider the following situations:
  i) $lab = (l, \varepsilon, \varepsilon)$. It is required the simultaneous enabling of the transitions labelled with $l^{=}$ belonging to other nets staying into the same place p' of the next upper level net. The firing of these transitions is simultaneous and all the (locally) synchronized nets remain into p'.
  ii) $lab = (\varepsilon, l, \varepsilon)$. It is required the enabling of the transitions labelled with $l^{\uparrow}$ belonging to other lower level nets into $\bullet t$. These transitions fire simultaneously and the lower level nets and symbols declared by $\pi((p, t), lab)_{<bt>}$ are removed.
  iii) $lab = (\varepsilon, \varepsilon, l)$. It is required the enabling of at least one of the $t' \in p'\bullet$, labelled with $l^{\downarrow}$, of the upper level net where the NET_i is contained. The firing of t provokes the transfer of NET_i and symbols declared into $\pi((p', t'), lab)_{<bt>}$.

A label may involve a combination of these clauses. So, (l, l, l) indicates that a transition must be synchronized locally, internally, externally respect to the symbol $l^{=\downarrow\uparrow}$. These situations apply to nets at all levels, except for the nets of levels 1 and n in which it is allowed only *internal synchronization* ($l^{\downarrow}$), and *local* and/or *external synchronization* ($l^{=}$, $l^{\uparrow}$, $l^{=\uparrow}$) respectively.

- ***Firing rule***

At all levels the firing of a transition t modifies the marking by removing $\pi((p, t), lab)_{<bt>}$ in all the input places and adding $\pi((t, p), lab)_{<bt>}$ to the output places. The binding $<b_t>$ is not necessary for nets of level n.

In figure 1, NET_1 is synchronized through the transition labelled with $a^{\downarrow}$, with $NET_{2,2}$, $NET_{3,2}$, $NET_{3,4}$ and $NET_{4,2}$ by mean of the transitions (locally synchronized) labeled with $a^{\uparrow}$; all these transitions must be enabled to fire. The simultaneous firing of the transitions removes these nets from the input places. $NET_{2,1}$, $NET_{3,1}$ and $NET_{4,1}$ are synchronized through the transitions labeled with $b^{\downarrow}$, $b^{\equiv}$, $b^{\uparrow}$ respectively; the firing of the transitions changes the marking of $NET_{2,1}$, and $NET_{3,1}$; $NET_{4,1}$ is removed from the place of $NET_{2,1}$. Also $NET_{3,3}$ is removed from the input place of $NET_{2,2}$, and $NET_{4,3}$ is removed from $NET_{3,3}$; this interaction is declared by $c^{\downarrow}$, $c^{\downarrow\uparrow}$, $c^{\uparrow}$, respectively.

3. Modelling of Interactive Mobile Robots

3.1 Mobile Robot Coordination

The coordination of multiple mobile robots evolving into a structured environment involves a wide variety of techniques and approaches that address problems related with the activities performed by robots namely sensing, planning, decision-making, interaction, and mission execution.

nLNS is used for specifying the coordination of these activities to facilitate the development of a distributed software coordination system. In this paper we focus on the modelling of robot tasks that require interaction with other robots. The proposed formalism allows representing in a modular way the main components of distributed software implementing a mission coordination system. The methodology for building coordination models that takes advantage of the modularity of nLNS is presented through a simple case study, modelled by three-level net system: the level 1 describes the robots environment, the level 2 models the general behaviour of the mobile robots, and the level 3 represents specific features of a given robot, namely missions, tasks, roadmaps, protocols, and resources.

3.2 Case Study

- ***System description***

Consider an automated discrete production system consisting of two manufacturing cells (C_1, C_2) and two warehouses, one for the raw material

(IW), and the other one for packed finished products (OW). A set of transporting robots R_1, …, R_m perform a) the feeding of the first processing stage (executed into C_1), b) the transfer among cells of products, and c) the storage of the packed products. The scheme of figure 2 depicts the layout of the manufacturing system. The example is close to that presented in [3].

- ***Scenario***

Suppose that three robots have been assigned to transfer the finished products from the C_2 to OW. So the mission for these robots consists in picking the packages up, transporting, and storing the packages in OW. Assume that the packages may have different weight, being necessary that at most two robots be involved in the transportation task. Suppose that all the robots have the same basic capabilities, i.e. they can perform the same operations, but the tasks are individually programmed in every robot.

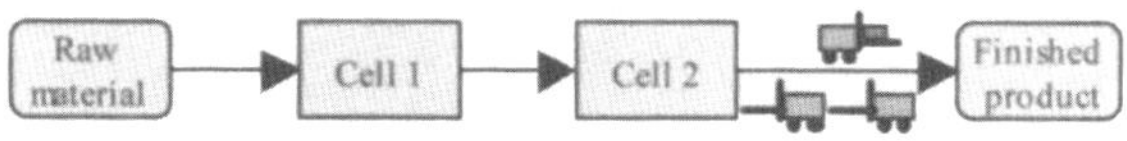

Fig. 2. System layout

3.3 Mission Coordination and Collaborative Task Models

- ***Layout model.***

The net of level 1 is straightforward obtained from the system layout; each place represents either a cell or a warehouse, and transitions represent the displacement of robots from one cell to another cell. Figure 3 shows the structure and the initial marking of this net for the proposed scenario.

Transitions have labels of the form R*i*C*k* and R*i*R*j*C*k*: the former represents the displacement of a robot *i* to the cell *k*, and the later the displacement of a robot *i* together with a robot *j* to the cell *k* (collaborative transporting). For readability it is only shown the labels of t_5 and t_6.

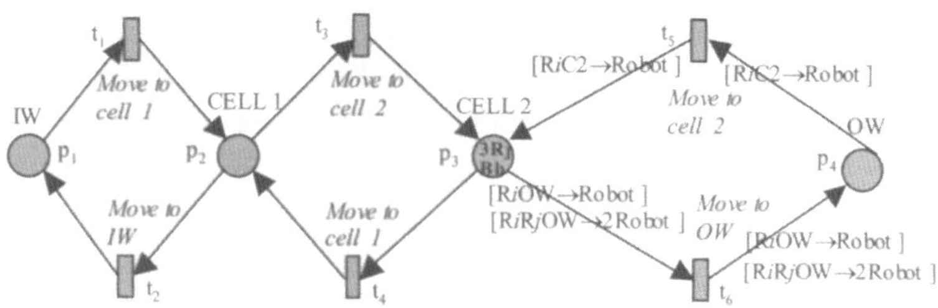

Fig 3. Layout model

The places of the layout model may be marked nets of level 2 representing the robots, and nets of level 3 representing resources; in this example it is only considered one type of level 2 for the robot, and one type of level 3 for the resources; these models are described below.

- ***Robot behaviour model.***

The behaviour of a mobile robot is described by a net of level 2 moving through the layout model; the robot model represents the possible states that the robot controller follows to execute its mission into the environment. The model presented herein handles token-nets of level 3 for representing the mission, the tasks the robot is able to perform, the interaction protocols, and the navigation roadmap. We first describe the robot model.

Robot model. Figure 4 shows the structure of the net of level 2 that describes the behaviour of the carrier robot. The task models are contained into p2, while the mission is marking p1 initially; the map model is contained into p6, while the nets specifying the protocols are marking p4; these token-nets are described below. The activities of the robot are exclusively navigating or executing tasks. t1 represents the displacement of the robot from one cell to another; its firing is synchronized with the transitions of both mission and map models through the label $\lambda(t_1)$.

When a robot enters to the cell where it must execute tasks, t2 fires by removing the mission and the pertinent task from p1 and p2 respectively; these token-nets are added into p3. In this place the token-net representing the task evolves by firing transitions that represent the operations of the tasks executed by the robot; when the task requires collaboration with another robot, t3 is fired, resuming a negotiation process that uses one protocol stored in p4 (a *contract net* protocol [4]) in order to determine which robot will help. When the negotiation stage finishes, t5 fires removing the task and protocol nets, and adding these nets into p3 and p4 respectively; this enables the robot the execution of the collaborative task.

Missions. The mission is modelled by a net of level 3 that describes the sequences of tasks to be executed by the robot; such tasks include navigation sequences. Figure 5 shows the mission each robot must complete.

If another robot is asking for help, the robot attempts to help; then a negotiation process is started: if the robot refuses to help or its offer is rejected, it restarts the mission. If no robot is asking for help, the robot lifts the package (alone or with the help of other robot), transports it to output warehouse, and deposits the package; then the robot returns to cell 2.

Tasks. The tasks are modelled by nets of level 3 that describe the operations to be performed by a robot into a cell; figure 6 shows the task each robot has to carry out. A robot tries to carry the package from cell 2 to OW; if this task must be performed in collaborative way, the negotiation process (coordinated with a negotiation protocol) is executed for selecting the collaborator; then the package is picked and transported by both robots. If there is no collaborator, the robot attempts to move another package.

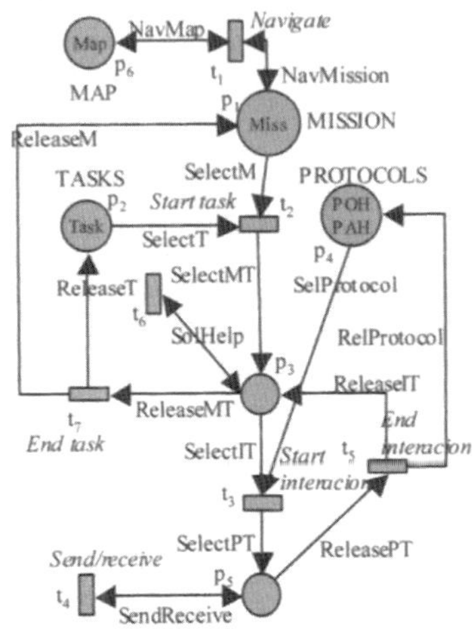

Fig. 4. Mobile robot model

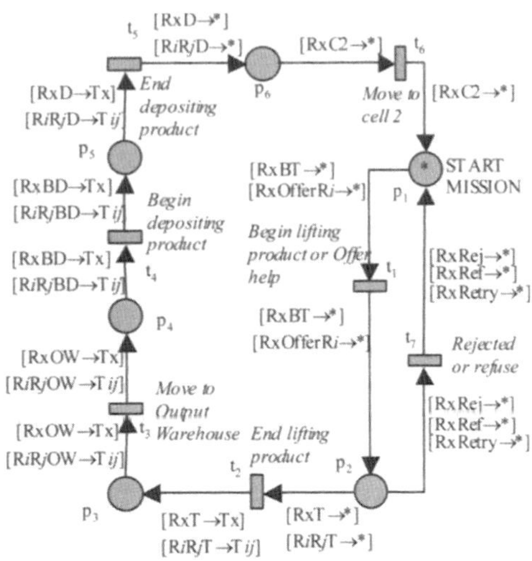

Fig. 5. Mission model

Maps. The navigation roadmap is a net of level 3 that specifies the permitted access of the robot to the cells; this allows specifying the mobility constraints of each robot. The roadmap net structure is a sub-graph of the layout net structure; if there are not mobility constraints the PN structure of this net is the same that that of the first level net.

Interaction protocols. Negotiation processes among the robots are performed according to *interaction protocols*. In this example a Contract Net protocol is used for determining the collaborator when a robot needs help to transport a package [4]. A robot may play either the roles of *manager* or *bidder* when it asks help or offers to collaborate respectively. Figures 7.a and 7.b show the token-nets representing the manager and bidder roles.

The manager sends the *call for proposals* (CFP), including the specification of the task, as well as any condition the manager is placing upon the execution of the task. The bidder can accept or refuse the CFP; if it accepts it sends a proposal. The manager receives the proposals and the refusing messages; then it evaluates the proposals, and selects the robot will help to perform the transportation task; then, it notifies its decision to all the robots, and the protocol ends. If no robot sends a proposal, the protocol ends, and the task starts again.

• Other third level models

A robot model interacts with other robot models and with the nets of level 3 representing resources or other abstract entities such as blackboards. In the example, a token-net residing into a place represents a blackboard, where the robots "write" a request for collaboration when they need help; also other robots find out who needs help. Moreover, this blackboard can be used to assure that the role of manager is assigned to just one robot at a time.

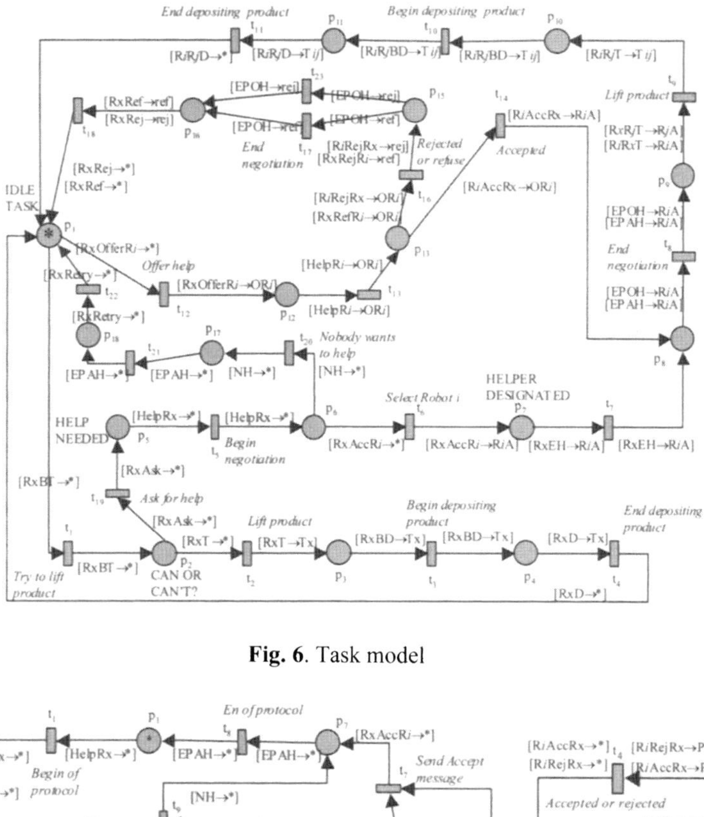

Fig. 6. Task model

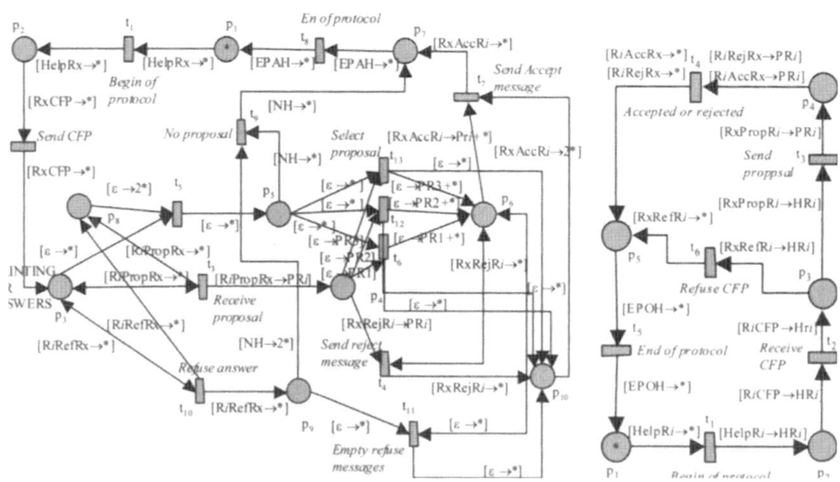

Fig. 7. Protocols for the manager and bidder role

4. Conclusions

This paper addressed the problem of high level specification of mobile robot systems. It is proposed a multi-level net system formalism allowing specifying the different activities performed by a mobile robot community evolving into a structured environment.

n-LNS suggest a modular and hierarchical modelling strategy allowing to define models for separate entities from the specification of tasks, missions, roadmaps, the environment, and robot capabilities; this feature permits to obtain systematically clear and compact nets, which are related through symbolic labelling of transitions, obtaining a global model of complex system. Such a model establishes an appropriate specification document for agent-based design of distributed control software.

References

1. Almeyda, H. I. (2002) "A three-level net system for the modelling of mobile agents" MSc.Thesis (*in Spanish*). Cinvestav-IPN, Guadalajara, México, September
2. DiCesare, F. G.Harhalakis, J.M.Proth, M.Silva, F.B.Vernadat. (1993) "Practice of Petri Nets in Manufacturing". Chapman & Hall.
3. Ferber, J. (1999) "Multi-Agent Systems. An introduction to Distributed Artificial Intelligence". Addison-Wesley. UK.
4. FIPA ORG. (2002) "FIPA Contract Net Interaction Protocol Specification". Document no. SC00029H.
5. Hiraishi K. (2000) "A Petri-net-based model for the mathematical analysis of multi-agent systems". Proc. of the IEEE ICSMC, Nashville, Tennessee, USA. October.
6. Jensen, K. (1981) "Coloured Petri Nets and the Invariant Method". Theoretical Computer Science Vol. 14 pp 317-336.
7. Kummer, O. (2001) "Introduction to Petri nets and Reference nets". Sozionikaktuell 1, pp.7-16
8. Lakos C. (1995) "From Coloured Petri Nets to Object Petri Nets". Proc. of 16th ICATPN, LNCS 935, pp 278-297, Torino, Italy, Springer-Verlag
9. Lomazova, I. (2000) "Nested Petri nets –a formalism for specification and verification of multi-agent distributed systems". Fundamenta informaticae 43 pp. 195-214. IOS Press
10. López, E., H. Almeyda (2003) "A Three-Level Net Formalism for the Modelling of Multiple Mobile Robot Systems". IEEE ICSMC, Washington, USA. Oct. pp.2733-2738
11. Sibertin-Blanc, C. (1994) "Cooperative Nets". Proc. of the 15th International Conference on Application and Theory of Petri Nets, LNCS 815, Zaragoza, Spain. June
12. Valk, R. (1998) "Petri nets as token objects: an introduction to elementary object nets" Proceedings of ICATPN, LNCS 1420 pp. 1-25, Springer-Verlag
13. Villanueva, N. , E. López, H. Almeyda (2003). "Three-level Modelling of Batch Processes". Proc. CESA 2003, Symposium on Discrete Events in Industrial and Manufacturing Systems. Lille, France, July

Part XI

Distributed Control

Lateral and Longitudinal Stability for Decentralized Formation Control

David J. Naffin[1], Mehmet Akar[2], and Gaurav S. Sukhatme[1]

[1] Robotic Embedded Systems Laboratory
Department of Computer Science
University of Southern California
Los Angeles, CA 90089
dnaffin@robotics.usc.edu
gaurav@robotics.usc.edu

[2] Communication Sciences Institute
Department of Electrical Engineering
University of Southern California
Los Angeles, CA 90089
akar@usc.edu

Summary. This paper analyzes the stability properties of a decentralized hybrid control system for maintaining formations. Utilizing only local sensing, the system assembles strings or "platoons" of robots that has each robot maintaining a fixed bearing to its nearest neighbor. Using these platoons, the system is able to construct more complicated geometries. A piecewise linear controller based on bidirectional controller design is utilized to ensure the stability of the system. The system is demonstrated in simulation as well as on a physical set on non-holonomic mobile robots.

1 Introduction

In a previous paper [8] we outlined a design for a decentralized control system that assembles as well as maintains formations of robots in simple geometric shapes. In this paper we examine the stability properties of this design. In particular we examine our solution to the string stability problem for both the longitudnal as well as the lateral control problem for each vehicle.

Recently there has been increased interest in assembling and maintaining formations of autonomous robots. Applications that would greatly benefit from robust formation control range from Automated Highway Systems (AHS) to clusters of satellites to formations of Unmanned Aerial Vehicles (UAVs) performing reconnaissance tasks.

Our formation controller design constructs larger formations from collections of smaller lines or "platoons" of robots. An important property of any formation controller is the ability to form stable configurations. In particular 'string stability' requires that all positional errors between robots when viewed from the lead vehicle be constant or decreasing. By positional errors we are referring to both the inter-robot spacing error (longitudinal errors) as well as each robot's bearing to its predecessor (lateral errors).

The rest of the paper is organized as follows. In Section II we address related works that have been published in the control, multi-agent and robotics community. Section III provides an overview of the system. Section IV defines the stability criteria and describes the basic approach we use for designing our control policies. Section V provides results of several simulations as well as a few physical robot results. Finally, section VI comments on our future works.

2 Related Works

Some of the earliest research on strings of moving vehicles was done by Levine and Athans [6]. They showed how a string of high speed moving vehicles could be controlled using a Linear Quadratic Regulator. Peppard [9] added to this work by showing how string stability could be obtained with PID control using both forward and rearward separation measurements.

Swaroop and Hedrick [11] are often cited as the first to give formal definitions for *string stable*, *exponentially string stable* and l_p *string stable*. Li et al [7] explored the effects of communication delays on string stability. Canudas de Wit and Brogliato [2] provided a detailed overview of string stability and how various control polices and inter-vehicle spacing strategies affect string stability. Seiler et al [10] analyzed various classes of linear controllers with regard to string stability.

In the area of non-holonomic mobile robot control, Aguiar et al [1] used Lyapunov functions to design a nonlinear controller that produces smooth trajectories. Desai et al [3] used input-output linearization to design a nonlinear controller to maintain robot formations. Vidal et [15] demonstrated the use of omni-directional cameras and a nonlinear control to maintain formations.

Input-to-state stability of formations, a more relaxed form of mesh stability, have been studied by Tanner, Pappas and Kumar [13], [12]. They also formalized a new metric for analyzing Leader-to-formation stability *LFS* [14]. By using graph laplacians, Fax and Murray [4] have developed a Nyquist-like criteria for vehicle formations.

3 The Approach

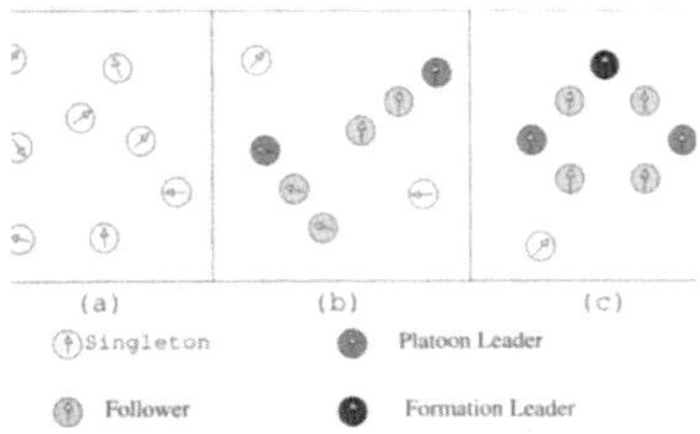

Fig. 1. Assembling Formations: (a) starting from a collection of singletons (b) coalescing into platoons (c) and finishing a diamond formation.

Our approach to assembling formations is to dynamically grow them from *singletons*, (i.e. single robots with no constraints on their motions) into *platoons* (i.e. line segments where each follower robot is constrained to follow its nearest neighbor) and finally into more complicated geometries (see figure 1). Our approach has several advantages over other approaches. The control graphs for most members of a formation only require sensing nearest neighbors. The exception to this rule are the platoon leaders. For some formation configurations, they will need to follow two leaders. However, for most formations, the number of platoon leaders is small when compared to the size of the formation. This minimizes the sensing requirements for each robot.

The system utilizes a hybrid control whereby robots switch between several behaviors, or modes of operation. The formation (global) state information is distributed among the robots in the two leader states. The rest of this paper addresses the issue of the stability of platoons within these formations. For the moment we will not address the issues of leader stability (i.e. mesh stability of the formation graph) nor the stability of behavior transitions.

4 String Stability and Linear Control

Simply stated, a system is considered *mesh stable* if a disturbance is attenuated as it propagates from one subsystem to the next. For the one-dimensional case, this property is refereed to as *string stability*. For multiple dimensional cases it is known as *mesh stability*.

Consider the interconnected system

$$\dot{x}_i = f(x_i, x_{i-1}, \cdots, x_1) \tag{1}$$

where $i \in \mathcal{N} = \{1, 2, \ldots, N\}$, $x_i \in \Re^n$, $f : \Re^n \times \cdots \times \Re^n \rightarrow \Re^n$, $f(0, \ldots, 0) = 0$, and $x = [x_1^T, \ldots, x_N^T]^T$.

String stability: The origin $x = 0$ is said to be *string stable* if for any $\epsilon > 0$, there exists a $\delta > 0$ such that

$$||x_i(0)||_\infty < \delta \Rightarrow sup_i||x_i(\cdot)||_\infty < \epsilon$$

Exponential string stability: The origin $x = 0$ is said to be *exponentially string stable* if it is string stable, and $x_i(t) \rightarrow 0$ asymptotically for all i. This property is considered stronger and therefore more desirable then just string stable.

l_p **string stability:** The origin $x = 0$ is said to be l_p *string stable* if given any $\epsilon > 0$, there exists a $\delta > 0$ such that

$$||x_i(0)||_p < \delta \Rightarrow sup_i \left(\int_0^\infty |x_i(t)|^p \right)^{1/p} < \epsilon$$

for all $p \in [1, \infty]$. This property is considered weaker than simple string stable. [11]

4.1 Longitudinal Control

We are assuming a string of N robots obeying identical kinematic and dynamic constraints. For this case study we will assume that each robot obeys the following constraint:

$$\dot{x} = u \tag{2}$$

where u is the control input to the system. All robots except the leader implement identical control policies. The design of these control policies can be categorized by the number and types of constraints they attempt to maintain. For the rest of this paper we will discuss three types of controller designs; *unidirectional*, *leader-centric* and *bidirectional* control.

Unidirectional Controllers: This type of controller implements only a single constraint. It attempts to minimize the following error:

$$e_i(t) = x_{i-1}(t) - x_i(t) - x_d \tag{3}$$

where x_d is the desired inter-robot spacing distance. Only a single local measurement of the inter-robot spacing from the robot immediately in front (i.e. toward the leader) is necessary to implement this control strategy. Since only a single local measurement is necessary, it is the most desirable of the three control strategies. Using the standard linear (PID) combinations of this error signal results in a control policy of:

$$U_i(s) = K(s)E_i(s) \tag{4}$$

where

$$K(s) = k_d s + k_p + \frac{k_i}{s} \tag{5}$$

Implementing a feedback system that utilizes this controller will result in the following transfer function:

$$\frac{E_i(s)}{E_{i-1}(s)} = \frac{k_d s^2 + k_p s + k_i}{(1+k_d)s^2 + k_p s + k_i} \tag{6}$$

In order for this system to Bounded-Input Bounded-Output (BIBO) stable, (6)'s two poles must be less than to zero (i.e. the LHP). The poles' locations are found by:

$$\frac{-k_p \pm \sqrt{k_p^2 - 4k_i - 4k_i k_d}}{2k_d + 1} \leq 0 \tag{7}$$

This requirement results in the following constraints on the selection of the gain parameters (k_p,k_i, and k_d):

$$k_p > 0 \tag{8}$$

$$k_i > 0 \tag{9}$$

$$k_d > -1 \tag{10}$$

In addition the system will need to meet the constraints necessary for string stability. In particular :

$$\left\| \frac{E_i(j\omega)}{E_{i-1}(j\omega)} \right\|_\infty = \left| \frac{k_i - k_d\omega^2 + jk_p\omega}{k_i - (1+k_d)\omega^2 + jk_p\omega} \right| \leq 1 \tag{11}$$

which simplifies to:

$$\frac{(k_i - k_d\omega^2)^2 + k_p^2\omega^2}{(k_i - (1+k_d)\omega^2)^2 + k_p^2\omega^2} \leq 1 \tag{12}$$

$$\omega \geq \sqrt{\frac{2k_i}{1+2k_d}} \tag{13}$$

In order for this system to be string stable for all frequency of $\omega \geq 0$ it will be necessary that the integral gain parameter (k_i) be zero. Therefore there is no choice of k_p, k_i and k_d that will result in a controller for each follower that will guarantee both BIBO stability as well as string stability. This result has been proved for systems that employ more complicated dynamics by many others (e.g. Peppard [9], Seiler et al[10], etc).

Leader-centric Controllers: Leader-centric controllers implement two constraints. They attempt to minimize both the inter-robot spacing error given by (3) as well as a new constraint of:

$$e_i^l(t) = x_0(t) - x_i(t) - ix_d \tag{14}$$

where $x_0(t)$ is the platoon leader's current position. This new constraint (14) requires a much more difficult to obtain global measurement between the platoon leader and itself. This additional measurement effectively decouples the followers from one another. Any disturbance in the leader's trajectory is immediately (or nearly immediately) sensed by each follower in the platoon. An instance of a linear controller for this type can be written as:

$$U_i(s) = K(s)[\beta E_i(s) + (1-\beta)E_i^l(s)] \tag{15}$$

where $0 \leq \beta \leq 1$ can be thought of as a measurement mixing factor. The special case in this system is the first robot after the leader. Since the $e_i(t)$ and $e_i^l(t)$ are the same measurement for this case this robot will end up implementing (4) control policy. From this controller we can derive the following error transfer function:

$$\frac{E_i(s)}{E_{i-1}(s)} = \frac{\beta(k_d s^2 + k_p s + k_i)}{(1+k_d)s^2 + k_p s + k_i} \tag{16}$$

The characteristic equation of (16) is identical to the characteristic equation of (6) and therefore the BIBO stability constraints on the various PID gains (8),(9) and (10) still apply. However, the mixing factor *beta* plays an important role in the system's string stability. The constraint for string stability for this system is given by:

$$\left\| \frac{E_i(j\omega)}{E_{i-1}(j\omega)} \right\|_\infty = \left| \frac{\beta(k_i - k_d\omega^2 + jk_p\omega)}{k_i - (1+k_d)\omega^2 + jk_p\omega} \right| \leq 1 \tag{17}$$

For any BIBO stable choice for the PID gains k_p,k_i,k_d there is always a choice for β that will satisfy this constraint. In the extreme case of β equal to zero, each follower is completely decoupled from its neighbors. In this case we have N independent systems and there are no string instabilities. In the case that β equals one, the system degrades into a unidirectional controller.

Bidirectional Controllers: Similar to the *leader-centric* approach, the bidirectional controller strategies employ two constraints. Like the unidirectional and leader-centric strategies, the bidirectional approach attempts to minimize the inter-robot spacing error (3) as well as the following:

$$e_{i+1}(t) = x_i(t) - x_{i+1}(t) - x_d \tag{18}$$

This type of controller utilizes two local measurements; the inter-robot spacing between itself and its immediate predecessor and between itself and its immediate successor (i.e. the robot ahead as well as behind). Since both constraints require only local measurements, this strategy is more desirable then the leader-centric approach. An instance of a linear controller for this type can be written as:

$$U_i(s) = K(s)[\beta E_i(s) + (1-\beta)E_{i+1}(s)] \tag{19}$$

Again where $0 \leq \beta \leq 1$ can be thought of as a measurement mixing factor. Implementing a feedback system that utilizes this controller will result in the following two transfer functions: (16) as well as

$$\frac{E_i(s)}{E_{i+1}(s)} = \frac{(1-\beta)(k_d s^2 + k_p s + k_i)}{(1+k_d)s^2 + k_p s + k_i} \tag{20}$$

(20) represents how errors propagate forward along the string. This error transfer function must also meet the requirements for string stability. In particular:

$$\left\| \frac{E_i(j\omega)}{E_{i+11}(j\omega)} \right\|_\infty = \left| \frac{(1-\beta)(k_i - k_d\omega^2 + jk_p\omega)}{k_i - (1+k_d)\omega^2 + jk_p\omega} \right| \leq 1 \tag{21}$$

One might wonder why there would be any error signals propagating from the back of the platoon toward the leader? First, the bidirectional strategy results in a fully connected system (i.e. all follower robots effect all other followers). Second, the last follower robot in the platoon is a special case and simply implements the unidirectional control law (4). When a forward propagating disturbance reaches the last follower, it has a "reflective effect" that creates a disturbance that propagates back toward the leader.

It should be noted that these two string stability constraints are opposing each other. The choice of the measurement mixing factor β is a design trade-off. The optimal mixing factor is affected not only by the choice of PID gains but also by the length of the platoon as well as the special case of the last follower. A value of 0.5 is the special case of equal attenuation of disturbances in both directions. Values greater than 0.5 attenuates the disturbances fast in the forward then in the backward direction while values less then 0.5 have the opposite effect.

The choice of β also effects the reactiveness of the platoon to the leader's changes. A smaller values of β has a "sluggish" effect resulting in larger overshoot and settling times for the first followers in the platoon. The extreme case of β equal to zero will result in the platoon ignoring all leader inputs! Larger values of β will result in a more reactive system but increasing it will eventually lead to string instabilities.

4.2 Lateral Control

The control law given in the previous section will maintain spacing for strings that requires each robot in the formation to maintain a zero bearing with the robot ahead of it. However our approach to formations require that the members of the platoons hold various bearings. Could these control laws be adapted for these cases as well and maintain their stability properties? The answer to this question is very dependent on the actuation model of the robot. If the robot has enough actuation to independently maintain separation, bearing as well as heading then the answer is yes. For robots equipped with holonomic actuation, the lateral control law for a bidirectional controller is given by

$$u_i(t) = h_1\bar{\psi}_i + h_2\int_0^t \bar{\psi}_i d\tau - h_3\bar{\psi}_{i+1} - h_4\int_0^t \bar{\psi}_{i+1} d\tau \tag{22}$$

Given that:

$$\bar{\psi}_i = \psi_i - \psi_d \tag{23}$$

where ψ_i is the bearing to the robot's target and ψ_d is the desired bearing to that robot.

Since all robots in a platoon must maintain the same heading as its platoon leader, it is assumed that this information is communicated to each robot in the platoon and that each robot has the ability to sense its global heading (via a compass, inertia measurements, etc.). If heading changes by the platoon leader are infrequent, then communication bandwidth can be maximized by only communicating heading changes.

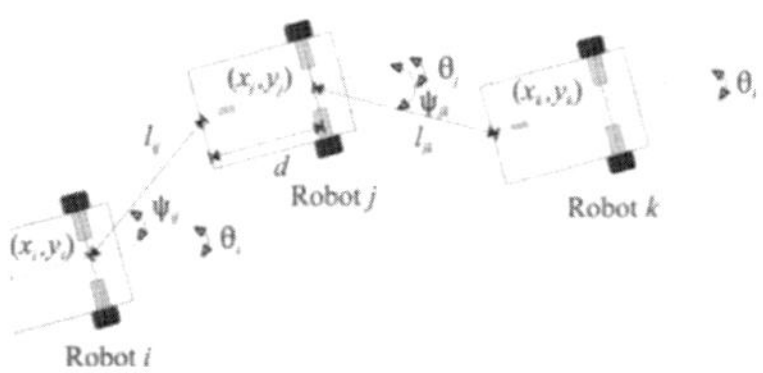

Fig. 2. Formation string

For under actuated robots the answer is yes as well, but with a few reservations. Given the following simplified kinematics model for a two-wheeled differential drive mobile robot:

$$\begin{bmatrix} \dot{x}_i \\ \dot{y}_i \\ \dot{\theta}_i \end{bmatrix} = \begin{bmatrix} \cos\theta_i & 0 \\ \sin\theta_i & 0 \\ 0 & 1 \end{bmatrix} \begin{bmatrix} v_i \\ \omega_i \end{bmatrix} \tag{24}$$

and the desired inter-vehicle spacing, bearing and orientation, l_d, ψ_d and θ_d respectfully, we can derive the error kinematics with a simple change of coordinates:

$$\begin{bmatrix} \dot{\bar{l}}_{ij} \\ \dot{\bar{\psi}}_{ij} \\ \dot{\sigma}_{ij} \end{bmatrix} = \begin{bmatrix} \cos\psi_{ij} & 0 \\ \frac{-\sin\psi_{ij}}{l_{ij}} & 1 \\ 0 & 1 \end{bmatrix} \begin{bmatrix} v_i \\ \omega_i \end{bmatrix} - \tag{25}$$

$$\begin{bmatrix} \cos(\sigma_{ij}+\psi_{ij}) & d\sin(\sigma_{ij}+\psi_{ij}) \\ \frac{-\sin(\sigma_{ij}+\psi_{ij})}{l_{ij}} & \frac{d\cos(\sigma_{ij}+\psi_{ij})}{l_{ij}} \\ 0 & 1 \end{bmatrix} \begin{bmatrix} v_j \\ \omega_j \end{bmatrix}$$

where

$$\bar{l}_{ij} = l_{ij} - l_d \tag{26}$$

$$\bar{\psi}_{ij} = \psi_{ij} - \psi_d \tag{27}$$

$$\bar{\theta}_i = \theta_i - \theta_d \tag{28}$$

$$\sigma_{ij} = \theta_i - \theta_j \tag{29}$$

and d is the distance between the robot's axis of rotation and the end of the robot (see figure 2). We can solve these sets of nonlinear equations with feedback linearization. Using the bidirectional controller design results in the following switched control laws:

$$u_j(t) = \begin{bmatrix} v_j \\ \omega_j \end{bmatrix} = \begin{cases} K_1, & \text{if } \bar{\theta}_j < \epsilon \\ K_2, & \text{otherwise} \end{cases} \tag{30}$$

where

$$K_1 : \begin{bmatrix} v_j \\ \omega_j \end{bmatrix} = \mathbf{M}_{jk} \begin{bmatrix} k_1 \bar{l}_{jk} + k_2 \dot{\bar{l}}_{jk} + k_3 \int_0^t \bar{l}_{jk} d\tau \\ h_1 \bar{\psi}_{jk} + h_2 \dot{\bar{\psi}}_{jk} + h_3 \int_0^t \bar{\psi}_{jk} d\tau \end{bmatrix} + \\ \mathbf{M}_{ij} \begin{bmatrix} k_4 \bar{l}_{ij} + k_5 \int_0^t \bar{l}_{ij} d\tau \\ h_4 \bar{\psi}_{jk} + h_5 \int_0^t \bar{\psi}_{jk} d\tau \end{bmatrix} \tag{31}$$

$$K_2 : \begin{bmatrix} v_j \\ \omega_j \end{bmatrix} = \begin{bmatrix} 0 \\ p\bar{\theta}_i \end{bmatrix} \tag{32}$$

and M_i and M_{jk} are given as:

$$\mathbf{M}_{ij} = \begin{bmatrix} \cos(\sigma_{ij} + \psi_{ij}) & -l_{ij} \sin(\sigma_{ij} + \psi_{ij}) \\ \frac{\sin(\sigma_{ij} + \psi_{ij})}{d} & \frac{l_{ij} \cos(\sigma_{ij} + \psi_{ij})}{d} \end{bmatrix} \tag{33}$$

$$\mathbf{M}_{jk} = \begin{bmatrix} \cos(\sigma_{jk} + \psi_{jk}) & -l_{jk} \sin(\sigma_{jk} + \psi_{jk}) \\ \frac{\sin(\sigma_{jk} + \psi_{jk})}{d} & \frac{l_{jk} \cos(\sigma_{jk} + \psi_{jk})}{d} \end{bmatrix} \tag{34}$$

K_1 is a bidirectional controller that attempts to minimize the spacing and bearing errors between the itself and the vehicle ahead as well as behind. However, this controller was designed under the assumption that all robots maintain the same heading as the leader. The K_2 control law switches on whenever this constraint is not met. Its purpose is to the keep the robot heading equalized with its leader. It is important to pick an ϵ that is not too small. Otherwise there will be too much trashing back and forth between the two control laws.

5 Simulations and Experiments

5.1 Experimental Methods

The various controllers were tested using USC's Player [5] robot server and Gazebo simulator. Physical robot experiments were performed using four Active Media Pioneer 2 DE mobile robots equipped with 802.11b wireless Ethernet and Sick LMS200 laser range finders.

	Unidirectional		*Bidirectional*	
No.	*Peak Value*	*% increase*	*Peak Value*	*% increase*
1	123	-	191	-
2	148	20.3	179	-6.3
3	184	24.3	142	-20.7
4	236	28.2	83	-41.5

Table 1. Comparison of peak overshoot values

5.2 Experimental Results

Figures 3 demonstrates the problem of string instability. Although all error signals have a steady-state error of zero (i.e. each robot is individually stable), the peak overshoot of each robot increases along the platoon. Table 1 list the peaks for each robot and the percentage increase from one robot to the next. Figure 4 shows the identical setup as figure 3, however the robots are utilizing a bidirectional controller. As revealed in Table 1 the initial disturbance of the lead robot's starting motion (a step response) is attenuated along the platoon.

The last six figures demonstrate the stability and performance of the bidirectional controller using both longitudinal and lateral feedback. In each case, four robots formed platoons at various bearings. Initially the robots are in their proper positions and at rest. The figures plot the separation and bearing errors generated by the leader's stop-and-go motion (i.e. the step response input). In each case the controller was able to attenuate the disturbance without any string stability problems.

6 Conclusion and Future Work

We have presented an approach to designing string stable formation controllers for certain modes of operations. In particular we have shown that linear bidirectional controller can be used to maintain platoons of robots and reject disturbances that may be introduced. We tested this design in simulation and for the longitudinal control on physical robots as well.

In the future we plan on studying the overall stability of the formation as well as addressing stability concerns regarding the switching of behaviors. We will continue testing our approach on non-holonomic robots (Pioneers) as well as holonomic platforms (model helicopters).

Acknowledgment

The authors would like to thank Srikanth Saripalli and Sameera Poduri for all their assistance. This work is supported in part by the DARPA MARS program under grants DABT63-99-1-0015 and 5-39509-A (via UPenn), and by NSF CAREER program under grant IIS-0133947.

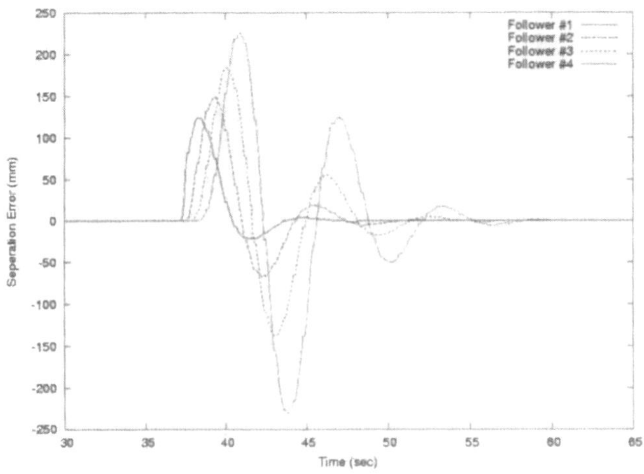

Fig. 3. Unidirectional Controller Demonstrating String Instabilities (Gazebo Simulation)

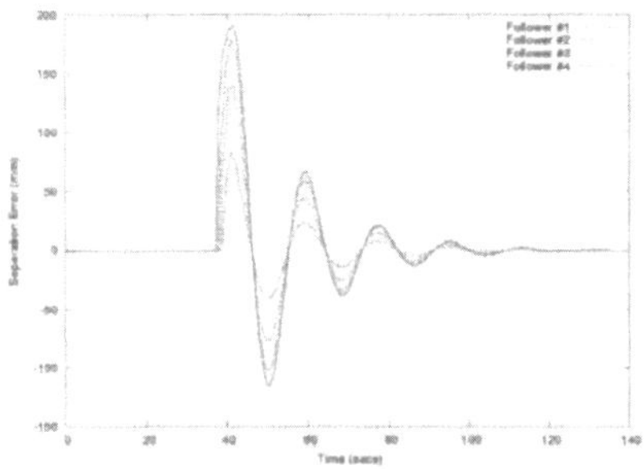

Fig. 4. Bidirectional Controller Demonstrating String Stabilities (Gazebo Simulation)

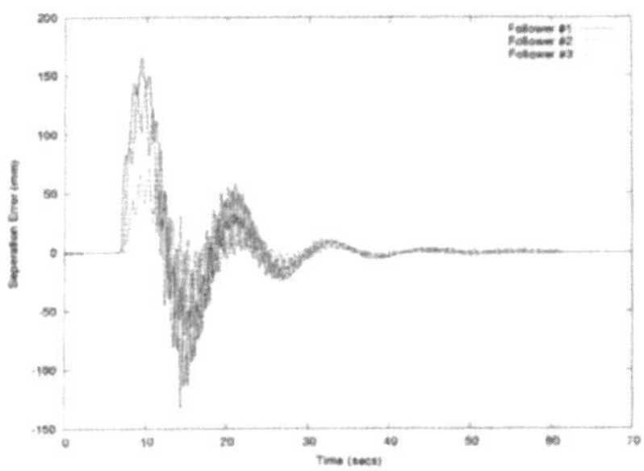

Fig. 5. Bidirectional Controller Demonstrating String Stabilities (Pioneer Robots)

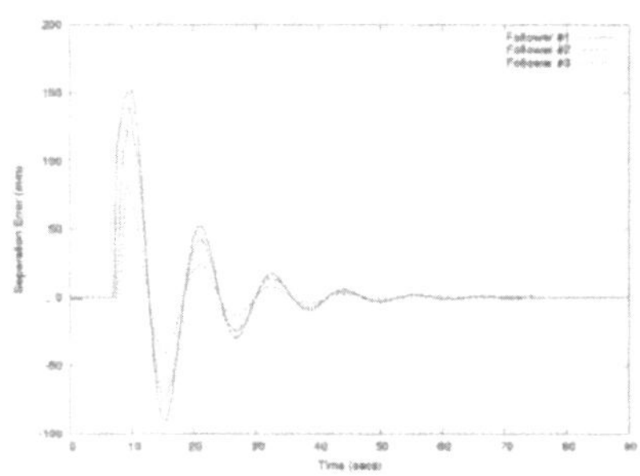

Fig. 6. Separation Errors for a Bidirectional Longitude Controller while maintaining a 30 degree bearing. (Gazebo Simulation)

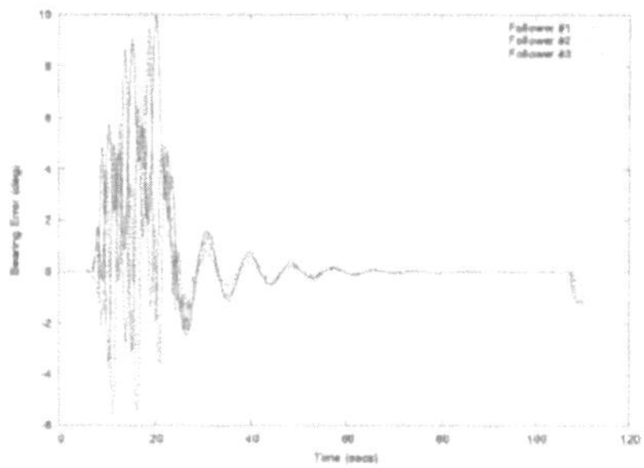

Fig. 7. Bearing Errors for a Bidirectional Longitude Controller while maintaining a 30 degree bearing. (Gazebo Simulation)

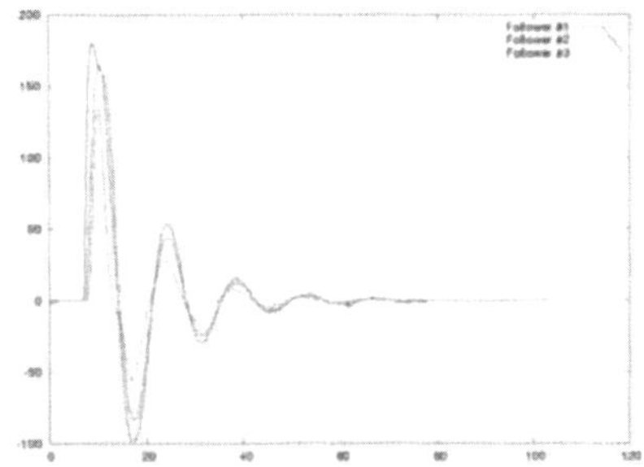

Fig. 8. Separation Errors for a Bidirectional Longitude Controller while maintaining a 45 degree bearing. (Gazebo Simulation)

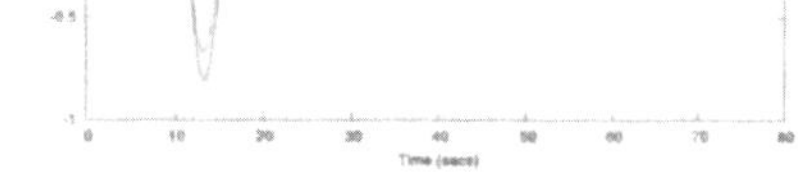

Fig. 9. Bearing Errors for a Bidirectional Longitude Controller while maintaining a 45 degree bearing. (Gazebo Simulation)

References

1. A. P. Aguiar, A. N. Atassi, and A. M. Pascoal. Regulation of a nonholonomic dynamic wheeled mobile robot with parametric modeling uncertainty using lyapunov functions. In *Proceedings of 39th IEEE Conference on Decision and Control (Sydney, Australia)*, December 2000.
2. Carlos Canudas de Wit and Bernard Brogliato. Stability issues for vehicle platooning in automated highway systems. In *Proceedings of the IEEE International Conference on Control Applications, (Hawaii USA)*, pages 1377–1382, August 1999.
3. J. P. Desai, J. Ostrowski, and V. Kumar. Controlling formations of multiple mobile robots. In *Proceedings of the IEEE International Conference on Robotics and Automation, (Leuven, Belgium)*, pages 2864–2869, May 1998.
4. J. Fax and R. Murray. Graph laplacians and stabilization of vehicle formations. In *Proceedings of the 15th IFAC Congress (Barcelona, Spain)*, July 2002.
5. B. Gerkey, R. T. Vaughan, K. Støy, A. Howard, G. S. Sukhatme, and M. J. Matarić. Most valuable player: A robot device server for distributed control. In *Proceedings of the IEEE/RSJ International Conference on Intelligent Robots and Systems (IROS'01)*, pages 1226–1231, October 2001.
6. W. S. Levine and M. Athans. On the optimal error regulation of a string of moving vehicles. *IEEE Transactions on Automatic Control*, 11:355–361, July 1966.
7. X. Liu, S.S. Mahal, A. Goldsmith, and J.K. Hedrick. Effects of communication delay on string stability in vehicle platoons. In *IEEE 4th International Conference on Intelligent Transportaion Systems (ITSC)*, 2001.
8. David J. Naffin and Gaurav S. Sukhatme. Negotiated formations. Submitted to The Eighth Conference of Intelligent Autonomous Systems (IAS-8), Amsterdam, March 2004.
9. Lloyd E. Peppard. String stability of relative-motion pid vehicle control system. *IEEE Transactions on Automatic Control*, pages 579–581, October 1974.
10. P. Seiler, A. Pant, and K. Hedrick. Disturbance propagation in large interconnect systems. In *Proceedings of the American Control Conference (Anchorage AK)*, May 2002.
11. D. Swarrop and J. K. Hedrick. String stability of interconnected systems. *IEEE Transactions on Automatic Control*, 41(3):349–357, March 1996.
12. H. Tanner, V. Kumar, and G. Pappas. Stability properties of interconnected vehicles. In *Proceedings of the 15 International Symposium on Mathematical Theory of Networks and Systems, (Notre Dame, IN)*, August 2002.
13. H. Tanner, G. Pappas, and V. Kumar. Input-to-state stability on formation graphs. In *Proceedings of the IEEE International Conference on Decision and Control, (Las Vegas, NV)*, December 2002.
14. Herbert G. Tanner, George J. Pappas, and Vijay Kumar. Leader-to-formation stability. In *IEEE International Conference on Robotics and Automation*, 2003.
15. Rene Vidal, Omid Shakernia, and Shankar Sastry. Formation control of nonholonomic mobile robots with omnidirectional visual servoing and motion segmentation. In *IEEE International Conference on Robotics and Automation*, 2003.

Decentralized Cooperative Object Transportation by Multiple Mobile Robots with a Pushing Leader

ZhiDong Wang, Yugo Takano, Yasuhisa Hirata, and Kazuhiro Kosuge

System Robotics Lab., Graduate School of Eng., Tohoku Univ.
Aobayama 01, Sendai 980-8579, JAPAN
{wang,takano,hirata,kosuge}@irs.mech.tohoku.ac.jp

Summary. We address a decentralized control method for object transportation in coordination by a leader-follower type multiple robot system. The proposed system consists of a pushing leader, a robot without grasping mechanisms, and multiple follower robots. During the object transportation, a desired trajectory is given to the leader robot only, and follower robots estimate the trajectory of the leader based on force/moment from the object. In the proposed system, a variable internal force is introduced to each robot's controller in decentralized style to let the follower's estimator work on not only the pushing but also the "pulling" case that the leader needs to slow down or stop the object. Finally, a robot system including three omni-directional mobile robots is presented and an experiment result is shown to illustrate the concept of the proposed control algorithm.

1 Introduction

The problem of single object transportation by using multiple mobile robots in coordination has been discussed and several control algorithms have been developed. Within them, as an important decentralized algorithm, leader-follower control algorithm shows its high potential on various aspects such as, efficiency on distributed control, applicability of the system with a large number of robots, independence of inter-robot communication, etc. We proposed a compliance control based leader-follower control system[15][16]. In the system, all robots are controlled to have the same compliant property when they are holding the object. A desired trajectory is given to the leader robot only. Other follower robots estimate the trajectory of the leader based on force/moment from the object and accomplish the object transportation task in coordination with the leader and other followers.

Basically, this control algorithm is designed under an assumption that each robot have a grasping mechanism, such as a griper, to hold the object firmly in motion. However, many robot systems, especially mobile robot systems, do not always equip grasping mechanism. Alternately, pushing is a useful strategy for object manipulation, which is familiar not only in operations by human being but also in performing some tasks by robot systems. Then a control strategy, which allows pushing, can be applied to almost all robots even without any grasping mechanism. Additionally, the strategy is easy to cope with the requirement on choosing or changing the contacting point.

In this paper, we proposed a decentralized control algorithm for multiple mobile robot system incorporating with a "pushing leader", a leader robot without grasping mechanism. The main difficulties on object manipulation via pushing are that the leader robot can not pull the object directly when it needs to slow down or stop the object, and can not inform follower robots these requirements through the object which is used in the original leader-follower control algorithm. To solve this problem, a variable internal force is introduced to each robot's controller in decentralized style to let the follower's estimator work on all cases, pushing and "pulling". The proposed algorithm is different from algorithms for human-robots cooperative system[8][9], because a pushing leader robot can be controlled to generate force including internal force precisely and have specific impedance character but human cannot.

2 Related Works

To perform the cooperative object-handling task, various distributed robots systems[3][13][22] have been proposed. But most of them need common model of the whole system or need explicit communication among robots. Some protocols for box pushing was demonstrated in distributed style on various topics, such as cooperative information invariants[5][6], sensor data sharing[21], fault tolerance[23], and reactive control system design for middle size robot team[19], but most of their cooperation strategies are static or quasi-static. Additionally because no tight connection exists between the object and robots, the robot team has to control the object with other extra components such as friction forces. Then appropriate pushing direction or path is necessary to be considered so that the object will not slide out from the pushing formation of the robot team. In dynamic cooperation control, a behavior-based cooperation strategy for multiple autonomous robots with a manager robot[29][30] is proposed to perform dynamic object transportation and assembly tasks. Some researchers are working on developing decentralized leader-follower mobile object manipulation system[11][15][16][22][26][27]. This type system has the advantage that it works without inter-robot communication, copes with slip of each robot's motion if the follower's control law is well designed. In this research, we are focusing on realizing the object transportation task in leader-follower control scheme with a pushing leader.

By considering closure conditions, conventional works on object manipulation can be categorized into three types: *grasping, caging, and conditional closure.* The grasping based manipulation is the most popular one, especially in multi-finger or multi-arm manipulation[4][20][24][28]. In this case, all robots are arranged so that the total robots system is grasping the object, and Form Closure or Force Closure condition should always be satisfied strictly. Most of them control robot system in centralized style. Recently, several research groups are working on developing control strategies for coping with distributed control requirements and distributed sensing errors[1][14][22][26][29][30]. As the second class of manipulation strategy, recently, caging based object transportation by using three or more mobile robots is discussed in [25][31][32] respectively. The key idea is to introduce a bounded movable area for the object during the object transportation. This strategy has the advantage of not using force control. Then using sequential or formation motion control of robot team, object transportation task can be achieved.

As the third class of manipulation strategy, *conditional closure manipulation* does not guarantee Form Closure or Force Closure. By including gravity force, inertia, or friction force, etc as an extra closure component, Force Closure is realized. Box pushing demonstration is the most typical example of *conditional closure* in 2D manipulation involving with friction between target object and supporting surface. Many cooperative control algorithms are proposed for systems with two robots[5][6][21][23] and more robots[19]. But closure conditions haven't been discussed directly in these researches. Performing an object lifting task in 3D case by a multi-robot team[2] is investigated as an example of the *conditional closure* which is using the gravity. Lynch and Mason[17] analyzed the closure condition carefully and showed results on controllability and stable pushing condition when the object is pushed by a flat plate equipped on the manipulator. They also proposed a method for designing stable pushing path. Later, they also worked on a task of throwing object[18] which could also be viewed as examples of *conditional closure* by using the gravity and inertial. But their research is limited in the single robot.

3 Cooperative Control of a Leader-Follower System

The leader-follower control system was proposed for controlling dual arms system[14] at first. Later, Bay's group [11] developed a leader-follower type control algorithms for multiple mobile robot system. In their research, the follower is controlled based on a compliance model with interaction to the object, and the cooperative control is realized by directly feedbacking errors of each follower's motion. In [15], Kosuge and his research group proposed a distributed control algorithm that the follower can perfectly estimate the motion of the leader and perform the cooperative transportation task by using this estimation. They demonstrated the algorithm in holonomic[15] and non-holonomic[27] mobile robots system. By incorporating a concept of virtual dual caster[7], they proposed cooperative control strategy without using position information of the follower's supporting point. Later, this algorithm is extended to mobile-manipulators systems[10][16] and human-robot cooperation system[8][9]. Here, we give a brief introduction on this distributed control algorithm for a system consisting of multiple holonomic mobile robots.

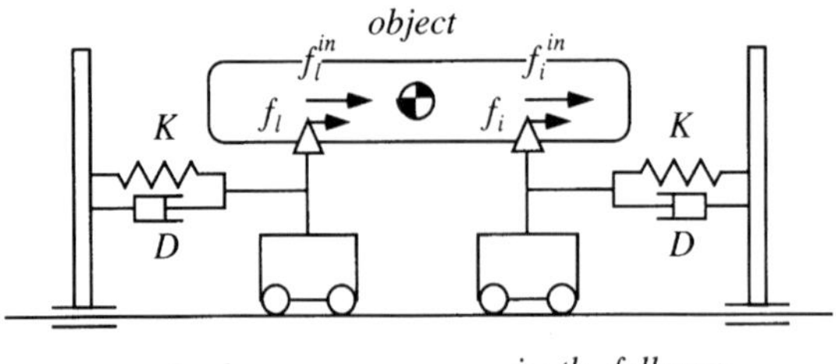

Fig. 1. Compliance based Leader-Follower Control for Cooperative Object Transportation.

In this algorithm, motion control of the leader and follower robots is designed so that their have the following characteristics. The system consists of a leader robot and n follower robots.

$$D\Delta\dot{\boldsymbol{x}}_l + K\Delta\boldsymbol{x}_l = \boldsymbol{f}_l - \boldsymbol{f}_l^{in} \tag{1}$$

$$D\Delta\dot{\boldsymbol{x}}_i + K\Delta\boldsymbol{x}_i = \boldsymbol{f}_i - \boldsymbol{f}_i^{in} \qquad i = 1, \cdots, n \tag{2}$$

where $\boldsymbol{x}_l$ and $\boldsymbol{x}_i$ are trajectories of the leader and ith follower respectively, $\boldsymbol{x}_{ld}$ and $\boldsymbol{x}_{id}$ are desired trajectories of robots, and motion errors of the leader and ith follower are defined as $\Delta\boldsymbol{x}_l = \boldsymbol{x}_l - \boldsymbol{x}_{ld}$ and $\Delta\boldsymbol{x}_i = \boldsymbol{x}_i - \boldsymbol{x}_{id}$ respectively. Because the sum of internal forces/moment($\boldsymbol{f}_l^{in}$ and $\boldsymbol{f}_i^{in}$ will not contribute to the motion of the object,

$$\boldsymbol{f}_l^{in} + \sum_{i=1}^{n} \boldsymbol{f}_i^{in} = 0 \tag{3}$$

. Also when the system reaches a steady state, forces applied to the object will balance.

$$\boldsymbol{f}_l + \sum_{i=1}^{n} \boldsymbol{f}_i = 0 \tag{4}$$

Then from Eq.1 and Eq.2, the following equation will be obtained in the steady state of the system.

$$\Delta\boldsymbol{x}_l + \sum_{i=1}^{n} \Delta\boldsymbol{x}_i = 0 \tag{5}$$

When the system consists of more than one follower robot, we set the leader and all other robots (robot j, $j = 1, \cdots, n$ and $j \neq i$) be a virtual leader to ith robot[15]. Then the system can be treated as a single leader single follower system (Fig.1). Then Eq.5, the errors in the steady state, can be written as:

$$\Delta\boldsymbol{x}_{li} + \Delta\boldsymbol{x}_i = 0 \tag{6}$$

where $\Delta \boldsymbol{x}_{li} = \sum_{j=1(j\neq i)}^{n} \Delta \boldsymbol{x}_j$. To the ith follower robot, if we set an estimator of leader's path be:

$$\Delta \boldsymbol{x}_{di} = \boldsymbol{x}_{dli} - \boldsymbol{x}_{ei} = 2\Delta \boldsymbol{x}_i \tag{7}$$

and design a transfer function matrix $\boldsymbol{G}_i$ as:

$$\boldsymbol{G}_i = \frac{a_i s + b_i}{s^2} \boldsymbol{I}_3 \tag{8}$$

, the system is stable and $\Delta \boldsymbol{x}_i$ will converge to zero and cooperative transportation of a single object is realized. The detailed discussion can be found in literature [15].

4 Pushing Model and Pushing Constraint

When a robotic mechanism is holding the object tightly, it can generate an arbitrary force and moment under the limitation of its maximum output to the object. However only force in a particular half force space can be generated from the robot mechanism to the object for pushing with a point contact. In a line contact case, the moment, which can be generated, is also limited, and depends on the magnitude of contacting force directly. Here, we consider that each robot contacts with the object by a line segment with length T.

For maintaining stable pushing, which means that the line contact between the object and the robot is maintained and no slip occurs on the contact. The force $\boldsymbol{f}$ generated by the robot should satisfies the following conditions.

$$\boldsymbol{f}^T \boldsymbol{e}_n \geq 0 \tag{9}$$

$$\boldsymbol{f}^T \boldsymbol{e}_n \geq \frac{1}{\sqrt{1+\mu^2}} ||\boldsymbol{f}|| \tag{10}$$

where, $\boldsymbol{e}_n$ is an unit inner normal vector of the contact edge of the object, and μ is the friction coefficient of the edge. Eq.9 indicates that only pushing force is feasible and Eq.10 provides non-slip condition of the contact.

For discussing the moment constraint, we first decompose the pushing force $\boldsymbol{f}$ applied to the object into two pure force vectors, $\boldsymbol{f}_L$ and $\boldsymbol{f}_R$, at two endpoints of the contacting line segment as follow.

$$\boldsymbol{f} = \boldsymbol{f}_R + \boldsymbol{f}_L \tag{11}$$

$$\boldsymbol{f}_R \geq 0, \qquad \boldsymbol{f}_L \geq 0 \tag{12}$$

Then the moment around the reference contact point, which is defined as the center of the contacting line segment, can be obtained as follow:

$$\boldsymbol{n} = \frac{T}{2} \boldsymbol{f}_R^T \boldsymbol{e}_n - \frac{T}{2} \boldsymbol{f}_L^T \boldsymbol{e}_n \tag{13}$$

Therefore, pushing constraints at each contact are:

$$\begin{cases} \boldsymbol{f}^T \boldsymbol{e}_n \geq \frac{1}{\sqrt{1+\mu^2}} ||\boldsymbol{f}|| \geq 0 \\ \frac{T}{2} \boldsymbol{f}^T \boldsymbol{e}_n \geq \boldsymbol{n} \geq -\frac{T}{2} \boldsymbol{f}^T \boldsymbol{e}_n \end{cases} \tag{14}$$

5 Pushing Leader based Decentralized Control

The cooperative control algorithm in section 3 should work under the condition of Eq.3. Introducing internal force into a cooperative control algorithm was proposed for dual arm control tasks at first[28][14]. A predetermined constant internal force is incorporated to the cooperative control of manipulators, a centralized control system in general. Later, this method was applied to cooperative control of distributed mobile system[15] for maintaining stable contacts with transported object. However, same with the dual arm control case, internal force in this system is predetermined constant during the manipulation also. Then the geometric relation, such as where robots are contacting with the object, should not be changed. This is because, to any change of the contact, such as that a robot changes its contacting position or leaves from the object transportation task, will break the condition of internal force (Eq.3) and make the resultant force of predetermined internal force of all robots do not be zero and affect the motion of the object. This means that the system is not reconfigurable if without introducing any other new method, e.g. using inter-robot communication, for passing information of the internal force.

In this study, we work on releasing this tight constraint among robots, and construct a distributed mobile robot system, which can reconfigure the contacting position of the robot during the manipulation. In this paper, we focus on developing a cooperative control strategy with a reconfigurable pushing leader. We assume that all follower robots are holding the object tightly and not any change on relative position and orientation between the object and each follower will occur. The leader robot contacts with the object by a line segment on its body, and can only push the object. We also assume that the pushing leader will only separate from the object when the object stops.

Since the object in transportation task will not be small and light like the object grasped by a multi-finger hand, forces or moment generated by some arm mechanism is too small in most of cases. Then, it can be said that our assumption of the pushing leader are reasonable because the robot generates the manipulation force by using its mobile platform. Also pushing the object by a line contact is a feasible, easy and low cost method to the most of robot system. For the same reason, the force on the direction perpendicular to the contacting edge can be set to any value if Eq.9 is satisfied. On the contrary, the force on the direction of the contact line comes from the friction and depends on magnitude of the force perpendicular to the edge. Then, we design the pushing action without using this friction force directly. By the action based on this pushing force perpendicular to the contacting edge and the moment applied by the edge, the object can be transported to its goal position.

The purpose of the internal force is for maintaining the stable pushing even the control error occurs or the leader needs to slow down the object by pushing contact. This means that we need to keep $\boldsymbol{f}_l$ always have minus value and $\sum \boldsymbol{f}_i$ always have plus value(following Eq.9) in dynamics of the leader(Eq.1) and followers(Eq.2). Under an assumption that acceleration of the desired path is relative small and up-bounded, up-bounds of the left side of both Eq.1 and Eq.2 are linearly relative to velocity of the leader robot. Then we design the internal force as follow:

$$f_l^{in} = \begin{cases} D_x \dot{x}_l & \dot{x}_l \geq 0 \\ 0 & \dot{x}_l < 0 \end{cases}, \quad f_i^{in} = \begin{cases} -\frac{1}{n} D_x \dot{x}_i & \dot{x}_i \geq 0 \\ 0 & \dot{x}_i < 0 \end{cases}, \tag{15}$$

where, D_x is a positive constant and $\dot{x}_l$ and $\dot{x}_i$ are velocity of the pushing operational point(Fig.2-(b)) measured by the pushing leader and by ith follower respectively. When pushing contact is maintained, these two velocities are exactly the same. Then our distributed control system can guarantee the Eq.3 without using inter-robot communication. Also this method makes the internal force be zero when the object stops, and allows the pushing leader to leave from the object-follower system. For reducing effect of follower's slip, Eq.2 is implemented by using the virtual caster model proposed in [7] with the variable internal force(Eq.15). This implementation makes the estimation of the leader's motion be an easy 1DOF estimation problem.

In the proposed system, the feasible moment that the pushing leader can apply to the object can be obtained as follow, and will increase when the object has higher velocity.

$$\frac{TD_x}{2} \dot{x}_l \geq n \geq \frac{-TD_x}{2} \dot{x}_l \tag{16}$$

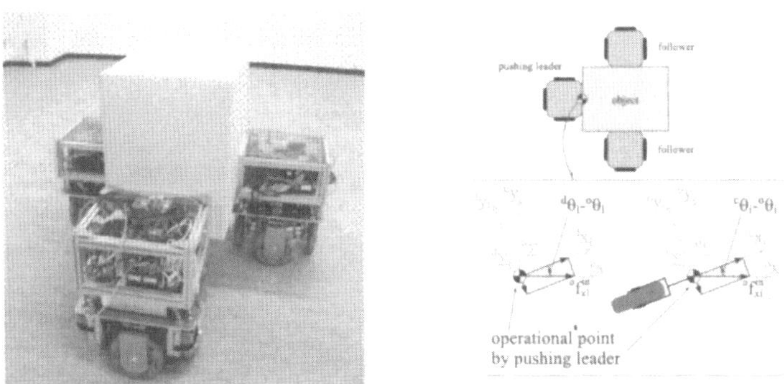

(a) robot system (b) virtual caster model

Fig. 2. Leader-Follower System with a Pushing Leader. Each follower is implemented in a virtual caster model, a 1DOF leader-path estimator and a variable internal force.

6 Experiment

We did experiments of transporting a single object using three omni-directional mobile robots, DR Helper robots(Fig.2-(a)), to illustrate the validity of the proposed algorithm. The control algorithm is implemented on each robot's onboard computer system with QNX realtime operating system. A body force sensor system is installed on each robot for measuring the pushing force.

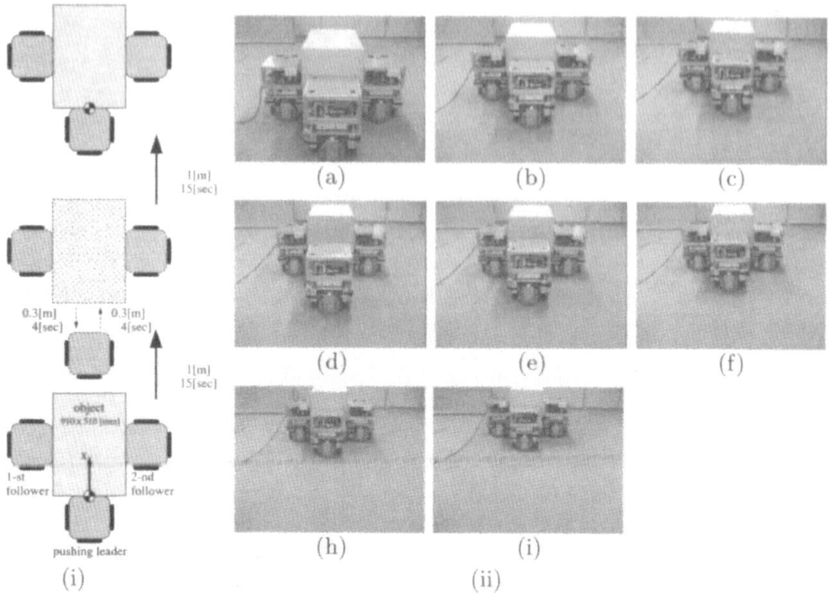

Fig. 3. Experiment Trajectory and Result. In (d), the internal force becomes zero and the pushing leader leaves from the object and the follower system. In (e), the

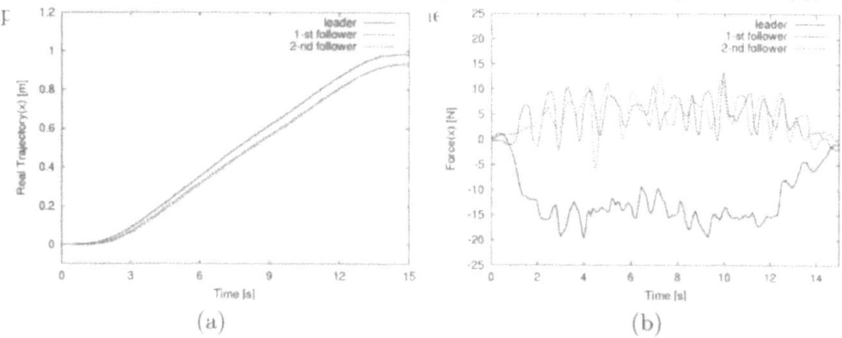

Fig. 4. Experiment Result: Trajectories and forces in the first pushing action.

In this experiment, an object transportation path is set to the pushing leader only (Fig.3-(i)). In the beginning, the leader pushes the object forward 1[m] (Fig.3-(a)-(c)), then stops and leaves from the object 0.3[m](Fig.3-(d)). At this moment, the object and followers successfully stop without inter-robot communication. Finally, the leader moves back and pushes the object forward again, and the object transportation task is achieved in coordination. Fig.4-(a) shows that two followers successfully estimate the leader's path and control themselves in the first pushing action. Fig.4-(b) shows that, even followers' force outputs are oscillating and are near to zero sometimes, force applied from the pushing leader is totally less than zero. Then stable pushing is maintained by using the proposed internal force decision algorithm successfully.

7 Conclusion

In this paper, we proposed a decentralized control algorithm for multiple mobile robot system incorporating with a leader robot without grasping mechanisms. A variable internal force is introduced to each robot's controller in decentralized style to make the follower's estimator work not only on pushing but also on "pulling" case that the leader robot needs to slow down or move back the object without any inter-robot communication. Since the designed internal force depends on the velocity of the transported object, the leader robot can leave from the object for changing the contacting point or moving away when the system stops. An experiment system, which consists of three holonomic mobile robots, is presented and some basic experiment results are shown for illustrating the validity and the effeteness of the proposed control algorithm. The main limitation of the system with pushing leader is that the system cannot follow an arbitrary path especially when the velocity is low. However, our system involving the active internal force concept can follow more possible path then the object pushing with passive internal force which is generated by friction or inertia force[17][18], etc. Realizing more complicated transportation motions is our future research issue.

References

1. Ahmadabadi M.N. and Nakano E., *A constrain and move approach to distributed object manipulation protocols*, IEEE Transaction on Robotics and Automation, 17(2), pp.157-172, 2001.
2. Ahmadabadi M.N., Rushan S.M., Wang Z.D., Nakano E., *A Constrain-Move based distributed cooperation strategy for four object lifting robots*, IROS2000, pp.2030-2035, 2000.
3. Asama H., Matsumoto A., and Ishida T., *Design of an Autonomous and Distributed Robot System*, IROS1989, pp.283-290, 1989.
4. Bicchi,A., and Kumar, V., *Robotic Grasping and Contact*, ICRA2000, pp.348-353, 2000.
5. Brown R.G. and Jennings J.S., *A Pusher/Steerer Model for Strongly Cooperative Mobile Robot Manipulation*, IROS1995, pp.562-568, 1995.
6. Donald B.R., Jenning J. and Rus D., *Information invariants for distributed manipulation*, Int. Journal of Robotics Research, 16(5), pp.673-702, 1997.
7. Hirata Y., Kosuge K., et al., *Coordinated Transportation of a Single Object by Multiple Mobile Robots without Position Information of Each Robot* IROS2000, pp.2024-2029, 2000.
8. Hirata Y., Kakagi T., et al., *Manipulation of a Large Object by Multiple DR Helpers in Cooperation with a Human* IROS2001, pp.126-131, 2001.
9. Hirata Y., Kume Y., Wang Z.D., and Kosuge K., *Decentralized Control of Multiple Mobile Manipulators Based on Virtual 3-D Caster Motion of Handling an Object In Cooperation with a Human* ICRA2003, pp.938-943, 2003.
10. Hirata Y., Kume Y., et al., *Handling of an Object by Multiple Mobile Manipulators in Coordination bas ed on Caster-like Dynamics*, ICRA2004, 2004.
11. Johnson P.J. and Bay J.S., *Distributed Control of Simulated Autonomous Mobile Robot Collectives in Payload Transportation*, Autonomous Robots, 2(1), pp.43-63, 1995.

12. Khatib O., Yokoi K., and et al., *Vehicle/Arm Coordination and Multiple Mobile Manipulator Decentralized Cooperation*, IROS1996, pp 546-553, 1996.
13. Kimura H., and Wang Z.D., *Huge-Object Manipulation in Space by Vehicle-Type Robots*, JSME Int. Journal, Series C, 38(3), pp. 543-551, 1995.
14. Koga M., Kosuge K., Furuta K., and Nosaki K., *Coordinated motion control of robot arms based on the virtual internal model*, IEEE Trans. on Robotics and Automation, Vol.1, No.1, pp. 77-85, 1992.
15. K. Kosuge, T. Oosumi, *Decentralized Control of Multiple Robots Handling an Object*, IROS1996, pp.318–323, 1996.
16. Kume Y., Hirata Y., Wang Z.D., and Kosuge K., *Decentralized Control of Multiple Mobile Manipulators Handling a Single Object in Coordination*, IROS2002, pp.2758-2763, 2002.
17. Lynch K.M. and Mason M.T., *Stable Pushing: Mechanics, Controllability, and Planning*, Int. J. Robotics Research, 16(6), pp.533-556, 1996.
18. Lynch K.M. and Mason M.T., *Dynamic nonprehensile manipulation: Controllability, planning, and experiments*, Int.J.Robotics Research,18(1),pp.64-92,1999.
19. Kube C.R. and Zhang H., *Task Modelling in Collective Robotics*, Autonomous Robots, 4(1), pp 53-72, 1997.
20. Nakamura Y., Nagai K., Yoshikawa T., *Dynamics and Stability in Coordination of Multiple Robotic Mechanisms*, Int.J.Robotics Research,8(2),pp.44-61,1989.
21. Mataric M.J., Nilsson M., and Simsarian K.T., *Cooperative Multi-Robot Box-Pushing*, IROS95, pp.556-561, 1995.
22. Ota J., Miyata N., Arai T., and et.al, *Transferring and Regrasping a Large Object by Cooperation of Multiple Mobile Robots*, IROS95, pp.543-548, 1995.
23. Parker L. E., *ALLIANCE: an architecture for fault tolerant multirobot cooperation*, IEEE Tran. on Robotics and Automation, 14(2), pp.220-240, 1998.
24. Rimon E. and Burdick J.W., *Mobility of Bodies in Contact -I: A New 2nd Order Mobility Index for Multiple-Finger Grasp.* IEEE Trans. Robotics and Automation, 14(5) pp.696-708, 1998
25. Sudsang A., Rothganger F., and Ponce J., *Motion planning for disc-shaped robots pushing a polygonal object in the pl ane*, IEEE Tran. on Robotics and Automation, 18(4). pp.550-562, 2002.
26. Sugar T., and Kumar V., *Multiple Cooperating Mobile Manipulators*, ICRA99, pp. 1538-1543, 1999.
27. Takeda H., Hirata Y., Wang Z.D., and Kosuge K., *Collision Avoidance Algorithm for Multiple Tracked Mobile Robots Transporting a Single Object in Coordination Based on Function Allocation Concept*, DARS 5, pp.155-164, 2002.
28. Uchiyama M. and Dauchez P., *A symmetric hybrid position/force control scheme for the coordination of two robots*, ICRA88, pp.350-355, 1988.
29. Wang Z.D., Nakano E., and Matsukawa T., *A New Approach to Multiple Robots' Behavior Design for Cooperative Object Manipulation*, DARS 2, pp.350-361, 1996.
30. Wang Z.D., Nakano E. and Takahashi T., *Solving Function Distribution and Behavior Design Problem for Cooperative Object Handling by Multiple Mobile Robots* IEEE Tran.on Systems, Man, and Cybernetics A, 33(5), pp.537-549, 2003
31. Wang Z.D. and Kumar V., *A Decentralized Test Algorithm for Object Closure by Multiple Cooperating Mobile Robots*, DARS 5, pp.165-174, 2002.
32. Wang Z.D., Kumar V., Hirata Y., and Kosuge K., *A Strategy and a Fast Testing Algorithm for Object Caging by Multiple Cooperative Robots*, ICRA2003,pp.938-943, 2003.

Part XII

Applications

Attentive Workbench: An Intelligent Production Cell Supporting Human Workers

Masao SUGI[1], Yusuke TAMURA[2], Jun OTA[2],
Tamio ARAI[2], Kiyoshi TAKAMASU[2], Kiyoshi KOTANI[1],
Hiromasa SUZUKI[2], and Yoichi SATO[3]

[1] School of Information Science and Technology, The University of Tokyo
sugi@prince.pe.u-tokyo.ac.jp
[2] School of Engineering, The University of Tokyo
{tamura,ota,arai}@prince.pe.u-tokyo.ac.jp
[3] Institute of Industrial Science, The University of Tokyo

Summary. We propose "Attentive Workbench (AWB)," a new cell production system in which an intelligent system supports human workers. Using cameras, projectors, self-moving trays driven by planar motors and so on, the system recognizes the worker's condition and intention and supports the workers from both physical and informational aspects. In this paper, AWB is outlined. The control system for multiple self-moving parts trays in AWB is proposed. The results of simulation are shown demonstrating the present control system.

1 Introduction

The constantly changing consumer trends have promoted innovations in a production system from time to time. Automated manufacturing lines were designed to produce specific products with special machines on a mass basis at the expense of high initial costs. To follow the changing consumer tastes, however, manufacturers have gradually replaced them with the flexible manufacturing system (FMS) since the 1970s. Consumers today are increasingly demanding, and the hot and emerging trends are being satisfied with a new idea called "cell production system". This is a system in which a single human worker assembles each product from start to finish almost manually [1, 2]. Using multiple skilled human workers, the cell production system, with lower level of automation, can accommodate diversified products and production quantity more flexibly than fully automated manufacturing system (i.e. conventional manufacturing line and FMS). There are dynamic changes in the age structure of the working populations. With negative and zero growth of the population, together with the tendency of young people avoiding manufacturing jobs, we will face a shortage of skilled workers, and hence a great difficulty in maintaining the cell production system.

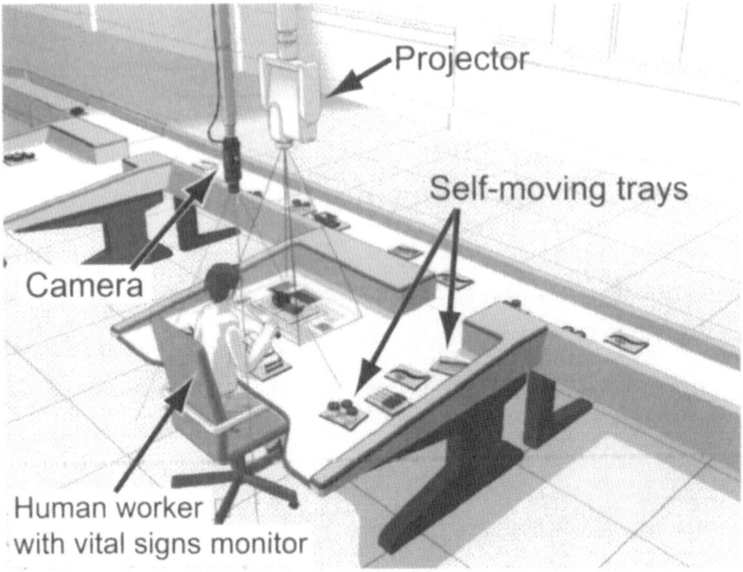

Fig. 1. Schematic view of Attentive Workbench (AWB).

To meet diverse needs with fewer labor force, we propose attentive workbench (AWB), which is shown in **Fig. 1**. Attentive workbench is an intelligent cell production system based on EnhancedDesk [3], an augmented human-computer interface. The system recognizes the intention and the condition of a human worker, and presents the information and supplies assembling parts to the worker. This may result in a higher yield rate and productivity, by reducing the failure and the time at the picking up of the parts. The system may also promote the less experienced people to enter the workforce.

As the related work, Roland DG Corp. (Hamamatsu, Japan) uses a production cell which is monitoring the assembly progress and presents the information about the next assembly process [1]. Reinhart and Patron [7] propose to apply augmented reality to the information presentation using semi-transparent head mount display. These researches deal with the support of human workers from the information side. As for the physical support, Sugano et. al [8] proposes an assembly supporting system in which a manipulator steadily holds the subassembly to help the human worker. Raven Engineering Inc. (Bloomington, IN, U.S.A.) invented a new production cell [2] with improved spatial layout to achieve a higher productivity.

In section 2, attentive workbench (AWB) is outlined. Components of AWB are presented. In section 3, we focus on the self-moving trays in AWB. After discussing on the function of the self-moving trays necessary for supporting human workers, a hierarchical architecture for controlling multiple trays is

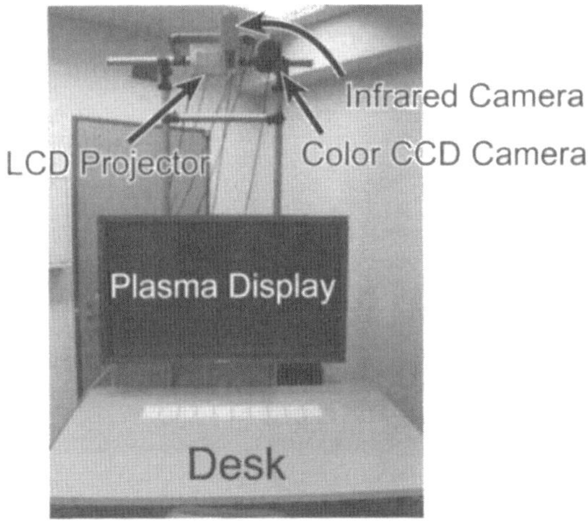

Fig. 2. Overview of EnhancedDesk[3].

proposed. In section 4, the demonstration of the proposed control method is shown. We conclude this paper in section 5.

2 Overview of Attentive Workbench (AWB)

2.1 System components of AWB

Major key technologies and the devices of the attentive workbench are the following three.

- **EnhancedDesk:** EnhancedDesk, shown in **Fig. 2**, is a desk-type human-computer interface with augmented reality proposed by Sato and Koike [3]. The user expresses his intention by hand gesture, and the system presents information using an LCD projector and a plasma display.
- **Self-moving parts trays:** In usual manufacturing systems, parts trays are transported by other manufacturing devices (e.g. belt conveyers), having no mobile mechanism in itself. We introduce self-moving trays driven by Sawyer planar motor [4] characterized with high speed (0.5 m/s at maximum) and high positioning accuracy (30μm).
  Figure 3 shows a prototype of the parts tray system. Each tray has square shape with a side of 8cm.
- **Estimation of the state of a worker based on bio-measurement technologies:** The state or intention of a human worker can be estimated from the heart rates and respirations of the worker that are measured by vital signs monitors, applying a method for analyzing respiratory sinus arrhythmia (RSA) with respect to respiratory phase, proposed by Kotani [6].

Fig. 3. Self-moving parts trays driven by sawyer planar motors and a passive steel platen.

2.2 System architecture

Figure 4 shows the architecture of AWB. System Manager at the top layer integrates the sensors and actuators at the second and the third layers.

The trays, belonging to the third layer, are connected to Tray Manager, which forms the second layer together with the other components, such as the projector and the camera. Tray Manager serves as an interface between the multiple trays and System Manager. The role of Tray Manager will be described in the next section.

System Manager in **Fig. 4** appropriately supports a human worker by interpreting intentional or unintentional messages from him using cameras and vital signs monitors. For example, checking the progress of assembly processes, AWB presents the assembly information (which parts are used in the next assembly process, where to assemble them, etc.) to the worker by the projector. The parts to be used are then supplied to the worker by self-moving parts trays. When AWB detects the fatigue of the worker through vital signs monitor, it advises the worker to take a rest in order to prevent failures or accidents.

3 Assembly Support by Self-Moving Trays

In an ordinary cell production system, the human worker picks up the assembling parts necessary for the current assembly process from the parts trays on the desktop shelf. Each of such holding-out motion brings about loss of time. Moreover, sometimes the worker mistakes the parts by picking up different (but similar) parts, which causes an assembling failure. If the assembly of a product is finished, the worker put the product away manually, which also

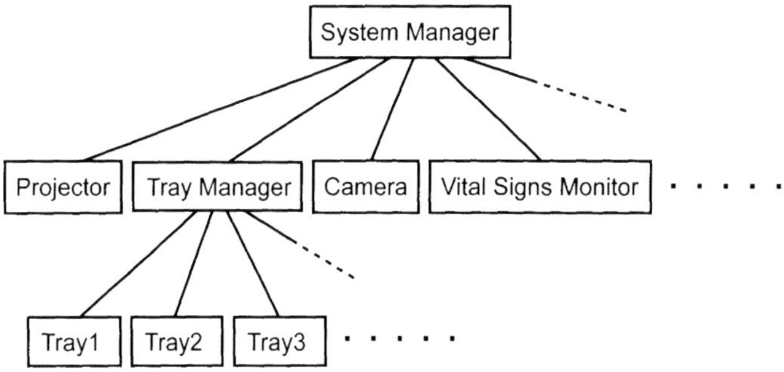

Fig. 4. Architecture of AWB.

spends time. Optimization of the spatial layouts of the parts trays can relax such problems, but cannot improve it radically. As another drawback of the cell production systems is the small workspace for human workers confined by the large desktop shelf.

To improve the fundamental defects of the cell production system mentioned above, we propose to support human workers by handing-over parts to them using self-moving trays. The necessary assembling parts are automatically delivered to the worker. The finished product is quickly cleared away. This will accelerate the speed of production and decrease the occurrence of assembling failures.

3.1 Tasks for Self-Moving Part Trays

Trays execute two kinds of tasks as follows,

- **Handing-over:**
  A tray carrying assembly parts moves toward the worker and supplies the parts to him.
- **Receiving:**
  An empty parts tray moves to the worker and receives a product or a subassembly from him.

A human worker requests these tasks using hand gestures. The camera module in **Fig. 4** recognizes it and informs it to System Manager. System Manager then informs the task request to Tray Manager.

Besides the above two items, we should allow for,

- **Standby:** When the parts tray does not execute any task, it moves back to its home position on the desk and stays there, preventing itself from colliding with other trays at work.

It is noted that standby is not a task, although it is associated with a movement of the tray like the above two tasks.

3.2 Roles of Tray Manager and Trays

The tasks requested by human workers are abstract and does not contain the concrete information, such as the necessary number of trays for a "receiving," the actual path of each trays, and so on. We introduce an interface agent "Tray Manager," which determines the detail of the task necessary for each tray to execute.

Tray Manager receives a task ("handing-over" or "receiving") from System Manager. Tray Manager checks the status of each tray (whether it is at work or not, what does it carry, etc.). Tray Manager then chooses an appropriate tray and allocates the task to the tray, informing it of the goal position of the task.

As another role, Tray Manager works as a server (or manager) in blackboard communication between trays. Each tray can know the position of others through Tray Manager.

Each tray executes the task allocated by Tray Manager. The tray determines its path to the goal position given by Tray Manager. Since the parts trays move in parallel and asynchronously, two trays sometimes collide with each other. We introduce a simple motion rule of each tray to solve the collision problem between trays, which is explained in the next subsection.

We can consider various schemes in dividing the roles of Tray Manager and each tray. For example, we can consider a simple centralized scheme, where Tray Manager determines the all, including the paths of all the trays, and the trays obey it. This method, however, imposes a large computational load on Tray Manager. Therefore it is difficult to manage the whole system in real-time when the number of trays increases. In our scheme, the path planning of each tray is responsible for itself. This is more suitable than the centralized scheme for handling large number of trays.

In the research field of autonomous decentralized mobile robots, sensors are necessary to each robot for the purpose of self-localization and collision detection. As a result, the hardware and the software of the each robot becomes massive and complex. In this paper, we suppose that each tray is powered by sawyer planar motor, which has high positioning accuracy. Each tray can know its accurate position by its odometry. Mutual collision therefore can be detected only by blackboard communication. Additional sensors is not necessary in our system.

3.3 Collision Avoidance between Two Trays

Each tray can know its current position and goal in the global coordinate system. Each tray can also know the position of the other trays through the communication with Tray Manager, which is the server of the blackboard communication. In the case of no collision, trays move toward their goal position. When two trays collide with each other, they should pass each other

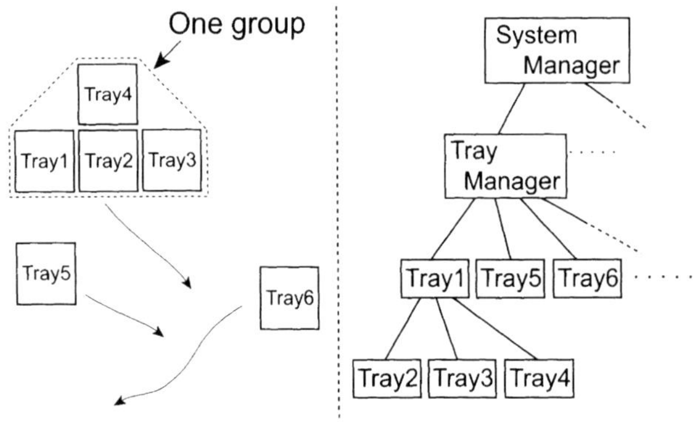

Fig. 5. Left: an example of the tray group composed of four trays. Right: The hierarchical architecture of AWB corresponding to the left figure.

according to the "keep-right" rule. Using this motion rule, two trays can avoid deadlock.

This motion rule is simple and practical, although the movement of trays generated by this rule is not so good in the view of efficiency. In order to realize a deadlock-free system with high efficiency of movement of trays, all the trays must be controlled through a centralized manner. This imposes a large computational load on Tray Manager, which is mentioned previously. It is beyond the scope of the present paper.

3.4 Group Formation of Multiple Trays

If an object (part or product) is larger in size than a single tray, which has square shape with a side of 8cm, a group of trays is subjected to a rigid body formation like a large single tray. This section deals with the way of forming a group of parts trays.

Tray Manager is given the shape of the object as a polygon by System Manager. Covering the given polygon by congruent squares, Tray Manager calculates the necessary number of trays and their conformation for carrying the object. Tray Manager gathers then the empty trays together according to the tray conformation calculated. The trays are now put into a group and treated as a single large tray. One of the trays involved in the group is chosen as the group master, which checks collision from outside, determines the path of the group, and controls all the trays in the group.

Figure 5 shows an example of tray group and the corresponding system architecture. The tray group in **Fig. 5** consists of four trays with Tray1 being the group manager. Tray2, Tray3 and Tray4 are not controlled by Tray Manager but by the group manager (Tray1 in the present case). The trays in the group except for the group manager are hidden from Tray Manager and any other tray outside the group.

When the trays, each represented by a square, are put together into a group, the resultant shape of the group is not always a convex polygon. In this research, the group manager calculates the minimal-area convex polygon covering the whole group, and uses it as the "shape" of the group in checking the collision from outside. This makes the collision check between trays easier.

Figure 5 shows an example. The original shape of the group is an octagon with two concaves. The minimal-area convex polygon covering the group is a hexagon, which is shown by the broken line.

4 Simulated Demonstration

We have made a simulator of the parts tray system. In the AWB system shown in **Fig. 4**, the trays and Tray Manager is realized in the simulator. Worker's requests of the assembly support, which will be presented by gesture in the implemented system, is substituted here by a keyboard and a mouse.

Figure 6 shows an example of the simulation, where the user requests Tray Manager to receive a large object. In **Fig. 6**(a), there are 20 trays, all being on standby at the home position. In (b), the user defines a large object and requests Tray Manager to receive it. Tray Manager calculates the number of trays and their conformation necessary for carrying the object, and in (b) to (e), Tray Manager gathers the trays one after another. In (e), all the necessary trays have been gathered. The trays are put together into a group, and the shape of the group is redefined as a convex polygon.

Through the simulations, the trays are found to move asynchronously in parallel, avoiding the mutual collisions. In the formative stage of the tray group, however, the deadlock is often caused by the collisions of three or more trays. To avoid this problem, in the present simulation, only one tray is moved at the same time in constructing the tray group (see **Fig. 6**). As a result, it takes long time to construct the group of trays after the object to be transported is given. To improve these problems, we will modify the motion rule in the next stage, introducing the order of priority of trays according to their status.

5 Conclusions

We have proposed attentive workbench (AWB), a new cell production system in which an intelligent system supports human workers from both information side and physical side. AWB consists of an augmented human-computer interface, vital signs monitors, and self-moving parts trays driven by planar motor. In this paper, AWB is outlined at first. Next, the control system of multiple self-moving trays is presented. The proposed control system is simulated, showing the main features of the system, i.e., parallel and asynchronous

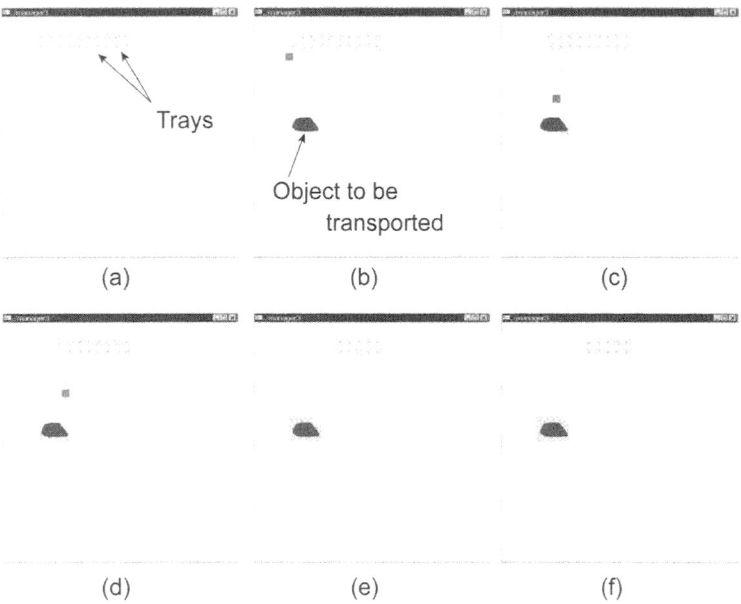

Fig. 6. An example of the simulation. A large object is received from the user to the tray system. (a) Initial condition. All the trays are in the home position. (b)–(e): Trays are gathered one by one. (f): The shape of the group is redefined as a convex polygon.

movement of trays, mutual collision avoidance of trays based on simple motion rule, and automatic group formation of trays.

In the next stage, we will integrate the proposed control architecture of trays with gesture-based interface, which is shown in **Fig. 7**, along with the prototype of the self-moving trays.

Acknowledgement

This research is partly supported by the 21st century COE program "Information Science and Technology Strategic Core" from the Ministry of Education, Culture, Sports, Science and Technology, Japan.

References

1. Shinichi Seki, "One by One Production in the 'Digital Yatai' — Practical Use of 3D-CAD Data in the Fabrication —," *Journal of the Japan Society of Mechanical Engineering*, Vol. 106, No. 1013, pp.32–36, 2003 (*in Japanese*).
2. Gary Vasilash, "Cell for Assembly," *Automotive Manufacturing & Production*, June 1999.

Fig. 7. Demonstration of self-moving trays. Using hand gestures, the human user controls the virtual trays which are displayed by a projector at the ceiling.

3. Yasuto Nakanishi, Yoichi Sato, Hideki Koike, "EnhancedDesk and Enhanced-Wall: Augmented Desk and Wall Interfaces with Real-Time Tracking of User's Motion," *Procedings of Ubicomp2002 Workshop on Collaborations with Interactive Walls and Tables*, pp.27–30, 2002.
4. B.A. Sawyer, "Magnetic Positioning Device," *US patent* 3,457,482, 1969.
5. X. Chen, K. Takamasu, M. Nikaidou, "Evaluation of Thrust Force and Positioning Accuracy of a New Linear Motor," *Proceedings of the 6th International Symposium on Measurement Technology and Intelligent Instruments*, 2003.
6. K. Kotani, I. Hidaka, Y. Yamamoto, S. Ozono, "Analysis of Respiratory Sinus Arrhythmia with Respet to Respiratory Phase," *Method of Information in Medicine*, Vol.39, pp.153-156, 2000.
7. C. Reinhart and C. Patron, "Integrating Augmented Reality in the Assembly Domain — Fundamentals, Benefits and Applications," *Annals of the CIRP*, Vol.52/1/2003, pp.5–8, 2003.
8. Yasuhisa Hayakawa, Ikuo Kitagishi, Shigeki Sugiano, "Human Intention Based Physical Support Robot in Assembling Work," *Proceedings of the 1998 IEEE/RSJ International Conference on Intelligen Rotobs and Systems*, pp.930–935, 1998.

Development of a Forward-Hemispherical Vision Sensor for Remote Control of a Small Mobile Robot and for Acquisition of Visual-Information

Jyun-ichi Eino[*1], Masakazu Araki[*1], Jun-ichi Takiguchi[*2], Takumi Hashizume[*1]

[*1] Waseda University, Shinjukku, Japan,
eino@power.mech.waseda.ac.jp, araki@power.mech.waseda.ac.jp, hasizume@waseda.jp
[*2] Mitsubishi Electric Corporation, Kamakura Works, Kamakura-shi, Japan,
Junichi.Takiguchi@kama.melco.co.jp

Abstract— **The target of this research is to develop a common-use sensor, which is useful for navigation and mission for rescue work. By combined use of ODV (OmniDirectional Vision) with a hemispherical forward-visual-field using a direct and reflection hybrid optical system, and MEMS (Micro Electro Mechanical Systems) IMU (Inertial Measurement Unit), the common-use sensor outputs a real-time image for remote control of the mobile robot and an environment map, including information on three-dimensional own-position, with a continuous panorama image for searching victims and for making a rescue plan.**

1 INTRODUCTION

The target of this research is to develop a common-use sensor, which is useful for navigation and rescue mission. By combined use of a hemispherical forward-visual-field using a direct and reflection hybrid optical system, and a MEMS IMU, the common-use sensor outputs a real-time image for remote control of a mobile robot and the panoramic integration map, including information on three-dimensional own-position, with a continuous panoramic image for searching victims and for making a rescue plan.

2 OPTICAL CONFIGURATION

Proposed sensor's appearance is shown in Fig.1 and its optical configuration is shown in Fig.2. The distinctive characteristic is its unique direct and reflection hybrid optical system, where a concave lens and an ODV featuring two mirrors with a contrived curvature for obtaining a panoramic image with low aberration are designed to obtain forward hemispherical view field. ODV's characteristics, comparison of the optical structure with a conventional hyperbolical mirror, reveals that the proposed one has a nicer geometric property: a) two axis-symmetric mirrors are designed to minimize astigmatism and are contributed to realize a clear omni-directional image with little blur, b) any ray of light is reflected such that its focal point is coincided on the focal plain of a CCD, hence a precise image in every field of vision can be obtained, c) center of the view field can be used to combine the forward view with a concave lens [1-5]. The comparison of geometric image formation between the proposed optical structure and the conventional hyperbolical mirror with a perspective camera is shown in Fig.3 [6-8].The illuminator unit is designed to illuminate hemispherical view field within a radius of 1000 [mm] with minimum light and shade. The final assembled sensor's appearance is shown in Fig.1 and its head's diameter is 160 [mm] and the length is 190 [mm].

Figure 1. Sensor appearance.

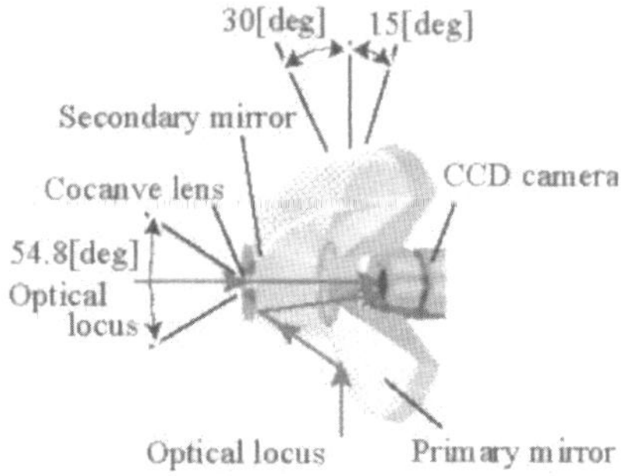

Figure 2. ODV optical structure.

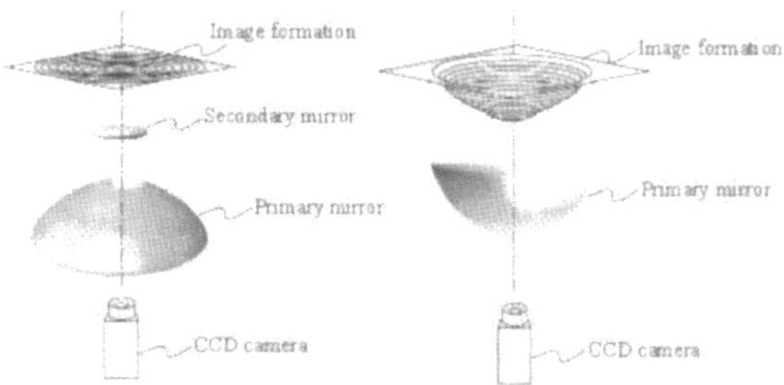

Figure 3. Image formation obtained by the ODV.

Fig.4 shows the obtained image. As most of the view field is occupied with side view field with low distortion, it is valuable for generating the column-shaped panoramic image with high resolution and useful for acquiring travel velocity by the parallactic to the environment. On the other hand, the forward view can be used for remote control of a mobile robot. Fig.5 shows the same image taken by a fish-eye lens [9,10]. Fig.4's image which is surrounded by two circles are corresponding to that of Fig.5. The radial resolution is 130[pixel] and 100[pixel] respectively. As Fig.4

efficiently occupies side-view field, the proposed optics proves to be effective for generating environmental map which should acquire side view rather than front view.

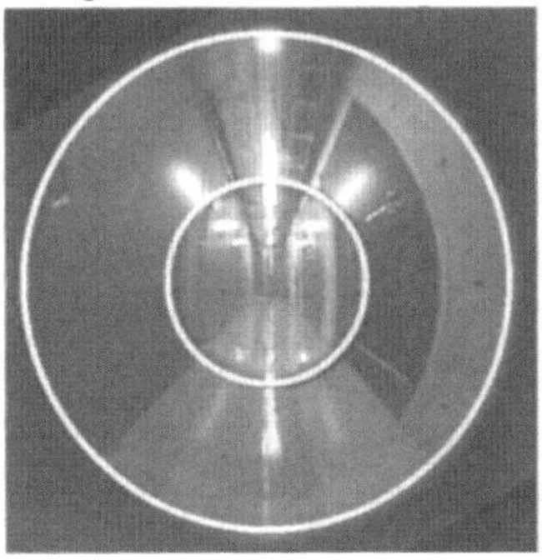

Figure 4. ODV image.

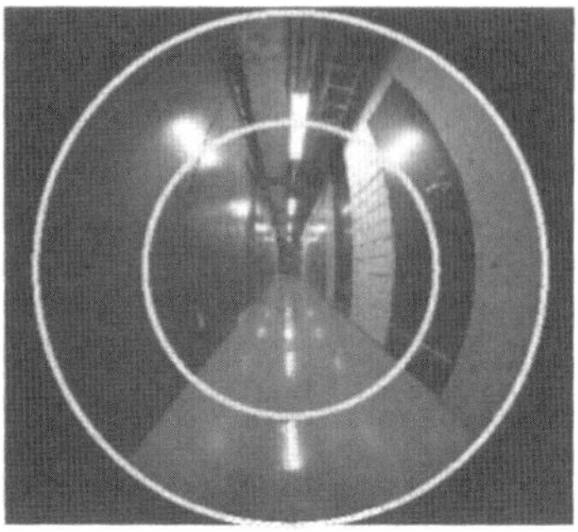

Figure 5. Fish-eye image.

3 PANORAMIC INTEGRATION MAP GENERATION

The proposed sensor consists of a forward-hemispherical vision, a MEMS IMU(Crossbow : AHRScc-100). It consists of three pairs of accelerometer, gyro and fluxgate compass to obtain Euler angles and three body-reference accelerations. The 3D self-position can be obtained from these signals by the strap down calculation. Suppose that the leveling error and the gyro random drift is negligible, the panoramic integration map which shows information on three-dimensional own-position with a continuous panoramic image by piling up the ODV's images on the horizontal plane with image coordinate transformation [11,12]. The system diagram is shown in Fig.6.

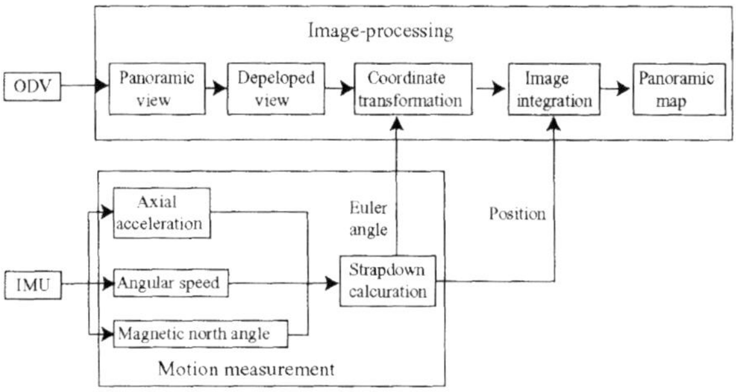

Figure 6. System block diagram.

Fig.7 shows an experimental remote-control mobile robot. The mobile robot is equipped with the proposed sensor on its head. Fig.8 shows an example of the panoramic integration map, which is taken in the floor of a laboratory. The robot is controlled so that it goes along the center of both walls. Because walls, the ceiling, and the floor can be clearly observed in Fig.8, it can be said that the proposed panoramic integration map can be very useful for surveillance and reconnaissance operation on the rescue campaign.

Figure 7. Experimental mobile robot.

Figure 8. Panoramic integration map.

4 IMAGE STABILIZATION

As the panoramic integration map is projected to a horizontal plane, the image is stabilized against pitching, rolling and yawing movement [13]. Pitch compensation scheme is shown in Fig.9. Each image pixel on the XYZ coordinate is transformed by (1). Roll compensation result is shown in Fig.10. Fig.11 shows the panoramic integration map without rotational compensation. The floor edge is distorted by pitch / roll disturbances. Fig.12 shows the panoramic integration map with rotational compensation. As the floor edge is reformed, it can be said that the proposed horizontal-plane based projection method successfully stabilizes rotational disturbances

$$\begin{pmatrix} X' \\ Y' \\ Z' \end{pmatrix} = \begin{pmatrix} \cos(-\phi) & -\sin(-\phi) & 0 \\ \sin(-\phi) & \cos(-\phi) & 0 \\ 0 & 0 & 1 \end{pmatrix} \begin{pmatrix} \cos(-\theta) & 0 & \sin(-\theta) \\ 0 & 1 & 0 \\ -\sin(-\theta) & 0 & \cos(-\theta) \end{pmatrix} \begin{pmatrix} 1 & 0 & 0 \\ 0 & \cos\psi & -\sin\psi \\ 0 & \sin\psi & \cos\psi \end{pmatrix} \begin{pmatrix} X \\ Y \\ Z \end{pmatrix} \quad (1)$$

where, ϕ , θ , ψ are roll angle, pitch angle, yaw angle , respectively.

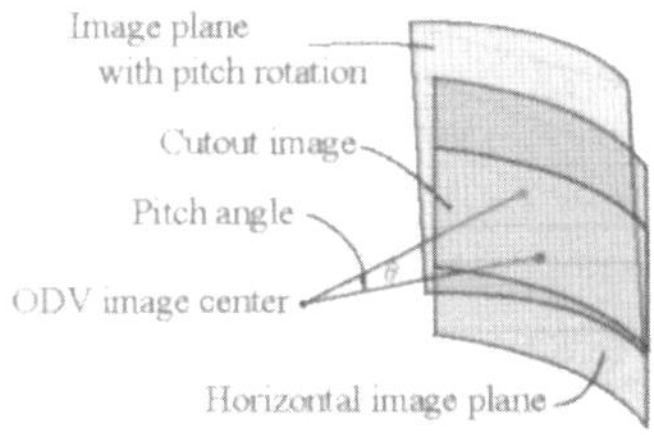

Figure 9. Image stabilization.

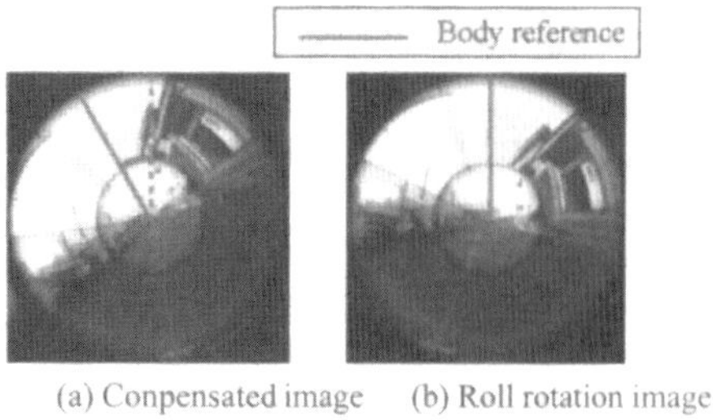

Figure 10. Roll compensation result.

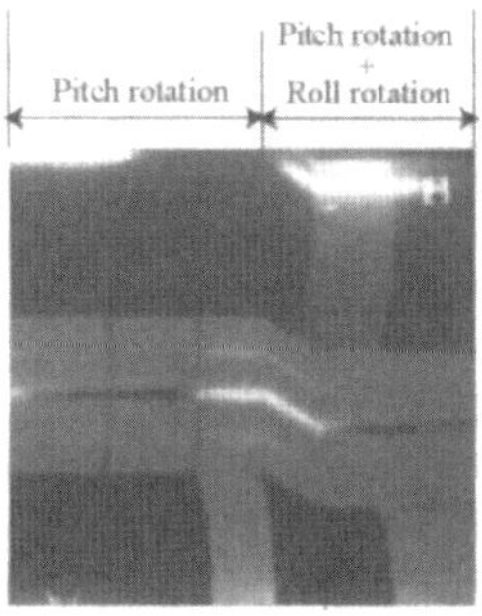

Figure 11. Panoramic map with Pitch/Roll disturbance.

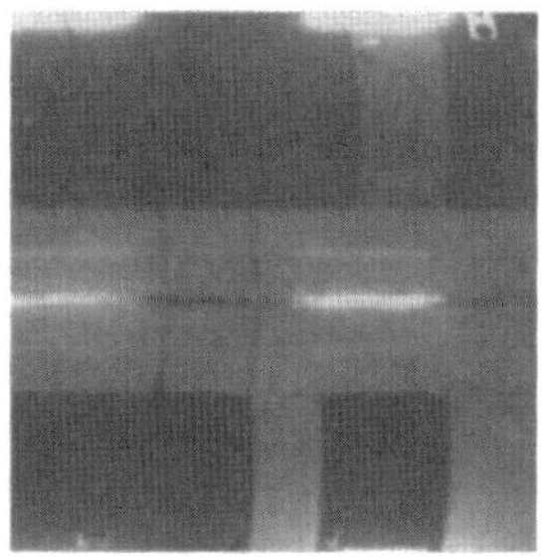

Figure 12. Compensated panoramic map.

5 EVALUATION OF THE PANORAMIC INTEGRATION MAP

The panoramic integration map is evaluated in a test field as shown in Fig.13. The corresponding panoramic integration map is shown in Figure .14.

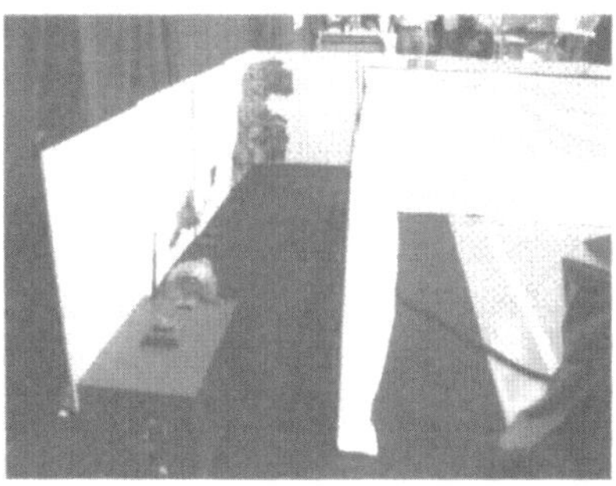

Figure 13. Evaluation field.

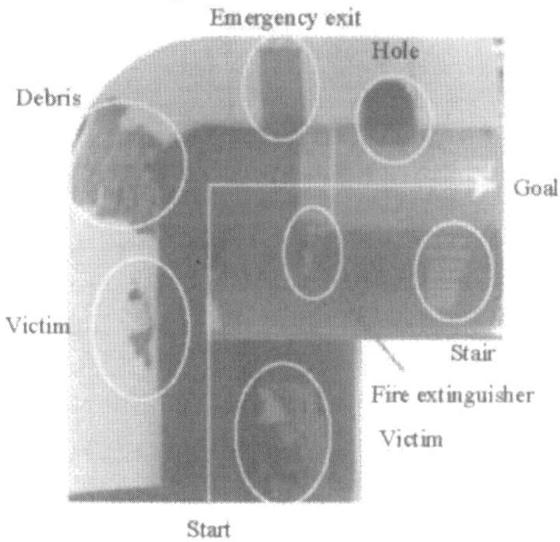

Figure 14. Corresponding panoramnic integration map.

In Fig.14, distinctive objects like a victim, a fire extinguisher and an emergency door can be easily identified. Theses information can be gathered by simply running a small mobile robot equipped with the proposed forward-hemispherical vision sensor without specific attention to control the camera view field, because the proposed sensor includes information on three-dimensional own-position, with a continuous 360 degrees panoramic image.

Thus, the effectiveness of the panoramic integration map for searching victims, and for making a rescue plan is proved.

6 RELATIVE VELOCITY ESTIMATION BY AN ODV IMAGE

In chapter 6 to 8, the ODV / IMU coupled self-positioning is discussed. In chapter 3, the IMU leveling error and the gyro random drift is neglected, but it is inevitable that these error cause serious problems for the self-positioning in practice. Proposed strategy is to apply the Kalman filter, which uses relative velocity as observation state for the accelerometer's integral error compensation. As the ODV can obtain high-resolution side view field with low distortion, distinctive image features like edges of a door, a window flame or the boundary between floor and wall can be easily tracked. Fig.15 shows an example of significant image feature of the environment. In this case, the perpendicularity between the vertical edge and the horizontal edge is used to identify the image feature. Several characteristic features can be tracked simultaneously with small tracking window to improve the velocity estimation accuracy.

Fig.16 shows the sensor coordinate definition. Suppose that the sensor's optical axis is 0-XYZ and the absolute coordinate is ENU. In Fig.16, relative angle between XYZ and ENU can be obtained from the IMU as Euler angle. Fig.17 shows the corresponding ODV image coordinate. Then, any image pixel "P_{ODV}" is obtained as P(n,e,u) shown below:

$$n = \frac{h_0 \tan(\alpha - (\alpha+\beta)\frac{r_P - r_S}{r_L - r_S})}{\cos(\tan^{-1}(\frac{x_c - x_{ODV}}{y_c - y_{ODV}}) - roll)} * \cos(yaw) + h_0 \tan(\tan^{-1}(\frac{x_c - x_{ODV}}{y_c - y_{ODV}}) - roll) * \sin(yaw) \tag{2}$$

$$e = -\frac{h_0 \tan(\alpha - (\alpha+\beta)\frac{r_P - r_S}{r_L - r_S})}{\cos(\tan^{-1}(\frac{x_c - x_{ODV}}{y_c - y_{ODV}}) - roll)} * \sin(yaw) + h_0 \tan(\tan^{-1}(\frac{x_c - x_{ODV}}{y_c - y_{ODV}}) - roll) * \cos(yaw) \tag{3}$$

$$u = -\frac{h_0 \tan(\alpha - (\alpha+\beta)\frac{r_P - r_S}{r_L - r_S})}{\cos(\tan^{-1}(\frac{x_c - x_{ODV}}{y_c - y_{ODV}}) - roll)} * \sin(pitch) \tag{4}$$

where, roll, pitch, yaw is Euler angle, and α is front side view angle, β is rear side view angle. x_c, y_c, r_P, r_L, r_S are corresponding to those constants in Fig.17.

Thus, the relative position can be calculated by tracking corresponding image features with each ODV's updated image. The objective relative velocity can be obtained as below:

$$(v_x, v_y, v_z) = \left(\frac{x_1 - x_2}{t_2 - t_1}, \frac{y_1 - y_2}{t_2 - t_1}, \frac{z_1 - z_2}{t_2 - t_1}\right) \tag{5}$$

where, t_n is the sampling timing, and (x_n, y_n, z_n) is the position of the significant feature.

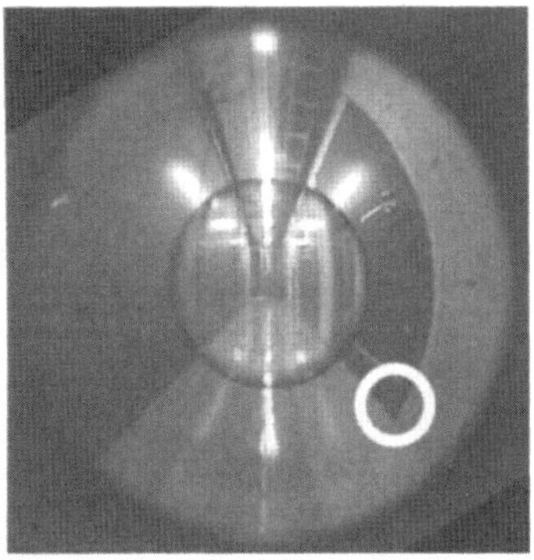

Figure 15. Significant feature example.

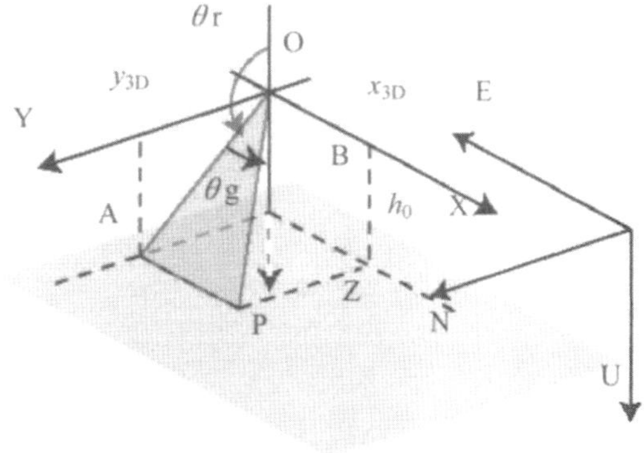

Figure 16. NEU coordinate system.

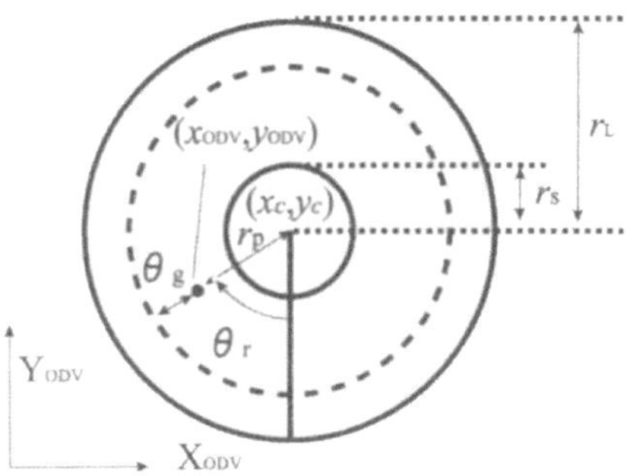

Figure 17. ODV coordinate system.

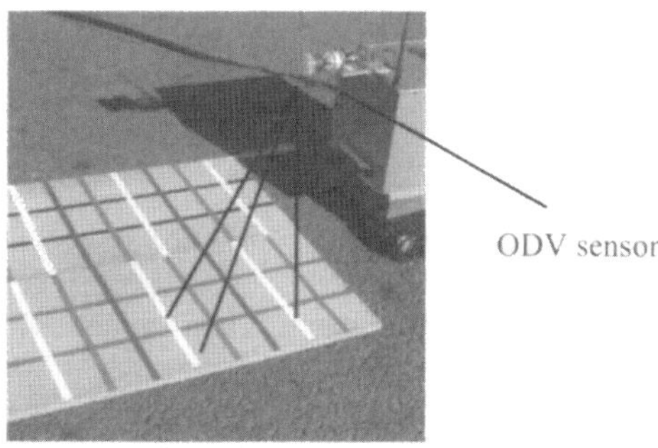

Figure 18. Experimental field for position measurement.

7 EVALUATION OF THE RELATIVE VELOCITY ESTIMATION

Fig.18 shows the experimental field. The intersection of the cross stripes are measured statically by the proposed sensor as shown in Table I. Dynamical velocity estimation accuracy is shown in Table II. Uni-point tracking means the result of one target tracking / velocity estimation, and multi-point tracking means the result of multiple target trackiNg. The average velocity of straight-line motion is 46.2[mm/s]. Thus, the velocity estimation accuracy achieves 3.7[%] as long as the system successfully tracks multiple image features.

TABLE I. STATIC POSITION MEASUREMENT RESULT.

	True value mm	Position error mm
Vertical accuracy	100	5.1
Horizontal accuracy	150	9.7

TABLE II. VELOCITY ESTIMATION RESULT.

	Average velocity error %	Average range error %
Uni-point tracking	4.9	7.2
Multi-point tracking	3.7	2.7

8 ODV / IMU COUPLED SELF-POSITIONING COMPENSATION WITH KALMAN FILTER

Fig.19 shows the proposed self-positioning system block diagram. It consists of the inertial strapdown calculation block and the KF (Kalman Filter), which uses the estimated velocity as observation state. Fig.20 shows the evaluation field. The mobile robot is controlled so that it can trace the line by way of A, B, C, D. Fig.21 shows the self-positioning result. Comparing the simple inertial strapdown calculation and the Kalman filtering compensation result, the velocity obtained from the ODV image can successfully compensate accelerometer's integral error. The remained error to the ideal locus heavily depends on the Euler angle error, which is caused in the leveling process, and gyro's random drift, which are not sensible to the velocity. It can be said that the proposed self-positioning method can be valid and effective.

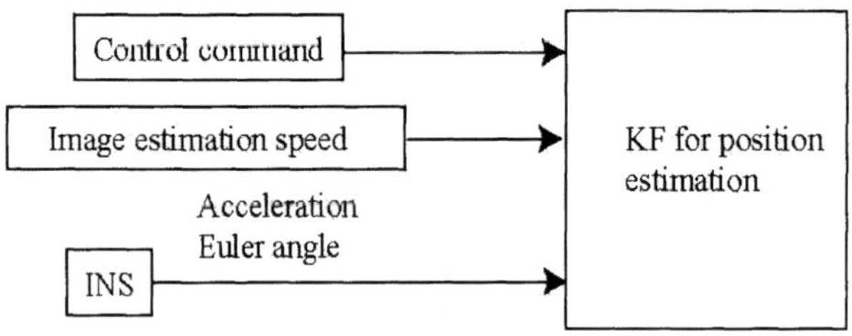

Figure 19. Self-positioning system block diagram.

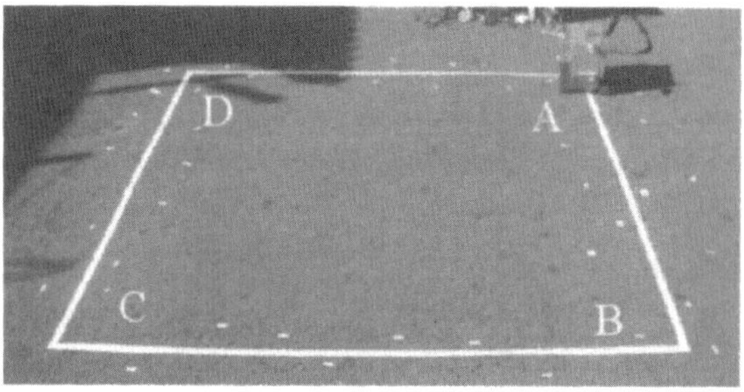

Figure 20. Experimental filed for self-positioning performance

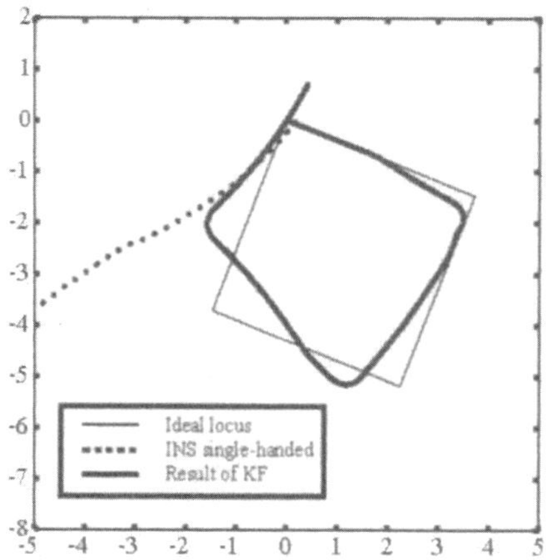

Figure 21. Self-positioning compensation result.

9 CONCLUSION

The target of this research is to develop a common-use sensor, which is useful for navigation and mission for rescue work. By combined use of a hemispherical forward-visual-field using a direct and reflection hybrid optical system, and a MEMS IMU, the common-use sensor outputs a real-time image for remote control of a mobile robot and the panoramic integration map, including information on three-dimensional own-position, with a continuous panoramic image. The evaluation of the panoramic integration map and the Kalman filter based self-positioning reveals that the proposed method is valid and effective.

ACKNOWLEDGMENT

This research is made under cooperation with an International Rescue System Institute and National Research Institute for Earth Science and Disaster Prevention as part of "Special project for earthquake disaster mitigation in urban areas".

REFERENCES

[1] A. Takeya, et al.,(1998). "Omnidirectional Vision System Using Two Mirrors," Novel Optical Systems Design and Optimization SPIE Vol. 3430, pp.50-60.

[2] T.Hasizume,et al.,"AStudy of Autonoumous Mobile System in Outdoor Environment (Part 14 Development of the self-positioning system with an ODV)", JSME, No.001-4 pp.98-99, 2000.3.11.

[3] T.Hasizume,et al.,"AStudy of Autonoumous Mobile System in Outdoor Environment (Part 15 Evaluation of the Self-positioning System with an ODV)",JSME,2P2-46-061,200.5.14.

[4] J. Takiguchi, et al.,"A Self-positioning System Using an Omnidirectional Vision", Transactions of the Japan Society of Mechanical Engineers(Part C), JSME, No68-673, pp.206-213, 2002.

[5] J. Takiguchi, et al., "High Precision Range Estimation from an Omnidirectional Stereo System", Proc. IEEE Int. Workshop Intelligent Robots & Systems, Switzerland, pp.263-268, 2002.

[6] Y.Yagi, et al.,(1990). "Panoramic scene analysis with conic projection", Proc. IEEE Int. Workshop Intelligent Robots & Systems, pp.181-187.

[7] J.Hong, et al.,(1991). "Image-based homing", Proc. IEEE Int. Conf. Robotics and Automation, pp.910-915.

[8] K.Yamazawa, et al.,(1993). "Omnidirectional imaging with hyperboloidal projection", Proc. Int. Conf. On Intelligent Robots and Systems, Yokohama, Japan, pp.1029-1034.

[9] Z.L.Cao,et al.,(1986). "Dynamic omnidirectional vision for mobile robots, " J. Robotic Systems, vol.3,No.1, pp.5-17.

[10] S.J.Oh,et al.,(1987). "Guidance of a mobile robot using an omnidirectional vision navigation system," Proc. Mobile Robots II,pp.288-300, SPIE 852.

[11] T.Hasizume,et al.,"AStudy of Autonoumous Mobile System in Outdoor Environment (Part 20 Outline of a Forward-looking Hemispheric Vision for Reconnaissance & Surveilance Operation)",JSME, pp259-260, 2002.12.19.

[12] T.Hasizume,et al .,"A Study of Autonoumous Mobile System in Outdoor Environment (Part 21 A Development of Intelligent Vision Sensor for a Remote-control Mobile Robot)", JSME, N2A1-2F-B3, 2003.5.25.

[13] T.Hasizume,et al.,"AStudy of Autonoumous Mobile System in Outdoor Environment (Part 22 Panoramic Map Generation by a Forward-Looking Hemispheric Vision)", JSME, 3J3-1,2003.12.21.

9784431358695

Printed in Japan

N0021979003